Geography of Climate Change

Climate change is one of the inescapable themes of current times. Climate change confronts society in issues as diverse as domestic and international political debate and negotiation, discussion in the media and public opinion, land management choices and decisions, and concerns about environmental, social and economic priorities now and for the future. Climate change also spans spatial, temporal and organizational scales, and has strong links with nature-society relationships, environmental dynamics, and vulnerability. Understanding the full range of possible consequences of climate change is essential for informed decision making and debate.

This book provides a collection of chapters that span geographical dimensions of environmental, social and economic aspects of climate change. Together the chapters provide a diverse and contrasting series that highlights the need to analyze, review and debate climate change and its possible impacts and consequences from multiple perspectives. The book also is intended to promote discussion and debate of a more integrated, inclusive and open approach to climate change and demonstrates the value of geography in addressing climate change issues.

This book was originally published as a special issue of *Annals of the Association of American Geographers.*

Richard Aspinall is a geographer with research interests in coupled human-environmental systems, especially in land use, environmental change, and ecosystem services, and in GIS. From 2006–2011 he was Director of the Macaulay Land Use Research Institute in Aberdeen, Scotland, an interdisciplinary institute addressing economic, social and environmental aspects of sustainability.

Geography of Climate Change

Edited by
Richard Aspinall

LONDON AND NEW YORK

First published 2012
by Routledge
2 Park Square, Milton Park, Abingdon, Oxfordshire OX14 4RN

Simultaneously published in the USA and Canada
by Routledge
711 Third Avenue, New York, NY 10017

First issued in paperback 2014

Routledge is an imprint of the Taylor and Francis Group, an informa business

This book is a reproduction of the *Annals of the Association of American Geographers*, vol. 100, issue 4. The Publisher requests to those authors who may be citing this book to state, also, the bibliographical details of the special issue on which the book was based.

British Library Cataloguing in Publication Data
A catalogue record for this book is available from the British Library

ISBN 978-0-415-69662-3 (hbk)
ISBN 978-1-138-85240-2 (pbk)

Typeset in Baskerville
by Taylor & Francis Books

Publisher's Note
The publisher would like to make readers aware that the chapters in this book may be referred to as articles as they are identical to the articles published in the special issue. The publisher accepts responsibility for any inconsistencies that may have arisen in the course of preparing this volume for print.

Contents

Geographical Perspectives on Climate Change

Richard Aspinall

Climate change is one of the inescapable themes of current times. Climate change confronts society in issues as diverse as domestic and international political debate and negotiation, discussion in the media and public opinion, land management choices and decisions, and concerns about environmental, social and economic priorities now and for the future, among others. Climate change also spans spatial, temporal and organizational scales and has links with many other strong and persistent geographic themes, including nature-society relationships, environmental dynamics, and vulnerability.

An indication of the increasing interest and importance of climate change is found in the remarkable growth in scientific publications, public awareness and political discussion of the many issues associated with climate change over the last 20 years. For example, summary of papers in ISI journals shows the rate of scientific publication on the subject of climate change has grown rapidly, the rate of increase being exponential since 1991 (Figure 1). The considerable and evolving knowledge on climate change and its possible impacts that is recorded in the international scientific literature is regularly synthesized and assessed in the reports of the Intergovernmental Panel on Climate Change, mostly recently the Fourth Assessment (Metz et al. 2007; Pachauri and Reisinger 2007; Parry et al. 2007; Solomon et al. 2007).

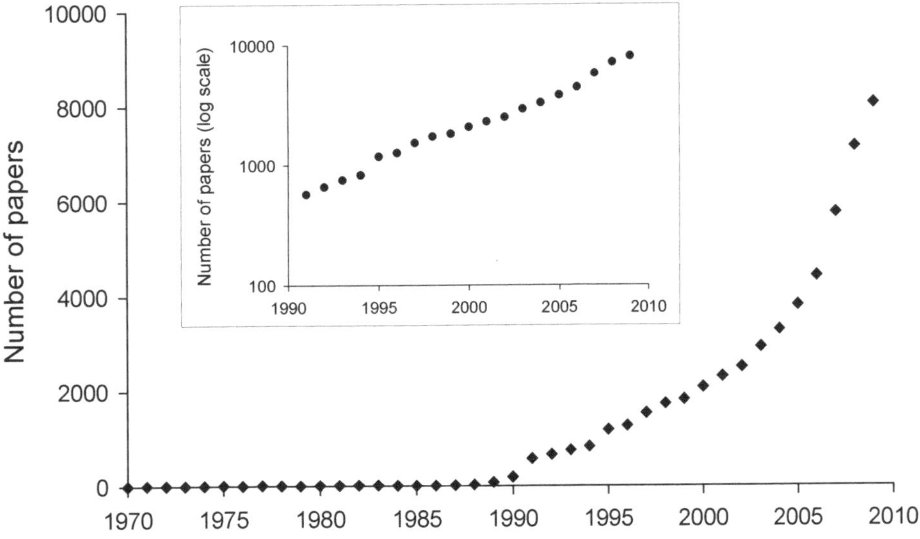

Figure 1. Number of papers per year with 'climate change' as either a keyword or in the abstract that are cited on ISI Web of Science. Data are for the period 1970 to 2009. The inset graph shows the number of papers from 1991 to 2009 using a log scale.

The issues and possible impacts of changes in climate reflect the interdependence of people, places and environment across geographic and political scales. Research can show how these global, regional, and local connections of the Earth's climate, and the extent to which human activity and climate are interrelated at multiple scales leading to local understanding of global changes (e.g., Corburn 2009; Furgal and Prowse 2009; Weare 2009). Science and policy both recognize the cross- and multi- scale nature of environmental changes and responses to climate changes (Cash and Moser 2000; Hedger, Connell and Bramwell 2006; Nelson, Adger and Brown 2007); science and policy also recognize a need for science to provide an input to the evidence base considered in policy and management decisions (Harman, Harrington and Cerveny 1998a, 1998b; Waterstone 1998; Cash and Moser 2000).

Although much of the debate about climate change is concerned with issues directly addressed by climate science, climate science alone is not sufficient to understand the full impact and importance of climate change to society and environment. Evidence that supports governmental, institutional and individual initiatives to mitigate and adapt to climate change similarly is not the exclusive preserve of climate science, but rather demands a diversity of disciplines to provide the knowledge, understanding and insights needed for society to answer its questions and respond to a changing global climate. Many disciplines recognize this, and both climate change and its impacts are regularly addressed in journals across the social (Lorenzoni, Pidgeon and O'Connor 2005; Lever-Tracy 2008), economic (Gowdy 2008; Gowdy and Julia 2010), behavioral (Bord, O'Connor and Fisher 2000; Sundblad, Biel and Garling 2007; Scannell and Grouzet 2010), biological (Swift 1999; Beaumont et al. 2007; Beale, Lennon and Gimona 2008; Sutherland et al. 2009) and environmental (Reed et al. 2006; Richards and Clifford 2008) sciences.

Climate change is additionally a subject that can also benefit from a systems-level understanding developed from multi- and interdisciplinary integrated approaches (Yarnal 1998, Steiner and Hatfield 2008, Schuepbach et al. 2009). It is here that geography and geographers can play a particularly important role. The diverse scope of the discipline and ability of geographers to communicate with other disciplines provide a platform for a successful contribution to integrative and interdisciplinary research focused on climate change.

Geography and Climate Change in the *Annals of the Association of American Geographers*

Geography contributes greatly to the understanding of changing climate, climate change and the impacts of changes. Climatology within physical geography has long addressed climate changes through its focus on dynamic and statistical climatology (Skaggs 2004). Publications in the *Annals* have reconstructed and analyzed different aspects of climatic fluctuations over timescales ranging from the postglacial (Kay and Andrews 1983; Liu and Lam 1985), to millennia (Liu, Shen and Louie 2001), centuries (Graumlich 1987), and shorter timescales during more recent periods (Carleton 1987; Hansen-Bristow, Ives and Wiison 1988; Bradbury, Keim and Wake 2003). Geographers have also contributed to numerical modeling used for simulation of climate in climate change studies (Balling Jr 2000). Studies of geomorphologic (Frei et al. 2002; Loaiciga 2003) and biogeographic (Shen et al. 2008) responses of environmental systems to changes in climate have also been published. Other geographers have highlighted issues of vulnerability and inequality (Liverman 1990, O'Brien and Leichenko 2003). The scientific, social and political construction of climate change and the need for a reflexive politics of climate change and of scientific knowledge that is based on active trust has also been discussed in the *Annals* (Demeritt 2001).

The chapters in this book were first published as a thematic special issue of the *Annals*. They develop a broad range of geographical perspectives on the subject of climate change. As for the 2009 special issue of the *Annals* on 'Geographies of Peace and Armed Conflict' edited by Audrey Kobayashi (Kobayashi 2009), the authors were asked to limit their contributions to about 5,000 words. This is a relatively short and challenging limit. It has the disadvantage of restricting the amount of theoretical development and other detail that can be included within each paper. However, it does have the benefit of allowing a larger number of papers to be included, which is important for coverage of a diverse and inclusive set of geographical perspectives on the wide range of material inherent in the theme of climate change. As for all papers in the *Annals*, selection was on the basis of quality assessed by peer review.

The papers do not sit comfortably within the four sections that are used to organize the research reported in the *Annals*, nor would they be expected to. Issues inherent in climate change and other themes based on human-environmental interactions span the sub-disciplinary interests of geographers and invite whole systems understanding. Of course, each of the papers can be read individually and in isolation. In keeping with the broad readership intended for the special issues of the *Annals* each paper may be used in a variety of ways to promote further questioning, research or debate. Additionally, as a collection of papers they provide a diverse and contrasting series that highlights not only the diversity of geographic contributions but also the need to analyze, review and debate climate change and possible impacts and consequences from multiple perspectives. Taken together, they are, therefore, intended to promote discussion and debate of a more integrated, inclusive and open approach to climate change and demonstrate the value of geography in addressing climate change issues.

Acknowledgements

Many individuals have contributed to this publication. The idea of a fifth issue of the *Annals* each year that concentrates on a topical issue is supported by the Council and Executive Committee of the Association of American Geographers, and is also supported by the publisher, Taylor & Francis. The editors of the *Annals* equally are all committed to the special issues and provide each other with considerable support. The experiences of Audrey Kobayashi in editing the 2009 Special Issue on 'Geographies of Peace and Conflict' were particularly informative and helpful for this second special issue. An initial call for abstracts for the special issue received 77 responses of which 41 were invited to submit manuscripts. All those who submitted abstracts and manuscripts are thanked and acknowledged, particularly for their cooperation with the timetable and its evolution. In addition, I also thank and acknowledge the *Annals* Editorial Board members and referees who provided timely feedback on abstracts and manuscripts. Finally, and profoundly, my thanks go to Robin Maier and Miranda Lecea in the Editorial Office at AAG who managed the whole process leading to production of the issue.

References

Balling Jr, R. C. (2000). The geographer's niche in the greenhouse millennium. *Annals of the Association of American Geographers* 90: 114.

Beale, C. M., J. J. Lennon and A. Gimona (2008). Opening the climate envelope reveals no macroscale associations with climate in European birds. *Proceedings of the National Academy of Sciences of the United States of America* 105: 14908–12.

Beaumont, L. J., A. J. Pitman, M. Poulsen and L. Hughes (2007). Where will species go? Incorporating new advances in climate modelling into projections of species distributions. *Global Change Biology* 13: 1368–85.

Bord, R. J., R. E. O'Connor and A. Fisher (2000). In what sense does the public need to understand global climate change? *Public Understanding of Science* 9: 205–18.

Bradbury, J. A., B. D. Keim and C. P. Wake (2003). The Influence of Regional Storm Tracking and Teleconnections on Winter Precipitation in the Northeastern United States. *Annals of the Association of American Geographers* 93: 544–56.

Carleton, A. M. (1987). Summer Circulation Climate of the American Southwest, 1945–1984. *Annals of the Association of American Geographers* 77: 619–34.

Cash, D. W. and S. C. Moser (2000). Linking global and local scales: designing dynamic assessment and management processes. *Global Environmental Change-Human and Policy Dimensions* 10: 109–120.

Corburn, J. (2009). Cities, Climate Change and Urban Heat Island Mitigation: Localising Global Environmental Science. *Urban Studies* 46: 413–27.

Demeritt, D. (2001). The Construction of Global Warming and the Politics of Science. *Annals of the Association of American Geographers* 91: 307.

Frei, A., R. L. Armstrong, M. P. Clark and M. C. Serreze (2002). Catskill Mountain Water Resources: Vulnerability, Hydroclimatology, and Climate-Change Sensitivity. *Annals of the Association of American Geographers* 92: 203.

Furgal, C. and T. Prowse (2009). Climate Impacts on Northern Canada: Introduction. *Ambio* 38: 246–47.

Gowdy, J. and R. Julia (2010). Global Warming Economics in the Long Run: A Conceptual Framework. *Land Economics* 86: 117–130.

Gowdy, J. M. (2008). Behavioral economics and climate change policy. *Journal of Economic Behavior & Organization* 68: 632–44.

Graumlich, L. J. (1987). Precipitation Variation in the Pacific Northwest (1675–1975) as Reconstructed from Tree Rings. *Annals of the Association of American Geographers* 77: 19–29.

Hansen-Bristow, K., J. D. Ives and J. P. Wiison (1988). Climatic Variability and Tree Response within the Forest-Alpine Tundra Ecotone. *Annals of the Association of American Geographers* 78: 505–19.

Harman, J. R., J. A. Harrington and R. S. Cerveny (1998a). Balancing scientific and ethical values in environmental science. *Annals of the Association of American Geographers* 88: 277–86.

______. (1998b). Reply: Values, ethics, and geographic research. *Annals of the Association of American Geographers* 88: 308–10.

Hedger, M. M., R. Connell and P. Bramwell (2006). Bridging the gap: empowering decision-making for adaptation through the UK Climate Impacts Programme. *Climate Policy* 6: 201–15.

Kay, P. A. and J. T. Andrews (1983). Re-evaluation of Pollen-Climate Transfer Functions in Keewatin, Northern Canada. *Annals of the Association of American Geographers* 73: 550–59.

Kobayashi, A. (2009). Geographies of Peace and Armed Conflict: Introduction. *Annals of the Association of American Geographers* 99: 819–26.

Lever-Tracy, C. (2008). Global warming and sociology. *Current Sociology* 56: 445–66.

Liu, K.-b., C. Shen and K.-s. Louie (2001). A 1,000-Year History of Typhoon Landfalls in Guangdong, Southern China, Reconstructed from Chinese Historical Documentary Records. *Annals of the Association of American Geographers* 91: 453.

Liu, K. B. and N. S. Lam (1985). Paleovegetational Reconstruction Based on Modern and Fossil Pollen Data–an Application of Discriminant-Analysis. *Annals of the Association of American Geographers* 75: 115–130.

Liverman, D. M. (1990). Drought Impacts in Mexico: Climate, Agriculture, Technology, and Land Tenure in Sonora and Puebla. *Annals of the Association of American Geographers* 80: 49–72.

Loaiciga, H. A. (2003). Climate change and ground water. *Annals of the Association of American Geographers* 93: 30–41.

Lorenzoni, I., N. F. Pidgeon and R. E. O'Connor (2005). Dangerous climate change: The role for risk research. *Risk Analysis* 25: 1387–98.

Metz, B., O. R. Davidson, P. R. Bosch, R. Dave and L. A. Meyer. 2007. Climate Change 2007: Mitigation of Climate Change. Contribution of Working Group III to the Fourth Assessment Report of the Intergovernmental Panel on Climate Change, 2007. Cambridge: Cambridge University Press.

Nelson, D. R., W. N. Adger and K. Brown (2007). Adaptation to environmental change: Contributions of a resilience framework. *Annual Review of Environment and Resources* 32: 395–419.

O'Brien, K. L. and R. M. Leichenko (2003). Winners and Losers in the Context of Global Change. *Annals of the Association of American Geographers* 93: 89–103.

Pachauri, R. K. and A. Reisinger. 2007. Climate Change 2007: Synthesis Report. Contribution of Working Group II to the Fourth Assessment Report of the Intergovernmental Panel on Climate Change, 2007. 104. Sqitzerland: IPCC.

Parry, M. L., O. F. Canziani, J. P. Palutikof, P. J. van der Linden and C. E. Hanson. 2007. Climate Change 2007: Impacts, Adaptation and Vulnerability. Contribution of Working Group II to the Fourth Assessment Report of the Intergovernmental Panel on Climate Change, 2007. Cambridge: Cambridge University Press.

Reed, P. M., R. P. Brooks, K. J. Davis, D. R. DeWalle, K. A. Dressler, C. J. Duffy, H. S. Lin, D. A. Miller, R. G. Najjar, K. M. Salvage, T. Wagener and B. Yarnal (2006). Bridging river basin scales and processes to assess human-climate impacts and the terrestrial hydrologic system. *Water Resources Research* 42.

Richards, K. and N. Clifford (2008). Science, systems and geomorphologies: why LESS may be more. *Earth Surface Processes and Landforms* 33: 1323–40.

Scannell, L. and F. M. E. Grouzet (2010). The metacognitions of climate change. *New Ideas in Psychology* 28: 94–103.

Schuepbach, E., E. Uherek, A. Ladstatter-Weissenmayer and M. J. Jacob (2009). Educating the next generation of atmospheric scientists within a European Network of Excellence. *Atmospheric Environment* 43: 5415–22.

Shen, C. M., K. B. Liu, L. Y. Tang and J. T. Overpeck (2008). Numerical Analysis of Modern and Fossil Pollen Data from the Tibetan Plateau. *Annals of the Association of American Geographers* 98: 755–72.

Skaggs, R. H. (2004). Climatology in American Geography. *Annals of the Association of American Geographers* 94: 446–57.

Solomon, D., D. Qin, M. Manning, Z. Chen, M. Marquis, K. B. Averyt, M. Tignor and H. L. Miller. 2007. Climate Change 2007: The Physical Science Basis. Contribution of Working Group I to the Fourth Assessment Report of the Intergovernmental Panel on Climate Change, 2007. Cambridge: Cambridge University Press.

Steiner, J. L. and J. L. Hatfield (2008). Winds of change: A century of agroclimate research. *Agronomy Journal* 100: S132–S152.

Sundblad, E. L., A. Biel and T. Garling (2007). Cognitive and affective risk judgements related to climate change. *Journal of Environmental Psychology* 27: 97–106.

Sutherland, W. J., W. M. Adams, R. B. Aronson, R. Aveling, T. M. Blackburn, S. Broad, G. Ceballos, I. M. Cote, R. M. Cowling, G. A. B. Da Fonseca, E. Dinerstein, P. J. Ferraro, E. Fleishman, C. Gascon, M. Hunter, J. Hutton, P. Kareiva, A. Kuria, D. W. MacDonald, K. MacKinnon, F. J. Madgwick, M. B. Mascia, J. McNeely, E. J. Milner-Gulland, S. Moon, C. G. Morley, S. Nelson, D. Osborn, M. Pai, E. C. M. Parsons, L. S. Peck, H. Possingham, S. V. Prior, A. S. Pullin, M. R. W. Rands, J. Ranganathan, K. H. Redford, J. P. Rodriguez, F. Seymour, J. Sobel, N. S. Sodhi, A. Stott, K. Vance-Borland and A. R. Watkinson (2009). One Hundred Questions of Importance to the Conservation of Global Biological Diversity. *Conservation Biology* 23: 557–67.

Swift, M. J. (1999). Integrating soils, systems and society. *Nature & Resources* 35: 12–20.

Waterstone, M. (1998). Better safe than sorry, or bettor safe, then sorry? In *Annals of the Association of American Geographers* 88: 297–300.

Weare, B. C. (2009). How will changes in global climate influence California? *California Agriculture* 63: 59–66.
Yarnal, B. (1998). Integrated regional assessment and climate change impacts in river basins. *Climate Research* 11: 65–74.

Beyond Adapting to Climate Change: Embedding Adaptation in Responses to Multiple Threats and Stresses

Thomas J. Wilbanks and Robert W. Kates

Climate change impacts are already being experienced in every region of the United States and every part of the world—most severely in Arctic regions—and adaptation is needed now. Although climate change adaptation research is still in its infancy, significant adaptation planning in the United States has already begun in a number of localities. This article seeks to broaden the adaptation effort by integrating it with broader frameworks of hazards research, sustainability science, and community and regional resilience. To extend the range of experience, we draw from ongoing case studies in the Southeastern United States and the environmental history of New Orleans to consider the multiple threats and stresses that all communities and regions experience. Embedding climate adaptation in responses to multiple threats and stresses helps us to understand climate change impacts, themselves often products of multiple stresses, to achieve community acceptance of needed adaptations as cobenefits of addressing multiple threats, and to mainstream the process of climate adaptation through the larger envelope of social relationships, communication channels, and broad-based awareness of needs for risk management that accompany community resilience. *Key Words: climate change adaptation, hazards, resilience, sustainability.*

气候变化的影响已经出现在美国和世界的每一个地区，其中最严重的就是北极地区，对此人们必须加以适应。虽然对于适应气候变化的研究仍处于起步阶段，美国已经开始在一些地方实施重要的适应规划。将适应气候变化的研究和更广泛的灾害研究的框架，可持续发展的科学工作，以及社区和地区性应变能力的研究相结合，籍此，本文旨在拓展适应气候变化的努力。为了扩大经验的范围，我们参考了美国东南部的当前个案以及新奥尔良的环境历史，以借鉴所有社区和地区所面临的多重威胁和压力。作为对多重威胁和压力的回应的一部分，对气候的适应有助于我们了解气候变化的影响，这些回应本身，往往造成了多种压力，作为应对多重威胁的双重好处，以实现社区接受那些必须的适应，通过一个更大的容器，包括社会关系，沟通渠道和对风险管理必须性的广泛意识，伴随着社会的适应能力，将气候适应的过程纳入主流。*关键词：适应气候变化，灾害，恢复力，可持续性。*

Los impactos del cambio climático ya se están sintiendo en cada región de los Estados Unidos y en todas partes del mundo—con mayor severidad en las regiones árticas—por lo que las adaptaciones a tal situación son impostergables. Aunque la investigación sobre adaptación a los cambios climátiicos se halla todavía en su infancia, ya se han inciado planes de adaptación significativos en un número de localidades de los Estados Unidos. Este artículo busca ampliar el esfuerzo de adaptación, integrándolo a esquemas de mayor amplitud de investigación sobre catástrofes, ciencia de la sustentabilidad y resiliencia comunitaria y regional. Con el ánimo de extender el ámbito de experiencia, nos apoyamos en estudios de caso en desarrollo en el sudeste de Estados Unidos y en la historia ambiental de de Nueva Orleans, para considerar las múltiples amenazas y estreses que experimentan todas las comunidades y regiones. Al incorporar la adaptación climática en la respuesta a múltiples amenazas y estrés nos ayuda a comprender mejor los impactos del cambio climático, que a la vez suelen ser producto de estreses múltiples, para lograr la aceptación de la comunidad a adeaptaciones necesarias, como beneficio derivado de tomar en cuenta amenazas múltiples, y para canalizar el proceso de adptación climática a través de la más grande envoltura de relaciones sociales, canales de comunicación y conciencia de base más amplia sobre la necesidad del manejo del riesgo, propia de la resiliencia comunitaria. *Palabras clave: adaptación al cambio climático, catástrofes, resiliencia, sustentabilidad.*

After a long period of neglect, adaptation to climate change impacts is finally getting a burst of attention, because it can no longer be ignored (Schipper and Burton 2009); examples include a May 2007 National Summit on Coping with Climate Change, a U.S. National Research Council study of "America's Climate Choices, 2008–2010," a U.S. National Climate Adaptation Summit in May 2010, and the scope of IPCC Working Group II's Fifth Assessment Report, 2010–2014, which will include four chapters on adaptation. In the United States, a recent synthesis study (Karl, Melillo, and Peterson 2009) found all regions at risk to impacts, although major impacts differed from region to region based on climate change exposure, climate-sensitive economic activities, and population vulnerabilities. Examples taken from the study are shown in Figure 1. As impacts emerge, affected peoples and systems must find ways to cope. We report on the adaptation planning in the United States that has already begun, but we also seek to go beyond these efforts to consider, in contexts of hazards research, sustainability science, and community and regional resilience, the multiple threats and stresses that all communities and regions experience, including but not limited to risks from climate change.

Adaptation to Climate Change

We begin with a brief discussion of climate change adaptation and then move on to our concept of multiple-threat adaptation. Vulnerabilities and risks associated with possible impacts of climate change can be prevented by reducing exposure to primary effects of climate change (warming, precipitation changes, extreme weather events, sea-level rise), reducing sensitivity to those effects, and increasing the capacity to cope with effects (Clark et al. 2000). Of course, the most important way to reduce exposure to climate change impacts is to limit climate change itself through mitigation. Using adaptations to sea-level rise as an example, exposure can also be limited by moving populations and land uses away from a vulnerable coastline, reducing sensitivity by constructing sea walls, or hardening coastal structures to flooding; coping capacities can be increased by strengthening emergency response capabilities.

Adapting to climate variability and change has been part of human experience for many millennia, and the historical record includes many cases of successful adaptations, often through migration (e.g., McIntosh, Tainter, and McIntosh 2000). Although research on adaptation to climate change in the modern context is still in its infancy, a number of the basic dimensions of adaptation are generally understood (Intergovernmental Panel on Climate Change [IPCC] 2001, 2007). For example, adaptation can either avoid costs or accept costs (Burton 1997). It can be autonomous (or spontaneous or voluntary, as systems react to observed changes, the prospect of changes, or market or other signals that incorporate general concerns about risks of change) or planned (encouraged or required by public policy interventions). It can be anticipatory, avoiding or moderating impacts by actions ahead of changes, or reactive, responding to impacts as they are experienced. It can be geographically widespread or localized. It can be sectorally focused, such as on agriculture or health, or crosscutting, such as the use of insurance to share risks (Schipper and Burton 2009).

Very little research attention has been given to adaptation costs and benefits or even to costs of failing to mitigate. One partial exception has been costs and benefits of major alternatives for adapting to coastal impacts from sea-level rise (Tol 2002a, 2002b; Tol et al. 2006); analyses of costs of recent severe weather events, such as hurricanes and floods, are often used to illustrate potential costs if such events were to become more intense or frequent (or shifted in location) because of climate change (IPCC 2007, chap. 7).

In its net costs and its capacities for action, however, adaptation to climate change is deeply and complexly linked with other economic and social goals. Thus, in many cases, adaptation can be done now at a relatively low cost, because of considerable cobenefits related to other aspects of sustainability and resilience. In fact, most adaptation actions in the near term to reduce impacts of climate change, most of which are relatively long term, need to be associated with other short-term benefits to be acceptable and sustainable. One example is a decision by the city of Boston in 1993 to raise a new waste disposal facility on an offshore island to protect it from sea-level rise (Moser 2009).

Adaptations also have physical, economic, or institutional limits, however. Physical limits to adaptation are identified by constraints such as the maximum height of levees and sea walls, the drawdown and rates of application of irrigation water, and the sizing and replacement of culvert capacity. Before their physical limits are reached, the costs of adaptations might exceed both customary usage and even future estimates of potential losses avoided. Institutional practice, determined by ignorance, uncertainty, custom, law, regulation, or competing agendas, further constrains adaptations. For example, coastal regions face uncertain increases in hurricane intensity and sea-level rise, although regulations are based on storm surges to 100-year flood plains, and losses are government subsidized for flood but not for wind or erosion damage.

Case Studies of Climate Change Adaptation in the United States

Examples of planned adaptations to climate change are still rare in the United States, aside from Alaska,

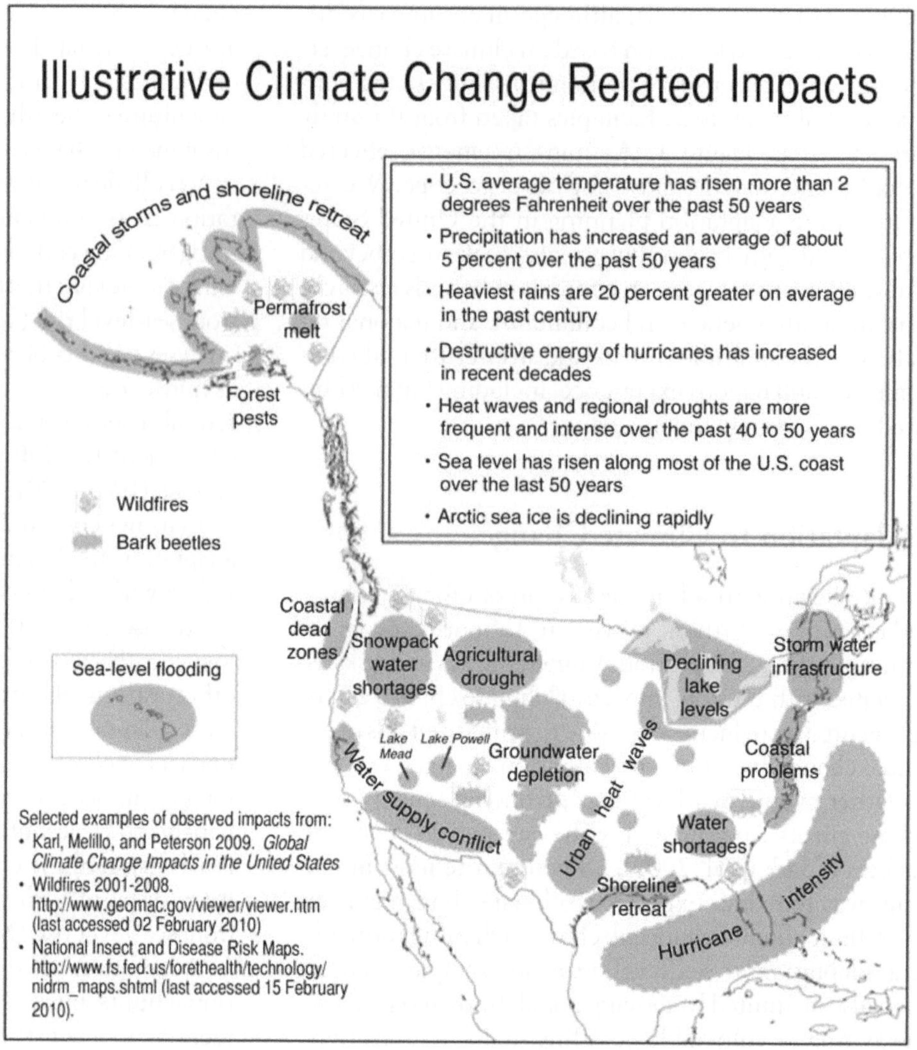

Figure 1. Climate change–related impacts (many have multiple causes) are already being experienced in every region of the United States as documented in a recent U.S. State of Knowledge Report (Karl, Melillo, and Peterson 2009). These generalized examples were selected to illustrate that although all regions are at risk for impacts, major impacts differ from region to region based on climate change exposure, climate-sensitive economic activities, and population vulnerabilities. Text box lists major observed climate changes in the United States.

because most climate change impacts lie in the future, shrouded in uncertainty. The risks are serious enough, however, that a number of states and localities have begun to take adaptation planning seriously (Moser 2009).

In most cases, planning for impacts of climate change at a regional or local level in the United States can be traced to the first U.S. National Assessment of Potential Consequences of Climate Variability and Change (2000), which included attention to possible regional impacts, with one of the regions a metropolitan area (New York City). Since then, a family of twenty-one summaries of climate change science, termed Synthesis and Assessment Products, has been commissioned by the U.S. Climate Change Science Program between 2005 and 2008 on the regional and more local implications of climate change (see http://www.climatescience.gov, last accessed 8 February 2010), with seven of the twenty-one specifically concerned with climate change impacts and adaptations (the reports numbered 4.1–4.7).

At a state scale, Alaska is a focus of action, not just planning, because climate change impacts are already proving to be significant (e.g., Sakakibara 2008). It has established an Alaska Climate Change Sub-Cabinet to prepare and implement an Alaska Climate Change Strategy (http://www.climatechange.alaska.gov/aag/aag.htm, last accessed 2 November 2009) and is seeking to implement plans to protect six native communities requiring immediate relocation (Immediate Action Workgroup 2009). Seven other states (California, Maryland, Oregon, Florida, Washington, Massachusetts, and New Hampshire) have begun climate change adaptation planning in parallel with discussions of mitigation issues (Pew Center on Global Climate Change 2008; Moser 2009). California has used climate change as a catalyst for addressing enormous environmental and economic challenges in the Sacramento–San Joaquin Delta, facing heavy water demands for agricultural and urban development in Southern California (Feldman and Jacobs 2008). Florida is considering adaptation challenges presented by prospects of intensified coastal storms combined with sea-level rise in a state where coastal amenities are keys to continued economic and social development.

Local adaptation planning has emerged as a focus more recently. Since 2006, New York City has organized an ambitious effort to prepare a climate change adaptation plan within the context of a broader sustainability and growth management initiative, PlaNYC. A community-wide Climate Change Adaptation Task Force was formed, provided with climate change projections, and asked to identify "acceptable levels of risk" and determine how to develop flexible adaptation pathways that would facilitate strategies to keep risks within those levels (New York Panel on Climate Change 2009). This adaptation planning effort was intended to contribute to the integration of PlaNYC's three challenge areas: growth, infrastructure, and environment not separate from those goals (http://www.nyc.gov/html/planyc2030/html/plan/plan.shtml, last accessed 8 January 2010).

This effort is not limited to large cities, however. A collaboration between Seattle/King County, Washington, with Local Governments for Sustainability (ICLEI) produced both an adaptation plan and a handbook to assist other cities and communities with adaptation planning (Snover et al. 2007). This handbook provided a tutorial for the ICLEI's new Climate Resilient Communities Program. The first of five pilot efforts to be completed is Keene, New Hampshire, a city of 23,000. Keene created a committee that included the senior city leadership as a focus for broad-based community participation. As the Keene committee identified climate change vulnerabilities, with support from ICLEI and help from the New England Integrated Sciences and Assessments, they grappled with difficulties in developing actions to address those vulnerabilities. They found it difficult to separate climate-related actions from more general sustainability and green economy issues and included all three in their adaptation plan, identifying a number of targeted actions. Since completion of the plan in 2007, some of the targeted actions have already been implemented and many more are to be "mainstreamed" by inclusion as part of a Community Master Plan to be completed in 2010 (Melissa Stults, ICLEI, personal communication, 26 May 2009; Mikaela Engert, Keene city planner, personal commuication, 27–28 May 2009).

Adaptation in a Broader Context of Sustainability and Resilience

These early experiments in planning for adaptation to impacts of climate change have approached climate change adaptation not as a narrow infrastructure or emergency preparedness assignment of traditional disaster planning but as an opportunity for broad-based participation by a wide range of stakeholders. Either as an initial objective or as an outcome of the participative process, these experiments have framed adaptation as an element of community or regional resilience and sustainability, related to current development stresses as well as longer term projections of climate change. In many cases, in fact, climate change has become the catalyst for more integrated attention to sustainability issues beyond climate change alone (Wilbanks 2003).

States, cities, and towns recognize that climate change is only one of many driving forces for global change that shape the sustainability of localities, regions, and nations. Its importance is wrapped up in how it interacts with other driving forces such as demographic change, global economic change, technological change, and institutional change (IPCC 2007, chap. 7). For example, climate change can mean exposure to more severe weather events, increased water scarcity, or sea-level rise; but these effects interact with changing population sizes and distributions, global and regional economies, public policies, and issues associated with resource consumption and waste disposal. Thus, climate change prevention and adaptation is important as a dimension of sustainability, not as an issue in itself.

They also recognize that most climate change impacts themselves are products of multiple stresses. For example, the coastal region from Mobile, Alabama, to Galveston, Texas, is now facing coastal retreat from the combined effects of increased hurricane intensity, subsidence, sea-level rise, wetlands destruction, and human settlement (Savonis, Burkett, and Potter 2008). The increased wildfire experience in the Mountain West is a combined product of previous fire management, drought, storm intensity, and human settlement. Thus, resilience to multiple threats and stresses might be an effective way to incorporate climate change adaptation into a wider effort for community and regional resilience.

The most common meaning of resilience is drawn from the engineering sciences, as the capacity to absorb disturbances and to return to a prior (relatively stable) state. An alternative meaning is drawn from the ecological sciences, where resilience is the capacity to both absorb disturbance and to reorganize into a system that still retains its previous functions (Gunderson 2008). Some interdisciplinary scientists use the term *resilience* to describe specific responses (adjustments, adaptations, coping actions, or adaptive capacity) used to reduce vulnerability to climate impacts. Sustainability scientists tend to use all of three of these concepts, including the capacity to absorb perturbations and return to previous states but beyond these the capacity to reorganize to move toward a state better than the previous state.

A current research project (the Community and Regional Resilience Initiative of the Community and Regional Resilience Institute [CARRI]) has defined community resilience as a community or region's capability to anticipate, prepare for, respond to, and recover from significant multihazard threats with minimum damage to public safety and health, the economy, and national security (Kates and Wilbanks 2009). As a concept for discussion and a goal to be sought, community resilience is real, but it is not simple. Through the choice of a community or region, resilience is place based, rooted in linked social, economic, and environmental systems that are always in some ways unique to a particular place. By addressing multihazard threats, including the geophysical, biological, and social, resilience means the capacity to address these often in combination, as well as dealing with surprises or threats that were not and could not have been anticipated. By espousing minimal damage as a criterion for success, the community commits itself to reduce the vulnerability of all parts of the community. By addressing concerns from local safety and health to national security, community resilience recognizes that no community is a self-sufficient island but is linked with other communities, its region, the nation, and indeed the world. Finally, as a measure of a community's capability, community resilience is a continuing process that adapts to changes in circumstances and learns from its (and others') experience as threats, vulnerabilities, and resources for response and recovery change through time.

Case Studies of Climate Change Adaptation in a Broader Context of Community Resilience

CARRI is an ongoing study involving the cities of Charleston, South Carolina; Memphis, Tennessee; and Gulfport, Mississippi. These case studies in the Southeast United States are intended to improve our understanding of how to enhance community resilience for the future. Threats being considered include natural and other disasters that might be associated with climate change, as well as earthquakes, economic changes, and exposures to health risks.

Charleston, South Carolina

Charleston, South Carolina, understands the idea of multiple threats to a community's well-being. Besides being threatened by hurricanes and other severe coastal storms, with associated flooding, wind damage, and other impacts (e.g., from Hurricane Hugo in 1989), and by earthquakes along its inland margins, its greatest threat in recent memory was the economic impact of the closure of the Charleston Naval Base and Shipyard in 1974. The economy of the city of Charleston itself is vulnerable because it is rather narrowly focused on tourism in contrast to its neighboring city of North Charleston, which is a regional commercial hub.

Climate change adaptation in Charleston would be expected to emphasize the possibility of increased risks of disruptive coastal storms, coupled with sea-level rise. Other climate-related vulnerabilities could include changes in exposures to health risks as pandemics arise and disease vectors shift location with climate change and changes in urban comparative advantage in such sectors as tourism, as competitors, markets, and sources of inputs in other areas are affected by climate change.

Viewing these sorts of vulnerabilities, along with others, through a resilience lens rather than a climate change lens, the Charleston tri-county metropolitan area (Berkeley, Charleston, and Dorchester) has concluded that its long-term well-being depends fundamentally on being ready for any of a wide variety of possible disasters (or combinations of them), adopting the CARRI philosophy that community resilience involves communication and cooperation across all parts of the community: not only government agencies but also business and industry, nongovernmental organizations (NGOs) and volunteer groups, neighborhoods, the media, and other components of a system of vulnerabilities and responses.

This effort is being coordinated by the Tri-County Council of Governments, focused initially on a set of resilience challenges that were identified through a series of community-wide focus group discussions (CARRI 2009b). Addressing these would contribute directly to climate change adaptation, such as transportation and mobility vulnerabilities and region-wide communication and information challenges.

One particular emphasis has been on improving the resilience of a large scattering of economically disadvantaged and rather isolated small municipalities east of the Cooper River, relying substantially on a network of faith-based organizations to be prepared to provide communications coordination and the distribution of goods and services in an emergency. This network not only offers resilience for possible disasters in the future but it pays benefits every month in coordinating dental, medical, educational, and training services that address chronic aspects of poverty in the area.

Clearly, these activities incorporate adaptation to possible climate change impacts, without any significant needs for a community climate change adaptation plan or external funding of adaptation actions. Maybe most important, they are perceived locally as providing benefits in current community operations, not just in the uncertain future event of a climate change–related disaster.

Memphis, Tennessee

The Memphis, Tennessee, area is a major focus of transportation systems of crucial national importance, from natural gas pipelines to Federal Express, and it is seriously threatened by earthquake hazards in the New Madrid zone along the course of the Mississippi River. Of more immediate concern to the community is an annual risk of severe tornadoes in the region, which in 2003 caused severe damage, along with its experience that in times of coastal hurricane exposure, the Memphis area becomes a destination for evacuees.

Climate change adaptation in the Memphis area would tend to focus on possible heat index increases associated with average warming, possible health effects, and effects associated with changes in the intensity or tracks of severe weather events, especially tornadoes. It would not include attention to such risks as earthquakes, and it would be weakened by the fact that existing climate models do not project tornado behavior.

The Memphis area has emphasized unprecedented networking among groups not previously in contact with each other, working toward a communication process that bridges diversity, layer after layer, evolving as leaders change and people move. The Shelby County Joint Economic and Community Development Board coordinates the Memphis resilience enhancement effort. It is also guided in its initial priorities, like Charleston, by the results of a series of community-wide discussions that identified focus areas including the following—again directly relevant to climate change adaptation (CARRI 2009c): identifying vulnerable residents, small business continuity and disaster recovery, and volunteer coordination.

A particular concern is with identifying and being prepared to address special needs, not only among the area's own population but also among evacuees who can arrive on very short notice in numbers that require creating an instant evacuee city. In addition, Shelby County has supported an innovative information and education program to increase community awareness of risks and vulnerabilities, using the theme "I'm ready!"

As in the Charleston case, the Memphis urban area experience with community resilience enhancement seems very likely to make this area more adaptable to climate change impacts through local initiatives, with near-term benefits that include improving relationships and communication structures across diverse groups within the community.

Gulfport, Mississippi

Gulfport, Mississippi, shares aspects of both the New Orleans case (discussed later) and the two other CARRI cases, because its perspectives on community resilience are dominated by its experience with Hurricane Katrina in late August 2005 and the months and years that followed. Responding to risks of similar events in the future is one of its challenges, seriously exacerbated by projections of land subsidence over the next half-century in the Gulf Coast region, associated with a projection that "apparent sea-level rise" is likely to be two to four feet by 2050 (Savonis, Burkett, and Potter 2008), along with the likelihood that coastal storms will become more intense with climate change. In addition, the area faces possible vulnerabilities to both international trade impacts and pandemics because of its proximity to and trade linkages with the Caribbean and Latin America.

Memories of the pain of Katrina for residents of the Gulfport area are still so vivid that thinking beyond that one kind of vulnerability has been difficult; at the same time, however, the community feels a strong need to get beyond such a close identification with hurricane risks that its prospects for tourism and other kinds of economic development are undermined. The main themes of multithreat resilience discussions in Gulfport have been the need to know the community and the need to come up with innovative solutions to problems that emerge in disaster conditions that were not anticipated. A community is not resilient because it has a plan; it needs to be prepared to listen, observe, adapt, address problems, and welcome unconventional partners in the emergency response.

Particular emphases in Gulfport include tracing out interdependencies among community functions and facilities, improving communication structures across the community to keep messages accurate and consistent, getting schools and businesses back in operation as quickly as possible (e.g., reopening local businesses to sell emergency materials and commodities, keeping the sales taxes within the community), and institutionalizing a local source of expertise regarding resilience knowledge (the University of Southern Mississippi has established a Center for Policy and Resilience in the area to serve as a local and regional focus).

The case of Gulfport is different from the other two CARRI partner cities because it starts with a focus on a particular climate-related disaster and then broadens to begin considering other threats as well, rather than beginning with a comprehensive frame and considering climate change impacts within that frame. If the projection of an apparent sea-level rise of two to four feet in forty years or less turns out to be accurate, this community is vulnerable to effects of intense coastal storms that could threaten its viability in the latter half of the twenty-first century. In this case, considering a combination of climate change and other environmental changes could push the limits of potentials for adaptation in situ, raising questions about contingency planning for the gradual relocation of some coastal land uses.

Gulfport, in fact, might be an exception to the more general CARRI observation that multithreat resilience should be the starting point, not climate change adaptation. Here, more attention to climate change risks and vulnerabilities might encourage the community to consider needs to adapt to longer term risks and vulnerabilities not limited to coastal hurricanes alone—and it is related to a potential for climate change–related risks to be a game changer for the community in coming decades, in which case climate change adaptation could mean a need to consider such structural changes as a relocation of population and socioeconomic activities.

General Lessons from the CARRI Experience

Lessons from these three community resilience enhancement cases include the following:

1. Community resilience means all-hazards planning—and also links with other community issues, such as poverty or economic growth. To sustain itself, resilience has to show that coordination offers benefits in daily operations, not just in the event of an emergency.
2. Resilience means that, in a community, people who need to respond together in an emergency know each other ahead of time. From Mayor A. C. Wharton of Shelby County, Tennessee: "A community that prepares together is going to stay together when something happens" (CARRI 2009b). In fact, the benefits from broader acquaintanceships in a community extend well beyond the immediate purpose of preparedness.
3. Timely communication structures that bridge community diversity are critical, especially nontraditional structures, as normal structures fail to operate during an emergency. NGOs and faith-based organizations are often adaptable gap-fillers if they are included in the resilient community network ahead of time.

It is instructive to compare lessons from the three CARRI cases with the experience of New Orleans with Hurricane Katrina. New Orleans has an extraordinary history of multihazard threats, experience, and resilience. Located on the on the subsiding delta of the lower Mississippi River, much of the city is below sea level. It has experienced twenty-seven major floods over the past 290 years (Kates et al. 2006), as well as nineteenth-century invasions, yellow fever epidemics, twentieth-century drinking water pollution, and a declining population and economy (Colten 2005). Hurricane Katrina accelerated the decline in the population and the economy. Today, flooding remains the most pressing concern, with future vulnerability increased by climate change increases in hurricane intensity, continued subsidence, loss of protective wetlands, and inadequate protection.

To deal with flooding, local and national institutions have combined to erect an extensive flood protection system, create river flood and hurricane forecasting, and develop evacuation plans. Exposure to flooding was relatively small in the most vulnerable locations until Hurricane Betsy in 1965. Following that storm, new levees and improved internal drainage encouraged new development in low-lying areas, increasing the most exposed population by 170,000 households across the metropolitan area.

When Hurricane Katrina arrived in August 2005, the storm overwhelmed the levee system and flooded 80 percent of the city, caused about 1,300 deaths, forced a long evacuation that led to the relocation (perhaps permanently) of 100,000 residents, damaged 70 percent of the city's residences, and caused an estimated monetary loss of $40 to $50 billion. Almost four years after Katrina, a population the equivalent of 70 percent of the prestorm population has returned, building permits for more than a third of residences have been issued, and the hospitality economy has been restored. Still, large areas of the city are still empty tracts, leading economic sectors in medicine and education have not recovered, organized reconstruction is just beginning, and some neighborhoods might have been lost forever.

Lessons learned from the New Orleans experience of the four key elements of resilience—anticipation, response, recovery, and reduced vulnerability—are detailed in a CARRI report (Colten, Kates, and Laska 2008). Six of these are especially relevant to climate change and multihazard adaptation:

1. Vulnerability grows from multiple causes, not just from climate. In New Orleans, geophysical vulnerability is characterized by its below-sea-level, bowl-shaped location, its accelerating subsidence, rising sea level, storm surges, and possible increased frequency of larger hurricanes from climate change. These are only partly natural phenomena and they have been made worse by settlement decisions, canal development, loss of barrier wetlands, extraction of oil and natural gas, and the design, construction, and failure of protective structures and rainfall storage. Social vulnerability grew as well, as new development in low-lying areas placed an additional 170,000 households at risk. Subsequent loss of population within the city (white flight) increased social vulnerability, followed by the Katrina failure to respond to the distinctive needs of the elderly, the poor, and households without autos.
2. Successful short-term adaptation might lead to larger long-term vulnerability. The forty-year period between Hurricanes Betsy and Katrina produced new and improved levees, drainage pumps, and canals—successfully protecting New Orleans against three hurricanes in 1985, 1997, and 1998. These same works, however, permitted the massive development of previously unprotected areas and, when the works themselves failed, became the major cause of the Katrina catastrophe.
3. Adaptation is a long-term process. In New Orleans, it took forty years to create an effective tracking and warning system and thirty-seven years to inform the community about the catastrophic threat. It also took forty years to reduce vulnerability with levees and drainage by a system that was only partly completed before Katrina and that subsequently failed. It will take at least six years to rebuild a reliable levee system to protect against a modest 100-year storm. The emergency response period following Katrina was the longest of any similar disaster in U.S. history (six weeks). To develop a community-acceptable reconstruction plan took twenty-one months and to reconstruct the city after Katrina will take at least a decade more.
4. The best available scientific and technological knowledge does not necessarily get used or widely disseminated. An extraordinary investment has been made in climate change research producing a growing body of scientific and technological knowledge. The New Orleans experience does not augur well for its utilization, however. The engineering designs for the new and improved protective works after Hurricane Betsy in 1965 took into account the effects of hurricane recurrence, storm surge, land subsidence, and rising sea level as measured at that time. These estimates were still being used nineteen years later, though, when sea level had risen by seven inches, storm waves and surges by similar amounts, subsidence had lowered the land surface by ten feet (U.S. Geological Survey 2004), and hurricane intensity increased from climate change. Moreover, the widely used risk assessments—in the form of Federal Emergency Management Agency maps of the 100-year floodplain—have never included sea-level rise or land subsidence effects.
5. Despite frequent references to partnerships, major response capacities and resources might be invisible to, refused by, or poorly used by the official emergency response structure. In every disaster there are unanticipated or unaddressed needs and "shadow responders" often emerge from households, friends and family, neighborhoods, NGOs, and voluntary organizations, businesses, and industry. In responding to Katrina, these emergent capabilities were sometimes refused or poorly used by government officials, even though they provided most of the initial evacuation capacity, sheltering, feeding, health care, and rebuilding and much of the search and rescue, cleanup, and post-Katrina funding.
6. Surprises should be expected. Every hazard event, climatic or otherwise, brings surprises and every disaster even more. Unanticipated events during Katrina included the massive breaches that flooded

80 percent of the city with more than twenty feet of water. In turn, these unanticipated events led to major failures in emergency response for events that had been anticipated. Thus, surprises come from unanticipated events, correctly anticipated events but failed responses, or wrongly anticipated events.

Climate Change Adaptation in a Broader Context

In summary, climate change impacts are real, and adaptation will be an unavoidable part of the response, but climate change adaptation is deeply and complexly linked with economic and social development paths and stresses. The experience to date in the growing efforts at adaptation planning finds states, cities, and towns moving beyond adaptations to climate change, seeking to be resilient to multiple threats and stresses and to achieve sustainability. In so doing, there are at least three main benefits: helping to understand climate change vulnerabilities as the products of multiple threats and stresses, achieving community acceptance of needed adaptations as cobenefits of addressing multiple threats, and mainstreaming the process of climate adaptation in the larger envelope of social relationships, communication channels, and broad-based awareness of needs for risk management that accompany true community resilience.

Acknowledgments

The research reported in this article was supported in part by funding from the U.S. Department of Homeland Security (DHS) to CARRI at the Oak Ridge National Laboratory (ORNL). The interpretations are those of the authors and do not necessarily speak for DHS, CARRI, or ORNL.

References

Burton, I. 1997. Vulnerability and adaptive response in the context of climate and climate change. *Climatic Change* 36:185–96.

Clark, W. C., J. Jaeger, R. Corell, R. Kasperson, J. J. McCarthy, D. Cash, S. J. Cohen et al. 2000. Assessing vulnerability to global environmental risks. Discussion Paper 2000–12, Environment and Natural Resources Program, Belfer Center for Science and International Affairs (BCSIA), Kennedy School of Government, Harvard University, Cambridge, MA.

Colten, C. E. 2005. *An unnatural metropolis: Wrestling New Orleans from nature*. Baton Rouge: Louisiana State University.

Colten, C., R. Kates, and S. Laska. 2008. Community resilience: Lessons from New Orleans and Hurricane Katrina. CARRI Research Report 3, Community and Regional Resilience Institute, Oak Ridge, TN.

Community and Regional Resilience Institute (CARRI). 2009a. *CARRI community forum white papers, volume 1*. Oak Ridge, TN: Community and Regional Resilience Institute. http://www.resilientus.org/library/Mayors_Panel_1258138040.pdf (last accessed 8 July 2010).

———. 2009b. *Charleston area case study*. Oak Ridge, TN: Community and Regional Resilience Institute.

———2009c. *Memphis urban area case study*. Oak Ridge, TN: Community and Regional Resilience Institute.

Feldman, D., and K. Jacobs. 2008. Making decision-support information useful, useable, and responsive to decision-maker needs. In *Decision-support experiments and evaluations using seasonal-to-interannual forecasts and observational data: A focus on water resources*, ed. N. Beller-Simms, 101–40. Asheville, NC: National Climatic Data Center.

Gunderson, L. 2008. Comparing ecological and human community resilience. CARRI Research Report 5, Community and Regional Resilience Institute, Oak Ridge, TN.

Immediate Action Workgroup. 2009. *Recommendations to the governor's sub-cabinet on climate change*. Juneau, AK: Immediate Action Workgroup.

Intergovernmental Panel on Climate Change (IPCC). 2001. *Climate change 2001: Impacts, adaptation, and vulnerability*. New York: Cambridge University Press.

———. 2007. *Climate change 2007: Impacts, adaptation and vulnerability*. New York: Cambridge University Press.

Karl, T., J. Melillo, and T. Peterson, eds. 2009. *Global climate change impacts in the United States*. Cambridge, MA: U.S. Global Change Research Program.

Kates, R., C. Colten, S. Laska, and S. Leatherman. 2006. Reconstruction of New Orleans after Hurricane Katrina: A research perspective. *Proceedings of the National Academy of Sciences* 103:14653–660.

Kates, R., and T. Wilbanks. 2009. Community resilience: What research tells us. Paper presented to the American Association for the Advancement of Science, Chicago, 15 February.

McIntosh, R., J. Tainter, and S. McIntosh, eds. 2000. *The way the wind blows: Climate, history, and human action*. New York: Columbia University Press.

Moser, S. 2009. *Good morning, America! The explosive U.S. awakening to the need for adaptation*. Santa Cruz, CA: Susanne Moser Research and Consulting.

National Assessment of Potential Consequences of Climate Variability and Change. 2000. *Climate change impacts on the United States: The potential consequences of climate variability and change*. Washington, DC: U.S. Global Change Research Program.

New York Panel on Climate Change. 2009. Climate risk information. New York: New York Panel on Climate Change. http://www.nyc.gov.html.om/pdf/2009/NPCC_CRI.pdf (last accessed 8 July 2010).

Pew Center on Global Climate Change. 2008. *Adaptation planning: What U.S. states and localities are doing (2008 update)*. Arlington, VA: Pew Center on Global Climate Change.

Sakakibara, C. 2008. Our home is drowning. *The Geographical Review* 98:456–75.

Savonis, M., V. Burkett, and J. Potter. 2008. *Impacts of climate change and variability on transportation systems and infrastructure: Gulf Coast study, Phase I*. Synthesis and Assessment Product 4.7, U.S. Climate Change Science Program, Washington, DC.

Schipper, E. L. F., and I. Burton, eds. 2009. *The Earthscan reader on adaptation to climate change*. London: Earthscan.

Snover, A. K., L. Whitely Binder, J. Lopez, E. Willmott, J. Kay, D. Howell, and J. Simmonds. 2007. *Preparing for climate change: A guidebook for local, regional, and state governments*. Oakland, CA: Local Governments for Sustainability (ICLEI).

Tol, R. S. J. 2002a. Estimates of the damage costs of climate change: Part I. Benchmark estimates. *Environmental and Resource Economics* 21:47–73.

———. 2002b. Estimates of the damage costs of climate change: Part II. Dynamic estimates. *Environmental and Resource Economics* 21:135–60.

Tol, R. S. J., M. Bohn, T. E. Downing, M. L. Guillerminet, E. Hizsnyik, R. Kasperson, K. Lonsdale et al. 2006. Adaptation to five metres of sea level rise. *Journal of Risk Research* 9:467–82.

U.S. Geological Survey. 2004. Subsidence and sea-level rise in southeastern Louisiana. http://coastal.er.usgs.gov/LA-subsidence/ (last accessed 8 July 2010).

Wilbanks, T. 2003. Integrating climate change and sustainable development in a place-based context. *Climate Policy* 3S1:147–54.

Changes in Annual Land-Surface Precipitation Over the Twentieth and Early Twenty-First Century

Elsa Nickl, Cort J. Willmott, Kenji Matsuura, and Scott M. Robeson

Time trends in annual land-surface precipitation during the twentieth and early twenty-first centuries and their spatial patterns are estimated from gridded (at a 0.5° × 0.5° spatial resolution) rain-gauge-based precipitation data sets available from the Climatic Research Unit (CRU), the Global Precipitation Climatology Centre (GPCC), and at the University of Delaware (UDel). Our analyses of these precipitation data sets make use of spatially weighted (geographic) percentiles as well as of join-point and simple linear regression. A consistent increase in annual land-surface-average precipitation (of approximately 0.2 and 0.5 mm/year) occurred during the first half of the twentieth century. This increase was followed by nearly a half-century (approximately forty-four years, from 1949 through 1993) of decreases in annual land-surface-average precipitation (on the order of 0.3 to 0.6 mm/year). Trends, once again, reversed themselves in the early 1990s and increased (at rates of approximately 0.75 to 2.1 mm/year) over the decade from 1992 through 2002. Maps of precipitation change during these alternating periods of increasing and decreasing precipitation show considerable spatial variability. *Key Words: climate change, precipitation, rain gauge data.*

依据从气候研究中心（CRU），全球降水气候学中心（GPCC），和特拉华大学（UDel）所得到的栅格雨量计降雨数据集（0.5 × 0.5 度的空间分辨率），本研究估算了二十世纪和二十一世纪早期的每年陆地表面降水的时间趋势以及它们的空间格局。我们对这些降水资料进行了空间百分比和联接点加权（地理）分析和简单线性回归分析。研究表明，在 20 世纪上半叶，年度陆表平均降水量产生了一致性的增长（大约 0.2 至 0.5 毫米/年）。这个增长之后是近半个世纪（大约 44 年，从 1949 年到 1993 年）年度陆表平均降水量的减少（依次为 0.3 至 0.6 毫米/年）。在二十世纪九十年代初，趋势再次逆转，从 1992 年到 2002 年的十年间重新出现了增长（约 0.75 至 2.1 毫米/年）。在这些降水增加和减少的交替时期，降水量地图展示出很大的空间变化性。*关键词：气候变化，降水，雨量计数据。*

Las tendencias temporales en precipitación anual sobre el terreno durante el siglo XX y comienzos del XI y sus patrones espaciales están calculados a partir de conjuntos de datos de precipitación basados en registros pluviométricos distribuidos en cuadrícula (a 0.5° x 0.5° de resolución espacial), disponibles en la Unidad de Investigación Climática (CRU, sigla en inglés), el Centro de Climatología para la Precipitación Global (GPCC) y la Universidad de Delaware (UDel). Nuestros análisis de estos conjuntos de datos de precipitación utilizan percentiles espacialmente ponderados (geográficos), lo mismo que regresiones *join-point* y linear simple. Se presentó un consistente incremento de la precipitación promedio anual sobre el terreno (de aproximadamente 0.2 y 0.5 mm/año) durante la primera mitad del siglo XX. A tal incremento siguió un período de cerca de medio siglo (aproximadamente cuarenta y cuatro años, de 1949 a 1993) de disminución de la precipitación promedio anual sobre el terreno (del orden de 0.3 a 0.6 mm/año). Nuevamente las tendencias se reversaron a sí mismas a principios de los 1990 y se incrementaron (a tasas de aproximadamente 0.75 a 2.1 mm/año) en la década que se extendió de 1992 a 2002. Los mapas del cambio de la precipitación durante estos períodos alternantes de incremento y disminución de la precipitación muestran considerable variabilidad espacial. *Palabras clave: cambio climático, precipitación, datos pluviométricos.*

A better understanding of the spatial and temporal variability of land-surface precipitation is indispensable for advancing climate change research, as well as for assessing the potential impacts of climate change on water resources. In addition to climate-model estimates of precipitation, measurement-based fields of precipitation and precipitation change are critical for evaluating the vicissitudes of Earth's climate. Our current measurement-based knowledge of land-surface precipitation variability and change over the last 100-plus years, however, is uncertain, as is evident in the differences between available gridded land-surface precipitation data sets. The Intergovernmental Panel on Climate Change (IPCC 2007), for example, reports substantial discrepancies among trend estimates derived from different data sets. According to the IPCC (2007) report, trends vary by regions and over time, although land-surface precipitation has generally increased north of 30°N over the past century and decreased over much of the tropics since the 1970s. Our purpose within this article is to explore, describe, and compare the patterns and historical trends in land-surface precipitation using three available high-resolution rain-gauge-based precipitation data sets.

Available Land-Surface Precipitation Data Sets

Estimated land-surface precipitation fields have tended to be of relatively coarse spatial resolution ($\geq 2.5°$), although efforts to develop higher resolution spatial and temporal (e.g., monthly or daily) data sets are increasing. Our analyses of the patterns and trends in annual land-surface precipitation make use of three recently available higher resolution land-surface precipitation data sets. Each of these data sets is derived primarily from in situ (rain-gauge) observations, has a monthly time step, and is gridded at (spatially interpolated to) a relatively high (0.5°) spatial resolution. The data sets are the Climatic Research Unit (CRU) archive (Mitchell and Jones 2005), the Global Precipitation Climatology Centre (GPCC) archive (Rudolf and Schneider 2005), and the recent University of Delaware (UDel) data set (Matsuura and Willmott 2009). Many of the same observational (rain-gauge) records inform all three of these data sets; however, each data set is not based on exactly the same set of rain-gauge records. It also should be noted that the CRU and GPCC archives do not contain estimates of precipitation over Antarctica; in turn, Antarctica is not included in our analyses of "land-surface" precipitation.

Each of the three precipitation data sets is based on a different method of spatial interpolation from the rain-gauge station locations to the nodes of the 0.5° grid. The CRU database is based on an angular-distance weighted (ADW) interpolation method (which bears conceptual similarity to the Shepard 1968 and Willmott, Rowe, and Philpot 1985 approaches). For each grid-node estimate, the ADW method weights each of the eight rain-gauge station precipitation observations that are nearest to the grid node by taking into account the distance from the grid node (using a correlation-decay distance [CDD]) and the directional (angular) isolation of each station. Interpolated fields are forced to a climatological mean value at grid points where there is no station within the CDD (New, Hulme, and Jones 2000; Mitchell and Jones 2005). As a consequence of this, estimated time series over some areas can be invariant for a number of years.

The GPCC data set (Rudolf and Schneider 2005) was spatially interpolated with the "SPHEREMAP" interpolation tool, which was developed at the University of Delaware. It is a spherical adaptation of Shepard's (1968) empirical weighting scheme, which takes into account the spherical distances from the nearby rain-gauge locations to the grid node (for a limited number of the nearest stations), the directional distribution of nearby rain gauges relative to the grid node (to avoid overweighting of clusters of stations), and spatial gradients within the rain-gauge data field (Willmott, Rowe, and Philpot 1985).

The UDel data set also is based on the Willmott, Rowe, and Philpot (1985) traditional-interpolation procedure as well as on Climatologically Aided Interpolation (CAI; Willmott and Robeson 1995; Matsuura and Willmott 2009), which employs a spatially high-resolution climatology to obtain a monthly precipitation difference at each station. These station differences then are interpolated to obtain a gridded field using a version of Shepard's algorithm (Willmott, Rowe, and Philpot 1985); finally, each gridded monthly difference field is added back onto the corresponding monthly climatology field to obtain monthly land-surface precipitation (Matsuura and Willmott 2009). These gridded monthly values then were integrated over the year to obtain an annual average for each year of interest.

The gridded monthly precipitation values contained within the three data sets analyzed here are not adjusted up for possible rain-gauge undercatch (Legates and Willmott 1990); therefore, our (and most other) estimates of annual precipitation are likely to be a little lower than actual precipitation. It is important to note, however, that this should have little or no effect on our estimates of time trends, as long as there was no meaningful time trend in rain-gauge undercatch.

Geographic Analyses of Annual Land-Surface Precipitation

Land-surface precipitation change since the beginning of the twentieth century is evaluated using spatially weighted (geographic) percentiles (Willmott, Robeson, and Matsuura 2007) and simple linear and join-point regression (Rawlins et al. 2006)—the latter is a type of change-point regression (Draper and Smith 1998)—of annual land-surface-average precipitation. An overall assessment of the longer term trends and variability apparent within the entire 100-plus years of record—contained in the CRU, GPCC, and UDel data sets—is made first. More detailed analyses of each of three major subperiods follow. The subperiods are the first half of the twentieth century (1902–1949 herein, for reasons explained later), 1950 through 1993, and 1993 through 2002. The year 1993 is a join-point estimate of the boundary between a long period of drying and more recent increases in land-surface precipitation (discussed later). Maps of change rates for each subperiod are presented and evaluated.

Because annual land-surface precipitation is intrinsically a geographic or spatial variable, we take explicit account of the sizes (areas) of the grid cells when performing our spatial analyses (Willmott, Robeson, and Matsuura 2007). As a consequence, each of our spatial percentiles has a fundamentally geographic interpretation. Our 75th spatial percentile, for instance, bounds that 25 percent of the land surface that contains annual precipitation values that are greater than the value of the 75th spatial percentile. Our join-point regression of land-surface-average precipitation is used to identify those year(s) when major precipitation change occurred; that is, since the beginning of the second half of the twentieth century. Join-point regression finds one or more "optimal" join-points in a time series by minimizing the sum-of-squared residuals of all possible join-point regressions. It helped us locate a year (1993) at or near the transition from more than forty years of drying to a decade or more of increasing precipitation.

Land-Surface Precipitation over the Twentieth and Early Twenty-First Centuries

Using the CRU, GPCC, and UDel data sets, time series of the annual spatial means as well as of the 5th, 25th, 75th, and 95th spatial percentiles are estimated and plotted for each data set (Figure 1). All of the data sets suggest that long-term land-surface average (rain-gauge-caught) precipitation is about 716 mm/year. The three data set average 5th, 25th, 75th, and 95th spatial percentiles are approximately 58, 234, 936, and 2,264 mm/year, which illustrates the right (positive) skew that typifies precipitation distributions. Over the entire period of record, but especially during the first half of the twentieth century, the three traces of the spatial means and spatial percentiles are remarkably similar, with a few exceptions. Prior to the 1930s, for example, the lower 25th spatial percentile of the GPCC trace is noticeably below the CRU and UDel traces, whereas during the second half of the twentieth century, the UDel trace becomes the lowest estimate of the higher spatial percentiles and the spatial mean. Variability, both within

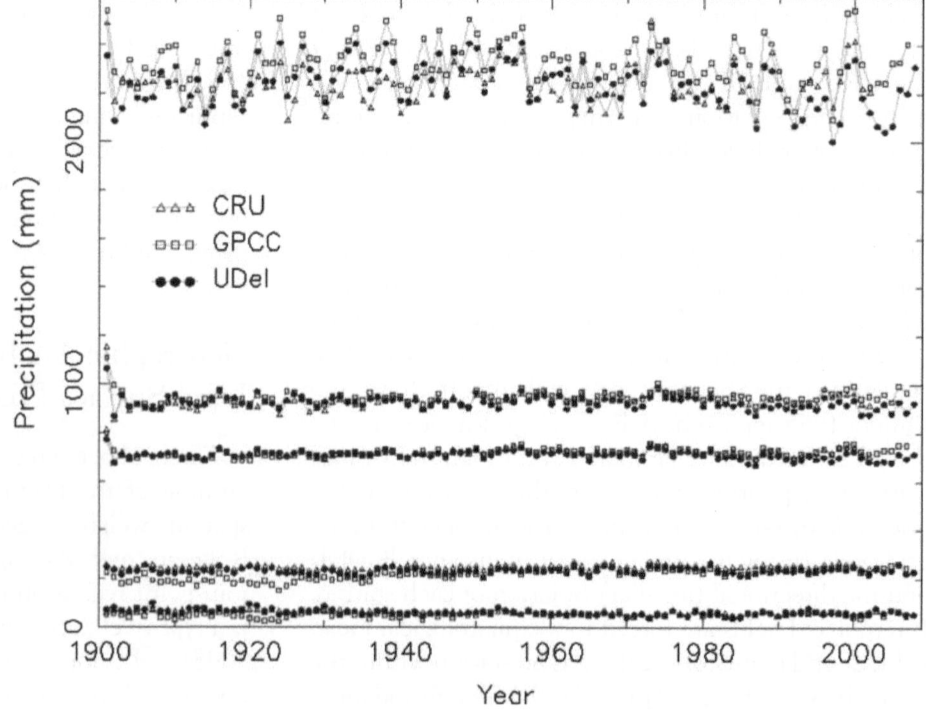

Figure 1. Time series (1901–2008) of annual land-surface precipitation (mm) estimated from the Climatic Research Unit (CRU), Global Precipitation Climatology Centre (GPCC), and University of Delaware (UDel) data sets. Annual spatial means as well as the 5th, 25th, 75th, and 95th spatial percentiles are plotted for each of the three data sets. Open triangles are used to represent CRU data, open squares indicate values derived from GPCC data, and blackened circles show estimates made from the UDel data set. There are five clusters of three curves each. The top set of three curves contains the estimates of the 95th spatial percentile of annual precipitation, the next set (moving down the graph) represents the 75th spatial percentiles, then comes the spatial means, then the 25th spatial percentiles, and the last (bottom) set indicates the 5th spatial percentiles.

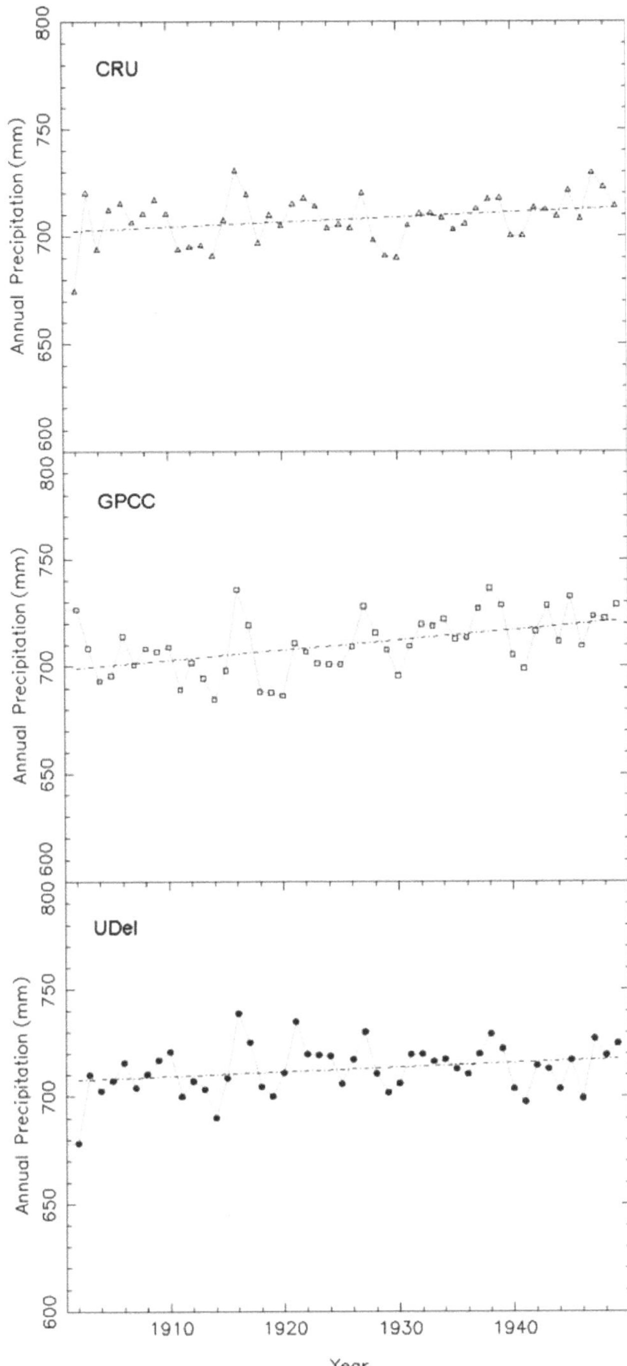

Figure 2. Three simple linear regressions of annual land-surface-average precipitation (over the period 1902–1949) are estimated from the Climatic Research Unit (CRU), Global Precipitation Climatology Centre (GPCC), and University of Delaware (UDel) data sets and plotted. The values of annual land-surface-average precipitation over the period 1902 to 1949 also are plotted. Open triangles are used to represent CRU data, open squares indicate values derived from GPCC data, and blackened circles show estimates made from the UDel data set.

and among the three traces, tends to increase after the 1970s. This appears to arise from increasing variability in the maxima, as suggested in the traces of the 95th spatial percentile.

There are several reasons why the UDel annual average precipitation measures fall increasingly below the comparable CRU and GPCC measures, as the present is approached (Figure 1). The UDel data set contains records from the Global Surface Summary of the Day (GSOD) archive, which improves the station-network coverage of the land surface, especially during the recent past and over the more harsh (drier and rugged regions) of the land surface. It follows that improved representation of drier regions would lessen (relatively speaking) the UDel estimates of land-surface-average precipitation. It also is true, however, that the GSOD archive contains a variety of unrealistic extreme values, including unbelievably long strings of zeros and a limited number of unusually high values. Although Matsuura and Willmott (2009) attempted to filter out only the truly unrealistic daily and monthly precipitation values from the GSOD station records, any erroneous zero value that was missed or an incorrectly removed maximum would tend to produce an underestimate.

Land-Surface Precipitation over the First Half of the Twentieth Century

It is clear that there was a general increase in land-surface-average precipitation during the first half of the twentieth century (1902–1949; Figure 2). Ignoring the estimated precipitation values for the year 1901, owing to probable station-network biases (Willmott, Robeson, and Feddema 1994), all three traces indicate an increase in the spatial mean, at estimated rates of between 0.2 and 0.5 mm/year. The spatial distribution of change over this period, however, is quite variable (Figure 3). Although there are many areas over which relatively small decreases in annual precipitation can be seen (e.g., over large areas of North America, North Africa, and Australia), these are outweighed by increases elsewhere. Large increases are evident over portions of the Amazon basin (especially within the GPCC and UDel fields), over the Maritime Continent (especially within the CRU and GPCC data sets), and, to a lesser extent, over northern India (within all three data sets). Spatially extensive but relatively small increases also appear over

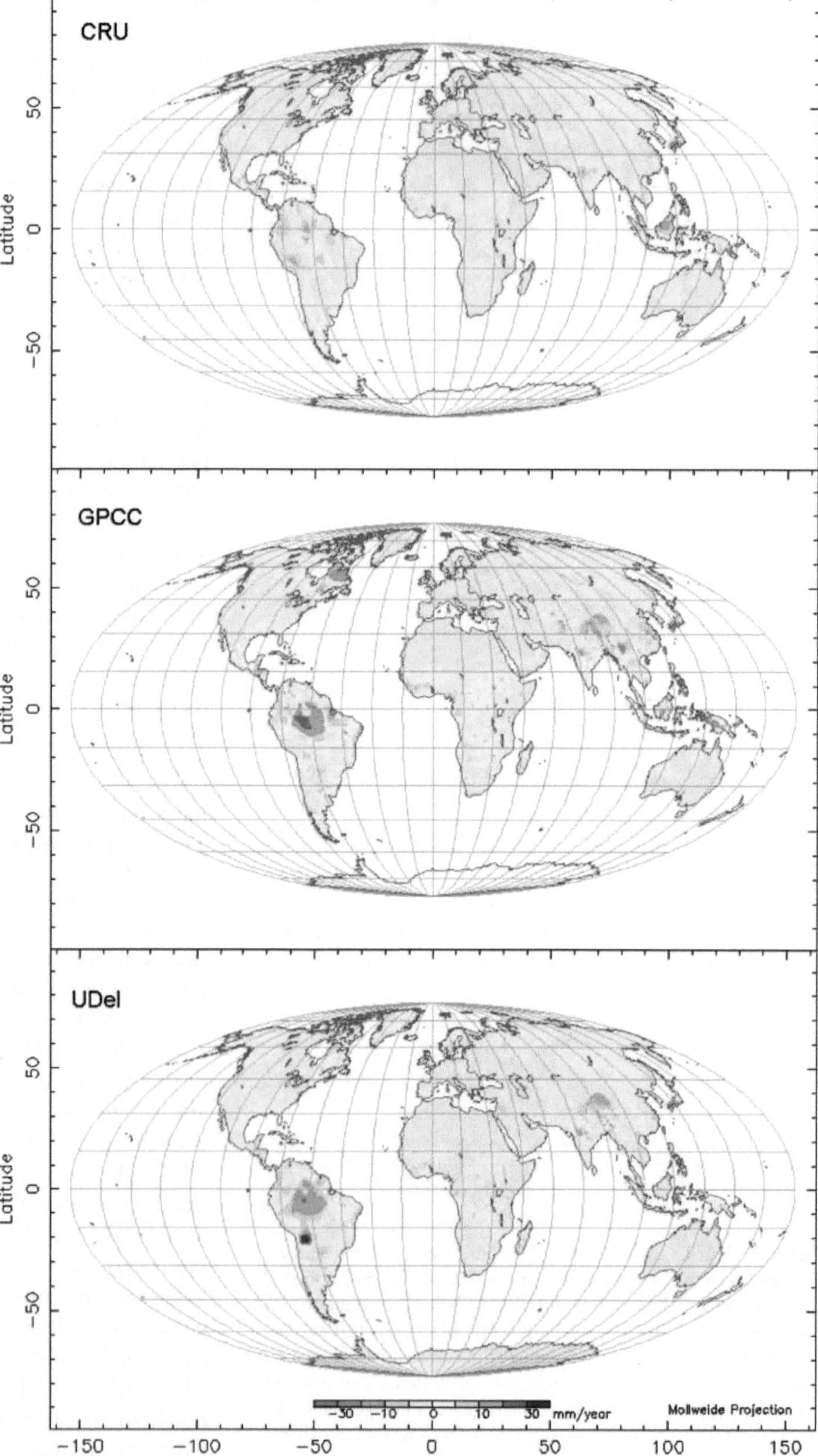

Figure 3. Spatial distribution of time trends (mm/year) in annual land-surface precipitation (1902–1949) apparent within the Climatic Research Unit (CRU), Global Precipitation Climatology Centre (GPCC), and University of Delaware (UDel) data sets.

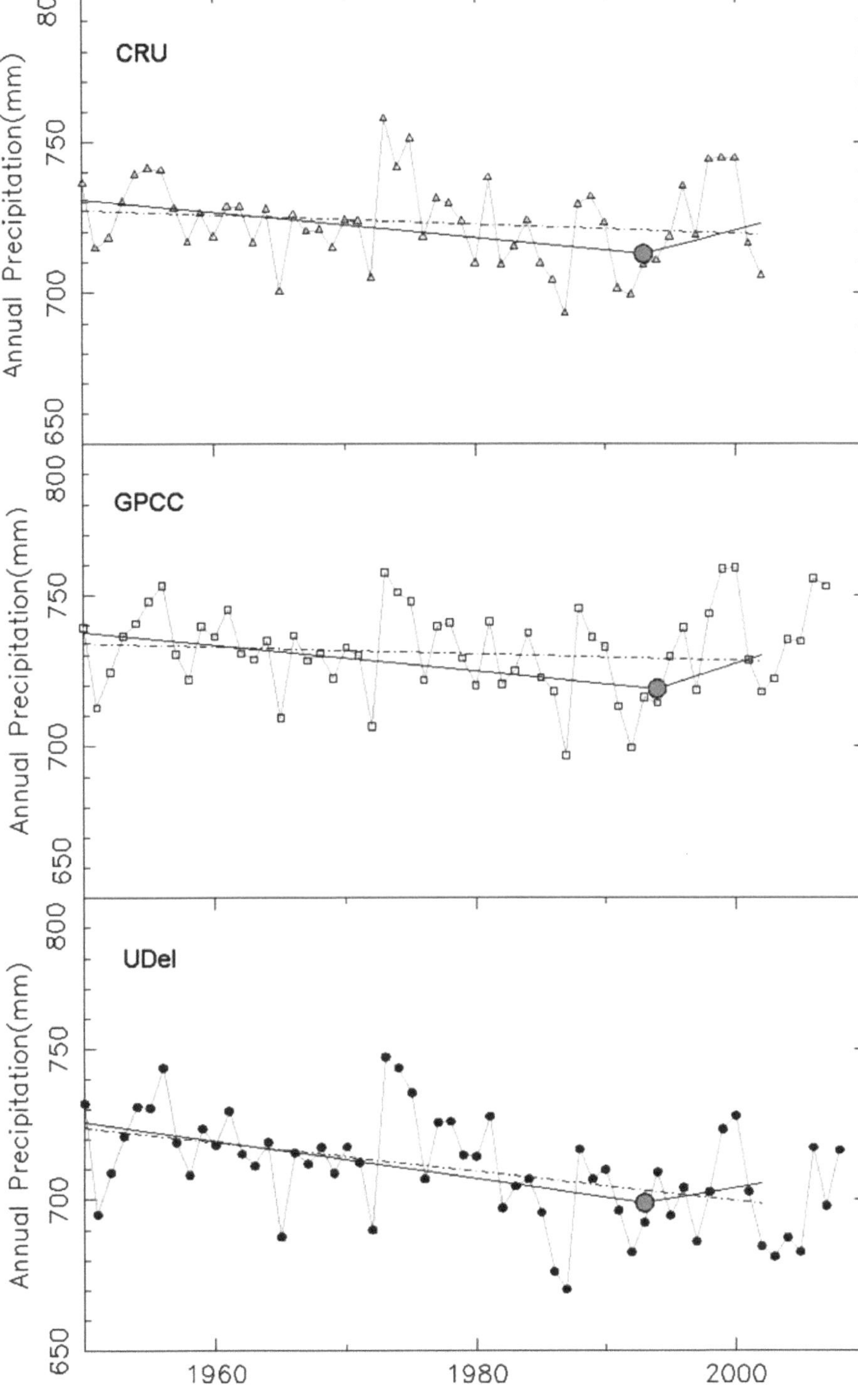

Figure 4. Three join-point and three simple linear regressions of annual land-surface-average precipitation (over the period 1950–2002) are estimated from the Climatic Research Unit (CRU), Global Precipitation Climatology Centre (GPCC), and University of Delware (UDel) data sets and plotted. The values of annual land-surface-average precipitation over the period 1950 through 2008 are also plotted. Open triangles are used to represent CRU data, open squares indicate values derived from GPCC data, and blackened circles show estimates made from the UDel data set.

much of the Arctic land surface and subtropical Africa. Annual land-surface-average precipitation during the first half of the twentieth century and into the 1950s was relatively stable (exhibiting limited variability) but increasing consistently.

Land-Surface Precipitation Since the Middle of the Twentieth Century

Graphs of the spatial means of annual land-surface precipitation from 1949 forward (especially in Figure 4)

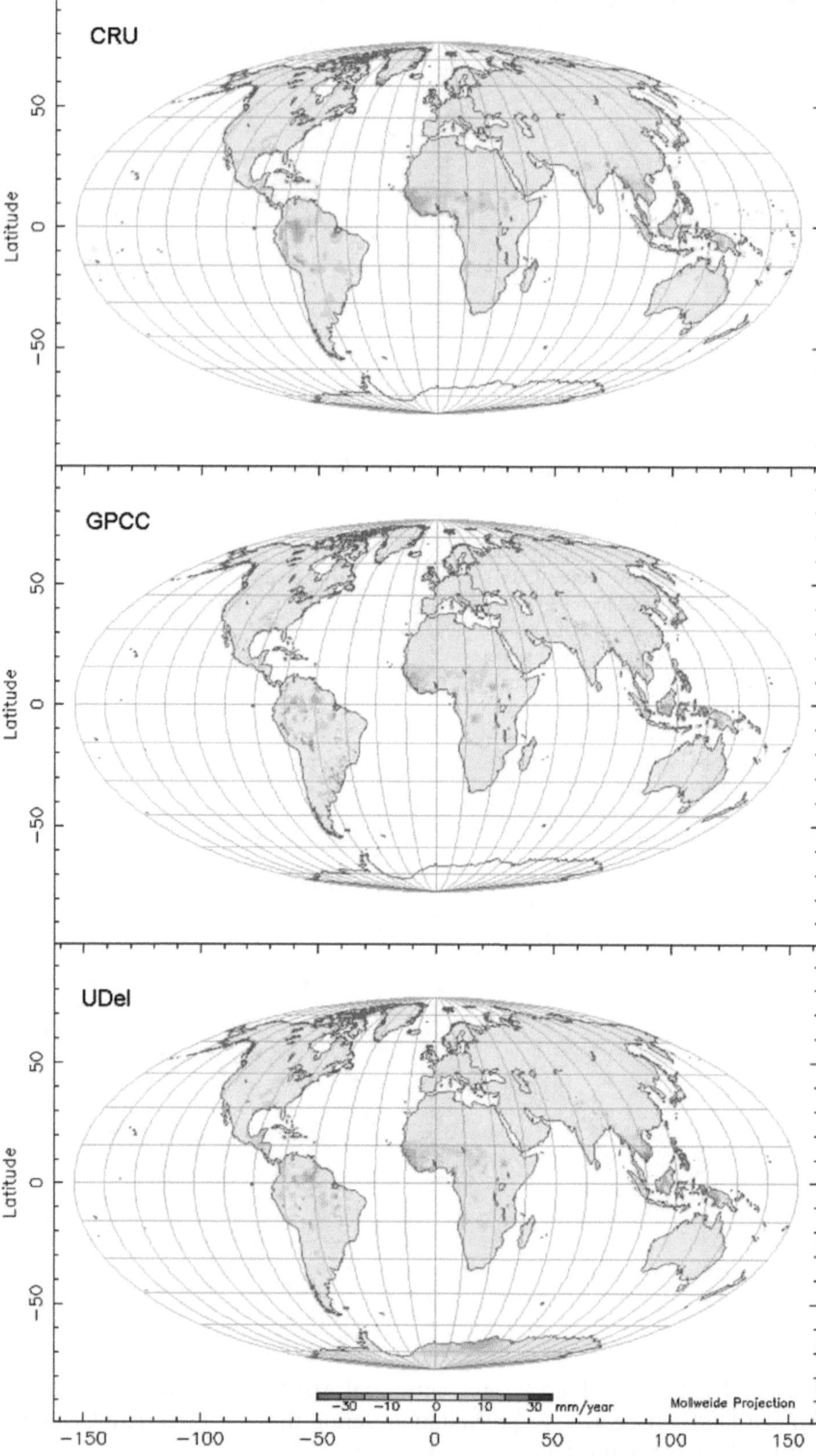

Figure 5. Spatial distribution of time trends (mm/year) in annual land-surface precipitation (1950–1993) apparent within the Climatic Research Unit (CRU), Global Precipitation Climatology Centre (GPCC), and University of Delaware (UDel) data sets.

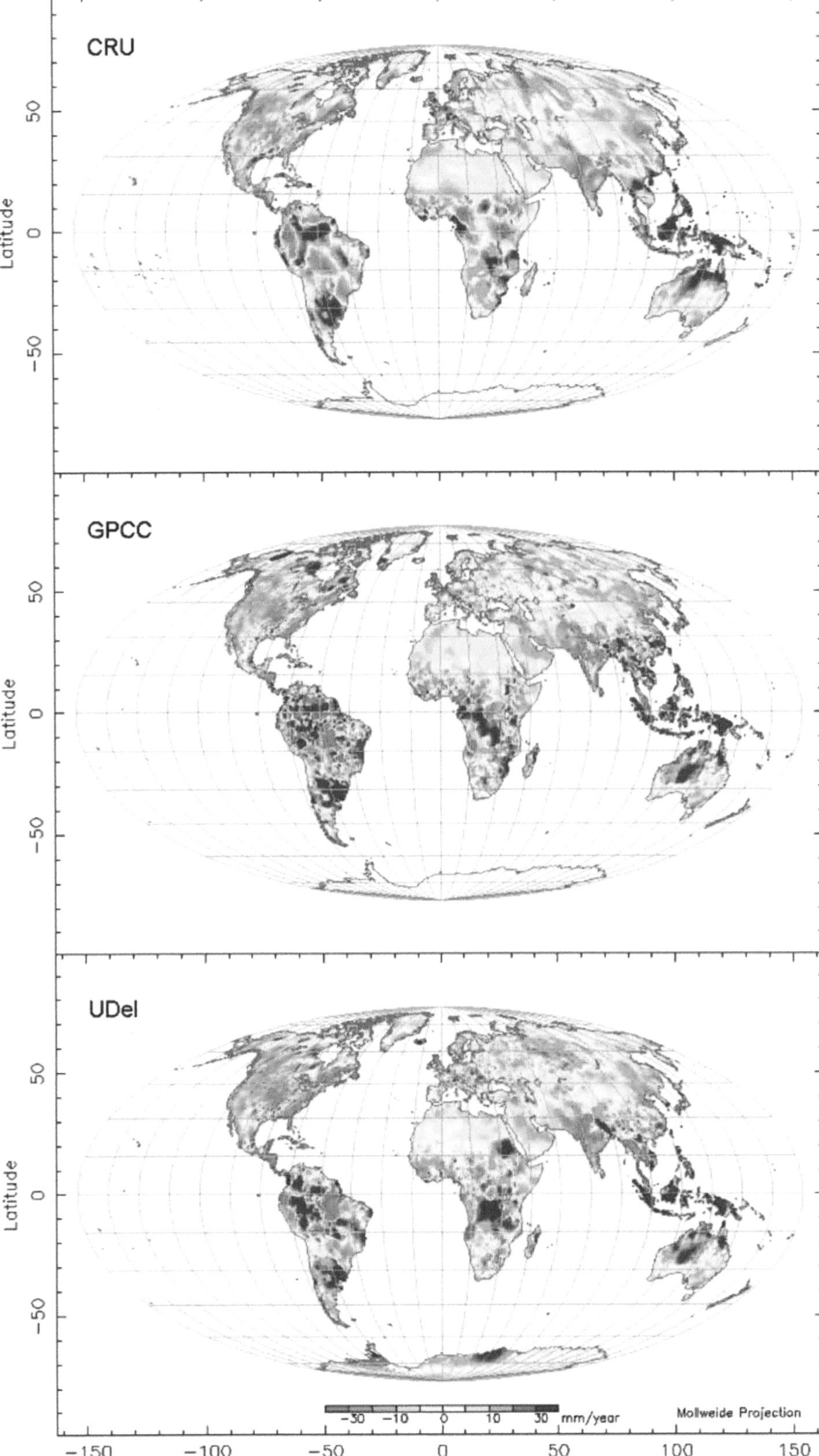

Figure 6. Spatial distribution of time trends (mm/year) in annual land-surface precipitation (1993–2002) apparent within the Climatic Research Unit (CRU), Global Precipitation Climatology Centre (GPCC), and University of Delaware (UDel) data sets.

indicate a dramatic decrease in land-surface precipitation as well as gradually increasing divergence among the three data sets. It is not entirely clear when the drying began but, on inspection (Figure 4), probably during the mid- to late 1950s. It is less clear exactly when it ended; therefore, we use join-point regression to help us make a determination. A single-point, join-point linear regression was fit to each of the three traces of the annual land-surface means over the period from 1949 through 2002 (Figure 4), and each suggests that the decrease ended during the early 1990s, most likely in 1993. Our regressions were applied only through 2002 because that is the last year for which CRU data were available. For comparison purposes, we also estimate and plot (Figure 4) a simple linear regression over the entire period from 1949 through 2002. It describes a general drying over the entire fifty-three-year period.

The reduction in land-surface-average precipitation from 1949 through 1993 was substantial, on the order of 0.3 to 0.6 mm/year or about 14 to 26 mm over the entire forty-four-year period. The steepest decline was estimated from the UDel data (Figure 4). To assess the spatial distribution of this land-surface drying, we estimate the linear time trend over the period from 1950 through 1993 at each land-surface grid node, and we map the rates of change (mm/year) obtained from each of the three data sets (Figure 5). Drying dominates Africa, especially over sub-Saharan Africa, as well as over the Maritime Continent, Southeast Asia, and the northwest Amazon basin. The strongest drying over sub-Saharan Africa and the northwest Amazon basin is estimated from the CRU data set. Areas wherein slight increases appear include North America, southeastern South America, western Australia, eastern Europe, and central Asia, although the signal over eastern Europe and central Asia is weaker within the UDel data set.

Reasons for the persistent, forty-four-year drying that occurred over extensive reaches of Earth's land surface, and especially over the lower latitudes, are not completely understood, but several contributing factors have been identified for the years after 1949 (e.g., see Marengo 2004; Hoerling et al. 2006). Marengo (2004), for example, associated circulation features typical of strong El Niño years with drier conditions over the northern Amazon basin region, while Aldrian and Djamil (2008) attributed decreases in precipitation over parts of the Maritime Continent (during the period from 1955–2005) to a waning of monsoonal dominance. Droughts also tend to develop over the Maritime Continent during strong El Niño years. The drying over much of Africa has been tied to relatively warm sea-surface temperatures atop the tropical oceans, including the Indian Ocean (Hoerling et al. 2006). Our sense is that these and perhaps other factors, such as deforestation—especially within the Amazon region—and global dimming, contributed to the drying.

Land-surface-average precipitation trends then reversed themselves, ostensibly beginning in 1993, and marked increases ensued over the decade from 1992 through 2002. Our second set of join-point regression lines (fit to the 1993–2002 land-surface-average precipitation values but constrained by the 1993 join point) suggest increases that are even more dramatic than the forty-four years of drying that occurred previously (Figure 4). These increases are on the order of 0.75 to 2.1 mm/year but, of course, the reduced length of record (ten years) and nontrivial year-to-year variability argue for cautious interpretation of these increases.

Once again, to assess the spatial distribution of these increases in yearly average precipitation, we estimate the linear time trend over the period from 1993 through 2002 at each land-surface grid node, and we map the rates of change (mm/year; Figure 6). All three data sets show significant drying over much of North America and south Asia as well as over much of North Africa and parts of South America. In contrast, substantial increases in annual average precipitation can be seen over parts of South America (especially over southern South America), southern Africa, northern Australia, the Maritime Continent, and parts of Southeast Asia. Regional dissimilarities in these recent trends emerge from the three data sets (compare the three geographies of trend estimates for South America, for example) and they underscore the need for more research into the finer scale nature of land-surface precipitation variability and change. Nonetheless, judging from the most recent values of estimated land-surface-average precipitation (2003–2008) obtained from the GPCC and UDel data (Figures 1 and 4), it seems likely that land-surface-average precipitation has continued to increase.

Summary and Conclusions

Our assessment of the changing patterns and trends in annual land-surface precipitation is based on three recently available monthly land-surface precipitation (rain-gauge-based) data sets. The data sets were assembled at the CRU, the GPCC, and UDel. Each of these data sets was gridded at (spatially interpolated to) a $0.5^\circ \times 0.5^\circ$ spatial resolution and we temporally integrated the monthly grid-node values to obtain the annual grid-node totals.

Our analyses of land-surface precipitation change made use of spatially weighted (geographic) percentiles, simple linear regression, and join-point regression. All of the precipitation data suggest that long-term average land-surface precipitation (rain-gauge-caught) is about 716 mm/year. There appears to have been a modest but consistent increase in land-surface-average precipitation (between 0.2 and 0.5 mm/year, approximately) during the first half of the twentieth century. This was followed by nontrivial reductions in annual land-surface-average precipitation (on the order of 0.3 to 0.6 mm/year), which were estimated over the period from 1949 through 1993. Trends in land-surface-average precipitation then reversed themselves again in the early 1990s and increased (at rates of approximately 0.75 to 2.1 mm/year) over the decade from 1992 through 2002. Recent data also suggest that land-surface-average precipitation has continued to increase until the present day.

The spatial distribution of the time trends in precipitation during these alternating periods of increasing and decreasing precipitation was explored with maps of the estimated trends, which showed considerable spatial variability. Over the ten years from 1992 through 2002, for instance, there was marked drying over much of North America and south Asia as well as over much of North Africa and parts of South America; at the same time, sizable increases in annual average precipitation occurred over parts of South America, southern Africa, northern Australia, the Maritime Continent, and Southeast Asia. A number of regional differences in trends estimated from the three data sets were apparent as well, especially in recent years.

Our findings argue for additional research into assessing the regional differences among the three sets of estimated trends that emerged from the three data sets; that is, into the finer scale spatial variability of land-surface precipitation. There also is a need to resolve the intra-annual variability of precipitation change. Our hope is that in situ–based precipitation fields, such as these, also will be used to evaluate and improve climate-model estimates of precipitation.

Acknowledgments

Portions of this work were made possible by NASA Grant NNG06GB54G to the Institute of Global Environment and Society and we are most grateful for this support.

References

Aldrian, E., and Y. S. Djamil. 2008. Spatio-temporal climatic change of rainfall in East Java Indonesia. *International Journal of Climatology* 28:435–48.

Draper, N. R., and H. Smith. 1998. *Applied regression analysis*. 3rd ed. New York: Wiley.

Hoerling, M., J. Hurrell, J. Eischeid, and A. Phillips. 2006. Detection and attribution of twentieth-century northern and southern African rainfall change. *Journal of Climate* 19 (16): 3989–4008.

Intergovernmental Panel on Climate Change (IPCC). 2007. *Climate change 2007: The physical science basis. Contribution of Working Group I to the Fourth Assessment Report of the Intergovernmental Panel on Climate Change*, ed. S. Solomon, D. Qin, M. Manning, Z. Chen, M. Marquis, K. B. Averyt, M. Tignor, and H. L. Miller, 1–996. New York: Cambridge University Press.

Legates, D. R., and C. J. Willmott. 1990. Mean seasonal and spatial variability in gauge-corrected, global precipitation. *International Journal of Climatology* 10:111–27.

Marengo, J. A. 2004. Interdecadal variability and trends of rainfall across the Amazon basin. *Theoretical and Applied Climatology* 78:79–96.

Matsuura, K., and C. J. Willmott. 2009. *Terrestrial precipitation: 1900–2008 gridded monthly time series* (Version 2.01). Newark: Center for Climatic Research, Department of Geography, University of Delaware. http://climate.geog.udel.edu/~climate/ (last accessed 15 July 2009).

Mitchell, T. D., and P. D. Jones. 2005. An improved method of constructing a database of monthly climate observations and associated high-resolution grids. *International Journal of Climatology* 25:693–712.

New, M., M. Hulme, and P. Jones. 2000. Representing twentieth-century space-time climate variability: Part II. Development of 1901–96 monthly grids of terrestrial surface climate. *Journal of Climate* 13: 2217–38.

Rawlins, M. A., C. J. Willmott, A. Shiklomanov, E. Linder, S. Frolking, R. B. Lammers, and C. J. Vörösmarty. 2006. Evaluation of trends in derived snowfall and rainfall across Eurasia and linkages with discharge to the Arctic Ocean. *Geophysical Research Letters* 33:L07403.

Rudolf, B., and U. Schneider. 2005. Calculation of gridded precipitation data for the global land-surface using in-situ gauge observations. In *Proceedings of the 2nd Workshop of the International Precipitation Working Group*, 231–47. Germany: Global Precipitation Climatology Centre.

Shepard, D. 1968. A two-dimensional interpolation function for irregularly-spaced data. In *Proceedings, 1968 ACM National Conference*, ed. R. B. Blue and A. M. Rosenberg, 517–23. New York: Association for Computing Machinery.

Willmott, C. J., and S. M. Robeson. 1995. Climatologically aided interpolation (CAI) of terrestrial air temperature. *International Journal of Climatology* 15 (2): 221–29.

Willmott, C. J., S. M. Robeson, and J. J. Feddema. 1994. Estimating continental and terrestrial precipitation averages from rain-gauge networks. *International Journal of Climatology* 14 (4): 403–14.

Willmott, C. J., S. M. Robeson, and K. Matsuura. 2007. Geographic box plots. *Physical Geography* 28 (4): 331–44.

Willmott, C. J., C. M. Rowe, and W. D. Philpot. 1985. Small-scale climate maps: A sensitivity analysis of some common assumptions associated with grid-point interpolation and contouring. *The American Cartographer* 12:5–16.

The Changing Geography of the U.S. Water Budget: Twentieth-Century Patterns and Twenty-First-Century Projections

C. Mark Cowell and Michael A. Urban

Persistent changes in temperature and precipitation patterns have dramatic effects on the availability of surface water for natural vegetation, streamflow, agricultural production, and human consumption. We use a combination of historical observational climate data and water budget equations to develop time-series and maps of twentieth-century water variables within the contiguous United States and compare these with anticipated twenty-first-century patterns projected by global climate models. The results graphically demonstrate regional variation in hydroclimatic trends: areas that experienced convergent actual (AET) and potential evapotranspiration (PET) rates during the twentieth century (such as the Great Lakes and Gulf South) witnessed long-term increases in available moisture, whereas areas with divergent rates (such as the Mid-Atlantic and Great Plains) had greater water deficits. Increasing temperatures through the twenty-first century will produce higher PET across the United States; areas where AET similarly escalates will maintain average moisture levels within twentieth-century ranges, but where AET does not correspondingly increase, as in much of the South and West, average conditions will be comparable to those of extreme twentieth-century droughts. The findings highlight the importance of a regional approach to environmental change, as the impacts of climate on water in the United States will be spatially uneven. *Key Words: climate change, United States, water budget.*

持续的气温和降水的变化模式对地表水的获取产生巨大的影响，包括自然植被，水流，农业生产和人类消费等各个方面。我们综合使用了气候历史观测数据和水预算方程，绘制了美国大陆 20 世纪水变量的时序图，并与依据全球气候模型所预测的 21 世纪的模式进行了对比研究。结果用图例证明了水文气候趋势的区域变化：那些经历了 20 世纪实际收敛蒸散量（AET）和潜在蒸散率（PET）的地区（例如大湖地区和海湾南部），证实了长期的可获得水分的增长，而那些具有发散蒸散率的地区（例如中大西洋和大平原）则经历了极大的水分缺乏。二十一世纪的温度升高会在美国各地产生更高的 PET，在那些 AET 类似地迅速上涨的地区，平均水分含量将会保持在二十世纪的水平，但在那些 AET 没有相应增加的地区，例如美国南部和西部地区，平均状况将和二十世纪最严重的干旱相当。这些结果突出体现了以区域方式研究环境变化的重要性，因为在美国，气候对水的影响将会是空间分布不均匀的。*关键词：气候变化，美国，水预算。*

La persistencia de cambios en los patrones de temperaturas y precipitaciones tiene efectos dramáticos sobre la disponibilidad de agua superficial para la vegetación natural, el flujo de las corrientes, la producción agrícola y el consumo humano. En nuestro estudio utilizamos una combinación de datos climáticos de las observaciones históricas y ecuaciones del balance hídrico, para desarrollar series de tiempo y mapas de las variables hidrológicas del siglo XX en los EE.UU. contiguos y compararlos con los patrones anticipados para el siglo XXI, según lo proyectado por modelos climáticos globales. Los resultados demuestran gráficamente la variación regional de las tendencias hidroclimáticas: áreas que experimentaron tasas convergentes reales (AET) y potenciales (PET) de evapotranspiración durante el siglo XX (como los Grandes Lagos y el Sur del Golfo) fueron testigos de incrementos de la humedad disponible a largo plazo, mientras que áreas con tasas divergentes (como el Medio-Atlántico y los Grandes Llanos) tuvieron déficits de agua mucho mayores. Los incrementos de las temperaturas durante el siglo XXI producirán PET más altas a través de los Estados Unidos; pero en las áreas donde las AET no se incrementen correspondientemente, como en la mayor parte del Sur y Oeste, las condiciones medias serán comparables con las de sequías extremas del siglo XX. Los descubrimientos del trabajo destacan la importancia de un enfoque regional para considerar el cambio ambiental, en cuanto que los impactos del clima sobre el agua en este país serán espacialmente desiguales. *Palabras clave: cambio climático, Estados Unidos, balance hídrico.*

Global climate change has received considerable scientific scrutiny in recent decades and has permeated the public consciousness. Much of this attention initially focused on demonstrating whether change is occurring, followed closely by scrutiny of its provenance. These broad-scale questions have been definitively answered by the scientific community (Rosenzweig et al. 2008). Among other evidence, empirical observations of average air and oceanic temperature patterns worldwide, melting of snow and ice in high latitudes and at high elevations, and rising global average sea levels led the Intergovernmental Panel on Climate Change (IPCC) to conclude that warming of the climate system is unequivocal (IPCC 2007).

Although average global temperatures are on the rise, regional patterns are more complex and variable. As such, attention is shifting from questions of change detection and attribution to investigating more localized patterns and impacts of short- and long-term change (Kintisch 2008). It is at these scales that many of the systemic consequences of climate change will present themselves, often mediated by the availability of water (Bates et al. 2008). Hydrological systems are likely to be altered not only by shifting precipitation patterns in the form of seasonal changes in the timing, intensity, magnitude, and phase state of precipitation but also by changes in surficial energy receipt, evaporation, and transpiration (IPCC 2007). Likewise, terrestrial ecosystems are affected by changes in moisture availability and ensuing disturbance, such as wildfire (Westerling et al. 2006; vanMantgem et al. 2009). Beyond direct health consequences (Patz et al. 2005), societal impacts will stem from modified environmental conditions that affect crop production (Battisti and Naylor 2009) and freshwater resources (Vörösmarty et al. 2000).

Climatic water budget analysis has long been used to identify spatial and temporal patterns in water utilization, flows, and storage at the Earth's surface (Mather 1991; Muller and Grymes 2005). This modeling approach integrates climatic controls with their resultant hydrological conditions (e.g., streamflow, soil moisture content) and therefore provides a useful means of testing how changes in energy, temperature, and precipitation alter the components of the hydrologic system. Since its inception, the water budget has been extensively applied to quantify global and regional hydroclimatic patterns (e.g., Willmott, Rowe, and Mintz 1985; Hay and McCabe 2002; Legates and McCabe 2005; Grundstein 2008) and their biogeographic (e.g., Stephenson 1998), fluvial (e.g., Flerchinger and Cooley 2000), and water resource (e.g., Frei et al. 2002) influences. Researchers have also demonstrated the usefulness of this approach in identifying regional consequences of past and anticipated climatic change (e.g., Mather and Feddema 1986; McCabe et al. 1990; Grundstein 2009).

This article seeks to further the goal of clarifying patterns of regional-scale climatic variability through quantification and visualization of change in water budget conditions within the contiguous United States. We use a combination of historical observational climate data and global climate model projections to identify spatial and temporal trends in twentieth-century water budget variables, which are then compared to conditions resulting from anticipated changes in temperature and precipitation by the end of the twenty-first century. Through this analysis we assess the extent to which projected water budget values are within the range of variability seen during the twentieth century and how this varies geographically.

Methods

As described by Mather (1978), a water budget accounts for what happens to the precipitation falling at a place and thus standardizes all of its climate-based measurements in terms of water. For this study we calculate the full array of water budget variables after the approach originally developed by Thornthwaite (1948; Thornthwaite and Mather 1955), focusing on (1) potential evapotranspiration (PET), a measure of the water that could be lost to the atmosphere given the (temperature-based) available energy; (2) actual evapotranspiration (AET), the water in fact moved to the atmosphere from soil and vegetation surfaces; (3) moisture deficit, the discrepancy between PET and AET, and thus representing the limits placed on evapotranspiration by water availability; (4) water surplus, the runoff from a place when water exceeds soil storage capacity; (5) soil moisture storage, which although highly dependent on localized soil characteristics, we assume to be saturated at 150 mm to standardize the climate-based reaction of this compartment; and (6) soil moisture change, which is termed *utilization* when negative, indicating loss of water from soil storage to compensate for a moisture deficit, or *recharge* when positive, indicating replenishment of soil storage.

To characterize twentieth-century conditions in the contiguous United States, we generated water budgets for each of more than 1,200 stations in the United States

Historical Climatology Network (USHCN; Williams et al. 2007) following the Thornthwaite-based methods adapted by Dingman (2002). This procedure estimates the suite of monthly water budget variables from inputs of monthly temperature and precipitation values; it determines PET via the Hamon (1963) equation and models soil water utilization as decreasing exponentially as available water storage declines. The benefit of this approach is the limited input requirements, which is important when incorporating a broad network of historical observations; drawbacks are the exclusion of variables such as wind and land cover that can have important impacts on PET, but Vörösmarty, Federer, and Schloss (1999) found that these estimates can perform well at the continental scale. We summarize the results of this analysis in two forms. First, we produce a twentieth-century water balance, calculated as an iterative mean for the period from 1901 to 2000, such that it contains one equilibrated set of monthly variables that represent the typical annual cycle, as conventionally portrayed by a water balance diagram. Second, we compute a continuous monthly time series for the period from 1901 to 2005 at each station, where water budget variables for each month are based on the water storage conditions inherited from the previous month. In successive drought years where deficits are continuous, for example, soil moisture loss is cumulative and carried forward into the next year, thus indicating the more extreme conditions that can occur during particular years. This time series thus represents water budget variability over time, as a complement to the mean conditions illustrated by the water balance.

The USHCN data set comprises high-quality meteorological observing stations selected by the National Climatic Data Center for maximal spatial coverage but low levels of missing data and potential biases such as station moves and adjacent urbanization. We utilized the USHCN mean monthly records of temperature and precipitation, which have undergone quality control measures to adjust for issues such as urban warming and missing records (i.e., "Filnet" and other procedures; Williams et al. 2007). Because computation of the continuous water budget depends on a complete series of monthly records, in the limited cases where temperature or precipitation observations were missing in the data set, we estimated placeholders for them as means of the ten closest years' values for that month but excluded these years' values from subsequent calculations of long-term or regional means. Many of the USHCN records predate 1901, but there are cases where the series began later (particularly in Western states) or ended early; only stations with a minimum of eighty-five years of complete data were utilized for the twentieth-century analyses.

To gauge the extent to which greenhouse gas–induced warming might alter the water budget in the contiguous United States, we develop projections of late twenty-first-century water balance conditions at each USHCN station from temperature and precipitation values produced by IPCC general circulation model (GCM) experiments. The IPCC (2007) Fourth Assessment utilizes a coordinated multimodel data set (MMD) of atmosphere–ocean GCMs to develop projections of climate change from a 1980 through 1999 baseline to 2080 through 2099 conditions (IPCC 2007; Meehl et al. 2007). Following the IPCC, we use the intermediate A1B scenario of greenhouse gas emissions and socioeconomic adaptation, wherein, for instance, CO_2 emissions peak around 2050 at 2.7 times 1990 levels and fall to about two times the 1990 level by 2100 (Nakicenovic and Swart 2000). Ensemble mean monthly late twenty-first-century temperature and precipitation change for the 612 grid cells (spaced at roughly 3° latitude/longitude intervals) centered on the contiguous United States were computed from the results of eighteen MMD GCMs and used to interpolate monthly change surfaces for both variables using ANUSPLIN (Hutchinson 2006; see later). Projected changes at each USHCN site were extracted from these surfaces and added to the twenty-year (1980–1999) means for all sites with at least nineteen complete years of data for that time period ($n = 965$) to create a projected 2080–2099 monthly temperature and precipitation data set. These inputs were then used to generate a water balance following the same method just described for the twentieth century. The nature of dynamic atmospheric models will always introduce a degree of predictive uncertainty to the accuracy of their outputs, but continuous refinement of these models has increased confidence in their estimates (IPCC 2007). The purpose of this study's projections is not to generate a precise forecast but to illustrate the potential magnitude of change over the coming century relative to the range of conditions during the twentieth century and to identify regions where the greatest consequences might be anticipated.

We portray spatial patterns of twentieth- and projected late twenty-first-century conditions throughout the contiguous United States by interpolating the USHCN point values to continuous surfaces with ANUSPLIN software. ANUSPLIN fits a thin plate-smoothing spline to data points as a function of independent variables with a technique comparable to standard multivariate regression (Hutchinson 1995, 2006). By incorporating latitude, longitude, and elevation as independent variables due to their key relationship with climatic gradients, ANUSPLIN has been shown to be a reliable means of more accurately estimating climatic surfaces from irregularly distributed stations than standard interpolation methods such as kriging (Price et al. 2000; McKenney et al. 2006). These independent variables are incorporated into the ANUSPLIN model through the use of a large-scale digital elevation model obtained from the U.S. Geological Survey. We present resultant maps of seasonal, rather than annual, water balance patterns because of important dimensions of spatio-temporal variability these indicate.

In addition to mapping the broad-scale continental patterns, we selected five regions to highlight details of historical and projected changes in the water budget. These regions are defined physiographically to represent environmentally distinctive portions of the United States with at least twenty-five USHCN stations: the Mid-Atlantic (coastal plain and piedmont settings from Massachusetts through southern Virginia), Great Lakes (central till plain and eastern lake sections), Gulf Coast (coastal plain sites from Texas through Georgia), Great Plains, and the Colorado River Basin. We derived regional mean water budget values from the set of USHCN sites within these areas to identify trends during the twentieth century (using simple linear regression) and compared these to patterns of interannual variability and projections for the late twenty-first century; the significance of these trends in relation to the interannual variability of the time series was tested with analysis of variance. Baseline early twentieth-century conditions are represented by the trend point values at 1901. Conventional water balance diagrams for one USHCN site typical of each region (representative of regional median twentieth-century temperature and precipitation) are presented for both twentieth-century and late twenty-first-century means as indicators of the nature and diversity of expected changes.

Results

Changes in the water budget reflect the interaction of varying temperature and precipitation. During the twentieth century, much of the United States experienced upward-trending temperatures, as southwestern and northern sections increased over 1°C, although the southeast ended the century about 0.5°C cooler (Figure 1; Karl, Melillo, and Peterson 2009). Simultaneously, annual precipitation remained generally consistent for many regions but increased by more than

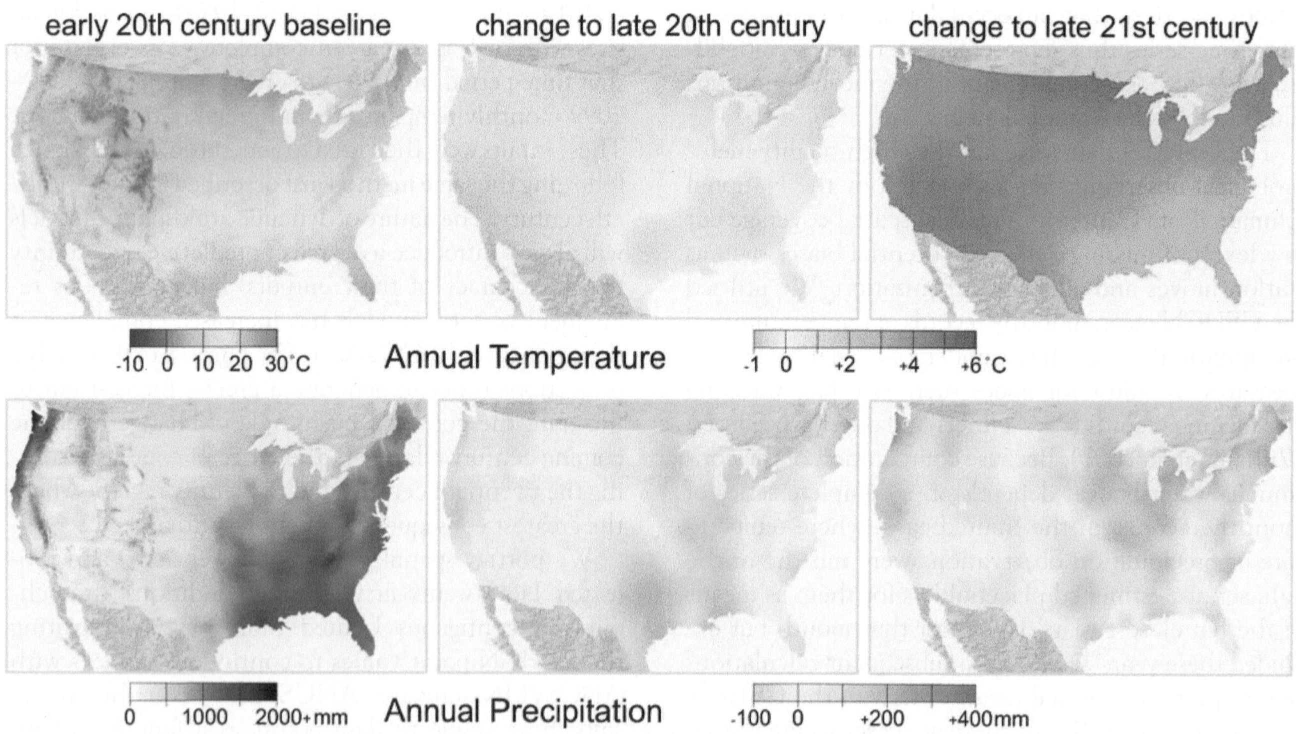

Figure 1. Early twentieth-century annual temperature and precipitation surfaces for the contiguous United States and departures from these baselines at the end of the twentieth century and as projected for the end of the twenty-first century.

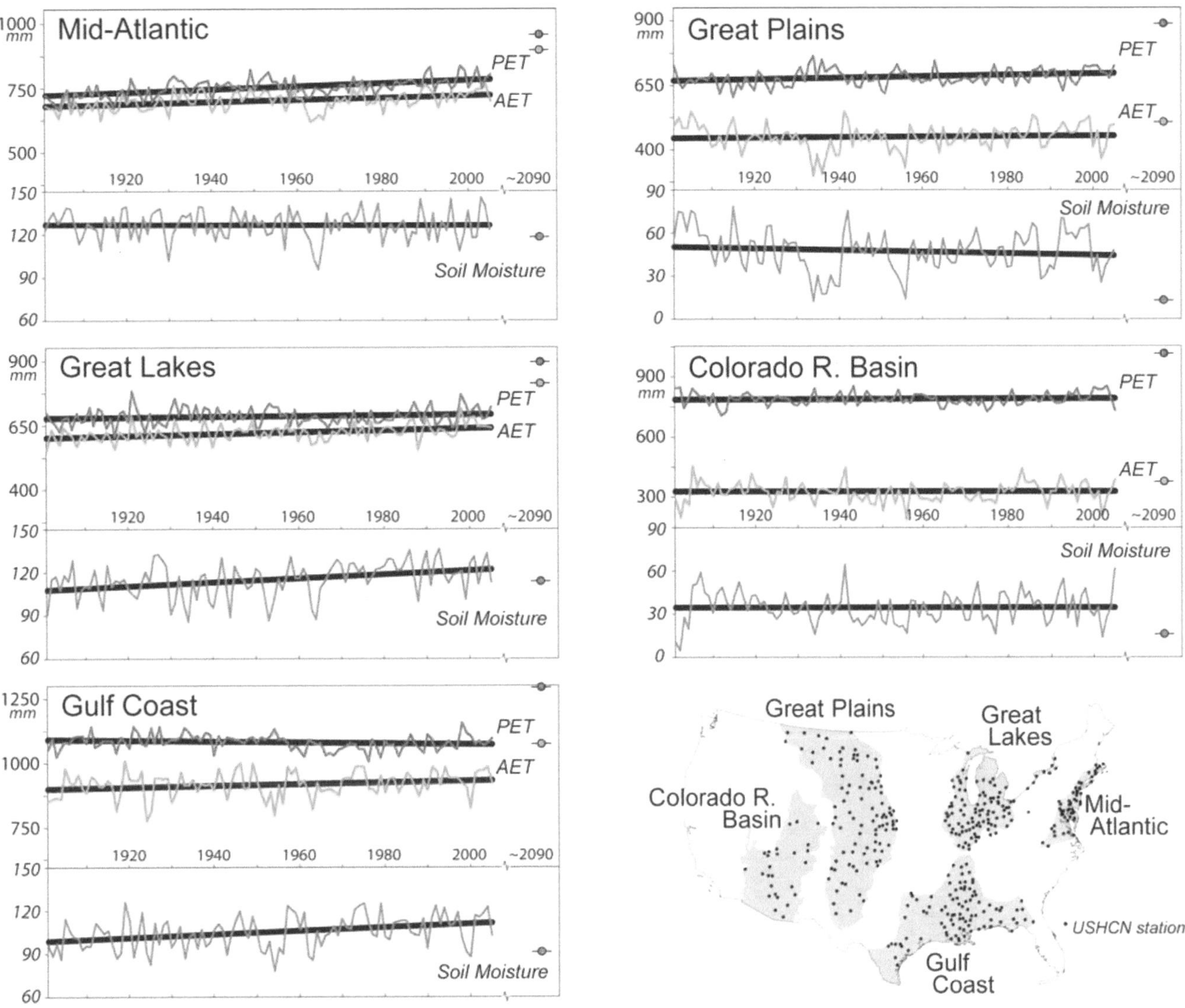

Figure 2. Regional trends in water budget conditions during the twentieth century and projected mean conditions for the late twenty-first century. PET = potential evapotranspiration; AET = actual evapotranspiration.

100 mm throughout much of the Gulf South and Great Lakes. GCM projections for the late twenty-first century indicate an intensification of both spatial patterns, with annual temperature increases of more than 4.5°C above the 1901 baseline in the southwestern and northern United States (with the southeastern and northwestern coasts nearer +3°C) and annual precipitation increasing little throughout much of the West and Southeast but by more than 100 mm in the Northeast and Northwest.

Temporal Change in Regional Water Budgets

Integration of these temperature and precipitation patterns in the water budget produces distinct regional responses (Figure 2, Table 1). Increased twentieth-century PET in most regions tracked temperature increases; where AET also increased (with precipitation) but by lower levels, as in the Mid-Atlantic and Great Plains, this divergence resulted in modest (2–7 percent) net long-term increases in annual moisture deficits and lower soil moisture. Where PET and AET trends converged, as in the Great Lakes and Gulf Coast, net reductions in deficits (of 24–34 percent) and increased soil moisture storage (by 11–12 percent) occurred. Consistent century-long levels for all water budget values were observed in the Colorado River Basin, although with high variability at shorter timescales. Indeed, ranges of interannual variability markedly exceed the twentieth-century trend in each of these regions.

Table 1. Mean regional trends in annual water budget values over the twentieth century and projections for change over the twenty-first century, indicating absolute and percentage changes from the 1901 baseline

Region	Time period	Temp (°C)	Precip (mm)	PET (mm)	AET (mm)	Deficit (mm)	Surplus (mm)	Soil moisture (mm)
Mid-Atlantic ($n = 53$)	1901 baseline	11.0	1122	720	677	42	444	128
	Twentieth-century Δ (1901–2000)	+1.2[a] (+11%)	+47 (+4%)	+58[a] (+8%)	+37[a] (+5%)	+21 (+49%)	+11 (+2%)	–2 (–2%)
	Twenty-first century Δ (1901–2090)	+4.2 (+39%)	+115 (+10%)	+235 (+33%)	+217 (+32%)	+17 (+41%)	–101 (–23%)	–9 (–7%)
Great Lakes ($n = 99$)	1901 baseline	9.6	872	679	601	78	272	108
	Twentieth century Δ (1901–2000)	+0.3 (+3%)	+99[a] (+11%)	+8 (+1%)	+35[a] (+6%)	–26[a] (–34%)	+64[a] (+24%)	+14[a] (+12%)
	Twenty-first century Δ (1901–2090)	+4.1 (+42%)	+166 (+19%)	+213 (+31%)	+210 (+35%)	+3 (+4%)	–44 (–16%)	+6 (+6%)
Gulf Coast ($n = 91$)	1901 baseline	18.5	1,258	1,091	904	187	355	99
	Twentieth century Δ (1901–2000)	–0.5[a] (–3%)	+144[a] (+11%)	–25[a] (–2%)	+20 (+2%)	–44[a] (–24%)	+122[a] (+35%)	+11[a] (+11%)
	Twenty-first century Δ (1901–2090)	+2.9 (+16%)	+83 (+7%)	+208 (+19%)	+167 (+18%)	+41 (+22%)	–84 (–24%)	–8 (–9%)
Great Plains ($n = 98$)	1901 baseline	9.0	483	667	446	221	37	50
	Twentieth century Δ (1901–2000)	+0.6[a](+6%)	–5 (–1%)	+23[a](+3%)	+9 (+2%)	+14 (+7%)	–13 (–36%)	–4 (–7%)
	Twenty-first century Δ (1901–2090)	+4.3 (+48%)	+23 (+5%)	+221 (+33%)	+60 (+14%)	+161 (+73%)	–37 (–100%)	–35 (–70%)
Colorado River Basin ($n = 25$)	1901 baseline	12.3	363	788	324	464	38	35
	Twentieth century Δ (1901–2000)	–0.3 (–2%)	+2 (+0%)	–4 (–1%)	+3 (+1%)	–7 (–2%)	–1 (–2%)	0 (0%)
	Twenty-first century Δ (1901–2090)	+3.7 (+30%)	+24 (+6%)	+227 (+29%)	+51 (+16%)	+176 (+38%)	–27 (–71%)	–19 (–54%)

Note: PET = potential evapotranspiration; AET = actual evapotranspiration.

[a]Twentieth-century trend line is significant in explaining the variability of the interannual time series ($p < 0.05$).

Projection of water budgets into the late twenty-first century can be similarly characterized by relative trends in PET and AET. Annual PET increases are consistent throughout all regions at over 200 mm; AET also generally increases but at levels that are far more spatially variable. For the Mid-Atlantic and Great Lakes, increases in AET of about 30 percent nearly match those of PET (i.e., considerable increases in both temperature and precipitation), with modest impacts on mean annual deficit or soil moisture—the latter remain near twentieth-century norms. In contrast, PET increases greatly outweigh those for AET in the other three regions, with correspondingly increased deficits and decreased soil moisture storage. The Great Plains, for instance, are shown to have average late twenty-first-century soil moisture conditions comparable to those during the extreme droughts of the mid-1930s Dust Bowl—about 70 percent lower than the early twentieth-century baseline. The convergence of PET and AET experienced by the Gulf Coast over the twentieth century reverses during the twenty-first century to return mean soil moisture to the lower levels seen around 1900 and the droughts of the 1950s, 1960s, and late 1990s. Late twenty-first-century annual water surpluses are projected to be lower than the 1901 baseline in all five regions, after increases during the twentieth century throughout the eastern United States and comparatively small decreases in the west. These projected mean decreases in surface water available for streamflow are on the order of 25 percent in the East and as high as 70 to 100 percent in the Western regions.

The magnitude and seasonality of these projected changes is evident by comparing twentieth- and twenty-first-century water balance diagrams for representative USHCN stations (Figure 3). For all sites, future increases in PET are relatively modest during cool months but much stronger (approx. 25 percent increase) during the summer to greatly expand warm-season moisture deficits. However, in more northern climates such as the Mid-Atlantic and Great Lakes, winter temperatures will be sufficiently high to more commonly exceed freezing (PET > 0), resulting in lower snowpack accumulation and therefore smaller and earlier snowmelt-induced spring surpluses. Reductions in spring surpluses in these two regions (62 percent for Westminster, MD; 16 percent for Rensselaer, IN) also accompany the higher levels of soil water recharge necessary during fall and winter months, despite higher precipitation then, to offset the greater soil water utilization predicted for the summer months. The Gulf Coast example follows a similar pattern of reduced water availability despite projections of relatively little change in the quantity or seasonality of precipitation, as the large summer rise in PET more than doubles the moisture deficit, thereby increasing soil water utilization and diminishing the spring surplus (24 percent for St. Joseph, LA) by extending the duration of soil recharge. The Great Plains and Colorado River Basin sites, which differ from the more humid examples by not experiencing spring surpluses, are nonetheless projected to experience similar drivers of water balance change. Because PET nearly always exceeds precipitation in these more arid climates, their soil moisture regimes are largely governed by precipitation during wintertime-low PET (often snow), as the amount of annual soil water utilization is limited to recharge at these times. Late twenty-first-century PET is projected to increase above zero for all months at both sites, thereby markedly diminishing their snowpack (41 percent in Imperial, NE; 84 percent in Blanding, UT) and spring recharge from twentieth-century levels, despite little precipitation change.

Projected Change in Continental-Scale Seasonal Patterns

The processes seen in the regional examples provide context for interpreting water budget patterns at the national scale. The mean twentieth-century seasonal progression of PET, AET, and consequent patterns of moisture deficit (Figure 4) highlight the contrast between the arid regime typical of the southwestern United States and the more humid east. Cool-season PET and AET are near zero for most of the United States, so moisture deficits are restricted to all but the southwestern deserts and subtropical Florida. As PET increases during warmer months, the area of deficits expands first to places with limited soil moisture storage, such as the southern Great Plains; by summer deficits are high throughout the intermountain West and Great Plains, with modest deficits also widespread throughout the Midwest and Southeast. Water surpluses and soil water recharge are highest during winter months in the Southeast and Pacific Northwest but near zero where precipitation is stored mainly as snowpack in the upper Midwest, Northeast, and many montane regions (Figure 5). The pattern of high runoff and soil recharge shifts northward with spring melt, and soil moisture nears or reaches capacity throughout humid regions. Soil water utilization begins in spring months throughout the Southeast (although the counterseasonal regime of peninsular Florida is clearly visible) and along the West Coast and becomes widespread during the

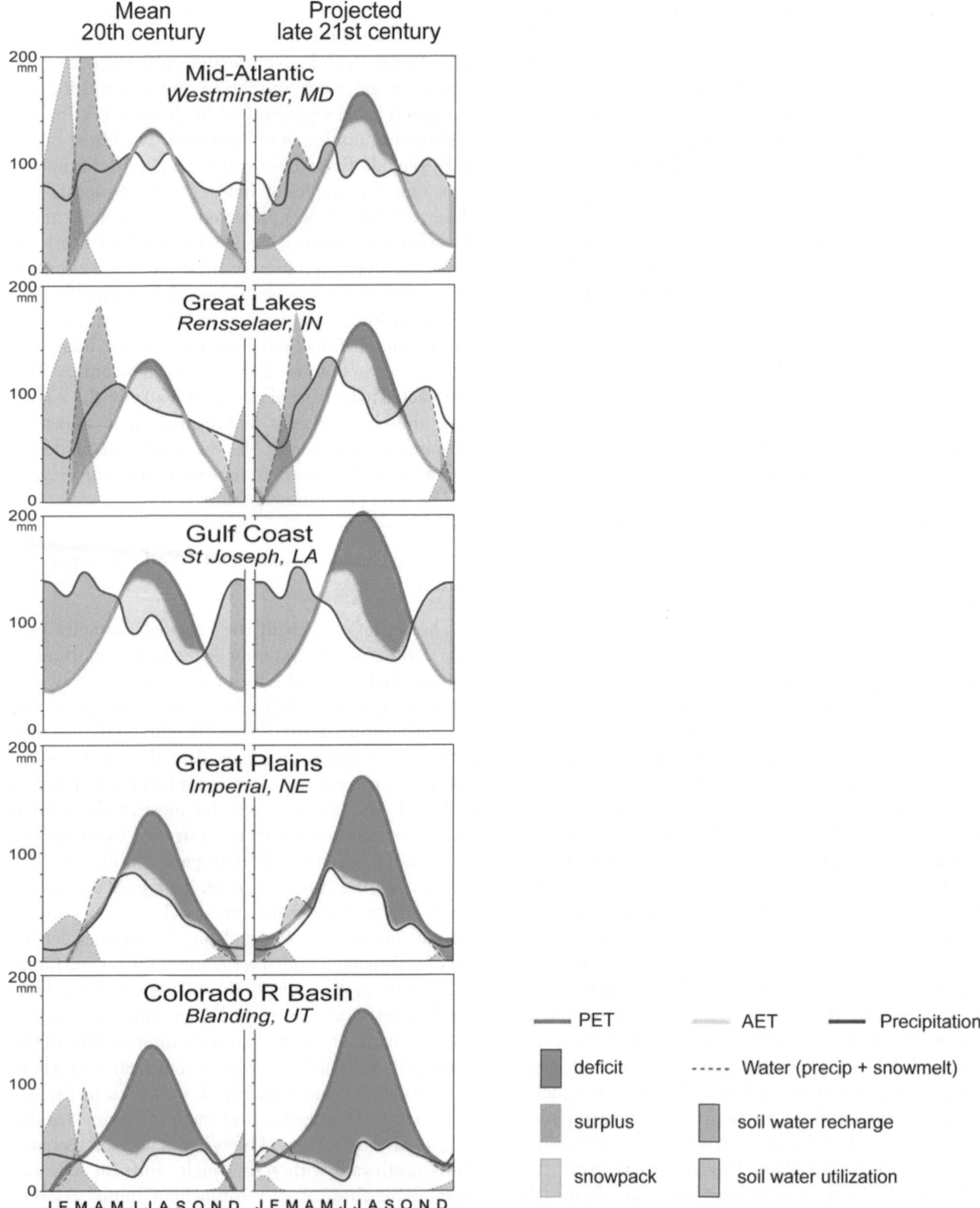

Figure 3. Water balance diagrams for representative regional United States Historical Climatology Network stations comparing mean twentieth-century patterns with those projected for the late twenty-first century. PET = potential evapotranspiration; AET = actual evapotranspiration.

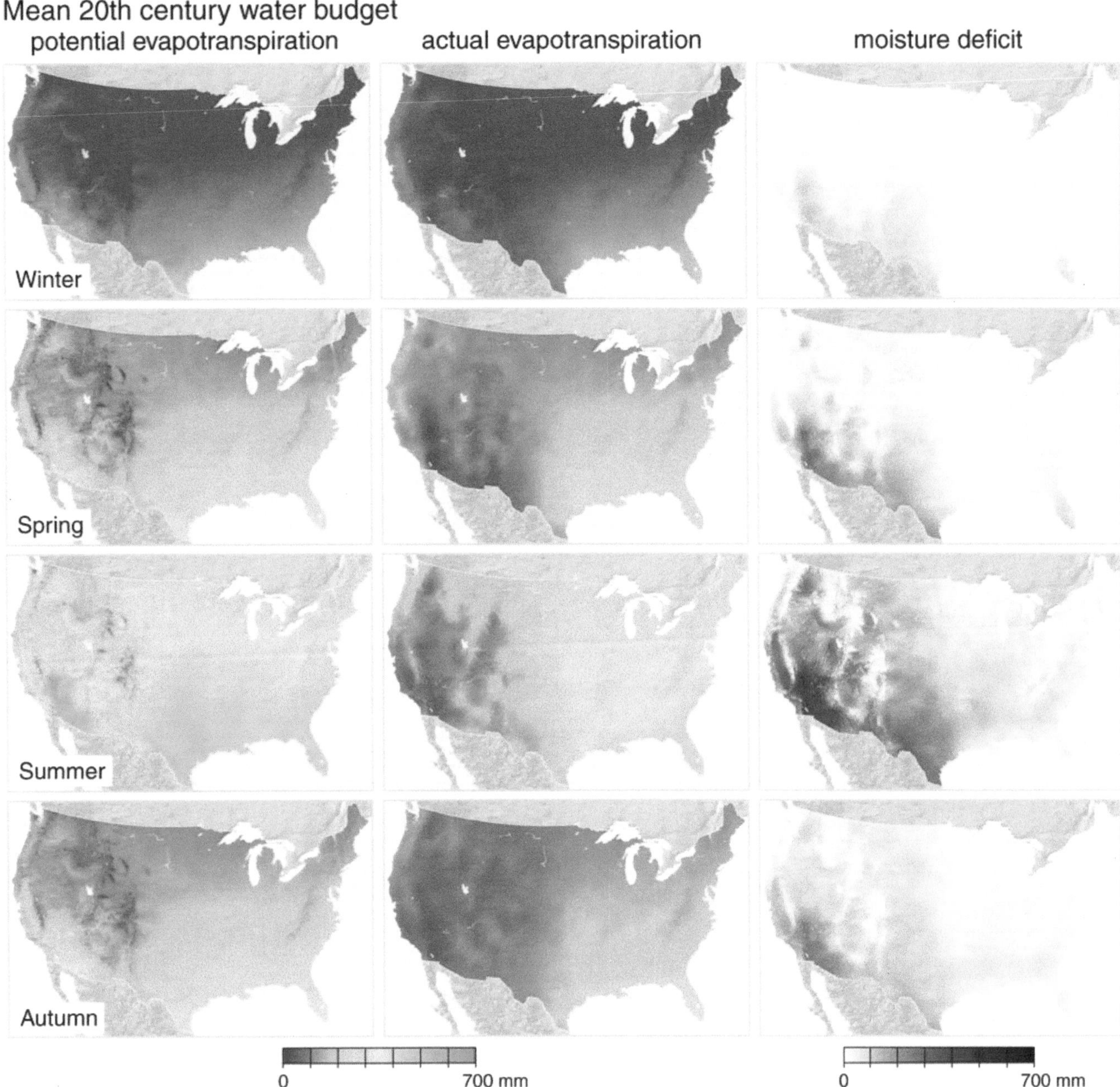

Figure 4. Mean twentieth-century seasonal water budget surfaces for the contiguous United States. Three-month total (mm) potential evapotranspiration, actual evapotranspiration, and moisture deficit during winter (December, January, February), spring (March, April, May), summer (June, July, August), and autumn (September, October, November).

summer, thereby reducing soil moisture. Reduced autumn PET allows a greater proportion of precipitation to go into soil recharge and small surpluses for much of the eastern United States and Pacific Northwest.

Projected patterns of change for the twenty-first century suggest continuity in the overall geography of the water budget but with modified seasonality and intensification of regional humid–arid contrasts. Milder winter temperatures result in a latitudinal band where PET and AET increase due to diminished snowpack and thus water available for evapotranspiration (Figure 6). Through much of the montane West and east-central United States, surpluses shift from spring to winter with the removal of the lag between winter precipitation (as snow) and spring runoff common there in the twentieth century (Figure 7). The relatively arid intermontane West and Great Plains, along with Florida, are projected to experience reduced winter surpluses and soil moisture,

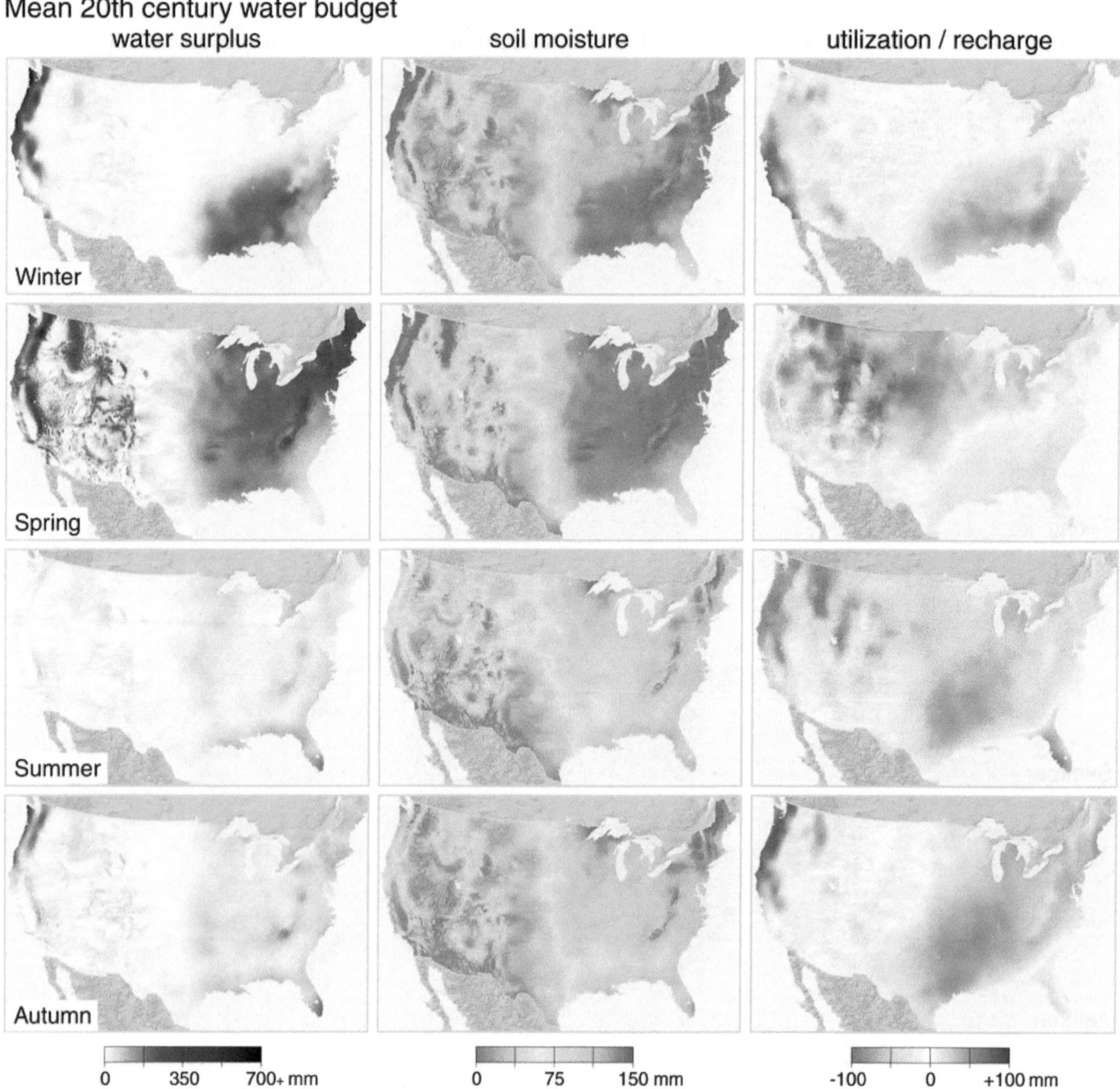

Figure 5. Mean twentieth-century seasonal water budget surfaces for the contiguous United States. Three-month total (mm) water surplus, average soil moisture, and total soil moisture change (+ = recharge, − = utilization).

which then continue throughout the year. Stronger moisture deficits occur throughout the Southwest in the spring and then spread widely throughout the West, Great Plains, and Gulf South during the summer when PET increases surge. In the humid Midwest and Northeast, summer AET nearly tracks the rise in PET, but soil moisture utilization is intensified and storage correspondingly decreased. Autumn PET increases are much less than summer nationwide, allowing precipitation to recharge soil moisture throughout the East; reduced snow accumulation in the Pacific Northwest and New England produce greater surpluses during these months than in the twentieth century.

The spatial consequences of these changes are best summarized by projected mean soil moisture conditions (Figure 8). Areas with marginal seasonally available soil moisture during the twentieth century, particularly the Great Plains (and to a similar extent, Florida), become dry year-round. The comparatively gradual transition zone between the 95th and 100th meridians on the

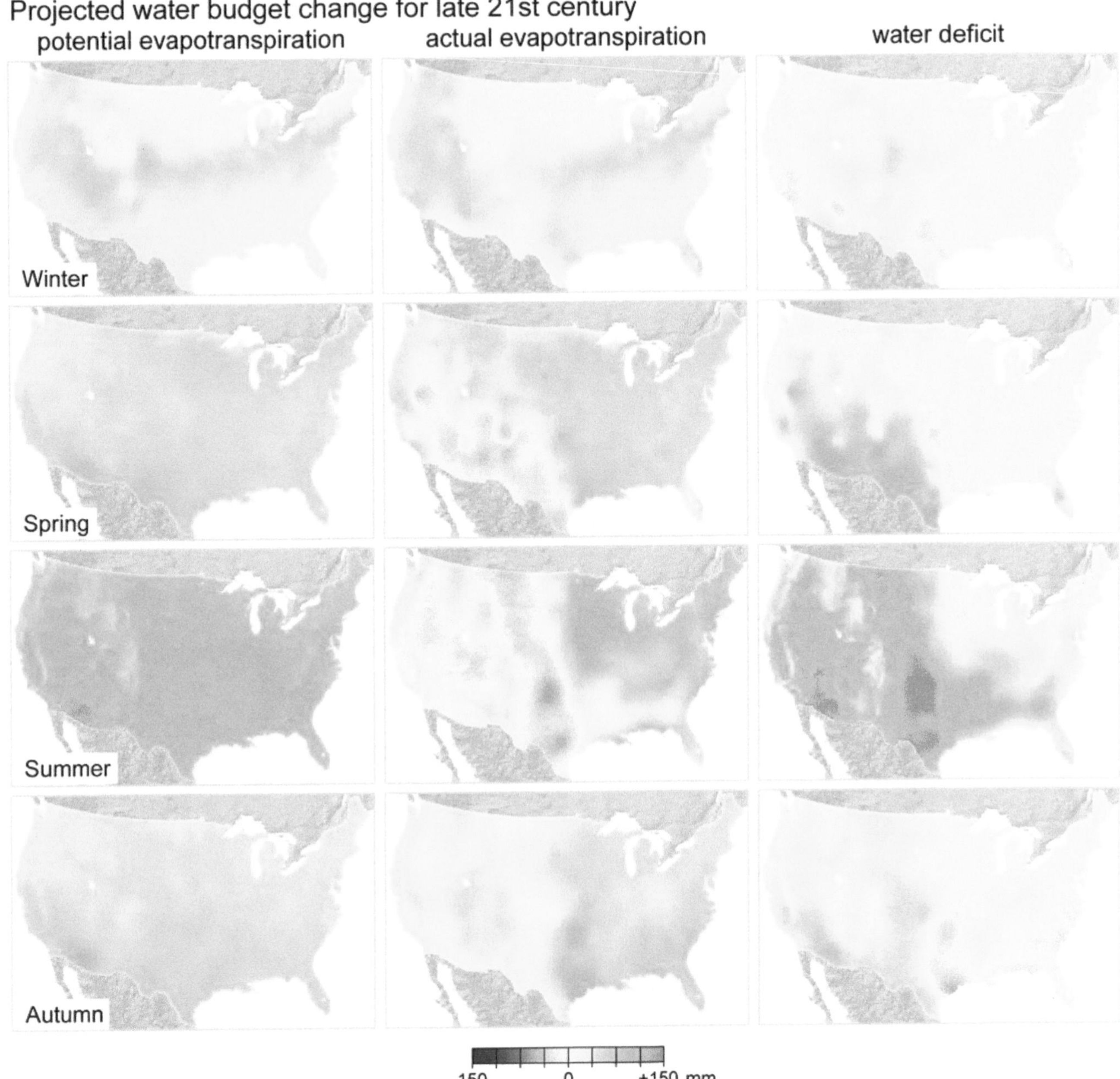

Figure 6. Projected change from twentieth-century mean to late twenty-first century (2080–2099) mean seasonal water budget conditions for the contiguous United States. Three-month total change (mm) in potential evapotranspiration, actual evapotranspiration, and moisture deficit.

twentieth-century map becomes a stark boundary from humid to arid at the end of the twenty-first century. In the humid regions of the East and Northwest, the extremes become greater during the year: more fully saturated over a broader extent during the winter and spring but more fully depleted during the summer and autumn.

Discussion

Just as twentieth-century temperature trends for the United States were not completely in step with global-scale increases (Rind 1991; Hansen et al. 2001), there was also considerable subregional variability in water budget trends within the United States. Projections

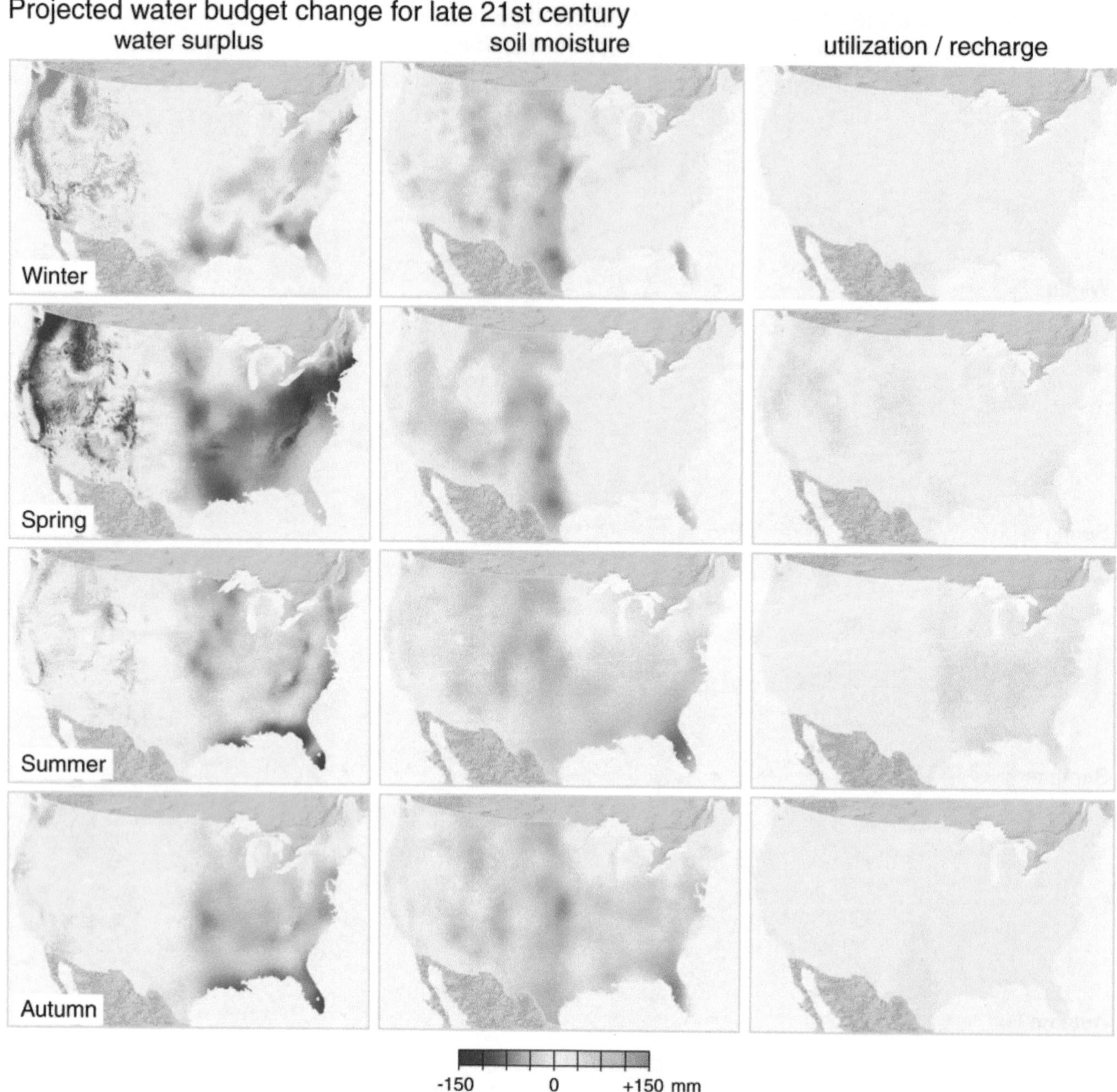

Figure 7. Projected change from twentieth-century mean to late twenty-first-century (2080–2099) mean seasonal water budget conditions for the contiguous United States. Three-month total change (mm) in water surplus, average soil moisture, and soil moisture recharge/utilization.

for twenty-first-century water budgets indicate shifts in these patterns of spatial variability, with much of this differentiation tied to the changing distribution of precipitation. This is due to the more local scales at which precipitation and associated AET will likely vary, relative to the more ubiquitous future temperature and PET increases, a pattern seen during the late twentieth century (Szilagyi, Katul, and Parlange 2001; Qian, Dai, and Trenberth 2007). With the expected expansion of summer subtropical dry conditions over much of the southern United States, but increased rainfall in northern regions (IPCC 2007), magnified summer deficits in the Southeast will reverse the twentieth-century trend of increasing moisture availability there, whereas the Northeast might maintain average annual moisture conditions consistent with the past century's ranges. Arid Western regions are less sensitive to changes in PET because of their low initial

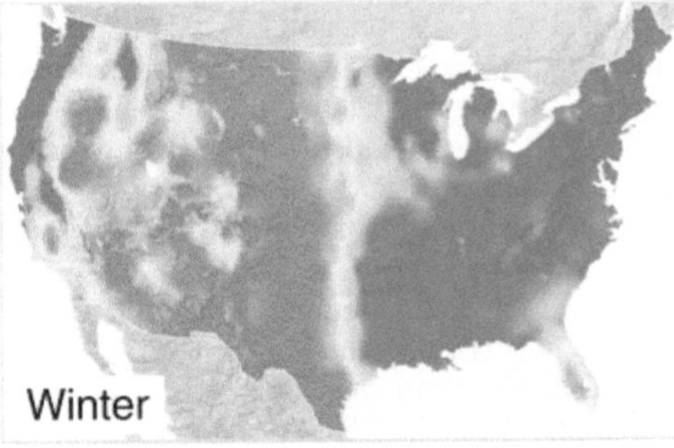

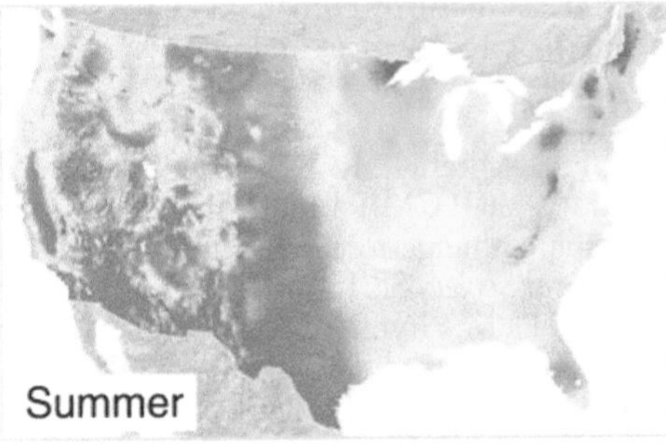

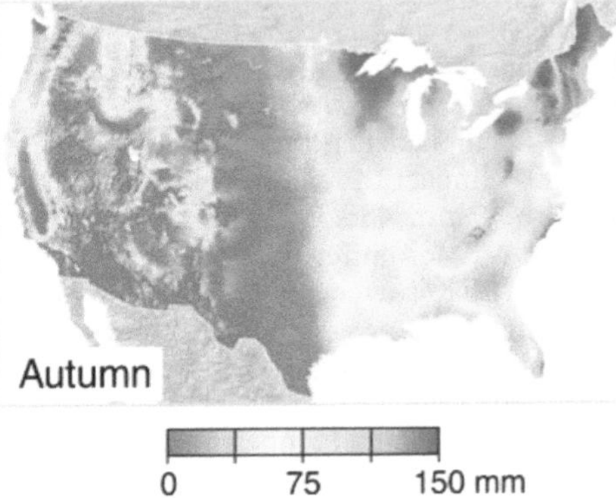

Figure 8. Projected average seasonal soil moisture conditions for the late twenty-first century in the contiguous United States.

levels of moisture (McCabe and Wolock 2002), but areas with these conditions are projected to increase in extent, creating sharper boundaries between humid and arid zones.

Although the magnitude of winter twenty-first-century temperature increase is anticipated to be much less than summer, this factor will be responsible for some of the more significant seasonal and geographic changes in water budgets. Reduced snowpack and modified soil recharge cycles are shown here to be tied to future decreases in water surplus and to cause earlier peaks in runoff, consistent with emerging patterns of increased winter streamflow in the later twentieth century (Lettenmaier, Wood, and Wallis 1994; Regonda et al. 2005; Stewart, Cayan, and Dettinger 2005). The hydrologic impacts of these changes will be greatest for the montane West and Northeast.

Our regional examples suggest that average annual water budget conditions will be near the dry extremes of the temporal range seen during the twentieth century, particularly in the Great Plains, Southeast, and Southwest. This analysis considers only late twenty-first-century means, and an important next step will be to assess the future interannual variablility projected by GCMs. Some models forecast that the frequency and magnitude of extreme events will shift dramatically, such as large increases in hot events and large decreases in cold events (Diffenbaugh et al. 2005), and therefore future droughts could certainly fall well outside the twentieth-century range. These projections are, of course, dependent on the accuracy of GCMs—which might in fact turn out to be conservative (Cox and Jones 2008; Kintisch 2009)—and are also contingent on forthcoming socioeconomic behavior.

An advantage of using the climatic water budget to assess the impact of past and future changes in climate is that it offers a unified set of measures that express the complex interconnections among temperature, precipitation, and available moisture in ways that can effectively portray their inherent spatial and temporal variation. The continuous bookkeeping procedure of water budget time series also captures the variability of soil moisture and water surpluses in a manner that is readily translatable to hydrological, ecological, and agricultural applications. As the focus of investigations continues to shift to addressing the impacts of climate change at regional scales, water budget approaches offer an effective tool to integrate understanding of the multiple physical processes involved and their geographic consequences.

Acknowledgments

We thank Claudia Tebaldi, Climate Central, for assistance with the GCM data used in this study.

References

Bates, B. C., Z. W. Kundzewicz, S. Wu, and J. P. Palutikof, eds. 2008. *Climate change and water: Technical report of the Intergovernmental Panel on Climate Change*. Geneva, Switzerland: IPCC Secretariat.

Battisti, D. S., and R. L. Naylor. 2009. Historical warnings of future food insecurity with unprecedented seasonal heat. *Science* 323:240–44.

Cox, P., and C. Jones. 2008. Illuminating the modern dance of climate and CO_2. *Science* 321:1642–44.

Diffenbaugh, N. S., J. S. Pal, R. J. Trapp, and F. Giorgi. 2005. Fine-scale processes regulate the response of extreme events to global climate change. *Proceedings of the National Academy of Sciences* 102:15774–778.

Dingman, S. L. 2002. *Physical hydrology*. Upper Saddle River, NJ: Prentice Hall.

Flerchinger, G. N., and K. R. Cooley. 2000. A ten-year water balance of a mountainous semi-arid watershed. *Journal of Hydrology* 237:86–99.

Frei, A., R. L. Armstrong, M. P. Clark, and M. C. Serreze. 2002. Catskill Mountain water resources: Vulnerability, hydroclimatology, and climate-change sensitivity. *Annals of the Association of American Geographers* 92:203–24.

Grundstein, A. 2008. Assessing climate change in the contiguous United States using a modified Thronthwaite classification scheme. *The Professional Geographer* 60:398–412.

———. 2009. Evaluation of climate change over the continental United States using a moisture index. *Climatic Change* 93:103–15.

Hamon, W. R. 1963. Computation of direct runoff amounts from storm rainfall. *International Association of Scientific Hydrology Publication* 63:52–62.

Hansen, J., R. Ruedy, M. Sato, M. Imhoff, W. Lawrence, D. Easterling, T. Peterson, and T. Karl. 2001. A closer look at United States and global surface temperature change. *Journal of Geophysical Research* 106:23947–63.

Hay, L. E., and G. J. McCabe. 2002. Spatial variability in water-balance model performance in the coterminous United States. *Journal of the American Water Resources Association* 38:847–60.

Hutchinson, M. F. 1995. Interpolating mean rainfall using thin plate smoothing splines. *International Journal of Geographic Information Systems* 9:385–403.

———. 2006. *ANUSPLIN version 4.36 user guide*. Canberra: Australian National University Center for Resource and Environmental Studies.

Intergovernmental Panel on Climate Change (IPCC). 2007. *Climate change 2007: The physical science basis. Contribution of Working Group I to the fourth assessment report of the Intergovernmental Panel on Climate Change*. Cambridge, UK: Cambridge University Press.

Karl, T. R., J. M. Melillo, and T. C. Peterson, eds. 2009. *Global climate change impacts in the United States*. Cambridge, UK: Cambridge University Press.

Kintisch, E. 2008. Impacts research seen as next climate frontier. *Science* 322:182–83.

———. 2009. Global warming: Projections of climate change go from bad to worse, scientists report. *Science* 323:1546–47.

Legates, D. R., and G. J. McCabe. 2005. A re-evaluation of the average annual global water balance. *Physical Geography* 26:467–79.

Lettenmaier, D. P., E. F. Wood, and J. R. Wallis. 1994. Hydroclimatological trends in the continental United States, 1948–88. *Journal of Climate* 7:586–607.

Mather, J. R. 1978. *The climatic water budget in environmental analysis*. Lexington, MA: Lexington Books.

———. 1991. A history of hydroclimatology. *Physical Geography* 12:260–73.

Mather, J. R., and J. J. Feddema. 1986. Hydrologic consequences of increases in trace gases and CO_2 in the atmosphere. In *Effects of changes in stratospheric ozone and global climate: Vol. 3. Climate change*, ed. J. G. Titus, 251–71. Washington, DC: U.S. Environmental Protection Agency.

McCabe, G. J., and D. M. Wolock. 2002. Trends and temperature sensitivity of moisture conditions in the conterminous United States. *Climate Research* 20:19–29.

McCabe, G. J., D. M. Wolock, L. E. Hay, and M. A. Ayers. 1990. Effects of climatic change on the Thornthwaite moisture index. *Water Resources Bulletin* 26: 633–43.

McKenney, D. W., J. H. Pedlar, P. Papadopol, and M. F. Hutchison. 2006. The development of 1901–2000 historical monthly climate models for Canada and the United States. *Agricultural and Forest Meteorology* 138:69–81.

Meehl, G. A., C. Covey, T. Delworth, M. Latif, B. McAvaney, J. F. B. Mitchell, R. J. Stouffer, and K. E. Taylor. 2007. The WCRP CMIP3 multi-model dataset: A new era in climate change research. *Bulletin of the American Meteorological Society* 88:1383–94.

Muller, R. A., and J. M. Grymes. 2005. Water budget analysis. In *Encyclopedia of world climatology*, ed. J. E. Oliver, 798–805. Dordrecht, The Netherlands: Springer.

Nakicenovic, N., and R. Swart. 2000. *Special report on emissions scenarios: A special report of Working Group III of the Intergovernmental Panel on Climate Change*. Cambridge, UK: Cambridge University Press.

Patz, J. A., D. Campbell-Lendrum, T. Holloway, and J. A. Foley. 2005. Impact of regional climate change on human health. *Nature* 438:310–17.

Price, D. T., D. W. McKenney, I. A. Nalder, M. F. Hutchinson, and J. L. Kesteven. 2000. A comparison of two statistical methods for spatial interpolation of Canadian monthly mean climate data. *Agricultural and Forest Meteorology* 101:81–94.

Qian, T., A. Dai, and K. E. Trenberth. 2007. Hydroclimatic trends in the Mississippi River basin from 1948 to 2004. *Journal of Climate* 20:4599–614.

Regonda, S. K., B. Rajagopalan, M. Clark, and J. Pitlick. 2005. Seasonal cycle shifts in hydroclimatology over the western United States. *Journal of Climate* 18:372–87.

Rind, D. 1991. Climate variability and climate change. In *Greenhouse-gas-induced climatic change: A critical appraisal of simulations and observations*, ed. M. E. Schlesinger, 69–78. New York: Elsevier.

Rosenzweig, C., D. Karoly, M. Vicarelli, P. Neofotis, Q. Wu, G. Casassa, A. Menzel, et al. 2008. Attributing physical and biological impacts to anthropogenic climate change. *Nature* 453:353–57.

Stephenson, N. L. 1998. Actual evapotranspiration and deficit: Biologically meaningful correlates of vegetation distribution across spatial scales. *Journal of Biogeography* 25:855–70.

Stewart, I. T., D. R. Cayan, and M. D. Dettinger. 2005. Changes toward earlier streamflow timing across western North America. *Journal of Climate* 18:1136–55.

Szilagyi, J., G. G. Katul, and M. B. Parlange. 2001. Evapotranspiration intensifies over the conterminous United States. *Journal of Water Resources Planning and Management* 127:354–62.

Thornthwaite, C. W. 1948. An approach toward a rational classification of climate. *Geographical Review* 38: 55–94.

Thornthwaite, C. W., and J. R. Mather. 1955. The water balance. *Publications in Climatology* 8:1–104.

vanMantgem, P. J., N. L. Stephenson, J. C. Byrne, L. D. Daniels, J. F. Franklin, P. Z. Fule, M. E. Harmon, A. J. Larson, J. M. Smith, A. H. Taylor, and T. T. Veblen. 2009. Widespread increase of tree mortality rates in the Western United States. *Science* 323:521–24.

Vörösmarty, C. J., C. A. Federer, and A. L. Schloss. 1999. Potential evaporation functions compared on US watersheds: Possible implications for global-scale water balance and terrestrial ecosystem modeling. *Journal of Hydrology* 207:147–69.

Vörösmarty, C. J., P. Green, J. Salisbury, and R. B. Lammers. 2000. Global water resources: Vulnerability from climate change and population change. *Science* 289:284–88.

Westerling, A. L., H. G. Hidalgo, D. R. Cayan, and T. W. Swetnam. 2006. Warming and earlier spring increase Western U.S. forest wildfire activity. *Science* 313:940–43.

Williams, C. N., Jr., M. J. Menne, R. S. Vose, and D. R. Easterling. 2007. United States Historical Climatology Network monthly temperature and precipitation data. ORNL/CDIAC-118, NDP-019. http://cdiac.ornl.gov/epubs/ndp/ushcn/usa_monthly.html (last accessed 1 October 2007).

Willmott, C. J., C. M. Rowe, and Y. Mintz. 1985. Climatology of the terrestrial seasonal water cycle. *Journal of Climatology* 5:589–606.

The Columbian Encounter and the Little Ice Age: Abrupt Land Use Change, Fire, and Greenhouse Forcing

Robert A. Dull, Richard J. Nevle, William I. Woods, Dennis K. Bird, Shiri Avnery, and William M. Denevan

Pre-Columbian farmers of the Neotropical lowlands numbered an estimated 25 million by 1492, with at least 80 percent living within forest biomes. It is now well established that significant areas of Neotropical forests were cleared and burned to facilitate agricultural activities before the arrival of Europeans. Paleoecological and archaeological evidence shows that demographic pressure on forest resources—facilitated by anthropogenic burning—increased steadily throughout the Late Holocene, peaking when Europeans arrived in the late fifteenth century. The introduction of Old World diseases led to recurrent epidemics and resulted in an unprecedented population crash throughout the Neotropics. The rapid demographic collapse was mostly complete by 1650, by which time it is estimated that about 95 percent of all indigenous inhabitants of the region had perished. We review fire history records from throughout the Neotropical lowlands and report new high-resolution charcoal records and demographic estimates that together support the idea that the Neotropical lowlands went from being a net source of CO_2 to the atmosphere before Columbus to a net carbon sink for several centuries following the Columbian encounter. We argue that the regrowth of Neotropical forests following the Columbian encounter led to terrestrial biospheric carbon sequestration on the order of 2 to 5 Pg C, thereby contributing to the well-documented decrease in atmospheric CO_2 recorded in Antarctic ice cores from about 1500 through 1750, a trend previously attributed exclusively to decreases in solar irradiance and an increase in global volcanic activity. We conclude that the post-Columbian carbon sequestration event was a significant forcing mechanism of Little Ice Age cooling. *Key Words: Americas, carbon dioxide, Early Anthropocene, fire history, Little Ice Age.*

1492 年的前哥伦布时代，在新热带区的低地，农民人数估计有 2500 万，其中至少有百分之八十是生活在森林生物群落里。现在人们公认，在欧洲人到来之前，为方便农业活动，新热带区内大量地区的森林被清除并焚烧。古生态和考古证据表明，人口增长对森林资源的压力（籍由人为地焚烧森林），在整个晚全新世一直稳步增加，在欧洲人到来的 15 世纪后期达到顶峰。旧世界疾病的引入导致经常性的疫病流行，造成整个新热带区前所未有的人口锐减。大约在 1650 年，人口总数的快速崩溃基本已经结束，那时，估计该地区所有原居民约有百分之九十五都已丧生。我们回顾了整个新热带区低地部分的火灾历史，总结了新的高分辨率碳记录和人口估计，这些研究共同支持下面的观点：新热带区的低地，从原来的净二氧化碳大气排放源，在遭遇哥伦布之后的几百年里，逐渐变成了净二氧化碳汇集源。我们认为，在哥伦布登陆之后，新热带区的森林再生造成了陆地生物圈的碳汇集，依次为 2 到 5 Pg C，从而支持了南极冰芯自 1500 年到 1750 年的记录，这些冰芯记录很好地展示了当时大气中二氧化碳的减少，这一减少趋势以前一直被完全归功于太阳辐射的减少和全球性火山活动的增加。我们的结论是：后哥伦布时代的碳吸存事件是小冰期变冷的一个重要影响机制。*关键词：美洲，二氧化碳，早期人类世，火的历史，小冰期。*

Hacia 1492, los cultivadores precolombinos de las tierras bajas neotropicales sumaban aproximadamente 25 millones, de los cuales por lo menos el 80 por ciento vivía dentro de los biomas de bosque. Ya se ha establecido que antes de la llegada de los europeos áreas importantes de las selvas neotropicales habían sido desbrozadas y quemadas para facilitar las actividades agrícolas. Las evidencias paleoecológica y arqueoecológica muestran que la presión demográfica sobre los recursos forestales—facilitada por la quema antropogénica—se incrementó constantemente a través del Holoceno Tardío, alcanzando el máximo cuando los europeos llegaron a finales del siglo XV. La introducción de enfermedades del Viejo Mundo se tradujo en epidemias recurrentes que llevaron a un colapso sin precedentes de la población en todas partes del neotrópico. La rápida catástrofe demográfica quedó concluida en gran medida para 1650, cuando se estima que cerca del 95 por ciento de todos los habitantes indígenas de la región habían perecido. Se revisaron los registros de la historia del uso del fuego en todas las tierras bajas neotropicales al tiempo que reportamos nuevos registros de alta resolución de carbón vegetal y cálculos demográficos que, conjuntamente, apoyan la idea de que las tierras bajas neotropicales pasaron de ser una fuente neta de suministro de CO_2 a la atmósfera, antes de Colón, a una caída neta de carbono durante varios siglos posteriores al encuentro colombino. Arguimos que el recrecimiento de los bosques neotropicales después del encuentro colombino condujo al secuestro del carbono biosférico terrestre del orden de 2 a 5 Pg C, contribuyendo así a la bien documentada disminución del CO_2 atmosférico registrado en núcleos de hielo antárticos, acumulados entre 1500 y 1750. Esta tendencia se le atribuyó antes exclusivamente a descensos de la irradiación solar y a un incremento del volcanismo global. Concluimos que el evento del secuestro de carbono poscolombino fue un mecanismo de forzamiento significativo en el enfriamiento global durante la Pequeña Edad del Hielo. *Palabras clave: Américas, dióxido de carbono, Antropoceno Temprano, historia del fuego, Pequeña Edad del Hielo.*

> Not even the more primitive nonagricultural folk may be considered as merely passive occupants of particular niches in their forest environment. By their continued presence and activity, they enlarged such niches against the forest.
>
> —Sauer (1958, 107)

The role of human beings in creating and expanding forest openings in the American tropics has been a topic of conjecture and debate for more than a half-century, and interest in the subject outside of academia has risen dramatically in recent years together with concerns about global warming. The Amazon rainforest in particular has widely been hailed as the "lungs of the world" due to its tremendous capacity for both oxygen production and carbon dioxide sequestration (Moran 1993). It has been estimated that the carbon flux from Neotropical forests (Figure 1) to the atmosphere totaled approximately 37 Pg from 1850 through 2000, which contributed measurably to the postindustrial CO_2 increase (Houghton 2003). Indeed, the collective roles of anthropogenic deforestation, burning (including accidental fires that are generally referred to as *leaked* or *escaped* fires), and agricultural expansion in the lowland Neotropics have been identified as major factors in both postindustrial and future global warming scenarios (Fearnside 2000; Mahli, Meir, and Brown 2002; Cramer et al. 2004; Malhi et al. 2008; Langmann et al. 2009), and yet until recently pre-Columbian land use has been considered to be negligible at the scale of global greenhouse forcing.

Widespread biomass burning and agricultural forest clearance predate the Industrial Revolution by several millennia in the Americas, Africa, Asia, Europe, and Australia, as well as on many oceanic islands (Sauer 1958; Crutzen and Andreae 1990; Goldhammer 1991; Chew 2001; Saarnak 2001; Williams 2003). The idea that preindustrial Holocene land use could have produced quantities of atmospheric CO_2 and CH_4 sufficient to impact the climate system was first outlined by Ruddiman's (2003) seminal paper, "The Anthropogenic Greenhouse Era Began Thousands of Years Ago." Although evidence mounts for a preindustrial Anthropocene (Ruddiman 2003, 2005, 2007; Faust et al. 2006; van Hoof et al. 2006; Nevle and Bird 2008; van Hoof et al. 2008; Vavrus, Ruddiman, and Kutzbach 2008), some critics maintain that human impacts in terms of climate forcing were negligible until the nineteenth century (Broecker and Stocker 2006; Olofsson and Hickler 2008; Elsig et al. 2009; Stocker, Strassmann, and Joos 2010).[1]

If preindustrial farmers did contribute measurably to the greenhouse effect via increased emissions of CO_2 and methane, only a massive and catastrophic collapse of agricultural populations could have led to significant decreases in anthropogenic emissions at any time. The post-Columbian encounter epidemics and pandemics were certainly the most rapid, thorough, and widespread to have occurred during the late Holocene (Crosby 1972; Lovell 1992), resulting in a loss of approximately 90 to 95 percent of the agricultural population throughout the Neotropics (Dobyns 1966; Lovell and Lutz 1995). The sixteenth- and seventeenth-century epidemics resulted in the abrupt abandonment of agricultural clearings in otherwise forested landscapes together with an unprecedented reduction in human fire ignitions, thus providing an ideal scenario for backcasting anthropogenic climate forcing before European contact. The widespread forest recovery that followed the native population crash after the Columbian encounter resulted in elevated biospheric sequestration of

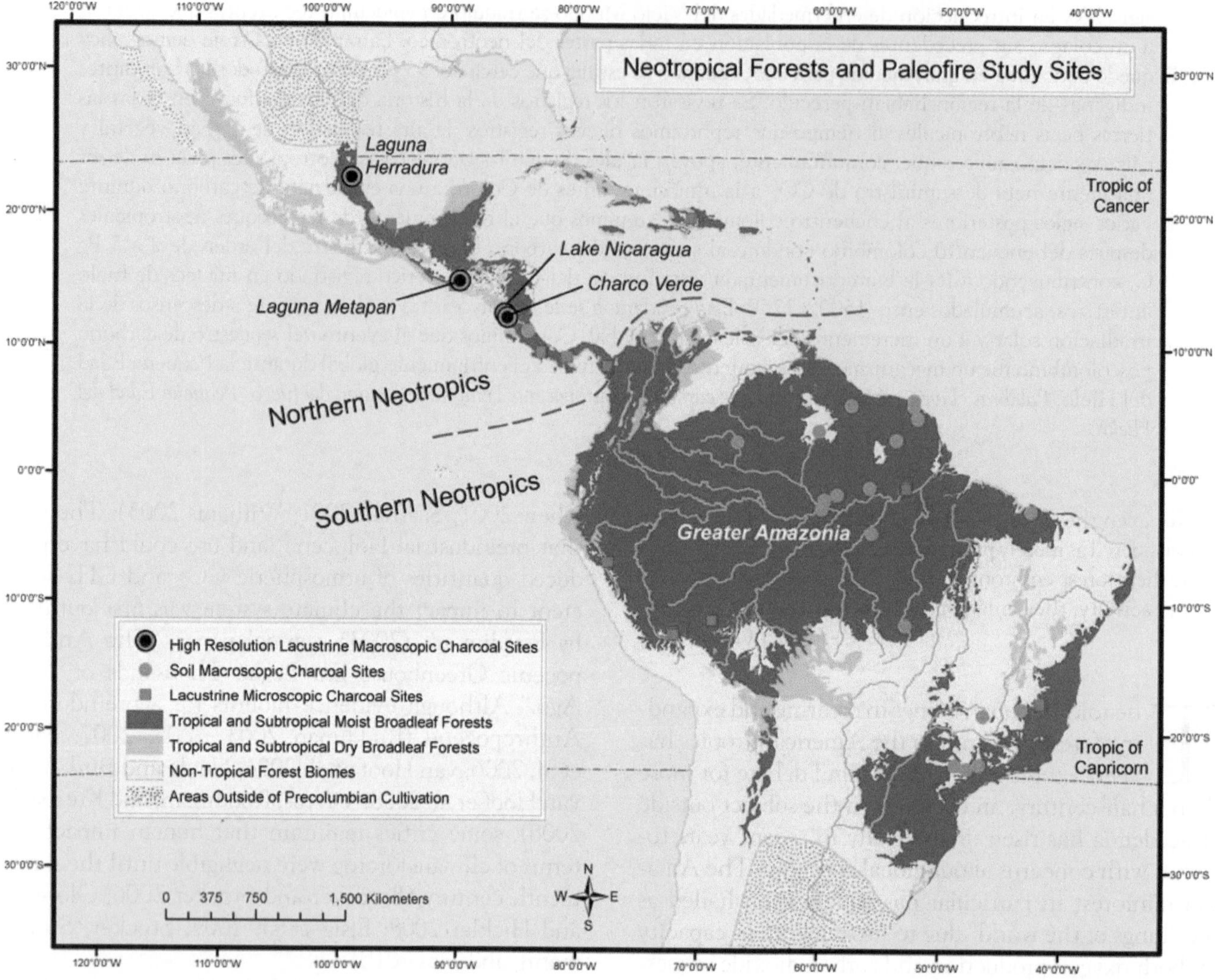

Figure 1. Map of study area illustrating distribution of Neotropical forest biomes, geographic regions discussed in text, and sites of charcoal records used to reconstruct biomass burning histories in this study (Figures 2 and 3A) and in Nevle and Bird (2008; Figures 3B–C). Cartography by Jon Lerner, Peter Dana, and Robert Dull.

atmospheric CO_2 in plant biomass because (1) forests rapidly reoccupied abandoned cultivated landscapes via secondary succession, and (2) existing forests became more carbon dense due to a reduction in wildfires related to anthropogenic fire ignitions, both intentional and accidental.

In this article we review the evidence for prehistoric anthropogenic biomass burning in the Neotropics and provide new data supporting the thesis that the aggregate carbon footprint of Neotropical farmers was sufficient to raise global temperatures via greenhouse forcing before the Columbian encounter. Furthermore, we argue that Little Ice Age (LIA) cooling and the attendant atmospheric CO_2 decrease can be explained in part by biospheric carbon sequestration following the native population collapse. The LIA is identified in surface temperature reconstructions of the past millennium as a global thermal anomaly of about $-0.1°C$ in which cooling was most pronounced from 1550 to 1750 AD in the Northern Hemisphere, particularly in northern Europe (Esper, Cook, and Schweingruber 2002; Jones and Mann 2004; Moberg et al. 2005). Atmospheric CO_2 concentration decreased by ~7 ppm during the same period (Meure et al. 2006). Previously, the LIA thermal anomaly and concomitant decrease in atmospheric CO_2 concentrations were attributed to solar–volcanic forcing (Joos et al. 1999; Hunt and Elliott 2002; Von Storch et al. 2004). Recent analyses, however, suggest that variations in solar luminosity are insufficient to drive significant climate variations on centennial to millennial timescales (Foukal et al. 2006), and were likely a negligible climate forcing factor at the onset of the LIA (Ammann et al. 2007). We conclude that the fifteenth- and sixteenth-century arrivals of Europeans in the Americas set into motion an unprecedented anthropogenic carbon sequestration event and contributed to the LIA climate anomaly. This event represents perhaps the best example of anthropogenic influence on Earth's climate system during the preindustrial period.

Neotropical Paleoecology: Pollen and Charcoal Records

Reconstructing pre-Columbian deforestation and the net forestation in the Neotropics following the Columbian encounter requires a thorough compilation of paleoecological data. Many pollen records from the Neotropics clearly show prehistoric anthropogenic forest clearance in tandem with fire and the rise of heliophytes, cultivars, and early successional pioneers (Tsukada and Deevey 1967; Rue 1987; Byrne and Horn 1989; Goman and Byrne 1998; Clement and Horn 2001; Anchukaitis and Horn 2005; Dull 2007; Lane et al. 2008). Although the evidence for agriculture-driven late Holocene deforestation is strongest for the northern Neotropics, records have been produced from a wide variety of settings across the American tropics (Piperno 2006), including what are today remote forested regions of the Petén (Islebe et al. 1996; Wahl et al. 2006), the Darién (Bush and Colinvaux 1994), and the Amazon (Colinvaux et al. 1988; Bush, Piperno, and Colinvaux 1989; Piperno 1990; Bush and Colinvaux 1994; Bush et al. 2000). Despite this rich and growing corpus of pollen data illustrative of prehistoric agricultural forest clearance, pollen analysis alone has several inherent limitations when it comes to reconstructing changes in past forest carbon density that do not involve dramatic shifts in forest composition or structure.[2]

Modern ecological studies from tropical regions demonstrate that most human impacts on tropical forests rarely result in a wholesale transformation from dense forest to open herbaceous parkland. Instead, what has been recorded are mosaics of cleared plots and successional plots of various ages together with patches of relatively mature closed-canopy forests, all in close proximity to one another and all with varying degrees of transience with regards to carbon storage and release (Laurance 2004; Fearnside 2005). Moreover, chronic understory burning via anthropogenic fire leakage can lead gradually to total biomass reductions of 40 percent or more without the loss of the forest canopy (Cochrane and Schulze 1999). Recognizing these landscape mosaics and gradual anthropogenic reductions in biomass in the pollen record can be exceedingly difficult due to inherent uncertainties related to pollen-vegetation representation, taxonomic precision, source area, and transport (Prentice 1985; Bush 1995).

Although the pollen record is crucial to our understanding of prehistoric land use practices, as are studies of stable carbon isotopes of organic matter in sediments (Dull 2004; Lane et al. 2009), we contend that the most compelling suite of available evidence in support of the pre-Columbian Anthropocene hypothesis comes from sedimentary charcoal records and the stable carbon isotopic signatures of atmospheric methane and CO_2 (Ferretti et al. 2005; Nevle and Bird 2008; Mischler et al. 2009). The decrease in pyrogenic methane from Law Dome, Antarctica indicates a decreased global burning of biomass from about 1500 through 1700 (Ferretti et al. 2005; Mischler et al. 2009) and fire history proxies from the American Tropics corroborate this trend (Carcaillet et al. 2002; Bush et al. 2008). That biomass burning decreased in the Neotropics during the LIA is widely acknowledged (Marlon et al. 2008; Nevle and Bird 2008); however, the causes of this decrease—and the establishment of its unique geography—remain at the crux of the present debate.

Ignition by lightning is rare in tropical moist forests and tropical seasonal forests, leaving human ignition as the dominant determinant of combustion (Janzen 1988; Stott 2000; Cochrane 2003). The more people burn, whether intentionally or accidentally, the more fire prone the forest becomes and the more likely it is to burn again when drier climatic conditions prevail (Field, van der Werf, and Shen 2009). Many anthropogenic fires escape their intended confines, agricultural or other. These accidental fires play a major role in the ecology of Neotropical forests today (Middleton et al. 1997) and contribute substantially to a positive fire feedback cycle; that is, more burning leads to more forest openings, more forest edges, drier fuel loads, higher wind speeds, and greater susceptibility to future flammability (Cochrane 2003; Laurance 2003, 2004; Fearnside and Laurance 2004). Conversely, the absence of anthropogenic fire in most tropical biomes leads to rapid increases in woody biomass (Bond, Woodward, and Midgley 2005) and a "hardening" of the forest against future fire incursions.

The El Niño droughts of 1997 and 1998 provide a relevant analog scenario for this study, demonstrating how both rural land use and population density contribute significantly to the geography of fire in the tropics (Cochrane and Schulze 1999; Cochrane 2003; Field, van der Werf, and Shen 2009) and by extension the geography of carbon flux from the biosphere to the atmosphere. Between ca. 3000 and 2000 calendar year before present (BP), the widespread adoption of plant cultivation as the primary means of food production in the Neotropics resulted in larger populations and as a consequence more fire, especially in the tropical moist forest biome where the anthropogenic-fire feedback is so essential to the fire regime. Because natural ignition is limited for tropical forest fires, the increased populations of pre-Columbian farmers spread across the lowland Neotropics would have facilitated large regional conflagrations via accidental ignitions when persistent drought conditions prevailed.[3]

Data Sources and Evidence for Holocene Paleofires in the New World Tropics

Charcoal records from the Neotropics, divided here into the northern Neotropics and the southern Neotropics (Figure 1), generally fall into two broad categories: lake and wetland charcoal, both microscopic and macroscopic; and soil charcoal, usually macroscopic (Carcaillet et al. 2002; Bush et al. 2007; Marlon et al. 2008; Nevle and Bird 2008). Of these, charcoal from lake and wetland sediment cores provides the most faithful reconstructions of past fire regimes (MacDonald et al. 1991; Long et al. 1998; Conedera et al. 2009). Several published reconstructions of Neotropical fire history have been produced by aggregating all charcoal data types to create composite paleofire reconstructions (Bush et al. 2007; Marlon et al. 2008; Power et al. 2008). Despite generally low chronological sampling resolutions (i.e., many sites with fewer than ten total samples), all of these fire history reconstructions show a dramatic decline in Neotropical burning after 1500 AD. The most recent of these studies concludes that the sixteenth century decreasing charcoal trend was consistent with global patterns of fire activity and was likely forced by LIA cooling with no evidence of preindustrial anthropogenic forcing of Neotropical fire regimes (Marlon et al. 2008).

Here we present a reevaluation of extant Neotropical charcoal data together with a compilation of four new high-resolution (<10 to <50-year sampling intervals) macroscopic charcoal records that indicate a dramatic fire regime shift took place in the tropical Americas at ca. 450 cal yr BP (AD 1500; Table 1, Figures 2 and 3). We have compiled previously unpublished macroscopic charcoal data from four lake sites in the northern Neotropics—Laguna Herradura, Veracruz, Mexico; Lago Metapan, El Salvador; and Charco Verde and Lago Nicaragua, Nicaragua (Table 1, Figures 1 and 2). Volumetric samples of 1 cm^3 were disaggregated and wet sieved with a 150 μ mesh and all

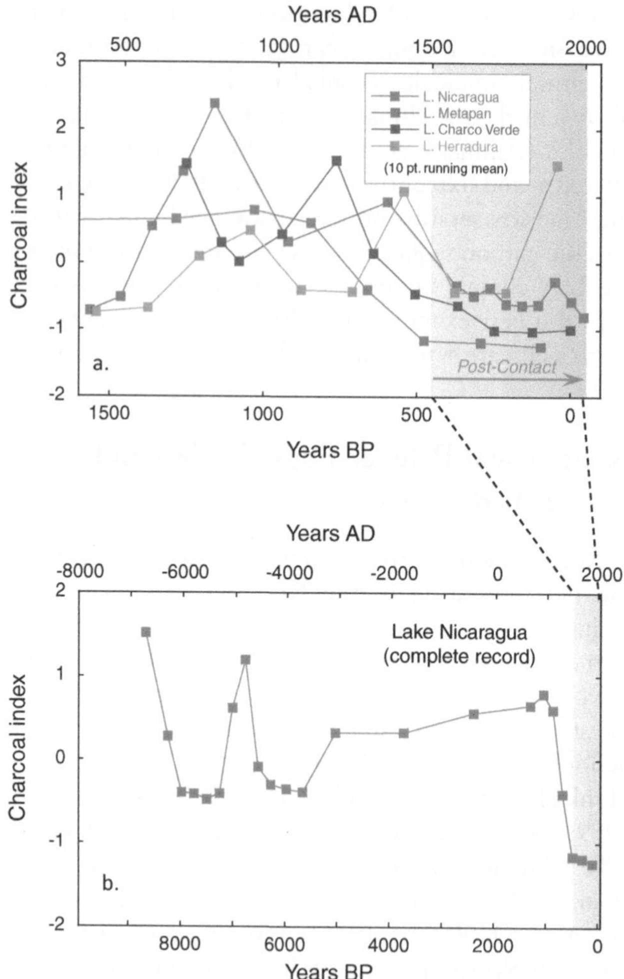

Figure 2. (A) Macroscopic (>150 μ) charcoal concentration records from four lakes in the northern Neotropics: Lago Herradura, Mexico; Lago Metapan, El Salvador; Laguna Charco Verde, Nicaragua; and Lago de Nicaragua, Nicaragua (partial record only). (B) Full charcoal concentration record for Lago de Nicaragua, Nicaragua. Data in Figures 2A and 2B reported as 10 pt. running mean of charcoal index calculated following method described by Nevle and Bird (2008). The charcoal index is a measure of the variation of the charcoal concentration, in units of standard deviation, about the mean charcoal concentration for each record. See Table 1 for characteristics of sedimentary records.

Table 1. Northern Neotropics lake sediment coring sites for macroscopic charcoal records reported in this article

Lake name	Location (latitude/ longitude)	Core length (m)	Basal age 2σ cal age	Total number of samples	Mean sampling interval (years/ sample)
Laguna Herradura, Mexico	22.01248N 98.155165W	2.00	*1677–1566 BP *AD 273–384	100	16.8
Laguna Metapan El Salvador	14.30071N 89.477829W	3.10	*1542–1414 BP *AD 408–536	155	9.8
Lake Nicaragua (LC-4), Nicaragua	11.76258N 85.872528W	4.39	*9015–8779 *7065–6829 BC	221	40.5
Charco Verde, Ometepe Island, Nicaragua	11.47615N 85.631777W	6.34	*1344–1297 BP *AD 606–653	108	12.7

Note: All ^{14}C dates calibrated with Calib 5.0 using the INTCAL 2004 data set (Reimer et al. 2004).

charcoal above that size fraction was tallied to calculate charcoal concentrations.[4] Basal ^{14}C dates are two sigma calibrated age ranges based on the calibration data set of INTCAL 2004 (Reimer et al. 2004). Data from two of these sites were reported in master's theses at the University of Texas at Austin (Lee 2006; Avnery 2007). All records are represented by at least 100 samples each, a sampling density not matched by any of the

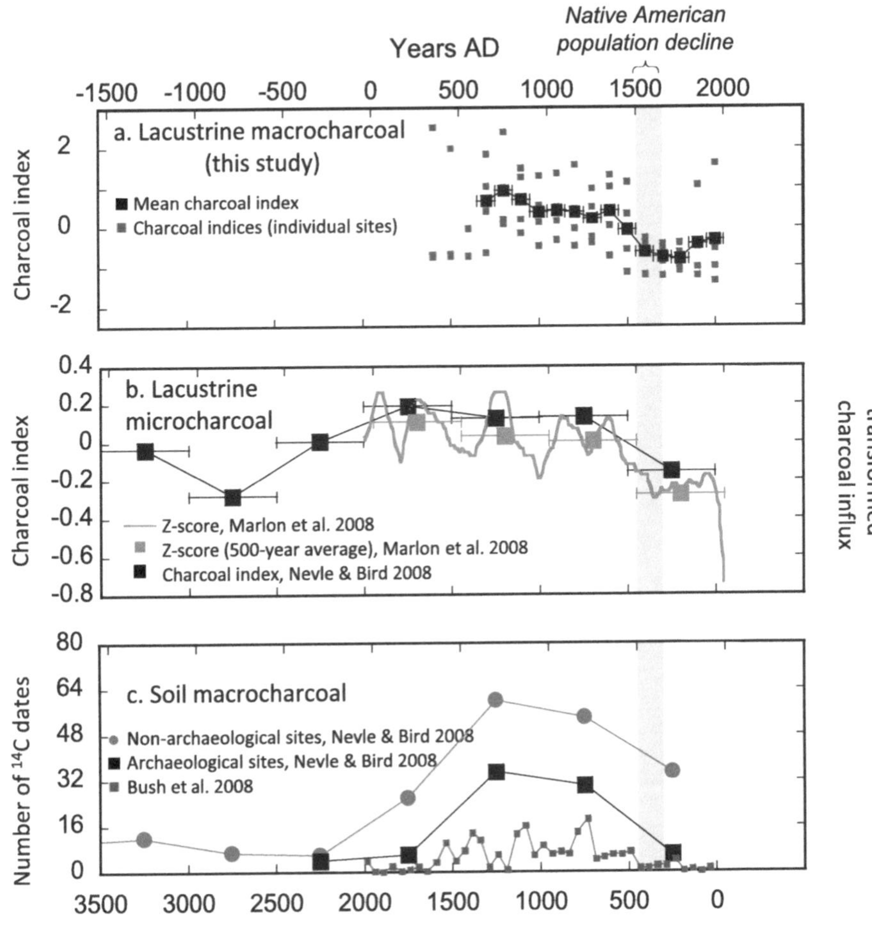

Figure 3. Comparison of Neotropical biomass-burning reconstructions. Biomass-burning reconstructions based on (A) lacustrine macroscopic charcoal records from this study (locations shown as blue/black symbols on Figure 1); (B) sedimentary charcoal records for the past 2,000 years (Marlon et al. 2008, red symbols) and 3,500 years (Nevle and Bird 2008, black symbols; locations shown as red symbols in Figure 1); and (C) soil–charcoal records (Bush et al. 2008; Nevle and Bird 2008; locations shown as pink symbols in Figure 1). Biomass-burning histories in A and B based on comparable numerical treatments of lacustrine charcoal records to reconstruct fire history. Methods used to reconstruct fire history indicate anomalies in regional variation of charcoal accumulation rates (charcoal index; black squares in A and B) and the charcoal influx (z score; red line in B), relative to their mean values for the duration of each reconstructed history. To permit direct comparison of the two fire histories based on lacustrine microscopic charcoal in B, 500-year averages of the z scores are also presented (red squares in B). Gray vertical bars represent the approximate duration of the Native American population decline (Dobyns 1966).

proxy fire data sets from the Neotropics reproduced in recent compilations (Marlon et al. 2008).

The new combined macroscopic charcoal index from the northern Neotropics (Figures 2 and 3A) is compared to several other composite fire records from the Neotropics constructed with microscopic charcoal data from lake sediments (Figure 3B) and macroscopic charcoal from terrestrial soils (Figure 3C). As evident in Figure 3C, the regional biomass burning reconstruction presented by Marlon et al. (2008) exhibits large amplitude oscillations from 2000 BP until ~500 BP and then decreases to values that remain below the minimum attained during the prior 1,500 years. The biomass burning index of Nevle and Bird (2008) is negative between 3500 and ~2500 BP and then increases and varies in a pattern nearly identical to that of the 500-year averages of the fire index from Marlon et al. 2008 (red squares). Frequencies of macroscopic soil charcoal ^{14}C dates (Bush et al. 2008; Nevle and Bird 2008)[5] obtained from both nonarchaeological and archaeological sites increase after 2500 years BP, obtain maxima after 1500 years BP, and decline markedly after 500 BP. All the biomass burning trends (Figure 3) are broadly similar despite differences in charcoal size fraction sampled, geographic distributions of sample sites, and numerical treatments of records used in each reconstruction (Bush et al. 2008; Marlon et al. 2008; Nevle and Bird 2008). The minima in the proxies of biomass burning after 500 years BP in the reconstructions shown in Figures 2 and 3 are synchronous with the absolute minimum in Holocene Neotropical charcoal accumulation in the 20,000-year reconstruction of Power et al. (2008).

Covariation between the biomass burning histories reconstructed from charcoal derived from both archaeological and nonarchaeological sites shown in Figure 3C suggests that anthropogenic activity controlled variations in the fire regime and vegetation cover in the Neotropical Lowlands prior to the Industrial Revolution. The data in Figures 2 and 3 are consistent with (1) increased biomass burning and deforestation during agricultural and population expansion in the Neotropics from ~2500 to ~500 years BP; and (2) declining anthropogenic use of fire due to pestilence-induced population collapse during the European conquest. The Late Holocene increase in biomass burning, including that indicated by the archaeological soil charcoal record (an unambiguous indicator of anthropogenic ignition), occurred as agriculture began to provide the dietary staple in the indigenous diet (Dull 2006; Rebellato, Woods, and Neves 2009). These observations suggest that human–landscape interaction profoundly influenced the Late Holocene Neotropical fire regime and implicate vegetation recovery during post-Contact demographic collapse as a potentially significant carbon sink.

Food Production, Population, and Anthropogenic Landscapes of the Neotropics

Although the term *Anthropocene* was coined just ten years ago (Crutzen and Stoermer 2000), there is now a broad consensus that the post-1850 rise in atmospheric CO_2—and its attendant 1°C rise in average global temperature—occurred largely because of increases in the burning of fossil fuels together with the conversion of forested lands to agriculture (Crutzen and Steffen 2003; Doney and Schimel 2007; Zalasiewicz et al. 2008). In the Americas, food production began in the early Holocene but was not widely adopted throughout the tropical lowlands until the late Holocene (Smith 2001; Dull 2006). The advent of food production in the Americas facilitated sedentary living, the rise of urban centers with populations numbering in the tens to hundreds of thousands, and widespread anthropogenic landscapes.

Population: Southern Neotropics

Greater Amazonia is the tropical lowland interior of South America, including considerable (~20 percent) savanna (Figure 1). Until recently this region has been seen as a hostile environment characterized by low soil fertility and meager protein resources beyond the fish and other animal sources common in riparian zones. As a result, pre-European population numbers were judged to be quite low and the consequent impacts on the environment correspondingly slight. For just the Amazon Basin, Meggers (1992) estimated a density of only 0.2 persons per square kilometer and a total of 1.5 to 2.0 million indigenous inhabitants. However, accumulating evidence indicates that there was both intensive and extensive environmental management and landscape modification; that subsistence systems varied in form and intensity; and that populations were dense along the rivers, in the wet savannas, and locally in the interior, with a total population of at least 5 to 6 million (Denevan 1992a, 2001, 2003; Heckenberger et al. 2003; Erickson 2008; Oliver 2008).

Recent research has examined the extent, ecology, chemistry, productivity, and significance of exceptionally fertile prehistoric, anthropogenic Dark Earth soils (*terra preta* in Brazil; Woods et al. 2009). Three different calculations based on maize and manioc production and the physical–chemical characteristics of the deposits indicate that populations of 3.1, 3.3, and 3.7 (average 3.4) million could have been supported by these soils, which total a conservative 0.2 percent of forested Amazonia (Woods, Denevan, and Rebellato forthcoming). With at least 5 million people in the rest of the region (Denevan 1992a, 2003), the total estimated population comes to 8.4 million for greater Amazonia, with an overall density of about 0.86 per square kilometer. For Greater Amazonia (ca. 9,770,000 km^2) about 78 percent was forested (Denevan 1992a). Proportionally, this reduces the prehistoric population estimate of 8.4 million to 6.6 million for the forested area.

A population estimate of this magnitude is supported by archaeological and historical evidence for large settlements, numbering in the thousands of people in each, in upper Amazonia, along the central Amazon and major tributaries, the uplands adjacent to the Rio Tapajós, on Marajó Island, the Upper Xingu region, and the Mojos savannas of eastern Bolivia (Denevan 1966, 1996, 2003; Roosevelt 1987; Heckenberger, Petersen, and Neves 1999; Petersen, Neves, and Heckenberger 2001; Erickson 2008). Many of the settlement sites, large as well as small, contain Amazonian Dark Earths, a clear indication of at least semipermanent habitation because these soils formed over long periods of time. Anthropogenic earthworks include mounds, ditches, causeways, canals, raised fields, fish traps, moats, and embankments. Complexes of huge geometrically shaped earthern berms (geoglyphs) have been exposed by recent deforestation in western Brazil (Pärssinen, Schaan, and Ranzi 2009). All of these are indicators of numerous people and forest clearing in the past.

The forested tropical coastal areas from central Ecuador to Lake Maracaibo in Venezuela were inhabited by sophisticated gold-working chiefdoms and contained a population of possibly 1.5 million (Denevan, unpublished). This 1.5 million plus 6.6 million for greater Amazonia gives an estimated total population of 8.1 million for the forested lowland southern Neotropics.

Population: Northern Neotropics

The mostly forested lowland northern Neotropics extended from Panama through Central America through Yucatan and the southern coasts of Mexico, plus the Caribbean.[6] The pre-Columbian population numbered an estimated 11.3 million. This includes 5.1 million in the Audiencia de Guatemala (Chiapas, Soconusco [western Chiapas], El Salvador, Honduras, Nicaragua, and Costa Rica; Lovell and Lutz 1995); 2.45 million in Panama, Belize, Yucatan, and Tabasco (Denevan 1992a); 500,000 in Veracruz (Sluyter 2002); possibly 250,000 on the southwest coast of Mexico (Denevan, unpublished); and 3.0 million in the Caribbean Islands (Denevan 1992a). Thus the total estimated pre-Columbian population for the forested lowlands of the Neotropics is 19.4 million (8.1 for the south and 11.3 for the north).

Demography of the Columbian Encounter

The late fifteenth-century arrival of the Spaniards in the New World unleashed a cascade of diseases (i.e., smallpox, typhus, diphtheria, mumps, measles, influenza, etc.) that swiftly swept the Americas (Dobyns 1966; Lovell 1992), resulting in an unprecedented demographic collapse referred to by some as the American Indian Holocaust (Thornton 1990). It has been estimated by several scholars that the population of the Americas was reduced by 90 percent or more (Dobyns 1966; Denevan 1992a),[7] a loss of approximately one fifth of the Earth's human inhabitants (Mann 2005; Figure 4C). The tropical lowlands of Central and South America were particularly hard hit, where the population collapse amounted to approximately 95 percent (Dobyns 1966). The more densely populated regions of Mesoamerica were gravely affected, but so were many regions in the Amazon basin, such as the Xingu drainage, where a population estimated in the tens of thousands in the early sixteenth century was reduced to approximately 500 individuals by the mid-twentieth century (Heckenberger and Neves 2009). Populations were decimated not only by epidemics but also the ancillary effects of colonialism such as warfare, slavery, starvation, and reduced fertility.

Estimating the Pre-Columbian Carbon Footprint in the Neotropics

The thesis that post-Columbian carbon sequestration in the Neotropics was partially responsible for the rapid decrease in atmospheric CO_2 concentration during the LIA requires a significant pre-Columbian carbon footprint for the American tropics. How much

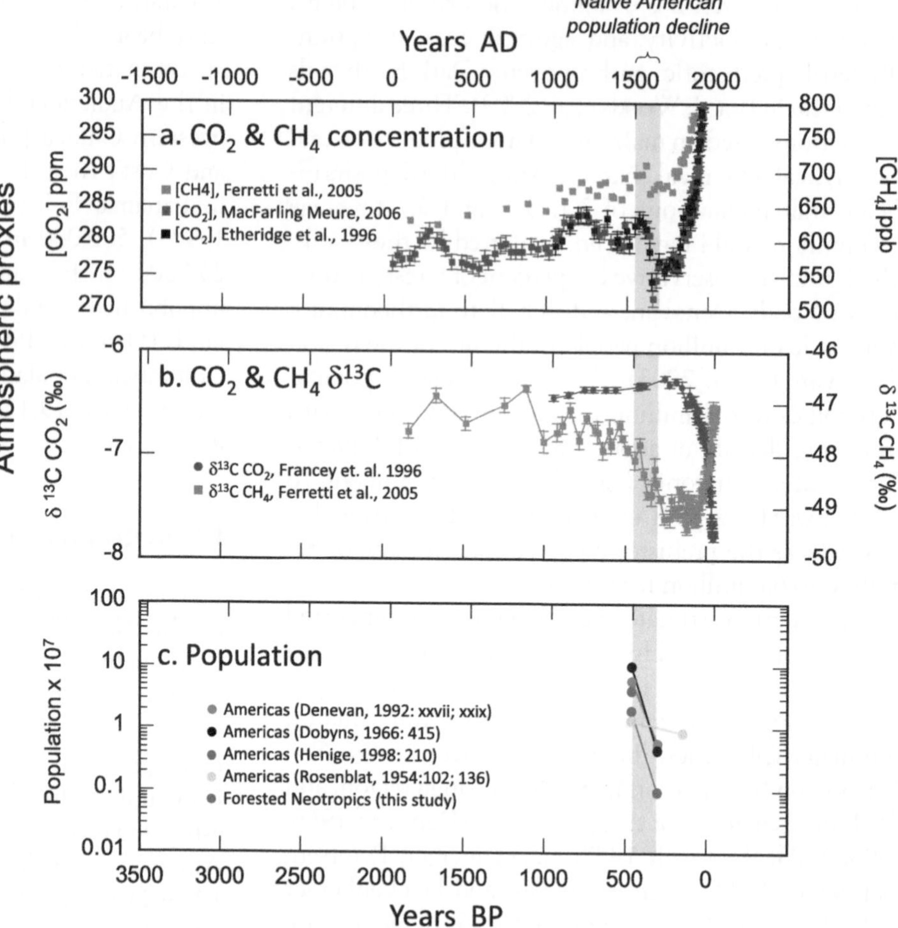

Figure 4. Comparison of atmospheric [CO_2], [CH_4], $\delta^{13}CCO_2$, $\delta^{13}CCH_4$, and population data for the Americas: (A) Concentration of atmospheric CO_2 (Etheridge et al. 1996; Meure et al. 2006; blue and black symbols) and CH4 (Ferretti et al. 2005; red symbols from Law Dome; (B) $\delta^{13}C$ of atmospheric CO_2 (Francey et al. 1999; blue symbols) and CH_4 (Ferretti et al. 2005; red symbols) from Law Dome; (C) population estimates for the Americas (Rosenblat 1954; Dobyns 1966; Denevan 1992a; Henige 1998) and Neotropics (this study). Gray vertical bars as in Figure 3.

biologically productive land was cleared or kept in early stages of succession by the land use practices of prehistoric Americans? The amount of land cleared for agriculture in prehistory has been estimated in several recent studies based on presumed populations (Olofsson and Hickler 2008; Pongratz et al. 2008; Pongratz et al. 2009), but the resulting maps are not at all consistent with the pollen and archaeological data, which indicate widespread cultivation and forest clearance, especially in the northern Neotropics (Whitmore and Turner 2001; Piperno 2006; Dull 2008).

The pre-Columbian carbon footprint consisted primarily of forest biomass cleared for farming and nonagricultural burning (primarily via fire leakage). Many other cultural practices, not explicit in the carbon sequestration calculations to follow, would have been common in the pre-Columbian Neotropics, such as fuel wood harvesting; establishment of habitations; weed, pest, and game management; maintenance of trails; and construction of monumental architecture, plazas, causeways, and other facilities.

The Farming Footprint

The total land area needed to provide for caloric needs varied widely across the prehistoric Americas, ranging from shifting cultivation[8] with mostly short fallows to highly productive semipermanent and permanent systems, such as those associated with Amazonian Dark Earths[9] (Denevan 2001; Whitmore and Turner 2001; Woods, Denevan, and Rebellato 2010). Per capita land needs for agricultural production in Latin America today range from approximately 0.2 to 0.3 ha $person^{-1}$ $year^{-1}$ for intensive agriculture on Amazonian Dark Earths to 2.2 ha $person^{-1}$ $year^{-1}$ in long-fallow swidden systems (Drucker and Heizer 1960; Beckerman 1987; Hecht 2003; Denevan 2004; Woods, Denevan, and Rebellato 2010; see Table 2). This range is consistent with those reported by Seiler and Crutzen (1980) of 0.4 to 0.6 ha $person^{-1}$ $year^{-1}$ for intensive agriculture and 2.0 to 3.2 ha $person^{-1}$ $year^{-1}$ for extensive long fallow systems. We estimate that a range of approximately 0.9 to 1.5 ha of land, including fallow,

Table 2. Representative agricultural land area per capita per year in the Neotropical lowlands

Crop system	Average fallow length (including ranges)	Land area needed per person/ year including fallows[a]	Sources
Long fallow, shifting cultivation (Amazonia)	15 years (10–20)[b]	2.2 ha	Beckerman (1987)[c]
Short fallow, shifting cultivation (Amazonia)	6 years (4–8)	1.0 ha	Denevan (2004)
Short fallow (Mesoamerica)	4 years (3–5)	1.5 ha	Drucker and Heizer (1960)
Semipermanent/permanent cultivation on superior soils (e.g., Amazonian Dark Earths)	2 years (0–4)	0.25 ha (0.2–0.3)	Woods, Denevan, and Rebellato (2010)
		1.2 ha (mean)	

[a]Based on average size of one family field plus the average length of fallow, divided by five, which is used here for the average size of family, which gives the amount of land needed to support one person.

[b]Long fallows can range up to fifty years or more, but most fall between ten and twenty years. Even a fifty-year fallow period would result in only about 50 percent biomass recovery.

[c]Beckerman (1987) gives field sizes for nineteen tribes, averaging 0.68 ha per field. With five people/family/field, the amount of cultivated land needed to feed one person for a year is 0.14 ha. With an additional fifteen fields in fallow, the amount needed per person for a year is 2.2 ha.

was needed per person to supply the majority of the caloric needs of the prehistoric population (Table 3).[10]

The Fire Footprint

The average global carbon emissions from fire today total amount to ~1.4 to 2.8 Pg C year^{-1}, with ~30% of global CO_2 fire emissions originating in the New World tropics (Schultz et al. 2008; Langmann et al. 2009). Even where forest regrowth and sequestration has been factored in, the "net tropical source" of CO_2 to the atmosphere during the recent era of tropical deforestation has been estimated to be as high as ~1.6 Pg C year^{-1} (Randerson et al. 1997). The association of fire with prehistoric land use and specifically agriculture in the forest biomes of the Neotropics is well established (Piperno and Pearsall 1998; Bush et al. 2008; Mayle and Power 2008). For example, Amazonian Dark Earths, which have been identified throughout the Amazon, were formed in part through repeated, intentional burning (Woods et al. 2009). This demonstrated use of fire in prehistoric agricultural systems has implications for the anthropogenic fire carbon footprint in the Prehistoric Americas.

Fire leakage is a major cause of chronic forest degradation in all tropical forests today, most notably in the Amazon (Laurance 2003; Aragão et al. 2007). For example, the 1997–1998 El Niño drought resulted in widespread fires that burned 20 million ha of tropical forests, most due to anthropogenic ignitions (Cochrane 2003), releasing ~0.7 to 0.9 Pg C to the atmosphere from the Neotropics alone, or about 20 percent of the CO_2 growth rate anomaly (van der Werf et al. 2004). Overall, roughly 2.3 to 3.0 Pg C was released from the tropics due to unintentional anthropogenic fire during the 1997–1998 El Niño, about one third from the Neotropics (Laurance 2003; van der Werf et al. 2004; Lewis 2006; Aragão et al. 2007). Unintentional fire leakage likely contributed substantially to pre-Columbian Neotropical biomass burning (Mayle and Power 2008), and consequently CO_2 flux to the atmosphere, but calculating average annual net CO_2 fluxes from this source in prehistory is nearly impossible without invoking modern fire emissions estimates from

Table 3. Summary of parameters used to calculate carbon sequestration potential of post-Contact Neotropical reforestation

Estimated pre-Contact population (millions)	Estimated 95% mortality by 1650 (millions)	Tropical forest carbon density (Mg C/ha)	Agricultural land clearance (ha/person)	Carbon sequestration potential from reforestation (Pg C)	Potential contribution to Little Ice Age CO_2 anomaly
19.4	18.4	120–190	0.9–1.5[a]	2–5	6–25%

[a]Represents average farming footprint of 1.2 ha/person/year (Table 2) ± 25 percent.

this region derived largely from remote sensing; no such explicit numbers are offered here.[11]

Postcontact Carbon Sequestration

Net primary production is highest globally in low-latitude forests, where it ranges from approximately 1,600 g m^{-2} $year^{-1}$ in tropical dry forests to 2,200 g m^{-2} $year^{-1}$ in tropical moist forests (Seiler and Crutzen 1980). Neotropical forests have played an especially crucial role in the carbon cycle because of their vast potential as a carbon sink. Several authors have suggested that the Neotropical lowlands became a massive sink for carbon following the Columbian encounter (Ruddiman 2005, 2007; Faust et al. 2006; Nevle and Bird 2008), with some suggesting that the effects of that demographic shift could still be echoing today via accelerated modern Amazon carbon sequestration rates (Phillips et al. 1998).

We calculate the carbon sequestration potential from reforestation of abandoned agricultural landscapes in the lowland Neotropics to be in the range of 2 to 5 Pg C, assuming 0.9 to 1.5 ha/person for agricultural production, 95 percent mortality of 19.4 million people, and above-ground (plant) carbon density of tropical forest of 120 to 190 Mg/ha (Prentice et al. 2001; Table 3). The 2 to 5 Pg C range represents (1) about 6 to 25 percent of terrestrial carbon sequestration (20–38 Pg; Joos et al. 1999; Faust et al. 2006) required to produce the atmospheric CO_2 anomaly of about −5 ppm from 1500 to 1750 AD; and (2) a minimum estimate for carbon sequestration associated with reforestation of humanized landscapes because indigenous people cleared land both intentionally for a variety of purposes besides agriculture and unintentionally through fire leakage.

Our order-of-magnitude calculations illustrate the potential influence of postcontact Neotropical reforestation on Earth's carbon budget and help explain anomalous variations in the concentration and $\delta^{13}C$ of greenhouse gases during the LIA. Anomalies in atmospheric [CO_2], [CH_4], $\delta^{13}C_{CO2}$, and $\delta^{13}C_{CH4}$ recorded in the Law Dome ice core (Etheridge et al. 1996; Francey et al. 1999; Ferretti et al. 2005; Meure et al. 2006; summarized in Figures 4A–B) are synchronous with Neotropical population decline (Figure 4C). The ~5 ppm decrease in CO_2 from 1500 to 1800 AD and simultaneous increase in the $\delta^{13}C$ of CO_2 (Figures 4A–B) require terrestrial biospheric carbon uptake (Francey et al. 1999; Joos et al. 1999), due to the strong discrimination against ^{13}C by the C_3 photosynthetic pathway. Although models of LIA cooling have suggested that the drop in CO_2 concentration during this period is tenuously linked to solar–volcanic forcing (Hunt and Elliott 2002; Von Storch et al. 2004), it has been demonstrated more recently that solar luminosity decreases were not a significant contributor to LIA cooling (Foukal et al. 2006). Moreover, the $\delta^{13}C$ of atmospheric CO_2 began to increase as CO_2 concentrations began to decrease about a century before the solar–volcanic forcing events commonly associated with the inception of LIA cooling (the Maunder Minimum in sunspot activity, 1645–1715 AD; Rind et al. 2004; and a cluster of major volcanic eruptions toward the end of the sixteenth century). This sequence indicates that factors besides solar–volcanic forcing contributed to the sequestration of CO_2 into terrestrial biosphere, which is consistent with the observation made by Siegenthaler et al. (2005) that cooling was unlikely to have independently produced the ~5 ppm decrease in atmospheric CO_2 that persisted for more than two centuries after 1500 AD (Figure 4B). Our mass balance calculations implicate Neotropical reforestation as a first-order contributor to changes in the atmospheric CO_2 concentration during the LIA.

Proxy records of atmospheric methane abundance and its carbon isotopic composition (Figures 4A–B) are also consistent with the hypothesis that postcontact changes in human–landscape interactions significantly influenced Earth's carbon budget. The $\delta^{13}C$ of CH_4 begins to decrease at about 1000 AD, with the rate of decrease accelerating at about 1500 AD. Ferretti et al. (2005) attributed the initial decrease in the $\delta^{13}C$ of CH_4 from 1000 to 1500 AD to natural climate change and the subsequent accelerated decrease in $\delta^{13}C$—as well as the decline in CH_4 (Figure 4B)—from 1500 to 1800 AD to reduced anthropogenic biomass burning coincident with rapid human population decline in the Americas (Figure 4C; Dobyns 1966; Henige 1998; Denevan 1992a; Dale and Adams 2003; Ruddiman 2005; Mischler et al. 2009). In summary, variations recorded by proxies of atmospheric CO_2 and CH_4 corroborate the hypothesis that preindustrial human–landscape interactions in the Neotropics influenced the regional fire regime and contributed to changes in Earth's atmospheric greenhouse gas budget.

Conclusion

The concept of anthropogenic forcing of climate before the Industrial Revolution has elicited a vigorous debate in recent years. We contend that by 1500 AD the global imprint of human land use on the carbon cycle was sufficient to produce measurable atmospheric warming and that the decrease in atmospheric CO_2 from 1500 to 1750 was in part caused by carbon sequestration in the lowland tropical forests of the Americas. We argue that the decline in atmospheric CO_2 must be evaluated not only in terms of land needed for food production but within the context of tropical fire ecology and the positive fire feedback loop created by increasingly pervasive agricultural land use and recurrent droughts during the late Holocene. The reduction in biomass burning in Neotropical regions that commenced 500 years ago cannot be explained on the basis of climatic factors alone. The sudden sixteenth-century decrease in human ignition sources contributed to an overall decrease in fire frequency—and a dramatic increase in woody biomass accumulation—in forest systems that have evolved where lightning ignition is rare.

The LIA was not caused by the Columbian encounter per se, but the evidence suggests that it was probably amplified measurably by the ecological effects of the demographic collapse. The estimates reported here of post-1500 carbon sequestration by the reforestation of lands previously cleared of tropical forests for agriculture represent a substantial terrestrial sink for CO_2 during the LIA, but these calculations only include the farming footprint with no quantification attempted of the potentially significant per capita anthropogenic fire footprint. Moreover, we have made no attempt to quantify the role of post-Conquest carbon sequestration in the Andes, the highland forests of Central America and Mexico, or the eastern deciduous forest of the United States. Future research on the climate forcing impacts of the Columbian encounter should seek to quantify a more geographically comprehensive per capita ecological footprint in the Neotropics and throughout the Americas.

Acknowledgments

We gratefully acknowledge J. W. C. White for providing Law Dome methane and carbon isotope data. We thank Bill Doolittle and Charles Mann for providing the base map of cultivated areas in Figure 1. We thank University of Texas students Kevin McKeehan and Maraigh Leitch for their work on the charcoal analyses. Funding to R. Dull was provided by the Mellon Foundation, the University of Texas Teresa Lozano Long Institute for Latin American Studies, the University of Texas Jackson School of Geosciences Initiative grant program, and the National Science Foundation Margins and Earth Science Programs (EAR 0440149) grant to Kirk McIntosh and Paul Mann. Funding for research by S. Avnery was provided by a Graduate Research Fellowship from the University of Texas College of Liberal Arts. We thank two anonymous reviewers for their constructive comments on an early version of this article.

Notes

1. Also see the book review by Turner (2006), a somewhat critical and yet balanced view.
2. Interpretations of the pollen data themselves have varied widely, with two recent studies from the Maya region supporting the notion of limited prehistoric forest clearance beyond agricultural fields (Ford and Nigh 2009; McNeil, Burney, and Burney 2010).
3. In the seasonal tropical dry forest regions, these "drought" conditions would have been achieved on a near annual basis, when dry season burning would have resulted in escaped fire.
4. Samples of 1 cc were soaked overnight in 50 mL of deflocculant: a 5 percent solution of sodium hexametaphosphate. The soaking was followed by repeated warm water washes the following day. Sieved residues were transferred to standard 100 × 15 mm Petri dishes and counted with 40× magnification.
5. Periods with relatively larger numbers of ^{14}C dates from charcoal fragments correspond with periods having relatively higher mean rates of biomass burning.
6. Several million more people lived in the nonforested lowland Neotropics. Most nonforest dwellers lived in savannas and coastal scrub environments. Although there are extensive savanna regions in South America, there were much smaller patches of savannas scattered throughout the northern Neotropics (Beard 1953).
7. Although we accept the estimate of Denevan (1992a) as being the best estimate, we also include several other estimates in Figure 4C (Rosenblat 1954; Dobyns 1966; Henige 1998).
8. Long-fallow systems, although quite common in the Amazon today, are the subject of some debate in regard to prehistory. A strong case has been made that shifting cultivation was a post-Conquest adaptation in the Amazon, facilitated by the introduction of metal tools that allowed for more effective and rapid clearing of forests (Denevan 1992b). Tree cutting with stone axes would not have been a viable means of forest clearance in prehistory.
9. The productivity of Amazonian Dark Earths or *terra preta* has been shown in modern studies to be extraordinarily high with 1 ha of rich terra preta capable supporting as many as 24.5 people in a maize-based economy and 29.6 people subsisting primarily on manioc. In such a system 80 percent of the farmer's land would be in a short bush fallow rotation (about two years cropping and three to four years fallow; Woods, Denevan, and Rebellato forthcoming).

10. We consider these numbers to be representative of average agricultural land consumption patterns based on the cited literature, but we do not assign percentages of Neotropical forest land area under each fallow system.
11. Using the 1997–1998 El Niño numbers of ~0.7 to 0.9 Pg C net Neotropical flux as an upper limit (van der Werf et al. 2004), and even if we assume average growth rate anomalies an order of magnitude or so lower than this (~0.05–0.10 Pg year^{-1}) during prehistory, significant cumulative net CO_2 fluxes could have built up over decades and centuries (~0.5–10+ Pg C) during the late Holocene as agricultural populations were increasing and human-mediated fire regimes were expanding.

References

Ammann, C. M., F. Joos, D. S. Schimel, B. L. Otto-Bliesner, and R. A. Tomas. 2007. Solar influence on climate during the past millennium: Results from transient simulations with the NCAR Climate System Model. *Proceedings of the National Academy of Sciences* 104 (10): 3713–18.

Anchukaitis, K. J., and S. P. Horn. 2005. A 2000-year reconstruction of forest disturbance from southern Pacific Costa Rica. *Palaeogeography, Palaeoclimatology, Palaeoecology* 221 (1–2): 35–54.

Aragão, L., Y. Malhi, R. M. Roman-Cuesta, S. Saatchi, L. O. Anderson, and Y. E. Shimabukuro. 2007. Spatial patterns and fire response of recent Amazonian droughts. *Geophysical Research Letters* 34 (7): L07701. http://www.agu.org/journals/ABS/2007/2006GL028946.shtml (last accessed 3 August 2010).

Avnery, S. 2007. 1400 years of natural and anthropogenic sources of environmental change on Ometepe Island, Lake Nicaragua. Master's thesis, Geography and the Environment, University of Texas, Austin, TX.

Beard, J. S. 1953. The savanna vegetation of northern tropical America. *Ecological Monographs* 23:149–215.

Beckerman, S. 1987. Swidden in Amazonia and the Amazon Rim. In *Comparative farming systems*, ed. B. L. Turner II and S. B. Brush, 55–94. New York: Guilford.

Bond, W. J., F. I. Woodward, and G. F. Midgley. 2005. The global distribution of ecosystems in a world without fire. *New Phytologist* 165:525–37.

Broecker, W. S., and T. F. Stocker. 2006. The Holocene CO_2 rise: Anthropogenic or natural? *Eos Transactions, American Geophysical Union* 87 (3): 27–29.

Bush, M. B. 1995. Neotropical plant reproductive strategies and fossil pollen representation. *The American Naturalist* 145 (5): 594–609.

Bush, M. B., and P. A. Colinvaux. 1994. Tropical forest disturbance—Paleoecological records from Darien, Panama. *Ecology* 75 (6): 1761–68.

Bush, M. B., M. C. Miller, P. E. De Oliveira, and P. A. Colinvaux. 2000. Two histories of environmental change and human disturbance in eastern lowland Amazonia. *The Holocene* 10 (5): 543–53.

Bush, M. B., D. R. Piperno, and P. A. Colinvaux. 1989. A 6,000 year history of Amazonian maize cultivation. *Nature* 340:303–5.

Bush, M. B., M. R. Silman, M. B. de Toledo, C. Listopad, W. D. Gosling, C. Williams, P. E. de Oliveira, and C. Krisel. 2007. Holocene fire and occupation in Amazonia: Records from two lake districts. *Philosophical Transactions of the Royal Society B: Biological Sciences* 362 (1478): 209–18.

Bush, M. B., M. R. Silman, C. McMichael, and S. Saatchi. 2008. Fire, climate change and biodiversity in Amazonia: A Late-Holocene perspective. *Philosophical Transactions of the Royal Society B: Biological Sciences* 363:1795–1802.

Byrne, R., and S. P. Horn. 1989. Prehistoric agriculture and forest clearance in the Sierra de los Tuxtlas, Veracruz, Mexico. *Palynology* 13:181–93.

Carcaillet, C., H. Almquist, H. Asnong, R. H. W. Bradshaw, J. S. Carrión, M.-J. Gaillard, K. Gajewski et al. 2002. Holocene biomass burning and global dynamics of the carbon cycle. *Chemosphere* 49:845–63.

Chew, S. C. 2001. *World ecological degradation: Accumulation, urbanization, and deforestation 3000 B.C.–A.D. 2000.* Walnut Creek, CA: Alta Mira Press.

Clement, R. C., and S. P. Horn. 2001. Pre-Columbian land-use history in Costa Rica: A 3000-year record of forest clearance, agriculture and fires from Laguna Zoncho. *The Holocene* 11 (4): 419–26.

Cochrane, M. A. 2003. Fire science for rainforests. *Nature* 421 (6926): 913–19.

Cochrane, M. A., and M. D. Schulze. 1999. Fire as a recurrent event in tropical forests of the Eastern Amazon: Effects on forest structure, biomass, and species composition. *Biotropica* 31 (1): 2–16.

Colinvaux, P. A., M. Frost, I. Frost, K. B. Liu, and M. Steinitz-Kannan. 1988. Three pollen diagrams of forest disturbance in the western Amazon basin. *Review of Palaeobotany and Palynology* 55:73–81.

Conedera, M., W. Tinner, C. Neff, M. Meurer, A. F. Dickens, and P. Krebs. 2009. Reconstructing past fire regimes: Methods, applications, and relevance to fire management and conservation. *Quaternary Science Reviews* 28 (5–6): 555–76.

Cramer, W., A. Bondeau, S. Schaphoff, W. Lucht, B. Smith, and S. Sitch. 2004. Tropical forests and the global carbon cycle: Impacts of atmospheric carbon dioxide, climate change and rate of deforestation. *Philosophical Transactions: Biological Sciences* 359 (1443): 331–43.

Crosby, A. W. 1972. *The Columbian exchange: Biological and cultural consequences of 1492.* Westport, CT: Greenwood Press.

Crutzen, P. J., and M. O. Andreae. 1990. Biomass burning in the tropics: Impact on atmospheric chemistry and biogeochemical cycles. *Science* 250 (4988): 1669–78.

Crutzen, P. J., and W. Steffen. 2003. How long have we been in the Anthropocene era? *Climatic Change* 61 (3): 251–57.

Crutzen, P. J., and E. F. Stoermer. 2000. The Anthropocene. *IGBP Newsletter* 41:17–18.

Dale, V. H., and W. M. Adams. 2003. Plant reestablishment 15 years after the debris avalanche at Mount St. Helens, Washington. *The Science of the Total Environment* 313:101–13.

Denevan, W. M. 1966. *The Aboriginal cultural geography of the Llanos de Mojos of Bolivia, Ibero-Americana: 48.* Berkeley: University of California Press.

———, ed. 1992a. *The native population of the Americas in 1492*. 2nd ed. Madison: University of Wisconsin Press.

———. 1992b. Stone vs. metal axes: The ambiguity of shifting cultivation in prehistoric Amazonia. *Journal of the Steward Anthropological Society* 20 (1–2): 153–65.

———. 1996. A bluff model of riverine settlement in prehistoric Amazonia. *Annals of the Association of American Geographers* 86 (4): 654–81.

———. 2001. *Cultivated landscapes of native Amazonia and the Andes*. Oxford, UK: Oxford University Press.

———. 2003. The native population of Amazonia in 1492 reconsidered. *Revista de Indias* 63 (227): 175–87.

———. 2004. Semi-intensive Pre-European cultivation and the origins of Anthropogenic Dark Earths. In *Amazonian Dark Earths: Explorations in space and time*, ed. B. Glaser and W. I. Woods, 135–43. Berlin: Springer.

Dobyns, H. F. 1966. Estimating aboriginal American population: An appraisal of techniques with a new hemispheric estimate. *Current Anthropology* 7:395–416.

Doney, S. C., and D. S. Schimel. 2007. Carbon and climate system coupling on timescales from the Precambrian to the Anthropocene. *Annual Review of Environment and Resources* 32:31–66.

Drucker, P., and R. F. Heizer. 1960. A study of the Milpa system of La Venta Island and its archaeological implications. *Southwestern Journal of Anthropology* 16:36–45.

Dull, R. A. 2004. A Holocene record of Neotropical savanna dynamics from El Salvador. *Journal of Paleolimnology* 32:219–31.

———. 2006. The maize revolution: A view from El Salvador. In *Histories of maize: Multidisciplinary approaches to the prehistory, biogeography, domestication, and evolution of maize*, ed. J. E. Staller, R. H. Tykot, and B. F. Benz, 357–66. San Diego, CA: Academic Press.

———. 2007. Evidence for forest clearance, agriculture, and human-induced erosion in Precolumbian El Salvador. *Annals of the Association of American Geographers* 97 (1): 127–41.

———. 2008. Unpacking El Salvador's ecological predicament: Theoretical templates and "long-view" ecologies. *Global Environmental Change* 18 (2): 319–29.

Elsig, J., J. Schmitt, D. Leuenberger, R. Schneider, M. Eyer, M. Leuenberger, F. Joos, H. Fischer, and T. F. Stocker. 2009. Stable isotope constraints on Holocene carbon cycle changes from an Antarctic ice core. *Nature* 461 (7263): 507–10.

Erickson, C. L. 2008. Amazonia: The historical ecology of a domesticated landscape. In *Handbook of South American archaeology*, ed. H. Silverman and W. H. Isbell, 157–83. Berlin: Springer.

Esper, J., E. R. Cook, and F. H. Schweingruber. 2002. Low-frequency signals in long tree-ring chronologies for reconstructing past temperature variability. *Science* 295 (5563): 2250–53.

Etheridge, D. M., L. P. Steele, R. L. Langenfelds, R. J. Francey, J. M. Barnola, and V. I. Morgan. 1996. Natural and anthropogenic changes in atmospheric CO_2 over the last 1000 years from air in Antarctic ice and firn. *Journal of Geophysical Research-Atmospheres* 101 (D2): 4115–28.

Faust, F. X., C. Gnecco, H. Mannstein, and J. Stamm. 2006. Evidence for the Postconquest demographic collapse of the Americas in historical CO_2 levels. *Earth Interactions* 10 (11): 1–14.

Fearnside, P. M. 2000. Global warming and tropical land-use change: Greenhouse gas emissions from biomass burning, decomposition and soils in forest conversion, shifting cultivation and secondary vegetation. *Climatic Change* 46 (1): 115–58.

———. 2005. Deforestation in Brazilian Amazonia: History, rates, and consequences. *Conservation Biology* 19 (3): 680–88.

Fearnside, P. M., and W. F. Laurance. 2004. Tropical deforestation and greenhouse-gas emissions. *Ecological Applications* 14 (4): 982–86.

Ferretti, D. F., J. B. Miller, J. W. C. White, D. M. Etheridge, K. R. Lassey, D. C. Lowe, C. Meure, M. F. Dreier, C. M. Trudinger, and T. D. Van Ommen. 2005. Unexpected changes to the global methane budget over the past 2000 years. *Science* 309 (5741): 1714–17.

Field, R. D., G. R. van der Werf, and S. S. P. Shen. 2009. Human amplification of drought-induced biomass burning in Indonesia since 1960. *Nature Geoscience* 2 (3): 185–88.

Ford, A., and R. Nigh. 2009. Origins of the Maya forest garden: Maya resource management. *Journal of Ethnobiology* 29 (2): 213–36.

Foukal, P., C. Frohlich, H. Spruit, and T. M. Wigley. 2006. Variations in solar luminosity and their effect on the Earth's climate. *Nature* 443 (7108): 161–66.

Francey, R. J., C. E. Allison, D. M. Etheridge, C. M. Trudinger, I. G. Enting, M. Leuenberger, R. L. Langenfelds, E. Michel, and L. P. Steele. 1999. A 1000-year high precision record of 13C in atmospheric CO_2. *Tellus B* 51 (2): 170–93.

Goldhammer, J. G. 1991. Tropical wild-land fires and global changes: Prehistoric evidence, present fire regimes, and future trends. In *Global biomass burning: Atmospheric, climatic, and biospheric implications*, ed. J. S. Levine, 83–91. Cambridge, MA: MIT Press.

Goman, M., and R. Byrne. 1998. A 5000-year record of agriculture and tropical forest clearance in the Tuxtlas, Veracruz, Mexico. *The Holocene* 8 (1): 83–89.

Hecht, S. B. 2003. Indigenous soil management and the creation of Amazonian Dark Earths: Implications of Kayapó practices. *Amazonian Dark Earths: Origin, properties, and management*, ed. J. Lehmann, D. C. Kern, B. Glaser, and W. I. Woods, 355–72. Dordrecht, Netherlands: Kluwer Academic.

Heckenberger, M. J., A. Kuikuro, U. T. Kuikuro, J. C. Russell, M. Schmidt, C. Fausto, and B. Franchetto. 2003. Amazonia 1492: Pristine forest or cultural parkland? *Science* 301 (5640): 1710–14.

Heckenberger, M., and E. G. Neves. 2009. Amazonian archaeology. *Annual Review of Anthropology* 38:251–60.

Heckenberger, M. J., J. B. Petersen, and E. G. Neves. 1999. Village size and permanence in Amazonia: Two archaeological examples from Brazil. *Latin American Antiquity* 10 (4): 353–76.

Henige, D. P. 1998. *Numbers from nowhere: The American Indian contact population debate*. Norman: University of Oklahoma Press.

Houghton, R. A. 2003. Revised estimates of the annual net flux of carbon to the atmosphere from changes in land use and land management 1850–2000. *Tellus B* 55 (2): 378–90.

Hunt, B., and T. Elliott. 2002. Mexican megadrought. *Climate Dynamics* 20 (1): 1–12.

Islebe, G. A., H. Hooghiemstra, M. Brenner, J. H. Curtis, and D. A. Hodell. 1996. A Holocene vegetation history from lowland Guatemala. *The Holocene* 6 (3): 265–71.

Janzen, D. H. 1988. Management of habitat fragments in a tropical dry forest: Growth. *Annals of the Missouri Botanical Garden* 75 (1): 105–16.

Jones, P. D., and M. E. Mann. 2004. Climate over past millennia. *Reviews of Geophysics* 42 (2): 1–42.

Joos, F., R. Meyer, M. Bruno, and M. Leuenberger. 1999. The variability in the carbon sinks as reconstructed for the last 1000 years. *Geophysical Research Letters* 26 (10): 1437–40.

Lane, C. S., S. P. Horn, Z. P. Taylor, and C. I. Mora. 2009. Assessing the scale of prehistoric human impact in the Neotropics using stable carbon isotope analyses of lake sediments. *Latin American Antiquity* 20 (1): 120–33.

Lane, C. S., C. I. Mora, S. P. Horn, and K. H. Orvis. 2008. Sensitivity of bulk sedimentary stable carbon isotopes to prehistoric forest clearance and maize agriculture. *Journal of Archaeological Science* 35 (8): 2119–32.

Langmann, B., B. Duncan, C. Textor, J. Trentmann, and G. R. van der Werf. 2009. Vegetation fire emissions and their impact on air pollution and climate. *Atmospheric Environment* 43 (1): 107–16.

Laurance, W. F. 2003. Slow burn: The insidious effects of surface fires on tropical forests. *Trends in Ecology & Evolution* 18 (5): 209–12.

———. 2004. Forest–climate interactions in fragmented tropical landscapes. *Philosophical Transactions of the Royal Society B: Biological Sciences* 359 (1443): 345–52.

Lee, J. L. 2006. Late Quaternary environmental change in the Huasteca cultural region of the Mexican Gulf Coast, geography and the environment. Master's thesis, University of Texas, Austin, TX.

Lewis, S. L. 2006. Tropical forests and the changing Earth system. *Philosophical Transactions of the Royal Society of London B: Biological Sciences* 361:439–50.

Long, C. J., C. Whitlock, P. J. Bartlein, and S. H. Millspaugh. 1998. A 9000-year fire history from the Oregon Coast Range, based on a high-resolution charcoal study. *Canadian Journal of Forest Research* 28 (5): 774–87.

Lovell, W. G. 1992. "Heavy shadows and black night": Disease and depopulation in Colonial Spanish America. *Annals of the Association of American Geographers* 82 (3): 426–43.

Lovell, W. G., and C. H. Lutz. 1995. *Demography and empire: A guide to the population history of Spanish Central America, 1500–1821*. Boulder, CO: Westview.

MacDonald, G. M., C. P. S. Larson, J. M. Szeicz, and K. A. Moser. 1991. The reconstruction of boreal forest fire history from lake sediments: A comparison of charcoal, pollen, sedimentological, and geochemical indices. *Quaternary Science Reviews* 10:53–71.

MacFarling Meure, C., D. Etheridge, C. Trudinger, P. Steele, R. Langenfelds, T. Van Ommen, A. Smith, and J. Elkins. 2006. Law Dome CO_2, CH_4 and N_2O ice core records extended to 2000 years BP. *Geophysical Research Letters* 33 (14): L14810. http://www.agu.org/journals/ABS/2006/2006GL026152.shtml (last accessed 3 August 2010).

Mahli, Y., P. Meir, and S. Brown. 2002. Forests, carbon and global climate. *Philosophical Transactions of the Royal Society of London* 360 (1797): 1567–91.

Malhi, Y., J. T. Roberts, R. A. Betts, T. J. Killeen, W. Li, and C. A. Nobre. 2008. Climate change, deforestation, and the fate of the Amazon. *Science* 319 (5860): 169–72.

Mann, C. 2005. *1491: New revelations of the Americas before Columbus*. New York: Knopf.

Marlon, J. R., P. J. Bartlein, C. Carcaillet, D. G. Gavin, S. P. Harrison, P. E. Higuera, F. Joos, M. J. Power, and I. C. Prentice. 2008. Climate and human influences on global biomass burning over the past two millennia. *Nature Geoscience* 1 (10): 697–702.

Mayle, F. E., and M. J. Power. 2008. Impact of a drier Early-Mid-Holocene climate upon Amazonian forests. *Philosophical Transactions of the Royal Society B-Biological Sciences* 363 (1498): 1829–38.

McNeil, C. L., D. A. Burney, and L. P. Burney. 2010. Evidence disputing deforestation as the cause for the collapse of the ancient Maya polity of Copan, Honduras. *Proceedings of the National Academy of Sciences* 107 (3): 1017–22.

Meggers, B. J. 1992. Prehistoric population density in the Amazon basin. *Disease and Demography in the Americas*, ed. J. W. Verano and D. H. Ubelaker, 197–205. Washington, DC: Smithsonian Institution Press.

Middleton, B. A., E. Sanchez-Rojas, B. Suedmeyer, and A. Michels. 1997. Fire in a tropical dry forest of Central America: A natural part of the disturbance regime? *Biotropica* 29 (4): 515–17.

Mischler, J. A., T. A. Sowers, R. B. Alley, M. Battle, J. R. McConnell, L. Mitchell, T. Popp, E. Sofen, and M. K. Spencer. 2009. Carbon and hydrogen isotopic composition of methane over the last 1000 years. *Global Biogeochemistry Cycles* 23 (4): GB4024. http://www.agu.org/journals/ABS/2009/2009GB003460.shtml (last accessed 3 August 2010).

Moberg, A., D. M. Sonechkin, K. Holmgren, N. M. Datsenko, and W. Karlén. 2005. Highly variable Northern Hemisphere temperatures reconstructed from low- and high-resolution proxy data. *Nature* 433 (7026): 613–17.

Moran, E. 1993. Deforestation and land use in the Brazilian Amazon. *Human Ecology* 21 (1): 1–21.

Nevle, R. J., and D. K. Bird. 2008. Effects of syn-pandemic fire reduction and reforestation in the tropical Americas on atmospheric CO_2 during European conquest. *Palaeogeography, Palaeoclimatology, Palaeoecology* 264 (1–2): 25–38.

Oliver, J. R. 2008. The archaeology of agriculture in ancient Amazonia. In *Handbook of South American archaeology*, ed. H. Silverman and W. H. Isbell, 185–216. Berlin: Springer.

Olofsson, J., and T. Hickler. 2008. Effects of human land-use on the global carbon cycle during the last 6,000 years. *Vegetation History and Archaeobotany* 17 (5): 605–15.

Pärssinen, M., D. Schaan, and A. Ranzi. 2009. Pre-Columbian geometric earthworks in the upper Purus: A complex society in western Amazonia. *Antiquity* 83:1084–95.

Petersen, J. B., E. G. Neves, and M. J. Heckenberger. 2001. Gift from the past: Terra preta and prehistoric

Amerindian occupation in Amazonia. In *Unknown Amazon: Culture in nature in ancient Brazil*, ed. C. MecEwan and E. G. Neves, 86–105. London: British Museum Press.

Phillips, O. L., Y. Malhi, N. Higuchi, W. F. Laurance, P. V. Núñez, R. M. Vásquez, S. G. Laurance, L. V. Ferreira, M. Stern, and S. Brown. 1998. Changes in the carbon balance of tropical forests: Evidence from long-term plots. *Science* 282 (5388): 439–42.

Piperno, D. R. 1990. Aboriginal agriculture and land usage in the Amazon Basin, Ecuador. *Journal of Archaeological Science* 17:665–77.

———. 2006. Quaternary environmental history and agricultural impact on vegetation in Central America. *Annals of the Missouri Botanical Garden* 93 (2): 274–96.

Piperno, D. R., and D. M. Pearsall. 1998. *The origins of agriculture in the lowland Neotropics*. San Diego, CA: Academic Press.

Pongratz, J., T. Raddatz, C. H. Reick, M. Esch, and M. Claussen. 2009. Radiative forcing from anthropogenic land cover change since AD 800. *Geophysical Research Letters* 36 (2): L02709. http://www.agu.org/journals/ABS/2009/2008GL036394.shtml (last accessed 3 August 2010).

Pongratz, J., C. Reick, T. Raddatz, and M. Claussen. 2008. A reconstruction of global agricultural areas and land cover for the last millennium. *Global Biogeochemical Cycles* 22 (3): GB3018. http://www.agu.org/journals/ABS/2008/2007GB003153.shtml (last accessed 3 August 2010).

Power, M. J., J. Marlon, N. Ortiz, P. J. Bartlein, S. P. Harrison, F. E. Mayle, A. Ballouche, R. H. W. Bradshaw, C. Carcaillet, and C. Cordova. 2008. Changes in fire regimes since the Last Glacial Maximum: An assessment based on a global synthesis and analysis of charcoal data. *Climate Dynamics* 30 (7): 887–907.

Prentice, I. C. 1985. Pollen representation, source area, and basin size: Toward a unified theory of pollen analysis. *Quaternary Research* 23:76–86.

Prentice, I. C., G. D. Farquhar, M. J. R. Fasham, M. L. Goulden, M. Heimann, V. J. Jaramillo, and H. S. Kheshgi. 2001. The carbon cycle and atmospheric carbon dioxide. In *Climate change 2001: The scientific basis. Contribution of Working Group I to the Third Assessment Report of the Intergovernmental Panel on Climate Change*, 183–237. Cambridge, UK: Cambridge University Press.

Randerson, J. T., M. V. Thompson, T. J. Conway, I. Y. Fung, and C. B. Field. 1997. The contribution of terrestrial sources and sinks to trends in the seasonal cycle of atmospheric carbon dioxide. *Global Biogeochemical Cycles* 11 (4): 535–60.

Reimer, P. J., M. G. L. Baillie, E. Bard, A. Bayliss, J. W. Beck, C. J. H. Bertrand, P. G. Blackwell, C. E. Buck, G. S. Burr, and K. B. Cutler. 2004. INTCAL 04 terrestrial radiocarbon age calibration, 0–26 cal kyr BP. *Radiocarbon* 46 (3): 1029–58.

Rind, D., D. Shindell, J. Perlwitz, J. Lerner, P. Lonergan, J. Lean, and C. McLinden. 2004. The relative importance of solar and anthropogenic forcing of climate change between the Maunder Minimum and the present. *Journal of Climate* 17 (5): 906–29.

Roosevelt, A. C. 1987. Chiefdoms in the Amazon and Orinoco. In *Chiefdoms in the Americas*, ed. R. D. Drennen and C. A. Uribe, 153–85. Lanham, MD: University Press of America.

Rosenblat, Á. 1954. *La población indígena y el mestizaje en América.Vol. 1, La población indigena, 1492–1950* [The indigenous and mestizo population of the Americas. Vol. 1, The indigenous population, 1492–1950]. Buenos Aires, Argentina: Editorial Nova.

Ruddiman, W. F. 2003. The anthropogenic greenhouse era began thousands of years ago. *Climatic Change* 61:261–93.

———. 2005. *Plows, plagues and petroleum: How humans took control of climate*. Princeton, NJ: Princeton University Press.

———. 2007. The early anthropogenic hypothesis: Challenges and responses. *Reviews of Geophysics* 45 (4): RG4001. http://www.agu.org/journals/ABS/2007/2006RG000207.shtml (last accessed 3 August 2010).

Rue, D. J. 1987. Early agriculture and early Postclassic Maya occupation in western Honduras. *Nature* 326:285–86.

Saarnak, C. F. 2001. A shift from natural to human-driven fire regime: Implications for trace-gas emissions. *The Holocene* 11 (3): 373–76.

Sauer, C. O. 1958. Man in the ecology of tropical America. *Proceedings of the Ninth Pacific Science Congress of the Pacific Science Association* 20:104–10.

Schultz, M. G., A. Heil, J. J. Hoelzemann, A. Spessa, K. Thonicke, J. G. Goldammer, A. C. Held, J. M. C. Pereira, and M. van het Bolscher. 2008. Global wildland fire emissions from 1960 to 2000. *Global Biogeochemical Cycles* 22 (2): GB2002. http://www.agu.org/journals/ABS/2008/2007GB003031.shtml (last accessed 3 August 2010).

Seiler, W., and P. J. Crutzen. 1980. Estimates of gross and net fluxes of carbon between the biosphere and the atmosphere from biomass burning. *Climatic Change* 2 (3): 207–47.

Siegenthaler, U., E. Monnin, K. Kawamura, R. Spahni, J. Schwander, B. Stauffer, T. F. Stocker, J. M. Barnola, and H. Fischer. 2005. Supporting evidence from the EPICA Dronning Maud Land ice core for atmospheric CO 2 changes during the past millennium. *Tellus* 57 (1): 51–57.

Sluyter, A. 2002. *Colonialism and landscape: Postcolonial theory and applications*. Oxford, UK: Rowman & Littlefield.

Smith, B. D. 2001. Documenting plant domestication: The consilience of biological and archaeological approaches. *Proceedings of the National Academy of Sciences* 98 (4): 1324–26.

Stocker, B., K. Strassmann, and F. Joos. 2010. Sensitivity of Holocene atmospheric CO_2 and the modern carbon budget to early human land use: Analyses with a process-based model. *Biogeosciences* 7:921–52.

Stott, P. 2000. Combustion in tropical biomass fires: A critical review. *Progress in Physical Geography* 24 (3): 355–77.

Thornton, R. 1990. *American Indian holocaust and survival: A population history since 1492*. Norman: University of Oklahoma Press.

Tsukada, M., and J. Deevey. 1967. Pollen analysis from four lakes in the southern Maya area of Guatemala and El Salvador. In *Quaternary paleoecology*, ed. E. J. Cushing and

H. E. Wright, 303–32. New Haven, CT: Yale University Press.

Turner, B. L. II. 2006. Review of *Plows, plagues, and petroleum* by William F. Ruddiman. *Geographical Review* 96 (3): 516–19.

van der Werf, G. R., J. T. Randerson, G. J. Collatz, L. Giglio, P. S. Kasibhatla, A. F. Arellano, Jr., S. C. Olsen, and E. S. Kasischke. 2004. Continental-scale partitioning of fire emissions during the 1997 to 2001 El Nino/La Nina period. *Science* 303 (5654): 73–76.

van Hoof, T. B., F. P. M. Bunnik, J. G. M. Waucomont, W. M. Kürschner, and H. Visscher. 2006. Forest re-growth on medieval farmland after the Black Death pandemic—Implications for atmospheric CO_2 levels. *Palaeogeography, Palaeoclimatology, Palaeoecology* 237 (2–4): 396–409.

van Hoof, T. B., F. Wagner-Cremer, W. M. Kürschner, and H. Visscher. 2008. A role for atmospheric CO_2 in preindustrial climate forcing. *Proceedings of the National Academy of Sciences* 105 (41): 15815–18.

Vavrus, S., W. F. Ruddiman, and J. E. Kutzbach. 2008. Climate model tests of the anthropogenic influence on greenhouse-induced climate change: The role of early human agriculture, industrialization, and vegetation feedbacks. *Quaternary Science Reviews* 27 (13–14): 1410–25.

Von Storch, H., E. Zorita, J. M. Jones, Y. Dimitriev, F. Gonzalez-Rouco, and S. F. B. Tett. 2004. Reconstructing past climate from noisy data. *Science* 306 (5696): 679–82.

Wahl, D., R. Byrne, T. Schreiner, and R. Hansen. 2006. Holocene vegetation change in the northern Peten and its implications for Maya prehistory. *Quaternary Research* 65 (3): 380–89.

Whitmore, T. M., and B. L. Turner. 2001. *Cultivated landscapes of middle America on the eve of conquest*. Oxford, UK: Oxford University Press.

Williams, M. 2003. *Deforesting the earth: From prehistory to global crisis*. Chicago: University of Chicago Press.

Woods, W. I., W. M. Denevan, and L. Rebellato. Forthcoming. Population estimates for anthropogenically enriched soils (terra preta) in Amazonia. In *Living on the land: The complex relationship between population and agriculture in the Americas*, ed. J. Wingard and S. Hayes. Boulder: University of Colorado Press.

Woods, W. I., W. G. Teixeira, J. Lehmann, C. Steiner, A. WinklerPrins, and L. Rebellato, eds. 2009. *Amazonian Dark Earths: Wim Sombroek's vision*. Berlin: Springer.

Zalasiewicz, J., M. Williams, A. Smith, T. L. Barry, A. L. Coe, P. R. Bown, P. Brenchley, D. Cantrill, A. Gale, and P. Gibbard. 2008. Are we now living in the Anthropocene? GSA *Today* 18 (2): 4–8.

Climate Change and Mountain Topographic Evolution in the Central Karakoram, Pakistan

Michael P. Bishop, Andrew B. G. Bush, Luke Copland, Ulrich Kamp, Lewis A. Owen, Yeong B. Seong, and John F. Shroder, Jr.

Mountain geodynamics represent highly scale-dependent interactions involving climate, tectonic, and surface processes. The central Karakoram in Pakistan exhibit strong climate–tectonic feedbacks, although the detailed tectonic and topographic responses to climate perturbations need to be systematically explored. This study focuses on understanding climate variations in relation to glacier erosion and relief production. Field data, climate modeling, remote sensing, geomorphometry, geochronology, glaciology, and geomorphological assessment are utilized to characterize climate change and geomorphic response. Climate simulations suggest that the region has experienced significant climate change due to radiative forcing over at least the past million years due to changes in Earth's orbital configuration, as well as more temporally rapid climate dynamics related to the El Niño Southern Oscillation. Paleoclimate simulations support geomorphological evidence of multiple glaciations and long-term glacier retreat. Mesoscale relief patterns clearly depict erosion zones that are spatially coincident with high peaks and rapid exhumation. These patterns depict extreme spatial and temporal variability of the influence of glacier erosion in the topographic evolution of the region. Results support the interpretation of high-magnitude glacial erosion as a significant denudational agent in the exhumation of the central Karakoram. Consequently, a strong linkage is seen to occur between global, or at least hemispheric, climate change and the topographic evolution of the Karakoram and the western Himalaya. *Key Words: central Karakoram, climate forcing, erosion, glaciation, landscape evolution.*

高山地球动力学代表与规模高度相关的，涉及气候，构造，和地面过程的相互作用。虽然还需要系统地探讨详细的构造和地形对气候扰动的响应，但是巴基斯坦中央喀喇昆仑已展现强有力的气候一构造反馈。本研究着重于了解有关气候变化与冰川侵蚀和生产救灾的相关性。利用现场数据，气候模型，遥感，地貌测量，年代学，冰川学，和地貌评估来表征气候变化和地貌反应。气候模拟表明，该地区在过去至少 100 年来，由于地球轨道构造变化的辐射力，已经经历了重大气候变化，以及与厄尔尼诺南方涛动有关的更快速的气候动态。古气候模拟支持多冰川和长期冰川退缩的地貌证据。中尺度救助格局图清楚地描绘了与高峰和险坡在空间上一致的侵蚀区。这些格局描述了在该地区的地形演化过程中冰川侵蚀影响的极端时空变异性。结果支持了高幅度冰川侵蚀是中央喀喇昆仑山峦一个重要剥离剂的解释。因此，能看到在全球，或者至少半球，气候变化和喀喇昆仑以及西部喜马拉雅地区的地形演变发展的密切联系。*关键词：中央喀喇昆仑山，气候强迫，侵蚀，冰川化，地貌演变。*

Las geodinámicas de montaña representan interacciones altamente dependientes de la escala en las que están involucrados procesos climáticos, tectónicos y topográficos. El Karakoram central de Pakistán muestra fuertes indicios de efectos climático–tectónicos, aunque las respuestas tectónicas y topográficas detalladas a las perturbaciones climáticas necesitan de una exploración sistemática. El presente estudio está enfocado a comprender las variaciones climáticas en relación con la erosión glaciar y la producción de relieve. Para caracterizar el cambio climático y la respuesta geomorfológica se utilizaron datos de campo, modelado climático, percepción remota, geomorfometría, geocronología, glaciología y análisis geomorfológico. Las simulaciones climáticas sugieren que la región ha experimentado cambio climático significativo por forzamiento radiativo durante por lo menos el pasado millón de años, causado por cambios en la configuración de la órbita terrestre, lo mismo que por la dinámica climática de temporalidad más rápida asociada con la Oscilación Meridional de El Niño. Las simulaciones paleoclimáticas refrendan la evidencia geomorfológica de múltiples glaciaciones y recesión glaciar de larga duración. Los patrones de relieve a escala mediana claramente muestran zonas de erosión espacialmente coincidentes con picos altos y rápida exhumación. Estos patrones muestran variabilidad temporal y espacial extremas de la influencia de erosión glaciar en la evolución topográfica de la región. Los resultados apoyan la interpretación de erosión glacial de alta magnitud como agente de denudación significativo en la exhumación del Karakoram central. En consecuencia, se nota que existe un lazo fuerte entre el cambio climático global, o por lo menos hemisférico, y la evolución topográfica del Karakoram y el Himalaya occidental. *Palabras clave: Karakoram central, forzamiento climático, erosión, glaciación, evolución del paisaje.*

Mountain topographic evolution has been traditionally explained by collisional tectonics (Bender and Raza 1995; Turcotte and Schubert 2002). Molnar and England (1990) suggested that climatic forcing in the Quaternary enhanced glacial erosion in the Himalaya, thereby causing isostatic uplift of peaks and relief production. Conversely, Raymo, Ruddiman, and Froelich (1988) and Raymo and Ruddiman (1992) indicated that enhanced chemical weathering of exposed rock in the Himalaya could reduce atmospheric CO_2, such that tectonic forcing caused climatic cooling and glaciation. Central to these forcing arguments is the recognition of scale-dependent interactions among climatic, surface, and tectonic processes (Koons et al. 2002; Bishop, Shroder, and Colby 2003).

Key issues revolve around climate change, glacier erosion, erosion patterns, exhumation and uplift patterns, and relief production (Montgomery 1994; Finlayson, Montgomery, and Hallet 2002; Bishop, Shroder, and Colby 2003; Spotila et al. 2004; Whipple 2009). Complex feedback mechanisms are expected, although it is difficult to study these interactions because of the operational scale dependencies of numerous processes and polygenetic topographic evolution (Bishop et al. 2002; Bishop, Shroder, and Colby 2003). Consequently, many parameters and process rates need to be included, characterized, and quantified in the context of polygenetic topographic evolution. Issues and debate include the importance of isostatic uplift (Gilchrist, Summerfield, and Cockburn 1994), exhumation rates (Whittington 1996), climate control and erosion distribution and timing (Clift, Giosan, et al. 2008; Clift, Hodges et al. 2008; Rahaman et al. 2009), glaciers and relief production (Harbor and Warburton 1992; Hallet, Hunter, and Bogen 1996; Whipple and Tucker 1999; Whipple, Kirby, and Brocklehurst 1999), focused erosion and spatial coincidence (Zeitler, Koons et al. 2001; Zeitler, Meltzer, et al. 2001; Finlayson, Montgomery, and Hallet 2002; Finnegan et al. 2008), geomorphometry and erosion modeling (Montgomery and Brandon 2002; Bishop and Shroder 2004), and equilibrium concepts such as topographic steady state (Tomkin and Braun 2002). Finding definitive field evidence of strong impacts of climate change on mountain building has been difficult (Whipple 2009). Glaciation's role is probably significant (Whipple 2009), although numerous issues are unresolved because erosion-process mechanics are not fully understood, glacier erosion is not adequately accounted for in numerous studies, the timing and mechanisms of climate-controlled erosion in the Himalaya are controversial (Pelletier 2008; Rahaman et al. 2009; van der Beek et al. 2009), and demonstration of a cause-and-effect relationship between rapid exhumation and precipitation remains elusive (Pratt-Sitaula et al. 2009; Whipple 2009).

Nevertheless, there are three dominant conceptual models that attempt to characterize the nature of mountain geodynamics and relief production in active orogens. Researchers have recognized that variations in fluvial erosion are caused by precipitation gradients wherein rapid river incision causes larger scale mass movements that increase slope angles and relief (Burbank et al. 1996; Egholm et al. 2009). Associated rapid rock uplift produces threshold slopes where relief is governed by rock strength. The proximity of high-discharge rivers to massifs, with high exhumation and recently metamorphosed rocks, represents a new conceptual model termed a *tectonic aneurysm* (Zeitler, Koons et al. 2001; Zeitler, Meltzer et al. 2001), where tectonic and isostatic rock uplift beneath a focused erosion zone advects hot and weak material from the midcrust toward the surface. Finally, the so-called glacial buzzsaw hypothesis suggests that the height of mountain ranges is primarily governed by glaciation and high-magnitude denudation, such that height and relief of mountain ranges are controlled by local climate rather than tectonic forces (Brozovic, Burbank, and Meigs 1997; Egholm et al. 2009).

The K2 mountain region in the central Karakoram in Pakistan represents an excellent natural laboratory for studying these concepts of landscape evolution, the role of glaciers, and relief production. The central Karakoram contains some of the highest peaks in the world and exhibits extreme relief and high uplift rates (Foster, Gleadow, and Mortimer 1994; Seong et al. 2008). The evolution of the K2 mountain region has been dominated by glacial and meltwater stream systems (Seong et al. 2007; Seong, Bishop et al. 2009). This is unlike other massifs such as Nanga Parbat and Namche Barwa, which exhibit localized exhumation and are spatially coincident with large-discharge rivers (Zeitler, Koons et al. 2001; Zeitler, Meltzer et al. 2001; Finnegan et al. 2008). However, a paucity of information exists regarding long-term and high-altitude climate conditions, glacier fluctuations, and erosion and uplift patterns. Furthermore, the influence of glacier erosion on relief production is difficult to characterize, as glacier erosion includes numerous processes including a nonlinear interaction with the overall climate. In addition, complications involving the coupling of climate-modulated erosion with rock rheology and strength have been recognized (Schmidt and Montgomery 1995; Koons et al. 2002). Although point estimates of river incision and exhumation in the western Himalaya of Pakistan do exist (e.g., Foster, Gleadow, and Mortimer 1994; Burbank et al. 1996; Zeitler, Koons et al. 2001; Zeitler, Meltzer et al. 2001; Seong et al. 2008), little is known about climate forcing dynamics and erosion patterns within the K2 region.

The objectives of this article are to summarize our fieldwork studies on landscape dynamics in the central Karakoram during the summer of 2005. Specifically we undertake (1) an assessment of historical, modern, and future climate conditions and effects; (2) an assessment of the geomorphic conditions and the timing of glaciations; (3) characterization of relief and erosion patterns; and (4) characterization of modern glacial conditions to provide insight into the magnitude of glacier erosion and relief production.

Study Area

The central Karakoram occurs at the western end of the trans-Himalaya along the border between China and Pakistan (Figure 1). This active margin is partially the result of the Indian–Asian continental–continental collision at a rate of ~4.4 cm per year^{-1} (Minster and Jordan 1978). The central Karakoram is currently underthrust from the south by the Indian plate and from the north by the Asian plate, which has resulted in a crustal thickness of > 65 km (Molnar 1988; Searle et al. 1990; Searle 1991; Foster, Gleadow, and Mortimer 1994). The Indus River, one of the world's largest rivers, separates the central Karakoram from the Greater Himalaya. The region is bounded to the north by the Karakoram Fault and to the south by the Main Karakoram Thrust (MKT) that trends southeast across the region. Our study region (Figure 1) is composed of rocks of the Karakoram metamorphic series, Karakoram granitoids, Gasherbrum sedimentary deposits, Gasherbrum diorite, and the K2 and Muztagh Tower gneisses (Searle et al. 1990; Searle 1991).

Intense denudation in the central Karakoram is associated with rapid and spatially variable uplift rates that have unroofed Neogene gneiss domes (Mahéo et al. 2004; Seong et al. 2008). Foster, Gleadow, and Mortimer (1994) estimated denudation rates of 3 to 6 mm per year^{-1} during the last 3 to 5 Ma, with ≥6 km of exhumation at an altitude of 6,000 m above sea level (asl). The region exhibits extreme relief with valley floors averaging 2,000 m asl and more than seventy peaks that rise above 7,000 m asl, including K2 (8,611 m asl), Gasherbrum I (8,068 m asl), Broad Peak (8,047 asl.), and Gasherbrum II (8,035 m asl.). More than 80 percent of the landscape is at an elevation of between 3,000 and 6,000 m asl.

Climate

The climate dynamics of the region are dominated by the midlatitude westerlies and the southwest Asian summer monsoon, whereas microclimatic conditions vary significantly given topographic controls on precipitation and the surface-energy balance (Owen 1988; Hewitt 1989). Most of the precipitation occurs in the spring from the westerlies, although two thirds of the snowfall occurs during the winter and spring. Summer snowfall from the southwest Asian summer monsoon occurs (Hewitt 1989), although the westerlies likely also make summer contributions depending on the phase of El Niño Southern Oscillation (ENSO), which, during its warm phase, significantly reduces monsoon strength (e.g., Bush 2001). Measured weather data in the Karakoram are meager (Pant and Kumar 1997), although a climate station exists in Skardu, near K2 (Mayer et al. 2006), and the Pakistani government has installed high-altitude stations. In general, annual precipitation exhibits a large vertical gradient, with maximal amounts at Biafo Glacier from 4,900 m to 5,400 m (Wake 1989). On the southern slopes of K2, precipitation is about 150 mm to 200 mm per year^{-1} at 2,000 m to 3,200 m, 1,600

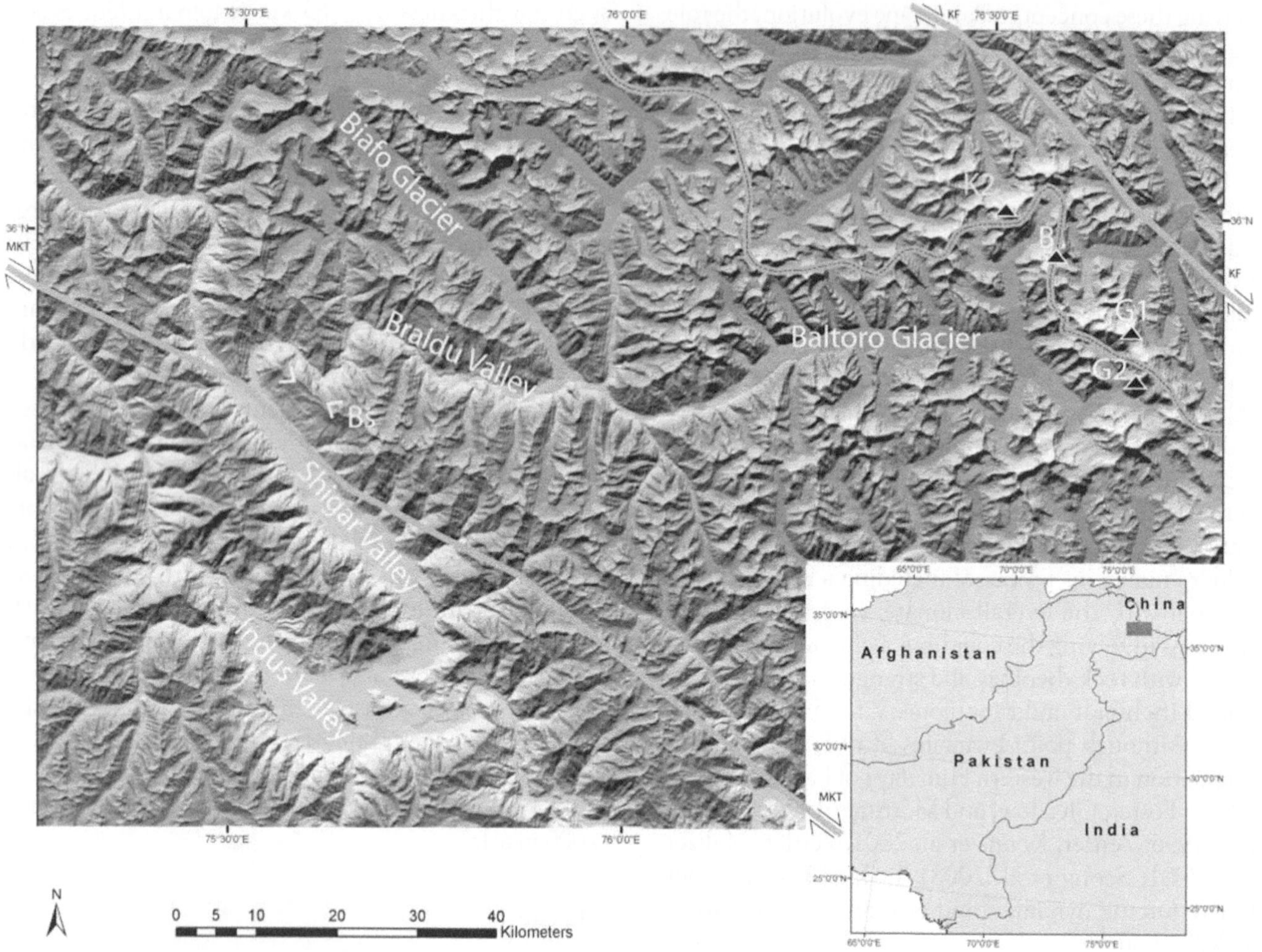

Figure 1. Shuttle Radar Topographic Mission shaded relief map of the K2 study region of the central Karakoram in northeastern Pakistan. Bs = Busper sackung; MKT = Main Karakoram Thrust; KF = Karakoram Fault; B = Broad peak; K2 = K2 peak; G1 = Gasherbrum 1 peak; G2 = Gasherbrum 2 peak.

mm per year^{-1} at 6,100 m, and about 2,500 mm per year^{-1} at 8,000 m (Mayer et al. 2006).

Surface Processes

The altitudinal zonation of the climatic-geomorphic conditions and their downslope relations have been previously recognized in terms of landforms and the sediment-transfer cascade (Hewitt 1989, 1993; Seong, Bishop et al. 2009). In this region, the process regimes alternate in concert with climate fluctuations. The overall sediment-transfer cascade involves glacier and fluvial incision, pervasive high frequency, moderate to high-magnitude mass movements onto and into glaciers and rivers, as well as catastrophic floods that remove valley fill out of the system (Shroder and Bishop 2000; Bishop et al. 2002).

The mass-movement types can be best considered in terms of regolith movement through common debris avalanches and flows, more deep-seated rock falls and rockslides, and massive sackung failures in which whole mountain ridges collapse internally along many interlinked and widely distributed shear planes. In all probability, debuttressing of rock slopes by glacier retreat during interglacials is likely to have produced many of the known rock falls and rockslides (Hewitt 1988, 1998a, 1999). In the Braldu River valley the huge Gomboro rockslide and the massive Busper sacking failure are seen to unroof the Neogene gneiss domes of Mahéo et al. (2004). Regional rates of erosion by mass-movement processes in the Himalaya are difficult to quantify, although their pervasive effects are obvious (Shroder 1998).

In the valley bottoms, fluvial and glacial processes are themselves the main erosional and transporting agents. Hydrological monitoring of river discharges is extremely limited in the K2 region. Nevertheless, high summer discharges result from monsoon precipitation and summer melt regimes. Monsoonal flood peaks dominate rivers at lower elevations with 60 to 80 percent of discharge from rain, whereas 50 to 70 percent of total discharge at higher altitudes is glacier and snowmelt (Wohl 2000).

Drainage area is commonly used as a surrogate for discharge, but drainage area–discharge relations are far from linear because of the glacier-melt addition. For example, the Upper Hunza River drains ~5,000 km^2 of somewhat lower mountains and small glaciers north of the highest Karakoram chain. The discharge of the Upper Hunza is then doubled by its confluence with the short Batura River that drains the ~300 km^2 Batura Glacier (Gerrad 1990; Wohl 2000).

Stream-channel characteristics in the western Himalaya are characteristically diverse, with alternating reaches of exposed bedrock as well as long sections of gravel-armored beds. Coarse bedload has two opposing roles in bedrock erosion (Sklar and Dietrich 1998, 2001) such that at low sediment concentrations, an increase in gravel supply can provide more tools with which water can erode the bedrock, whereas at higher concentrations the channel bed can be protected. Rockfalls from unbuttressed walls during interglacials in the western Himalaya (Hewitt 1998a, 1999) can cause valley blocking and upstream sedimentation above the bedrock. Incision rates appear to be highest at intermediate levels of sediment supply and transport capacity. On channels steeper than about 20 percent, incision is probably dominated by episodic debris flows.

River incision into bedrock links topography to tectonics and climate but is poorly understood and is typically modeled using the so-called stream-power law (Howard 1980; Sklar and Dietrich 1998), which is appealing and has been used by many because it has a minimum number of parameters and can be empirically calibrated from topographic data. Nevertheless, it is not an actual physical law, is not directly process related, and must be applied with care (Hancock, Anderson, and Whipple 1998). In contrast, direct measures of incision rates based on terrestrial cosmogenic nuclide (TCN) surface exposure dating of strath terraces on the Indus and Braldu Rivers (Burbank et al. 1996; Seong et al. 2008), and other rivers in the Himalaya, might produce the most reliable data concerning incision rates. Along a 100-km reach, incision rates were found to vary from ~1 mm per year^{-1} to 12 mm per year^{-1}, which is quite commensurate with inferred high uplift rates there (Zeitler 1985). Combined glacial and fluvial incision rates on Nanga Parbat appear even higher at ~22 ± 11 mm per year^{-1} (Shroder and Bishop 2000) but are highly episodic and unlikely to be sustained over long periods. Nevertheless, highly focused river incision can affect crustal structure in mountain belts by changing the distribution of stress in the crust (Beaumont and Quinlan 1994; Wohl 2000; Koons et al. 2002). Rapid exhumation results from advecting crust, which is known as the tectonic aneurysm. This is considered the case at Nanga Parbat in Pakistan and Namche Barwa in Tibet (Zeitler, Koons et al. 2001; Zeitler, Meltzer et al. 2001; Finnegan et al. 2008).

Low-frequency, high-magnitude catastrophic floods in the western Himalaya and trans-Himalaya might be the single most important erosion process there because they scour the bedrock and transport such large quantities of sediment. Mass movement and glacial dams can be many hundreds of meters high, with the result that lakes impounded behind the barriers can become quite large. Eventually the impounded water will overtop most landslide barriers, or float most glacial ice, with the result that the large breakout floods can erode and transport considerable quantities of sediment (Cenderelli and Wohl 2003; Korup and Montgomery 2008). Imbricated flood boulders of up to 15 m in size serve as evidence of this process in the K2 region (Seong, Bishop et al. 2009).

Glaciers

The glaciers in this region over the last century were much larger than they are today (Mayewski and Jeschke 1979). Evidence of this is found in moraine deposits at high altitude, high-altitude erosion surfaces, and large U-shaped valleys that exhibit striated and ice-polished bedrock surfaces (Figure 2). Currently, there are more than thirty glaciers over 20 km in length in the central Karakoram. The glaciers in this region, including the Batura, Hispar, Biafo, and Baltoro Glaciers, comprise some of the longest midlatitude ice masses in the world. The surface gradients of the glaciers range from very steep to gentle, with the largest glaciers often having ablation areas with very low surface slope angles (e.g., Baltoro Glacier has an average surface slope angle of about 2 degrees over its lowermost 35 km). Supraglacial debris covers most glaciers in their ablation zone. The thickness of this supraglacial debris is highly variable, although there is a general increase down-glacier, with maximum depths greater than 5 m.

Figure 2. Ice-polished bedrock surfaces on the south side of the Braldu Valley.

Thick debris effectively insulates (Hagg et al. 2008), thereby reducing ablation rates and allowing glaciers to exist at lower altitudes. Most of the glaciers are of winter accumulation type, reflecting the dominant influence of the midlatitude westerlies, although significant monsoon-related summer snowfall does occur. Snow and ice avalanching is significant given the extreme relief and feeds many glaciers. Ice velocities have been documented to increase by up to double in the summer compared to the winter due to increased meltwater (Copland et al. 2009; Quincey et al. 2009), and many glaciers actively incise the landscape (Figure 3).

The Karakoram is known for its many surging glaciers, and new data indicate an increase in the number of surging and advancing glaciers (Hewitt 1969, 1998b, 2005, 2007; Copland et al. 2009). Glaciers in this region are thought to be more responsive to change in precipitation than in temperature (Derbyshire 1981; Shi 2002; Owen et al. 2005), although systematic ablation studies are limited to short-term measurements

Figure 3. Urdukas West Glacier is a tributary glacier of the Baltoro Glacier system on the south side of the valley. The glacier has vertically incised into the landscape generating steep valley slopes and extreme relief. The width of the glacier is approximately 261 m.

such as Mihalcea et al. (2006). Unfortunately, individual glacier and regional mass balance trends are not currently known with any certainty.

Climate Modeling

To better understand climate variations, we conducted seventy-year simulations with a fully coupled atmosphere–ocean general circulation model (see Bush and Philander [1999] for details on the model configuration). Specifically, we examine paleoclimate conditions for the Last Glacial Maximum (LGM ~21 ka), 16 ka, 9 ka, and 6 ka, and compared results to modern climate conditions. We also simulated a future climate scenario assuming CO_2 forcing double that of the present day. Insolation data in all simulations are from Berger (1992) and ice-sheet topography is from Peltier (1994). The atmospheric model is spectral and has an equivalent spatial resolution of ~2.250 of latitude by 3.750 of longitude. The ocean model has comparable resolution.

Topographic heights are included in the model by spectrally decomposing digital elevation data and then truncating the resulting wave-number expansion at the smallest wave-number resolved by the model (wave-number thirty using a rhomboidal truncation scheme). Any resulting Gibbs oscillations are damped by a spectral smoothing technique (Navarra, Stern, and Miyakoda 1994). Although the model resolution is relatively fine from a global perspective, it is quite coarse when examining results on spatial scales of hundreds of kilometers. For this reason, spatial averages over multiple grid cells are commonly used to describe regional climate fluctuations because the averaging procedure produces statistically more significant and robust results. Nevertheless, the global model is incapable of generating detailed results on spatial scales smaller than a few hundred kilometers, and alternative methods such as statistical downscaling (e.g., Easterling 1999; von Storch and Navarra 1999; Schoof and Pryor 2001) or nested modeling methodologies (e.g., Christensen et al. 2001; Raisanen, Rummukainen, and Ullerstig 2001) are required to produce climate data in such fine spatial detail. We specifically focus on temperature and snow accumulation, as these parameters are most significant in governing glacier fluctuations and erosion.

Our numerical simulations reveal a monotonic decrease in the amount of snow accumulation averaged over the Himalayan area of 60–105°E, 24–44°N from the LGM to today, and this trend continues most dramatically in a doubled CO_2 climate (Figure 4). Simulated snow cover over the study region in the central Karakoram, averaged over 70–80°E, 30–37°N, indicates similar trends. As the atmosphere warms from a glacial to interglacial state (and beyond), the amount of snow accumulation in the western Himalaya decreases.

The spatial pattern of difference in snow accumulation indicates large changes along the front range of the Himalaya (Figure 5). From the LGM to 6 ka, snow accumulation was higher than today along the entire length of the front range that faces the influence of the summer monsoon and its orographic precipitation. The accumulation along the southwestern front of the Himalaya was highest, however, in the 16 ka simulation despite the fact that orbital obliquity, and hence monsoon strength, did not reach its maximum until 11 ka. Dramatic decreases in snow accumulation in the doubled CO_2 simulation occur across the entire study region. Our numerical simulation results indicate that the area around and to the south of Nanga Parbat has been in the past more susceptible to change than the central Karakoram (Figure 5B–E). In a doubled CO_2 climate, however, both regions are affected to a similar degree in terms of reduced snow accumulation (Figure 5F).

Summertime radiative forcing in the region increases between the LGM and 11 ka due to the increase in Earth's angle of obliquity. This strengthens the summertime monsoon and should, given the increase in temperature, increase precipitation during the summer months. The decline in snowfall is indicative of a change in type of precipitation, with more rainfall occurring rather than snowfall in a warmer climate.

Interannual variability associated with ENSO is also a major factor regulating precipitation in this region. As best as can be determined, during the LGM, ENSO was frequent but of small amplitude (Bush 2007) and would likely not have had as much influence on the summertime monsoon as it does today. The competition between monsoon southwesterly winds and the midlatitude westerlies that flow over Europe and the Caspian Sea is a delicate balance that determines the type of precipitation in this study region. If the monsoon is weak, the midlatitude westerlies are able to penetrate further east, bringing colder winds and more snowfall to the region. If the monsoon is strong, however, then warm, moist air dominates the region and snowfall is reduced but rainfall is enhanced.

Compared with the climate through the last glacial cycle, the most dramatic change simulated is the one between today and that of a climate with doubled atmospheric CO_2. For the study region, the simulated change between today and a possible future climate is

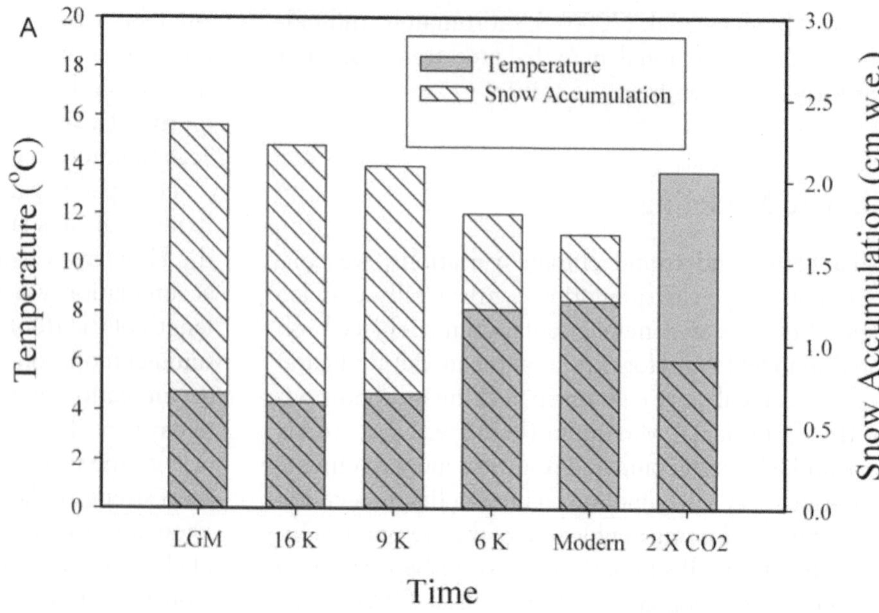

Figure 4. Annual mean temperature and snow accumulation averaged over (A) 60–105°E, 24–44°N; (B) 70–80°E, 30–37°N.

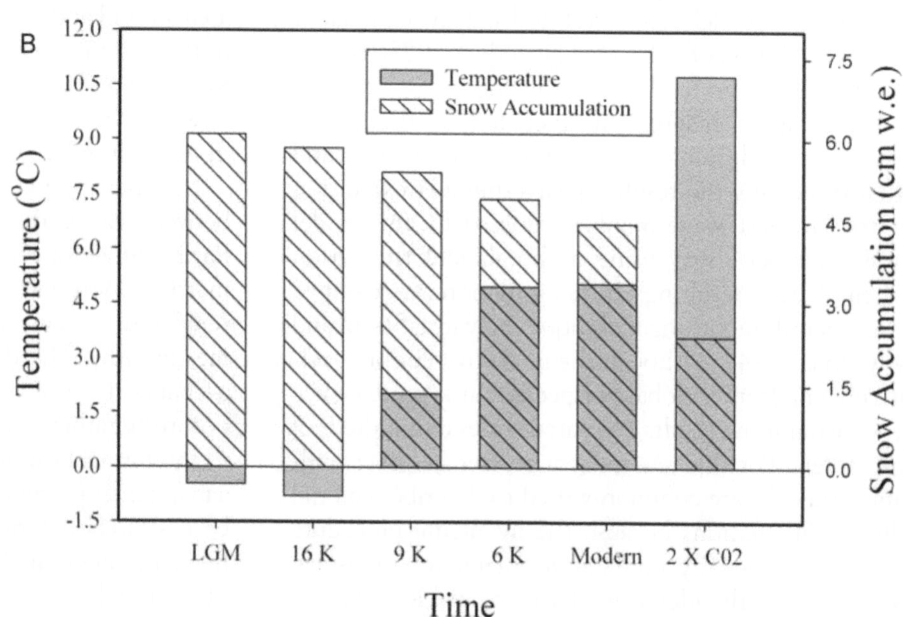

greater than the change between today and the LGM. This reflects the fact that a doubling of atmospheric CO_2 places our climate outside the bounds of the naturally varying glacial and interglacial climates of the Late Quaternary. Implications for future glacial mass balance and freshwater resources for the countries in this region are profound.

Timing and Style of Quaternary Glaciation

We examined the Quaternary glacial history of the central Karakoram using geomorphic mapping of landforms and sediments and ^{10}Be TCN surface exposure dating of boulders on moraines and glacially eroded surfaces (Seong et al. 2007). We mapped landforms (strath terraces, flood deposits, and landslides) and assessed ^{10}Be TCN ages to better understand glaciations and landscape evolution (Seong et al. 2008; Seong, Bishop et al. 2009). Details of our methods related to sampling and laboratory analysis for ^{10}Be TCN dating are provided in Seong et al. (2007, 2008; Seong, Bishop et al. 2009).

We recognized four glacial stages in the Skardu Basin and the Shigar and Braldu Valleys of the central Karakoram, which were defined as Bunthang glacial stage (>0.7 Ma), Skardu glacial stage (Marine Oxygen

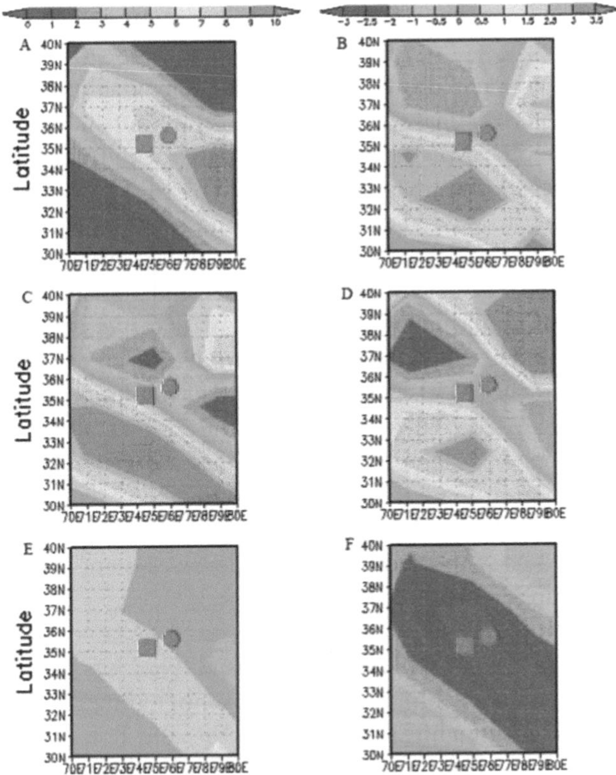

Figure 5. Annual mean snow accumulation (cm w.e.) over 73–80E, 32–37N. K2 is depicted by a red circle and Nanga Parbat is represented by a red square. (A) Modern climate simulation (control), with color scale range depicted overtop. (B–F) Last glacial maximum, 16 ka, 9 ka, 6 ka, and 2× CO_2, respectively, with color scale range depicted over B.

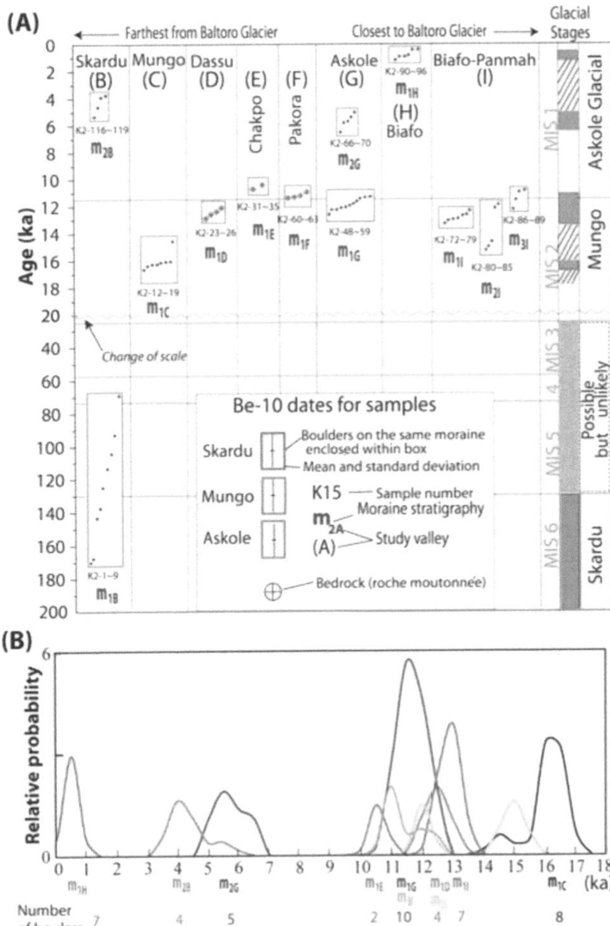

Figure 6. ^{10}Be TCN surface exposure ages for boulders on the moraines dated in the central Karakoram (from Seong et al. 2007). (A) Scatter plots of ages on each moraine. Each box encloses one landform. Marine oxygen isotope stages (MIS; after Martinson et al. 1987) are shown in light blue. The vertical scale changes from 20 ka and its boundary is marked by a light gray undulating line. (B) Probability distributions for ^{10}Be TCN surface exposure ages of each glacial advance after 18 ka. Each different color represents a landform formed during a different glacial stage.

Isotope Stage [MIS] 6 or older); Mungo glacial stage (MIS 2); and Askole glacial stage (Holocene). The glacial geologic evidence shows that glaciers oscillated several times during each glacial stage. Glacial advances during the oldest stage, the Bunthang, were not well defined because of the lack of preservation of landforms and sediments. In contrast, the glacial geologic evidence for the Mungo and Askole stages is abundant throughout the study region and we were able to define glacial advances that likely occurred at ~16, ~11–13, ~5, and ~0.8 ka (Figure 6). Furthermore, our data show that the extent of each progressive glaciation throughout the region became increasingly more restricted over time (Seong et al. 2007; Seong, Bishop et al. 2009). In the Braldu and Shigar valleys, glaciers advanced more than 150 km during the Bunthang and Skardu glacial stages, and glaciers advanced more than 80 km beyond their present positions during the Mungo glacial stage. In contrast, during the Askole glacial stage, glaciers only advanced a few kilometers from present ice margins. We calculated the equilibrium-line depression for the Mungo glacial stage to be ~500 m (Seong, Bishop et al. 2009).

Seong et al. (2007) argued that glaciers in the central Karakoram likely responded to the same forcing that caused changes in Northern Hemisphere oceans and ice sheets. These changes are likely teleconnected via the midlatitude westerlies and also to changes in monsoon intensity. Owen et al. (2008) highlighted the fact that the extent of glaciation between adjacent regions within the Himalayan–Tibetan orogen can vary considerably. This is particularly well illustrated across northern Pakistan and northern India for the Late Glacial at about 14 to 16 ka, coincident with the Mungo glacial stage of northern Pakistan. During this time in the central Karakoram an extensive valley glacier system extended more than 80 km from the present ice margin, whereas in the Hunza valley to the northeast and Muztag Ata-Kongur Shan to the north, for example, glaciers only advanced a few kilometers from their present position (Owen et al. 2002; Seong et al. 2007; Owen et al. 2008; Seong, Owen et al. 2009). In Ladakh, to the southeast of these regions, there is little evidence of a glacier advance at this time, and when glaciers did advance they were restricted to a few kilometers from their present ice margins (Owen et al. 2006; Owen et al. 2008). Furthermore, to the south of Ladakh, in the Lahul Himalaya, glaciers advanced more than 100 km beyond their present positions at this time (Owen et al. 2001; Owen et al. 2008). These contrasts in the extent of glaciation within a relatively small region of the Himalaya and trans-Himalaya highlight the important local climatic gradients and the strong topographic controls on climate forcing and glaciation.

Nevertheless, by recalculating all of the TCN ages for moraine boulders and glacially eroded surfaces throughout the Himalayan–Tibetan region, Owen et al. (2008) were able to make broad statements regarding the synchroneity of glaciation. In essence, glaciers throughout monsoon-influenced Tibet and the Himalaya and the trans-Himalaya (including the central Karakoram) appear to have responded in a similar fashion to changes in monsoon-driven and Northern Hemisphere cooling cycles alone. In contrast, glaciers in the far western regions of the Himalayan–Tibetan orogen are asynchronous with the other regions and appear to be dominantly controlled by the Northern Hemisphere cooling cycles and by the influential interaction between the relative strengths of the westerlies and the summer monsoon.

Geomorphometry

Such dramatic glacier fluctuations generate extreme relief and U-shaped valleys. Relief, however, might not always be related linearly to erosion and erosional efficiency. For glacier erosion, research has indicated that glaciation can limit relief up to some altitude approaching the equilibrium-line altitude (ELA) and that erosion fluctuations at the ELA can generate extreme relief (Brozovic, Burbank, and Meigs 1997; Bishop, Shroder, and Colby 2003). Given the anisotropic nature of relief, as a function of a variety of surface processes and deformation, it is necessary to quantitatively characterize the topography and compare high-relief regions, as the topography inherently represents the interplay among climate, tectonics, and surface processes (Wobus et al. 2006).

We conducted an analysis of the topography over the western Himalaya of Pakistan using Shuttle Radar Topographic Mission (SRTM30) data, acquired from the Spaceborne Imaging Radar-C. The SRTM30 data were constructed at thirty arc-second spacing, similar to GTOP030 data. GTOP030 data points were used where SRTM data were invalid (i.e., radar shadows). We projected the data set (Universal Transverse Mercator) and resampled (bilinear interpolation) it to generate an SRTM30 data set with a grid resolution of 1 km. Given the coarse resolution of the data set, altitude estimates and slope angle magnitudes are generalized and underestimated. The projected SRTM30 data set, however, adequately characterizes the relief structure of the landscape and permits a first-order assessment of the topography.

Geomorphometric analysis consisted of analyzing and comparing topographic regions (14,400 km^2) that were centered on high-altitude peaks including Batura Mustagh, Distaghil, K2, Nanga Parbat, and Rakaposhi (Figure 7A). Hypsometric analyses permitted an examination of the altitude/area relationship and the relative amount of mass removed from each region. It is important to note that the relative patterns of hypsometric curves are generally related to erosion potential, although the hypsometric integral does not represent an absolute magnitude of erosion given systematic and nonsystematic biases in the digital elevation model (DEM).

Scale-dependent analysis of the topography was used to characterize the relief structure of each region. Specifically, we performed semivariogram analysis and compared every grid cell to every other grid cell within each region. Within an $n_x \times n_y$ grid there will be N point pairs, where $N = n_x^2\ (n_y^2 - 1)/2$. The variance component is $\Delta z = (z_1 - z_2)^2$, and the horizontal distance is $\Delta x = [(x_1 - x_2)^2 + (y_1 - y_2)^2]^{0.5}$. The variance was summed over a binned horizontal distance interval (1 km).

We also conducted spatial analysis to assess the degree of landscape concavity and convexity. The approach represents a hemispherical analysis of the topography to measure the angular relation between the surface relief and the multidirectional horizontal distances. Yokoyama, Shirasawa, and Pike (2002) referred to this measure as *openness* and provided a quantitative description of positive openness (P) and negative openness (N). We computed P and N images using a radius distance of 30 km around each grid cell

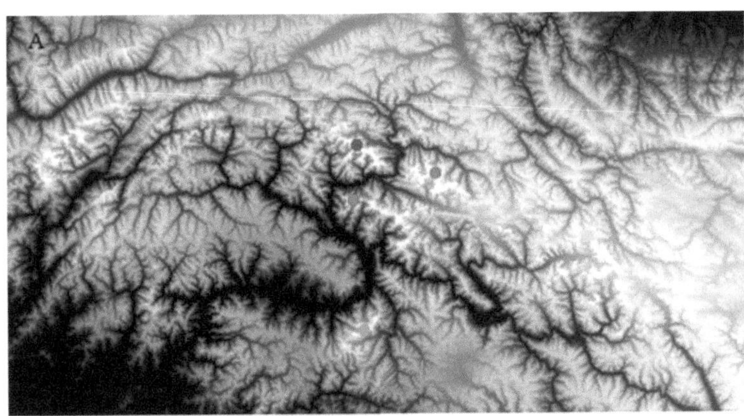

A

B

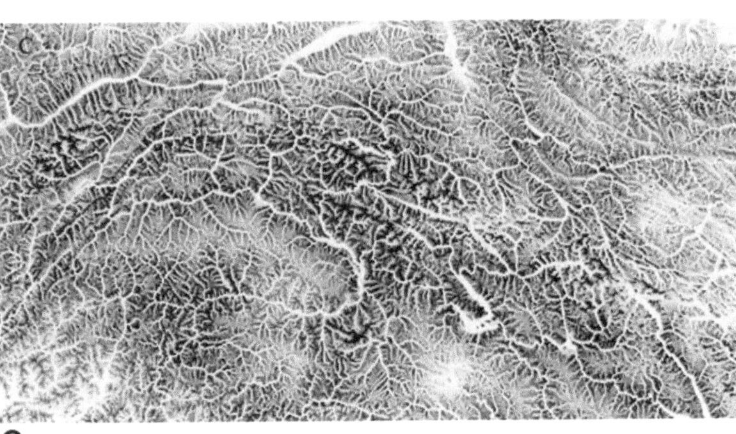

C

Figure 7. Subsection of the Shuttle Radar Topographic Mission (SRTM30) data set that was used for geomorphometric analysis. (A) Geomorphometric analysis of 120 × 120 km regions included Nanga Parbat (green dot), Rakaposhi (red dot), Batura (blue dot), Distaghil (magenta dot), and K2 (yellow dot). (B) Positive-openness measure. Dark gray tones represent concavity, and lighter gray tones depict convexity. (C) Negative-openness measure. Light gray tones represent concavity, and darker gray tones represent convexity.

(Figure 7B–C). We wanted to investigate the potential of landscape convexity and concavity to be used for assessing spatial variations in relief.

Hypsometric analysis revealed that the majority of the K2 massif (88 percent) occurs between 4,000 and 6,000 m asl, with an estimated mean elevation of 5,064 m asl. Average altitudes for the Nanga Parbat (3,598 m asl), Rakaposhi (3,895 m asl), and Distaghil (4,609 m asl) regions depict a systematic north and easterly increase in average altitude toward K2.

The hypsometric curve for the K2 region is similar in shape to Nanga Parbat's depicting similar relief at high altitudes due to deep glacier erosion, although less pervasive river incision at lower altitudes accounts for more mass there at K2 (Figure 8). In contrast, the Rakaposhi curve reveals more high-altitude land mass and a

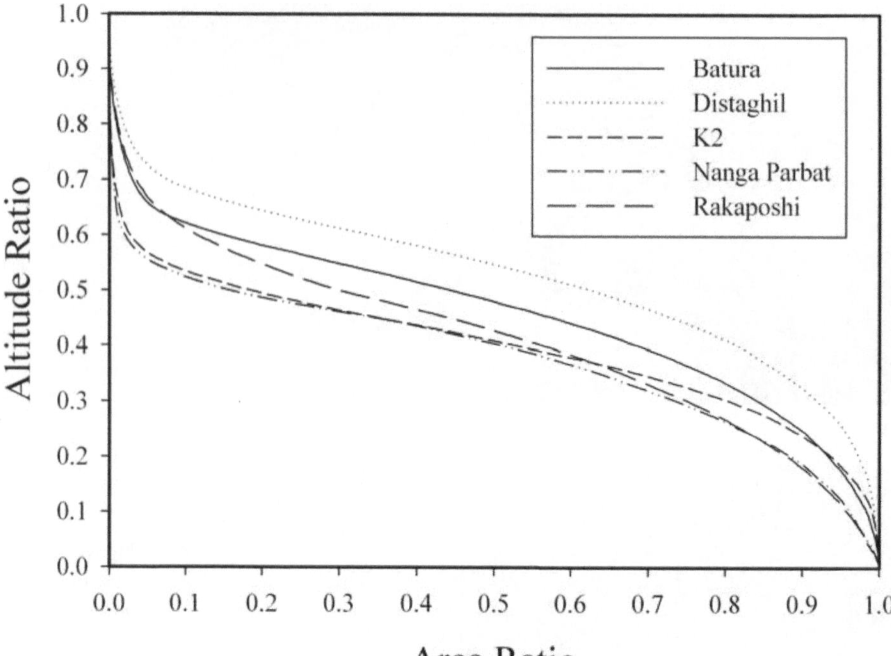

Figure 8. Hypsometry of the Batura, Nanga Parbat, Rakaposhi, Distaghil, and K2 regions. Each region was centered on each peak and represents 14,400 km^2.

similar low-altitude mass distribution, whereas the Distaghil region exhibits relatively more land mass at all altitudes. This indicates a difference in the surface processes responsible for erosion at various altitudes and the degree to which incision and hillslope processes dominate mass removal.

These results can be viewed from a denudational unloading perspective by comparing the hypsometric integrals (HI). Nanga Parbat exhibited the lowest HI (0.3789). Going north, Rakaposhi and Batura are characterized by HI values of 0.4125 and 0.4586. To the west of K2, Distaghil exhibits a value of 0.5268, and the K2 value is 0.3986. This suggests that denudational unloading is greatest at Nanga Parbat and K2, compared to the Hunza region of the Karakoram. In addition, these two regions exhibit the hypsometric signature of the glacial buzzsaw model. Given the relative high mean altitude of the K2 region and relatively large glaciers, we might expect K2 to exhibit the lowest integral value, although the removal of glaciers with respect to DEM altitude values would produce a lower value.

Assuming that greater relief is related to the magnitude of erosion, our variogram analysis reveals a different pattern. The Rakaposhi region exhibits the greatest relief at all lag distances (Figure 9). This is consistent with the coupled river-incision and slope-failure model. At lag distances ≤18 km, the Distaghil region exhibits greater mean relief than Nanga Parbat, but Nanga Parbat exhibits greater relief at lag distances ≥18 km. The relatively high mean relief characterizes the influence of the size of the fluvial drainage network. K2 exhibits the least amount of relief at all lag distances and the relief signature is best explained by the glacial buzzsaw model.

Analysis of the concavity and convexity patterns in the topography of the western Himalaya provides insight into the nature of erosion. The positive-openness measure uses the minimum zenith angle to characterize concavity and convexity (Figure 7B). The magnitude of the concavity is related to the mesoscale relief. The spatial patterns of concavity clearly show regions of relief where erosion has removed more lithospheric mass. Spatially contiguous concavity zones include Nanga Parbat and the Hunza region (Rakaposhi, Batura, Distaghil). These patterns explain the increased scale-dependent relief at Nanga Parbat, Rakaposhi, Batura, and Distaghil. K2 does exhibit significant mesoscale relief immediately surrounding the mountain, although it exhibits less of a spatial predominance of extreme relief compared to the aforementioned regions. Furthermore, areas of low relief include the Kohistan region to the west of Nanga Parbat and the Deosai plateau to the east of Nanga Parbat.

The negative-openness measure of concavity and convexity provides additional insight into the nature of erosion (Figure 7C). This measure uses the maximum zenith angle to characterize concavity and convexity.

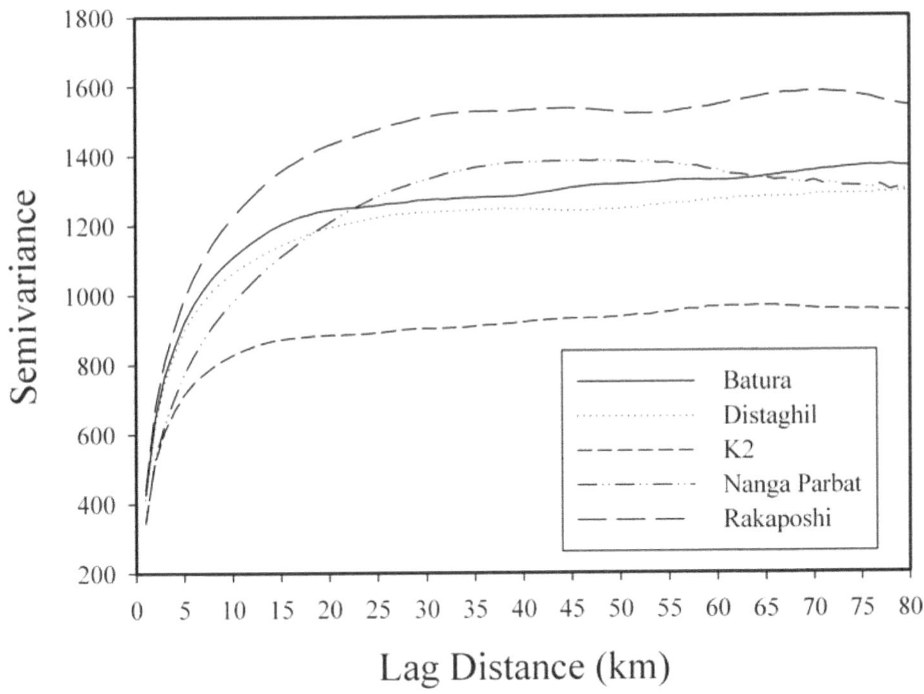

Figure 9. Semivariograms for the Batura, Nanga Parbat, Rakaposhi, Distaghil, and K2 regions of the central Karakoram.

The spatial patterns and magnitude of concavity identify the influence of river incision and the drainage network on the landscape. For example, the relatively high-magnitude concavity identified the rapid river incision associated with the Indus and Astor Rivers near Nanga Parbat, where rapid uplift is known to occur (Zeitler, Koons et al. 2001; Zeitler, Meltzer et al. 2001). Collectively, these results demonstrate the polygenetic nature of erosion and relief production, as a particular landscape will exhibit multiple geomorphometric signatures that characterize different landscape evolution models.

Glaciological Observations

Glaciological measurements in the K2 region of the central Karakoram in summer 2005 focused along a ~50 km length of Godwin Austen and Baltoro Glaciers below K2, as well as the nearby Biafo Glacier. To date even basic information such as ice depths and surface velocities are poorly known for these ice masses, yet this information is crucial for understanding glacier response to climate change and the role of glacier erosion in landscape evolution.

Surface Velocities

We made surface velocity measurements from the repeated positioning of a total of seventeen fixed markers with a Trimble R7 differential Global Positioning System (dGPS) unit. The interval between measurements varied between twenty-eight days for Biafo Glacier and six to twenty-two days for Baltoro Glacier (Quincey et al. 2009), although velocities are expressed in units of m per year^{-1} here for standardization. Permanent dGPS base stations do not exist in this region, so a temporary one was established in Skardu and run for the entire summer. We used the Precise Point Positioning technique and Trimble Geomatics Office software to process the dGPS data.

Velocities at the terminus of Biafo Glacier were 1.0 m per year^{-1} for a point located ~300 m from the terminus and 5.0 m per year^{-1} for a point located ~500 m from the terminus (Figure 10). These low values are characteristic for the lower parts of glaciers, although they also indicate that the glacier is active all the way to its terminus. This is further confirmed by Copland et al. (2009), who used feature tracking to determine velocities across the entire Biafo Glacier and found that velocities were low (<15 m year^{-1}) across most of the lower terminus but averaged 100 to 150 m per year^{-1} across most of the main ablation zone.

Surface velocities on Baltoro Glacier were generally much higher than Biafo, with a peak of ~240 m per year^{-1} recorded at two stakes at Gore II, ~12 km downglacier from Concordia (Quincey et al. 2009). Below this point the velocities gradually decreased over a distance of 23 km to a minimum of 10 m per year^{-1} at the glacier terminus. Velocities in the Concordia and K2 regions above Gore II varied between 125 and 168 m

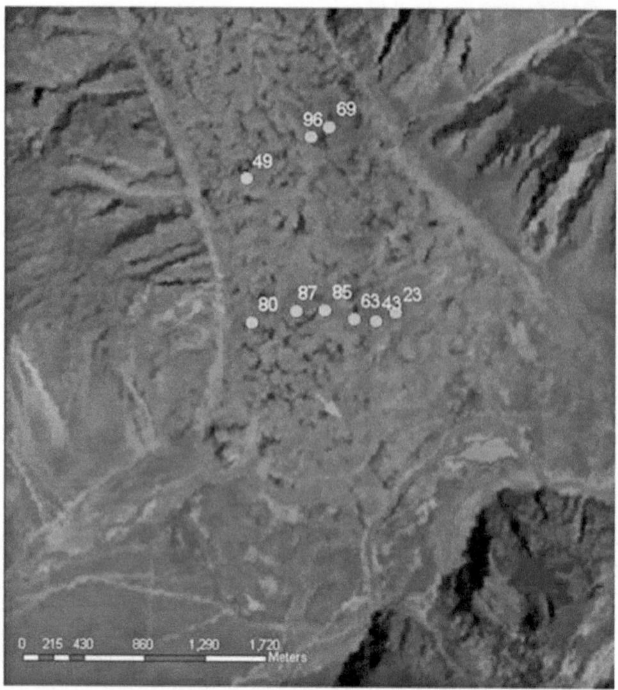

Figure 10. Ice depths in meters (green dots) and velocities in meters per year^{-1} (red arrows) measured across the terminus of Biafo Glacier in summer 2005.

Figure 11. View of the lowermost differential Global Positioning System measurement location on Baltoro Glacier (lower right foreground), with the tributary Uli Biaho Glacier visible in the background. Supraglacial debris thicknesses across the terminus of Baltoro Glacier are typically >1 m.

per year^{-1}. Comparisons with velocities derived from Synthetic Aperature Radar satellite imagery, feature tracking, and other field measurements (Mayer et al. 2006; Copland et al. 2009; Quincey et al. 2009) indicate generally good agreement for summer patterns but that the dGPS velocities are much higher than winter velocities over most of the glacier > 10 km upglacier from the terminus. These patterns suggest that motion of the lowermost ~10 km of the glacier is largely driven by ice deformation, whereas motion over the remainder of the ablation zone is dominated by basal sliding, with summer speed-ups of 200 percent or more compared to winter patterns (Quincey et al. 2009).

Surface Melt Rates

To sample a few surface melt rates, we placed ablation stakes at bare ice locations across Baltoro Glacier and remeasured them over periods of seven to nineteen days. The rates decreased with altitude, varying between 6.52 cm per day^{-1} at 3,907 m, 5.91 cm per day^{-1} at 4,055 m and 3.03 cm per day^{-1} at 4,837 m during a period with generally sunny weather and few clouds. These patterns agree well with those of Mihalcea et al. (2006), who also found a clear relationship between elevation and surface melt rates at Baltoro Glacier. Mihalcea et al. (2006) also found strong relationships between surface melt rates and surface debris thicknesses, with surface melt rates < 2.0 cm per day^{-1} when debris depths were > 0.10 m, regardless of elevation. Given the thick supraglacial debris cover over the lower part of Baltoro Glacier (Figure 11), it is likely that little melting occurs there.

The surface melt patterns provide an explanation for the observed velocity patterns, as the regions where the highest seasonal velocity variations occur are located at the lowest elevations where there is a bare ice surface. In these locations, large volumes of surface meltwater can make their way to the glacier bed via moulins and crevasses (Quincey et al. 2009) and produce high summer basal sliding rates. In contrast, the thick debris cover over the lowermost 10 km of the glacier occurs in a region where there is generally slow motion and a dominance of internal ice deformation. In this region the surface melt occurs only very slowly, producing little direct meltwater input to the glacier bed. The large volumes of meltwater from upglacier are presumably carried in hydraulically efficient subglacial channels that can remain open due to the low ice depths over the terminus.

Ice Depths

We measured ice depths with a monopulse ground-penetrating radar (GPR) system with a transmitter based on the design of Narod and Clarke (1994). The receiver consisted of an airwave-triggered Tektronix digital oscilloscope connected to a handheld computer. We typically used a center frequency of 10 MHz, although in locations where the ice was particularly deep and the bed could not be seen at 10 MHz, we used 5 MHz. Errors in the ice-depth measurements are typically quoted as 1/10 of the transmitted wavelength (Bogorodsky, Bentley, and Gudmandsen 1985), which equates to ±1.7 m at 10 MHz and ±3.4 m at 5 MHz. The radiowave velocity of the GPR signals through the ice was assumed to be 0.168 m per ns^{-1}, an average value for temperate ice (Macheret, Moskalevsky, and Vasilenko 1993). We used a handheld Garmin eTrex GPS unit to locate the position of each GPR measurement to within ±10 m horizontally.

We measured ice depths at two profiles across Biafo Glacier, at distances of ~1 km and ~2 km from the ice front at surface elevations of ~3,200 m asl (Figure 10). The depths varied between 23 m and 85 m across the lower transect and 49 m and 96 m across the upper transect, in a part of the glacier that is heavily debris covered (average debris thickness ~1 m). The only other previous GPR measurements in this region were completed by Hewitt et al. (1989), who measured centerline ice depths of ~500 to 700 m at a transect ~25 km upglacier from the terminus (4,100 m asl) and depths up to ~1,400 m at the equilibrium line ~45 km upglacier from the terminus (4,650 m asl). These authors cautioned, however, that the measurements at the equilibrium line had a high degree of uncertainty.

On Baltoro Glacier, the greatest ice depth of 171 m occurred in proximity to the stakes with highest surface velocities at Gore II, with depths generally decreasing to a minimum of ~40 m at the glacier terminus (Figure 12). Ice was 141 to 155 m thick in the region of K2 base camp, although no successful measurements were made in the Concordia region. We suspect that the ice there was deeper, or perhaps the presence of more basal meltwater precluded our measurements with the GPR system, particularly because the ice is likely warm based. Desio, Marussi, and Caputo (1961) and Caputo (1964) reported measurements of ice depths on Baltoro Glacier that were determined using a gravimetric method in the mid-1950s for three locations:

1. At Urdukas, gravimetric ice depths were determined to be ~170 to 400 m, compared to depths up to 144 m in our study.
2. At Concordia, gravimetric ice depths were measured at up to ~850 m.

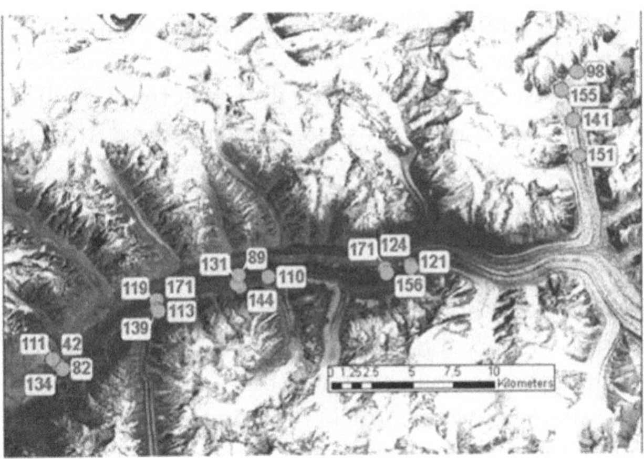

Figure 12. Ice depths (m) measured along Baltoro Glacier for summer 2005.

3. At K2 base camp, gravimetric ice depths were ~150 to 250 m, compared to 155 m in this study.

The gravimetric method exhibits a high degree of uncertainty and is therefore no longer used in glaciology, although the lower range of numbers just quoted are broadly comparable to the GPR measurements at Urdukas and K2 base camp. Given that the gravimetric method appears to have overestimated most depths, however, it is quite possible that ice depths at Concordia are lower than those quoted by Desio, Marussi, and Caputo (1961) and Caputo (1964). Based on repeat terrestrial photography of Concordia in 1912 and 2004, and the terminus of Baltoro Glacier in 1954 and 2004 (Mayer et al. 2006), there is little evidence that ice surface lowering has caused the differences in measured ice depths.

Discussion

Our climate simulations reveal a systematic pattern of increasing annual temperature and decreasing snow accumulation in the central Karakoram since the LGM. This result is consistent with long-term glacier retreat in the region (Mayewski and Jeschke 1979) and geomorphological evidence of extensive glaciations that generated high-altitude glacier erosion surfaces, U-shaped valleys exhibiting extreme relief, and extensive landslides that were most likely the result of debuttressing due to glacier retreat (Seong et al. 2007; Seong, Bishop et al. 2009). Furthermore, our paleoclimate simulations support our geochronology results of major glacier advances at ~16 and 11 ka, as snow accumulation was greater than modern-day conditions, and an enhanced monsoon in the Holocene would have contributed much more summer snowfall compared to contemporary conditions.

Extensive glaciations during the Quaternary resulted in much greater ice depths than were measured in 2005, and our glacial reconstructions indicate that the ice was ~1 km thick in the Baltoro Valley near Biafo Glacier and much more extensive (Seong et al. 2007). Therefore, based on our current understanding of glacier erosion (ice depth and basal sliding velocity are the primary controls on abrasion), the magnitude of glacier erosion was much higher than contemporary erosion rates given more spatial coverage, thicker ice, and positive ice discharge. Furthermore, given the long-term retreat pattern caused by increasing temperatures and decreasing snow accumulation, glacier erosion during the retreat phases might have been greater than during advances, as increasing temperatures would produce more meltwater that governs basal water pressure, sliding velocity, and glacio-fluvial erosion. Koppes and Hallet (2006) indicated that there is a strong correlation between glacial retreat rates and glacial sediment yields.

Our geomorphometry results clearly indicate that relief is not linearly correlated with the HI in Pakistan. Although the K2 region of the central Karakoram exhibits the highest altitudes and high exhumation rates, it exhibits the lowest relief, yet the region has experienced significant erosion and exhumation (Foster, Gleadow, and Mortimer 1994; Seong et al. 2007, 2008). This apparent contradiction can be explained by nonlinear relationships between erosion and relief production.

Scale-dependent analysis revealed that in the absence of local high-discharge rivers and a spatially dense drainage network, variation in relief is dramatically reduced. As the K2 region of the central Karakoram has been dominated by glaciation and glacierization, glacier erosion can limit the relief at intermediate altitudes (Brozovic, Burbank, and Meigs 1997; Bishop, Shroder, and Colby 2003), thereby decreasing the variability in scale-dependent relief. Relatively low slope angles at intermediate altitudes document the erosion and redistribution of material by glaciers from 4,000 to 6,000 m asl. These low slope angles represent actual glacier surfaces or high-altitude erosion surfaces from past glaciations. These are classic signatures of a glacial buzzsaw. It is clear that the K2, Nanga Parbat, and Hunza regions have been significantly affected by glaciation, although the subsequent influence of local river incision and mass movement can increase local relief.

The highest scale-dependent relief occurs in the Hunza and Nanga Parbat regions and mesoscale relief patterns indicate that these regions are active erosion zones currently dominated by large-scale river incision and mass movement. The Hunza region exhibits more scale-dependent relief than Nanga Parbat because of an extensive fluvial drainage network. Our climate simulations suggest that the Hunza erosion zone, as depicted by mesoscale concave relief patterns, is the result of climate forcing. The region exhibits a negative snow accumulation anomaly in paleoclimate simulations that represent increased rainfall due to an intensification of the monsoon, where increased precipitation and associated river incision and mass movement explain the high scale-dependent relief. Furthermore, spatial patterns of precipitation have been found to be strongly controlled by topography in the Himalaya, and considerable variability exists at scales of ~10 km (Anders et al. 2006; Barros et al. 2006).

The K2 region of the central Karakoram is dominated by large glaciers and has less drainage network to produce additional local relief. This indicates that large-discharge rivers might not be required to focus erosion and produce uplift zones. At intermediate to higher altitudes, deep valley-glacier erosion and headwall erosion can effectively remove rock and sediment, whereas protective cold-based ice and permafrost on the highest peaks permits ridge and peak formation. Glacier mass-balance gradients determine the availability of meltwater for river incision and hillslope adjustment down-valley at lower altitudes. Consequently, in high mountain environments, glaciation both directly and indirectly alters the relief structure of the landscape differently with altitude. These differences can be attributed to the degree of temporal overprinting and spatial overlap of glacial events, such that at intermediate altitudes, the degree of spatial overlap is less than at high altitudes.

Therefore, with respect to glacier erosion and relief production, we should not expect relief to be related linearly to the magnitude of glacier erosion, as relief might not scale with erosional efficiency. Whipple (2009) indicated that erosional efficiency should be associated with a decrease in relief. Numerous parameters such as rock strength, topographic-induced stress fields, basal water pressures, basal sliding velocity, ablation rates related to climate and supraglacial debris loads, and other erosion processes determine the magnitude of glacier erosion, and isostatic and tectonic uplift operate at completely different spatio-temporal scales. Furthermore, glacier-erosion efficiency is most likely highly spatially variable in the Himalaya, as characterized by research on regional precipitation variability (e.g., Anders et al. 2006), our geomorphometric analysis, and mesoscale

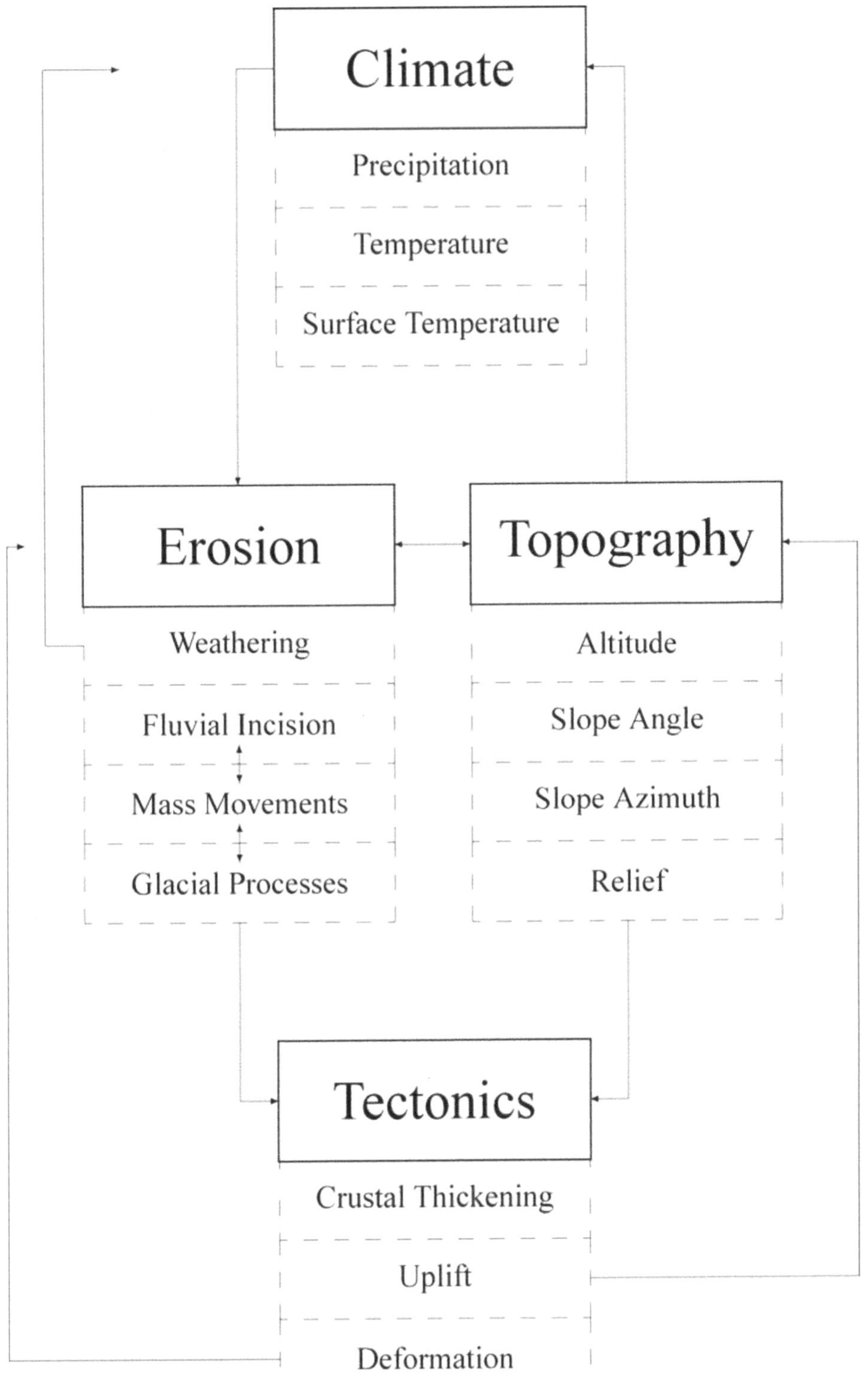

Figure 13. Interactions and selected feedback pathways for climate, erosion processes, and tectonics. Many parameters of the system are not included and the magnitudes of the linkages change with time. Positive and negative feedbacks exist for selected pathways.

relief variations that show little relief over the Kohistan region and Deosai plateau. Recent work by van der Beek et al. (2009) indicated that these two regions represent preserved remnants of the Eocene Tibetan plateau that have been glaciated and exhibit slow denudation. This demonstrates extreme spatio-temporal variability in the influence of glacier erosion in this region. Furthermore, large glaciers once flowed down all the major valleys in this region, significantly contributing to relief production. Such variability in climate change and erosion patterns indicates that the landscape is unlikely to reach equilibrium in terms of mass or topographic steady state.

Modern-day glaciological conditions indicate that the glaciers are highly active and responding to climate forcing. The region is currently experiencing an increase in precipitation according to European reanalysis and Tropical Rainfall Measuring Mission data, and this appears to be associated with an increase in glacier surging and an expansion of nonsurging glaciers, as observed in satellite imagery. Furthermore, abundant meltwater and high glacier ice velocities indicate that glacier erosion potential might be increasing (Quincey et al. 2009).

Our more recent regional climate modeling simulations suggest that future increases in precipitation might be related to the strength of the westerly winds that are correlated with ENSO conditions. If this can be verified, these results would be in direct opposition to the hypothesis of an enhanced monsoon due to global warming (Anderson, Overpeck, and Gupta 2002). It is yet unclear whether an intensification of ENSO precipitation events or increases in atmospheric temperature will facilitate further glacier expansion or result in negative mass balance that will regulate the magnitude of glacier erosion. Our global climate simulation results suggest a future regional negative mass-balance trend.

Clearly, climate forcing has had a significant influence on landscape evolution in the western Himalaya of Pakistan. Large glaciers and glacier fluctuations have removed significant mass, thereby causing isostatic and tectonic uplift. The climate-versus-tectonic forcing hypotheses are difficult to relate directly, as numerous feedback mechanisms are involved, and direct linkages among climate, erosion, and tectonics must be established to demonstrate a tectonic response to climate change. Another important issue is whether active orogenic zones are carbon sinks or sources. Silicate weathering and the drawdown of atmospheric CO_2 could cause global cooling; however, climate simulations and radiative forcing explain glacier fluctuations in the Quaternary. Furthermore, chemical analyses of hot springs in the Himalaya suggest that active orogenic zones are sources of CO_2 caused by metamorphic reactions (Gaillardet and Galy 2008).

Climate forcing would intensify the geomorphic influence on crustal scale processes such that enhanced erosion causes the upward flow of hot, low-viscosity crustal rock into areas or localized uplift zones (Koons et al. 2002; Whipple 2009). Rapid erosion generates topographic stresses that reduce rock strength and intensify erosion in a positive feedback, thereby generating a strong crustal temperature and pressure gradient. Decompression melting and metamorphism is associated with rock flow along the gradient. Such erosional–rheological coupling can explain the presence of active metamorphic massifs (Koons et al. 2002). Metamorphic degassing of CO_2 would increase atmospheric concentrations. If this is the case, climate forcing and mountain building would inject CO_2 and warm Earth, resulting in limited glacier expansion and deglaciation. The hypothesis of orogenic CO_2 source contradicts the tectonic-forcing hypothesis and is dependent on the thermal history of the orogenic zone (Gaillardet and Galy 2008). Consequently, the climate-forcing hypothesis is predicated on whether mountain ranges are a net source or sink of atmospheric CO_2, and we must account for different processes and time scales such as the consumption of CO_2 by rock weathering (silicate and carbonate), the balance of organic matter burial and oxidation, and the fluxes of CO_2 degassing. It is also imperative that we better characterize the topography with respect to erosional efficiency and tectonic response so that the patterns of erosion and tectonics can be more directly related to climate forcing parameters.

Conclusions

Mountain geodynamics represent highly scale-dependent interactions involving climate, tectonic, and surface processes (Figure 13). The western Himalaya in Pakistan potentially exhibits strong climate–tectonic feedbacks, although the tectonic and topographic responses to climate perturbations have not been systematically explored. Nevertheless, the region exhibits unique patterns of topographic complexity, erosion, and exhumation.

Our GCM and regional-climate simulations suggest that the region has experienced extreme climate change due to radiative forcing and regional system response (i.e., westerlies, southwestern monsoon) to ENSO conditions. Paleoclimate simulations support geomorphological evidence of multiple glaciations and overall glacier retreat since the Holocene. High-magnitude glacier erosion has resulted in extreme relief, U-shaped valleys, high-altitude erosion surfaces, and extensive mass movements.

Our analysis of the topography indicates that relief is not related linearly to mass removal and erosional efficiency, as different surface processes and process couplings generate different topographic conditions that vary with altitude and geologic structure. Consequently, in some locations the topography contains overprinting patterns regarding the polygenetic evolution of the landscape. This is not the case where high erosion and uplift rates generate threshold topography that might not sustain relief. Furthermore, mesocale relief patterns clearly depict erosion zones that are spatially coincident with precipitation anomalies, high peaks, and rapid exhumation. Mesoscale relief patterns demonstrate extreme spatial and temporal variability in the influence of glacier erosion in the topographic evolution of the region. Collectively, our results demonstrate a strong linkage between climate change and the topographic evolution of the central Karakoram in Pakistan. Nevertheless, Figure 13 clearly depicts that although climate and erosion play a significant role, the system is complex, with multiple feedback pathways and mechanisms operating at a multitude of spatio-temporal scales, such that the dominance of any particular component, process, or parameter of the system is likely to be highly variable and difficult to know with certainty. It is essential that we better characterize the topography with respect to erosional efficiency and tectonic response, so that the patterns of erosion and tectonics can be more directly related to climate forcing parameters.

Acknowledgments

We would especially like to acknowledge the long-term and highly fruitful relationship with the late Syed Hamidullah, former Director of the Centre of Excellence at Peshawar University, who worked so much to help facilitate this project. We would also like to thank his students, Faisal Khan and Mohammad Shahid, for their excellent assistance in the field. This research was supported by funding from the National Geographic Society and the U.S. National Science Foundation (Grant BCS-0242339) to the University of Nebraska–Omaha and the University of Cincinnati.

References

Anders, A. M., G. H. Roe, B. Hallet, D. R. Montgomery, N. J. Finnegan, and J. Putkonen. 2006. Spatial patterns of precipitation and topography in the Himalaya. In *Tectonics, climate, and landscape evolution*, ed. S. D. Willett, N. Hovius, M. T. Brandon, and D. M. Fisher, 39–53. Boulder, CO: Geological Society of America.

Anderson, D. M., J. T. Overpeck, and A. K. Gupta. 2002. Increase in the Asian southwest monsoon during the past four centuries. *Science* 297:596–99.

Barros, A. P., S. Chiao, T. J. Lang, D. Burbank, and J. Putkonen. 2006. From weather to climate: Seasonal and interannual variability of storms and implications for erosion processes in the Himalaya. In *Tectonics, climate, and landscape evolution*, ed. S. D. Willett, N. Hovius, M. T. Brandon, and D. M. Fisher, 17–38. Boulder, CO: Geological Society of America.

Beaumont, C., and G. Quinlan. 1994. A geodynamic framework for interpreting crustal-scale seismic-reflectivity patterns in compressional orogens. *Geophysics Journal International* 116:754–83.

Bender, F. K., and H. A. Raza. 1995. *Geology of Pakistan*. Berlin, Germany: Borntraeger.

Berger, A. 1992. Orbital variations and insolation database. IGBP PAGES/World Data Center-A for Paleoclimatology Data Contribution Series # 92–007, NOAA/NGDC Paleoclimatology Program, Boulder, CO.

Bishop, M. P., and J. F. Shroder, Jr. 2000. Remote sensing and geomorphometric assessment of topographic complexity and erosion dynamics in the Nanga Parbat massif. In *Tectonics of the Nanga Parbat syntaxis and the western Himalaya*, ed. M. Khan, P. J. Treloar, M. P. Searle, and M. Q. Jan, 181–200. London: Geological Society London.

———. 2004. *Geographic information science and mountain geomorphology*. Chichester, UK: Praxis-Springer.

Bishop, M. P., J. F. Shroder, Jr., R. Bonk, and J. Olsenholler. 2002. Geomorphic change in high mountains: A western Himalayan perspective. *Global and Planetary Change* 32:311–29.

Bishop, M. P., J. F. Shroder, Jr., and J. D. Colby. 2003. Remote sensing and geomorphometry for studying relief production in high mountains. *Geomorphology* 55:345–61.

Bogorodsky, V. V., C. R. Bentley, and P. E. Gudmandsen. 1985. *Radioglaciology*. Dordrecht, The Netherlands: Reidel.

Brozovic, N., D. W. Burbank, and A. J. Meigs. 1997. Climatic limits on landscape development in the northwestern Himalaya. *Science* 276:571–74.

Burbank, D. W., J. Leland, E. Fielding, R. S. Anderson, N. Brozovic, M. R. Reid, and C. Duncan. 1996. Bedrock incision, rock uplift and threshold hillslopes in the northwestern Himalayas. *Nature* 379:505–10.

Bush, A. B. G. 2001. Pacific sea surface temperature forcing dominates orbital forcing of the early Holocene monsoon. *Quaternary Research* 55:25–32.

———. 2007. Extratropical influences on the El Nino Southern Oscillation through the late Quaternary. *Journal of Climate* 20:788–800.

Bush, A. B. G., and S. G. H. Philander. 1999. The climate of the Last Glacial Maximum: Results from a coupled atmosphere–ocean general circulation model. *Journal of Geophysical Research* 104:509–25.

Caputo, M. 1964. Glaciology. In *Geophysics of the Karakorum*, ed. A. Marussi. Leiden, The Netherlands: E. J. Brill.

Cenderelli, D. A., and E. E. Wohl. 2003. Flow hydraulics and geomorphic effects of glacial-lake outburst floods in the Mount Everest region, Nepal. *Earth Surface Processes and Landforms* 28:385–407.

Christensen, J. H., J. Raisanen, T. Iversen, D. Bjorge, O. B. Christensen, and M. Rummukainen. 2001. A synthesis of regional climate change simulations: A Scandinavian perspective. *Geophysical Research Letters* 28:1003–6.

Clift, P. D., L. Giosan, J. Blusztajn, I. H. Campbell, C. Allen, M. Pringle, A. R. Tabrez, et al. 2008. Holocene erosion of the Lesser Himalaya triggered by intensified summer monsoon. *Geology* 36:79–82.

Clift, P. D., K. V. Hodges, D. Heslop, R. Hannigan, H. V. Long, and G. Calves. 2008. Correlation of Himalayan exhumation rates and Asian monsoon intensity. *Nature Geoscience* 1:875–80.

Copland, L., S. Pope, M. P. Bishop, J. F. Shroder, Jr., P. Clendon, A. B. G. Bush, U. Kamp, Y. B. Seong, and L. A. Owen. 2009. Glacier velocities across the central Karakoram. *Annals of Glaciology* 50:41–49.

Derbyshire, E. 1981. Glacier regime and glacial sediment facies: A hypothetical framework for the Qinghai-Xizang Plateau. In *Proceedings of the Symposium on Qinghai-Xizang (Tibet) Plateau*, vol. 2, 1649–56. Beijing: Science Press.

Desio, A., A. Marussi, and M. Caputo. 1961. Glaciological research of the Italian Karakorum Expedition 1953–1955. *IASH Publication* 52:224–32.

Easterling, D. R. 1999. Development of regional climate scenarios using a downscaling approach. *Climatic Change* 41:615–34.

Egholm, D. L., S. B. Nielsen, V. K. Pedersen, and J. E. Lesemann. 2009. Glacial effects limiting mountain height. *Nature* 460:884–88.

Finlayson, D. R., D. R. Montgomery, and B. Hallet. 2002. Spatial coincidence of rapid inferred erosion with young metamorphic massifs in the Himalayas. *Geology* 30:219–22.

Finnegan, N. J., B. Hallet, D. R. Montgomery, P. K. Zeitler, J. O. Stone, A. M. Anders, and L. Yuping. 2008. Coupling of rock uplift and river incision in the Namche Barwa–Gyala Peri massif, Tibet. *Geological Society of America Bulletin* 120:142–55.

Foster, D. A., A. J. W. Gleadow, and G. Mortimer. 1994. Rapid Pliocene exhumation in the Karakoram (Pakistan), revealed by fission-track thermochronology of the K2 gneiss. *Geology* 22:19–22.

Gaillardet, J., and A. Galy. 2008. Himalaya: Carbon sink or source? *Science* 320:1727–28.

Gerrad, A. J. 1990. *Mountain environments*. Cambridge, MA: MIT Press.

Gilchrist, A. R., M. A. Summerfield, and H. A. P. Cockburn. 1994. Landscape dissection, isostatic uplift, and the morphologic development of orogens. *Geology* 22:963–66.

Hagg, W., C. Mayer, A. Lambrecht, and A. Helm. 2008. Sub-debris melt rates on Southern Inylchek Glacier, Central Tian Shan. *Geografiska Annaler Series A: Physical Geography* 90:55–63.

Hallet, B., L. Hunter, and J. Bogen. 1996. Rates of erosion and sediment evacuation by glaciers: A review of field data and their implications. *Global and Planetary Change* 12:213–35.

Hancock, G. S., R. S. Anderson, and K. X. Whipple. 1998. Beyond power: Bedrock river incision process and form. In *Rivers over rock: Fluvial processes in bedrock channels*, ed. K. J. Tinkler and E. E. Wohl, 35–60. Washington, DC: American Geophysical Union.

Harbor, J., and J. Warburton. 1992. Glaciation and denudation rates. *Nature* 356:751.

Hewitt, K. 1969. Glacier surges in the Karakoram Himalaya (Central Asia). *Canadian Journal of Earth Sciences* 6:1009–18.

———. 1988. Catastrophic landslide deposits in the Karakoram Himalaya. *Science* 242:64–77.

———. 1989. The altitudinal organisation of Karakoram geomorphic processes and depositional environments. *Zeitschrift für Geomorphologie* 76:9–32.

———. 1993. Altitudinal organization of Karakoram geomorphic processes and depositional environments. In *Himalaya to the sea*, ed. J. F. Shroder, Jr., 159–83. London and New York: Routledge.

———. 1998a. Catastrophic landslides and their effects on the Upper Indus streams, Karakoram Himalaya, northern Pakistan. *Geomorphology* 26:47–80.

———. 1998b. Glaciers receive a surge of attention in the Karakoram Himalaya. *Eos* 79:104–5.

———. 1999. Quaternary moraines vs. catastrophic avalanches in the Karakoram Himalaya, northern Pakistan. *Quaternary Research* 51:220–37.

———. 2005. The Karakoram anomaly? Glacier expansion and the "elevation effect," Karakoram Himalaya. *Mountain Research and Development* 25:332–40.

———. 2007. Tributary glacier surges: An exceptional concentration at Panmah Glacier, Karakoram Himalaya. *Journal of Glaciology* 53:181–88.

Hewitt, K., C. P. Wake, G. J. Young, and C. David. 1989. Hydrological investigations at Biafo Glacier, Karakorum Range, Himalaya: An important source of water for the Indus River. *Annals of Glaciology* 13: 103–8.

Howard, A. D. 1980. Thresholds in river regimes. In *Thresholds in geomorphology*, ed. D. R. Coates and J. D. Vitek, 227–58. London: Allen and Unwin.

Koons, P. O., P. K. Zeitler, C. P. Chamberlain, D. Craw, and A. S. Meltzer. 2002. Mechanical links between erosion and metamorphism in Nanga Parbat, Pakistan Himalaya. *American Journal of Science* 302:749–73.

Koppes, M., and B. Hallet. 2006. Erosion rates during rapid deglaciation in Icy Bay, Alaska. *Journal of Geophysical Research* 111:F02023.

Korup, O., and D. R. Montgomery. 2008. Tibetan plateau river incision inhibited by glacial stabilization of the Tsangpo gorge. *Nature* 455:786–90.

Macheret, Y. Y., M. Y. Moskalevsky, and E. V. Vasilenko. 1993. Velocity of radio waves in glaciers as an indicator of their hydrothermal state, structure and regime. *Journal of Glaciology* 39:373–84.

Mahéo, G., A. Pecher, S. Guillot, Y. Rolland, and C. Delacourt. 2004. Exhumation of Neogene gneiss domes between oblique crustal boundaries in south Karakorum (northwest Himalaya, Pakistan). In *Gneiss domes in orogeny*, ed. D. L. Whitney, C. Teyssier, and C. S. Siddoway, 141–54. Boulder, CO: Geological Society of America.

Mayer, C., A. Lambrecht, M. Belo, C. Smiraglia, and G. Diolaiuti. 2006. Glaciological characteristics of the ablation zone of Baltoro Glacier, Karakoram, Pakistan. *Annals of Glaciology* 43:123–31.

Mayewski, P. A., and P. A. Jeschke. 1979. Himalayan and Trans-Himalayan glacier fluctuations since AD 1812. *Arctic and Alpine Research* 11:267–87.

Mihalcea, C., C. Mayer, G. Diolaiuti, A. Lambrecht, C. Smiraglia, and G. Tartari. 2006. Ice ablation and meteorological conditions on the debris-covered area of Baltoro Glacier, Karakoram, Pakistan. *Annals of Glaciology* 43:292–300.

Minster, J. B., and T. H. Jordan. 1978. The present-day plate motions. *Journal of Geophysical Research* 83:5331–54.

Molnar, P. 1988. A review of the geophysical constraints on the deep structure of the Tibetan Plateau, the Himalaya and the Karakoram, and their tectonic implications. *Philosophical Transactions of the Royal Society of London, Series A* 326:33–88.

Molnar, P., and P. England. 1990. Late Cenozoic uplift of mountain ranges and global climatic change: Chicken or egg? *Nature* 46:29–34.

Montgomery, D. R. 1994. Valley incision and uplift of mountain peaks. *Journal of Geophysical Research* 99:913–21.

Montgomery, D. R., and M. T. Brandon. 2002. Topographic controls on erosion rates in tectonically active mountain ranges. *Earth and Planetary Science Letters* 201:481–89.

Narod, B. B., and G. K. C. Clarke. 1994. Miniature high-power impulse transmitter for radio-echo sounding. *Journal of Glaciology* 40:190–94.

Navarra, A., W. F. Stern, and K. Miyakoda. 1994. Reduction of the Gibbs oscillation in spectral model simulations. *Journal of Climatology* 7:1169–83.

Owen, L. A. 1988. Wet-sediment deformation of Quaternary and recent sediments in the Skardu Basin, Karakoram Mountains, Pakistan. In *Glaciotectonics: Forms and Processes*, ed. D. Croot, 123–48. Rotterdam, The Netherlands: Balkema.

Owen, L. A., M. W. Caffee, K. Bovard, R. C. Finkel, and M. Sharma. 2006. Terrestrial cosmogenic surface exposure dating of the oldest glacial successions in the Himalayan orogen. *Geological Society of America Bulletin* 118: 383–92.

Owen, L. A., M. W. Caffee, R. C. Finkel, and Y. B. Seong. 2008. Quaternary glaciations of the Himalayan-Tibetan orogen. *Journal of Quaternary Science* 23:513–32.

Owen, L. A., R. C. Finkel, P. L. Barnard, M. Haizhhou, K. Asahi, M. W. Caffee, and E. Derbyshire. 2005. Climatic and topographic controls on the style and timing of Late Quaternary glaciations throughout Tibet and the Himalaya defined by ^{10}Be cosmogenic radionuclide surface exposure dating. *Quaternary Science Reviews* 24:1391–1411.

Owen, L. A., R. C. Finkel, M. W. Caffee, and L. Gualtieri. 2002. Timing of multiple glaciations during the Late Quaternary in the Hunza Valley, Karakoram Mountains, Northern Pakistan: Defined by cosmogenic radionuclide dating of moraines. *Geological Society of America Bulletin* 114:593–604.

Owen, L. A., L. Gualtieri, R. C. Finkel, M. W. Caffee, D. I. Benn, and M. C. Sharma. 2001. Cosmogenic radionuclide dating of glacial landforms in the Lahul Himalaya, Northern India: Defining the timing of Late Quaternary glaciation. *Journal of Quaternary Science* 16:555–63.

Pant, G. B., and K. R. Kumar. 1997. *Climates of South Asia*. Chichester, UK: Wiley.

Pelletier, J. D. 2008. Glacial erosion and mountain building. *Geology* 36:591–92.

Peltier, W. R. 1994. Ice age paleotopography. *Science* 265:195–201.

Pratt-Sitaula, B., B. N. Upreti, T. Melbourne, A. Miner, E. Parker, S. M. Rai, and T. N. Bhattarai. 2009. Applying geodesy and modeling to test the role of climate controlled erosion in shaping Himalayan morphology and evolution. *Himalayan Geology* 30:123–31.

Quincey, D. J., L. Copland, C. Mayer, M. P. Bishop, A. Luckman, and M. Belo. 2009. Ice velocity and climate variations for the Baltoro Glacier, Pakistan. *Journal of Glaciology* 55:1061–71.

Rahaman, W., S. K. Singh, R. Sinha, and S. K. Tandon. 2009. Climate control on erosion distribution over the Himalaya during the past ~100 ka. *Geology* 37:559–62.

Raisanen, J., M. Rummukainen, and A. Ullerstig. 2001. Downscaling of greenhouse gas induced climate change in two GCMs with the Rossby Centre regional climate model for northern Europe. *Tellus Series A* 53:168–91.

Raymo, M. E., and W. F. Ruddiman. 1992. Tectonic forcing of late Cenozoic climate. *Nature* 359:117–22.

Raymo, M. E., W. F. Ruddiman, and P. N. Froelich. 1988. Influence of late Cenozoic mountain building on ocean geochemical cycles. *Geology* 16:649–53.

Schmidt, K. M., and D. R. Montgomery. 1995. Limits to relief. *Science* 270:617–20.

Schoof, J. T., and S. C. Pryor. 2001. Downscaling temperature and precipitation: A comparison of regression-based methods and artificial neural networks. *International Journal of Climatology* 21:773–90.

Searle, M. P. 1991. *Geology and tectonics of the Karakoram Mountains*. Chichester, UK: Wiley.

Searle, M. P., R. R. Parrish, R. Tirrul, and D. C. Rex. 1990. Age of crystallization and cooling of the K2 gneiss in the Baltoro Karakoram. *Journal of the Geological Society of London* 147:603–6.

Seong, Y. B., M. P. Bishop, A. B. G. Bush, P. Clendon, L. Copland, R. C. Finkel, U. Kamp, L. A. Owen, and J. F. Shroder, Jr. 2009. Landforms and landscape evolution in the Skardu, Shigar and Braldu Valleys, Central Karakoram. *Geomorphology* 103:251–67.

Seong, Y. B., L. A. Owen, M. P. Bishop, A. B. G. Bush, P. Clendon, L. Copland, R. C. Finkel, U. Kamp, and J. F. Shroder, Jr. 2007. Quaternary glacial history of the Central Karakoram. *Quaternary Science Reviews* 26:3384–405.

———. 2008. Rates of fluvial bedrock incision within an actively uplifting orogen: Central Karakoram Mountains, northern Pakistan. *Geomorphology* 97:274–86.

Seong, Y. B., L. A. Owen, C. Yi, R. C. Finkel, and L. Schoenbohm. 2009. Geomorphology of anomalously high glaciated mountains at the northwestern end of Tibet: Muztag Ata and Kongur Shan. *Geomorphology* 103:227–50.

Shi, Y. 2002. Characteristics of late Quaternary monsoonal glaciation on the Tibetan Plateau and in East Asia. *Quaternary International* 97–98:79–91.

Shroder, J. F., Jr. 1998. Slope failure and denudation in the western Himalaya. *Geomorphology* 26:81–105.

Shroder, J. F., Jr., and M. P. Bishop. 2000. Unroofing of the Nanga Parbat Himalaya. In *Tectonics of the Nanga Parbat syntaxis and the western Himalaya*, ed. M. Khan, P. J.

Treloar, M. P. Searle, and M. Q. Jan, 163–79. London: Geological Society of London.

Sklar, L. S., and W. E. Dietrich. 1998. River longitudinal profiles and bedrock incision models: Stream power and the influence of sediment supply. In *Rivers over rock: Fluvial processes in bedrock channels*, ed. K. J. Tinkler and E. E. Wohl, 237–60. Washington, DC: American Geophysical Union.

———. 2001. Sediment and rock strength controls on river incision into bedrock. *Geology* 29:1087–90.

Spotila, J. A., J. T. Buscher, A. J. Meigs, and P. W. Reiners. 2004. Long-term glacial erosion of active mountain belts: Example of the Chugach-St. Elias Range, Alaska. *Geology* 32:501–04.

Tomkin, J. H., and J. Braun. 2002. The influence of alpine glaciations on the relief of tectonically active mountain belts. *American Journal of Science* 302:169–90.

Turcotte, D. L., and G. Schubert. 2002. *Geodynamics*. Cambridge, UK: Cambridge University Press.

van der Beek, P., J. Van Melle, S. Guillot, A. Pecher, P. W. Reiners, S. Nicolescu, and M. Latif. 2009. Eocene Tibetan plateau remnants preserved in the northwest Himalaya. *Nature Geoscience* 2:364–68.

von Storch, H., and A. Navarra. 1999. *Analysis of climate variability*. New York: Springer.

Wake, C. P. 1989. Glaciochemical investigations as a tool for determining the spatial and seasonal variation of snow accumulation in the central Karakoram, northern Pakistan. *Annals of Glaciology* 13:279–84.

Whipple, K. X. 2009. The influence of climate on the tectonic evolution of mountain belts. *Nature Geoscience* 2:97–104.

Whipple, K. X., E. Kirby, and S. H. Brocklehurst. 1999. Geomorphic limits to climate-induced increases in topographic relief. *Nature* 401:39–43.

Whipple, K. X., and G. E. Tucker. 1999. Dynamics of the stream-power river incision model: Implications for height limits of mountain ranges, landscape response timescales, and research needs. *Journal of Geophysical Research* 104:17661–74.

Whittington, A. G. 1996. Exhumation overrated. *Tectonophysics* 206:215–26.

Wobus, C., K. X. Whipple, E. Kirby, N. Snyder, J. Johnson, K. Spyropolou, B. Crosby, and D. Sheehan. 2006. Tectonics from topography: Procedures, promise and pitfalls. In *Tectonics, climate, and landscape evolution*, ed. S. D. Willett, N. Hovius, M. T. Brandon, and D. M. Fisher, 55–74. Boulder, CO: Geological Society of America.

Wohl, E. 2000. Mountain rivers. Water Resources Monograph 14, American Geophysical Union, Washington, DC.

Yokoyama, R., M. Shirasawa, and R. J. Pike. 2002. Visualizing topography by openness: A new application of image processing to digital elevation models. *Photogrammetric Engineering and Remote Sensing* 68:257–65.

Zeitler, P. K. 1985. Cooling history of the NW Himalaya. *Tectonics* 4:127–51.

Zeitler, P. K., P. O. Koons, M. P. Bishop, C. P. Chamberlain, D. Craw, M. A. Edwards, S. Hamidullah, et al. 2001. Crustal reworking at Nanga Parbat, Pakistan: Metamorphic consequences of thermal-mechanical coupling facilitated by erosion. *Tectonics* 20:712–28.

Zeitler, P. K., A. S. Meltzer, P. O. Koons, D. Craw, B. Hallet, C. P. Chamberlain, W. S. F. Kidd, et al. 2001. Erosion, Himalayan geodynamics and the geomorphology of metamorphism. *Geological Society of America Today* 11:4–8.

Climate Change and Tropical Andean Glacier Recession: Evaluating Hydrologic Changes and Livelihood Vulnerability in the Cordillera Blanca, Peru

Bryan G. Mark, Jeffrey Bury, Jeffrey M. McKenzie, Adam French, and Michel Baraer

Climate change is forcing dramatic glacier mass loss in the Cordillera Blanca, Peru, resulting in hydrologic transformations across the Rio Santa watershed and increasing human vulnerability. This article presents results from two years of transdisciplinary collaborative research evaluating the complex relationships between coupled environmental and social change in the region. First, hydrologic results suggest there has been an average increase of 1.6 (± 1.1) percent in the specific discharge of the more glacier-covered catchments (>20 percent glacier area) as a function of changes in stable isotopes of water ($\delta^{18}O$ and $\delta^{2}H$) from 2004 to 2006. Second, there is a large (mean 60 percent) component of groundwater in dry season discharge based on results from the hydrochemical basin characterization method. Third, findings from extensive key interviews and seventy-two randomly sampled household interviews within communities located in two case study watersheds demonstrate that a large majority of households perceive that glacier recession is proceeding very rapidly and that climate change–related impacts are affecting human vulnerability across multiple shifting vectors including access to water resources, agro-pastoral production, and weather variability. *Key Words: climate change, glacier recession, hydrology, livelihoods, vulnerability.*

在秘鲁的科迪勒拉布兰卡地区，气候变化正在导致急剧的冰川物质损失，造成了里约圣塔流域的水文变化，并增加了人类的脆弱性。本文介绍了在该地区进行了两年之久的跨学科的合作研究成果，以评价环境和社会变革的之间的复杂关系。首先，水文结果表明，从 2004 到 2006 年，根据水的稳定同位素（δ18O 和 δ2H）的变化结果，该地区具有较高冰川覆盖的蓄水流域区（大于百分之二十的冰川面积）产生了平均增长百分之 1.6（± 1.1）的单位流量增长。第二，基于水化学盆地分析方法的结果，存在一个较大的（平均百分之六十）旱季流量的地下水体。第三，本文对位于上述两个个案流域内的社区进行了广泛的重点访谈和随机的 72 户抽样访问，研究结果表明：大部分的家庭感觉到冰川衰退地非常迅速，与气候变化有关的影响非常广泛，影响到人类的脆弱性，包括水资源的利用，农牧业生产和天气变化。*关键词：气候变化，冰川衰退，水文，生计，脆弱性。*

El cambio climático está causando una dramática pérdida de masa en los glaciares de la Cordillera Blanca, Perú, generando transformaciones hidrológicas en toda la cuenca del Río Santa e incrementando la vulnerabilidad humana. Este artículo presenta los resultados de dos años de investigación colaborativa transdisciplinaria para evaluar las complejas relaciones entre los cambios ambientales y sociales en la región. Primero, los resultados hidrológicos sugieren que ha habido un incremento promedio del 1.6 (±1.1) por ciento en la descarga específica de los desagües con mayor cobertura de glaciares (>20 por ciento de área glaciada), como una función del cambio en isótopos estables del agua (δ18O and δ2H) entre 2004 y 2006. Segundo, hay un gran componente (media de 60 por ciento) de agua subterránea en el descargue de la estación seca basado en resultados del método de caracterización de la cuenca hidroquímica. Tercero, los descubrimientos derivados de detalladas entrevistas a informantes claves y setenta y dos entrevistas de muestra aleatoria administradas a hogares de comunidades pertenecientes a dos estudios de caso de cuencas, demuestran que la gran mayoría de la gente intuye que la recesión de los glaciares está avanzando rápidamente y que los impactos relacionados con cambio climático afectan la vulnerabilidad humana por medio de muchos vectores cambiantes, incluyendo el acceso a los recursos hídricos, producción agro-pastoral y variación meteorológica. *Palabras clave: cambio climático, recesión de glaciares, hidrología, medios de vida, vulnerabilidad.*

Rapid glacier recession in the tropical Peruvian Andes due to recent climate change exemplifies environmental changes in mountain regions that have critical water supply implications for millions of people globally (Intergovernmental Panel on Climate Change 2007; Viviroli et al. 2007). Modern glacier recession is strongly correlated with a significant rising trend in atmospheric temperatures, and climate models predict enhanced future rates of glacier loss with increased warming at higher tropical altitudes (Barry and Seimon 2000; Bradley et al. 2009). This recession will significantly affect the future availability and use of water resources because human populations are highly dependent on glacial melt water to buffer the seasonally arid climate of the central Andes (Vuille et al. 2008). Although progress has been made in tracing glacier changes over time and space, new and more integrated research is needed to evaluate these ongoing social and hydrological transformations (Mark 2008).

The Cordillera Blanca, Peru, is a unique location to study these issues as more than 25 percent of the world's tropical glaciers are located in the range. There is also a long history of environment–society dynamics in this region that has been densely settled for many centuries and the spatial extent of human activities is often conditioned by water availability. The range has also been the epicenter of some of the most disastrous hazards in recorded history (Carey 2010). Finally, the differential glacial coverage of watersheds in the Cordillera Blanca region provides a natural continuum to compare hydrologic and human responses to climate change across subcatchments simultaneously as a proxy for sequential changes in time. This article presents results from our ongoing collaborative research project in the region. Working as a transdisciplinary team, we are evaluating the regional impact of glacier melt on seasonal and interannual water availability and assessing human vulnerability and governance shifts related to the political economy of climate change.

Setting

The Cordillera Blanca contains the world's largest concentration of tropical glaciers, most of which flow westward toward the Santa River (Figure 1 and Table 1). According to glacial inventories conducted in the 1970s, 35 percent of Peru's glacierized area and 40 percent of its glacial reserves (by volume) are stored in the more than 700 glaciers located in the range (Ames et al. 1989). The Santa River flows northwest over 300 km, draining a total watershed of 12,200 km^2 to become the second largest and most regular flowing Peruvian river to reach the Pacific Ocean (Mark and Seltzer 2003). The upper Santa River watershed, or Callejón de Huaylas, has an area of 4,900 km^2 defined by the outflow at the La Balsa station (Figure 1). Within the Callejón de Huaylas, tributary streams to the Santa River from the Cordillera Blanca define subcatchments with different percentages of glacial coverage. For some of these tributaries historical discharge records are available for forty years. Monthly average discharge from these gauged tributary streams is higher during the months of October through April, closely reflecting the seasonality of precipitation where greater than 80 percent of precipitation falls between October and May, and the austral winter months of June to September are known as the dry season.

Table 1. Characteristics of the Cordillera Blanca watersheds that have available historical discharge data

Stream	Period of record	Watershed area (km^2)	Percentage glaciated area
Llanganuco	1953–1997	85	36
Chancos	1953–1997	221	22
Quilcay	1953–1983	243	17
Olleros	1970–1997	175	10
La Balsa	1954–1997	4,768	8
Pachacoto	1953–1983	194	8
Quitaracsa	1953–1997	383	7
Querococha	1956–1996	62	3

Note: The case study watersheds are in bold type, and the La Balsa discharge station is italicized to indicate that it is located on the Rio Santa.

According to our subdistrict spatial analysis of the 2007 Peruvian Census data, the population of the Callejón de Huaylas is approximately 267,000 inhabitants. Major urban and periurban clusters along the valley include the cities of Huaraz (96,000), Caraz (13,000), Yungay (8,000), Carhuaz (7,200), Recuay (2,700), and Catac (2,400), and approximately 1,500 small rural settlements (INEI 2007). Our population estimate is roughly 25 percent higher than previous estimates, which significantly heightens the potential social risks of recent climatic change in the region (Byers 2000; Young and Lipton 2006). The watershed is also a critical source of water for urban centers, agricultural activities, and several hydroelectric power plants that account for approximately 10 percent of the

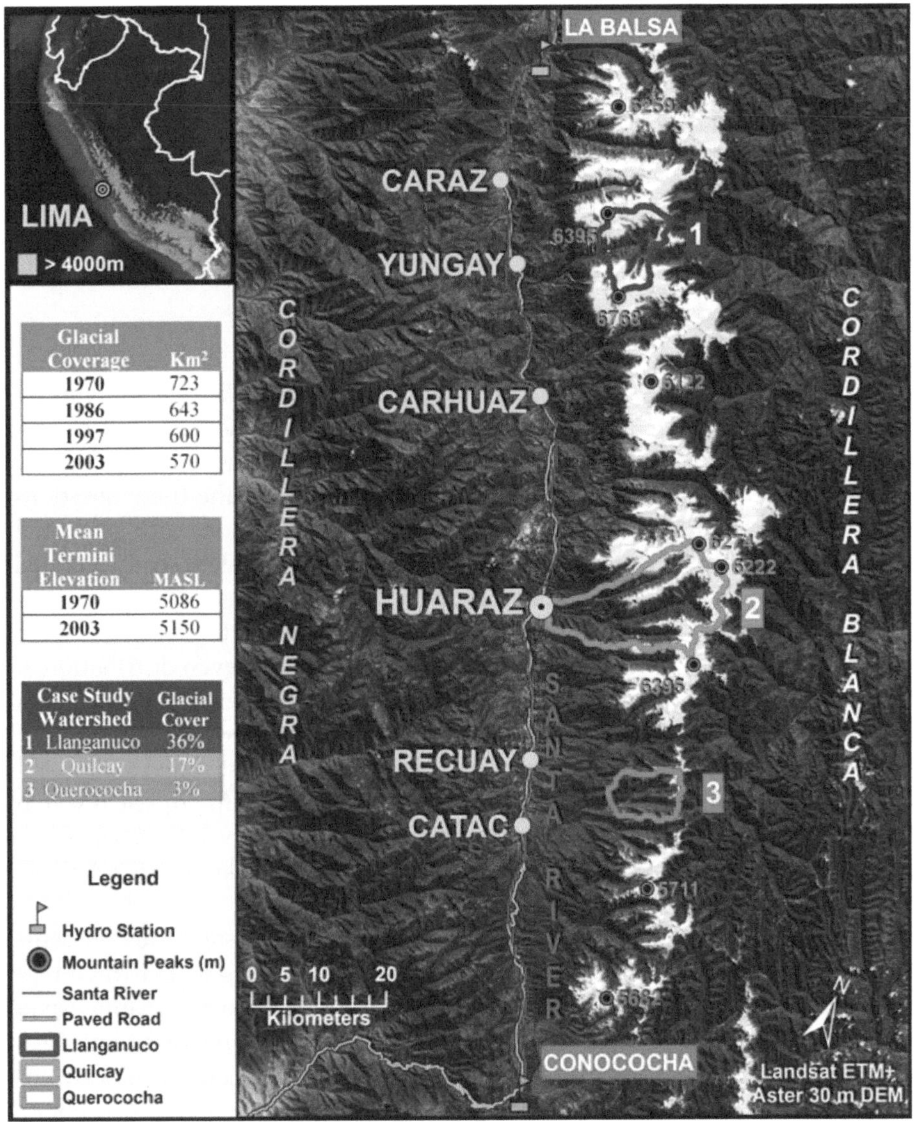

Glacial Coverage	Km²
1970	723
1986	643
1997	600
2003	570

Mean Termini Elevation	MASL
1970	5086
2003	5150

Case Study Watershed	Glacial Cover
1 Llanganuco	36%
2 Quilcay	17%
3 Querococha	3%

Figure 1. Map of the Cordillera Blanca, with study sites indicated. Inset tables include data on glacier coverage (Ames et al. 1989; Racoviteanu, Arnaud, and Williams 2008).

country's hydroelectric capacity (Ministry of Energy and Mines 2008).

Livelihoods in the region are generally dependent on access to water and other natural resources for agricultural and livestock production, as more than 80 percent of the population of the region is engaged in smallholder production (INEI 2007). Since the late 1990s, agricultural production along the Santa River has been shifting toward water-intensive crops for consumption outside the region and, similar to trends across the country, several new transnational polymetallic mining facilities have recently begun operating along the Cordillera Negra (Bury 2005; Bebbington and Bury 2009). However, more than 50 percent of the population still lives in conditions of poverty (defined as lacking more than one basic necessity) and 33 percent of the rural population is illiterate (INEI 2007).

Background and Rationale

Observations of continued glacier recession exist throughout the Andes, but the Cordillera Blanca is one of only a few locations in the tropics where research has quantified changes in the hydrologic regime due to glacier volume loss on a scale relevant to human impact. The range has lost one third (>30%) of its glacierized area since maximum extensions of the nineteenth century, with clear evidence of significant climate change over the twentieth century (Vuille et al. 2008). Multiscale studies have traced climate forcing and the hydrologic impacts of glacier volume changes over time (Mark 2008). Mark and Seltzer (2003) estimated that 35 to 40 percent of the mean annual discharge in the Querococha watershed during 1998 and 1999 was supplied by nonrenewed melt from the Yanamarey glacier. This increased to 58 percent over the observation period from 2001 to 2004 (Mark, McKenzie, and Gómez 2005). The Yanamarey glacier is projected to disappear within a decade and our recent research has been directed toward understanding the implications of this late-stage glacial retreat for hydrologic processes and downstream communities (Bury et al. forthcoming).

Changes in the stable isotopes of water observed over recent years in glacier-fed streams demonstrate relative increases in glacier melt water (Mark and McKenzie 2007). However, this enhanced melt contribution will diminish as glacier mass disappears, causing the streams to have smaller dry season flow and increased variability (Kaser et al. 2003; Mark and Seltzer 2003; Juen, Kaser, and Georges 2007). Glacier recession also modifies watersheds by forming wetlands, lakes, and groundwater reservoirs that alter the surface drainage. There is an outstanding need to understand and quantify hydrologic processes in these dynamic and transforming landscapes, particularly in view of the potentially severe water stress impacts of glacier loss highlighted in future climate change scenarios for Peru.

Complementing research on glacier recession and hydrologic change, our research also builds on recent environment–society research examining the adaptive behavior of human communities to environmental change in the Peruvian Andes (Young and Lipton 2006; Postigo, Young, and Crews 2008). Geographic research in the Andes has long focused on human responses and adaptation to diverse processes of environmental and socioeconomic change (Sauer 1941; Knapp 1991; Zimmerer 1991; Denevan 1992; Bebbington 2000; Bury 2004). As Sauer (1941) noted, the complex shifts that climatic change induces across landscapes have been an enduring theme in geography.

This research also has theoretical links to research on human vulnerability and adaptation to global change. In geographic studies, vulnerability research has evolved significantly from its foundations in early risk-hazard studies (White 1942; Burton, White, and Kates 1978) and subsequent critiques of this work for a lack of attention to political economy and structural conditions (Hewitt 1983; Watts and Bohle 1993). More recent vulnerability studies have developed integrative approaches that examine how biophysical, social, economic, and political factors interact with and feed back on one another across scales of analysis (Liverman 1990; Cutter 2003; Turner et al. 2003; Polsky 2004; Eakin 2006; Leichenko and O'Brien 2008). Much of this work makes a case for the transdisciplinary collaborations, place-based analyses, and mixed methodological approaches that we are currently utilizing in our research.

Methodology

The linked foci of this research are on shifts in regional water resources and the impacts of such changes on household livelihood strategies and vulnerability in the Cordillera Blanca. To examine these questions we have integrated in situ observations with geospatial analyses, hydrochemical mixing models, semistructured household surveys, and key interviews. Our methodology also identifies and analyzes patterns across nested spatial and temporal scales. The uneven distribution of modern glaciers in the tributary watersheds of the Santa River provides a natural continuum over which to evaluate changes in hydrologic processes related to different amounts of glacier coverage. Furthermore, the research is focused on the dry season when water resources are more stressed and glacier melt production is relatively more important.

Using geographic information systems, remote sensing techniques, and site visits we selected three representative tributary watersheds to the Santa River with different glacial coverage, variable hydrological characteristics and diverse livelihood pursuits to understand and measure hydrologic processes, calibrate hydrochemical mixing models, and evaluate the spatial distribution of household resource use patterns. These case study watersheds are Llanganuco, Querococha, and Quilcay (Figure 1). Specific digital products utilized to identify glacierized areas, generate digital elevation models, delimit watershed areas, and evaluate human activities include satellite imagery from the Advanced Spaceborne Thermal Emission and Reflection Radiometer (ASTER), Landsat, and Quickbird platforms.

Historic discharge records in the Callejón de Huaylas are available for the Santa River at La Balsa and for eleven tributaries of the Santa River, dating from the 1950s through the end of the twentieth century when observations were largely disbanded due to the privatization of the state-run water resources office (Mark and Seltzer 2003). To compare trends over time, normalized anomalies of annual discharge were generated for La Balsa and seven glacierized, noncontrolled (i.e., no dams) tributary watersheds with the longest records (Table 1). Discharges were normalized by subtracting the series mean from the annual value, then dividing that difference by the series standard deviation. We computed anomalies for both the total annual discharge and the dry half of the year (May–October).

To characterize the temporal and spatial variability of glacier melt water discharge in the case study watersheds, we measured dissolved ion concentrations and isotopic signatures ($\delta^{18}O$ and δD) of water samples collected between 1998 and 1999 and 2004 and 2009 (after Mark and Seltzer 2003; Mark and McKenzie 2007). We used the hydrochemical basin characterization method (HBCM), based on a multicomponent mass balance approach to identify hydrologic end member's relative or absolute contributions at a sample point (Baraer et al. 2009). Dry season discharge measurements and hydrochemistry results from 1998 and 1999 and 2004 to 2008 at the Querococha watershed were used to calibrate the HBCM and examine interannual variations of the melt water contribution from a single glacier (Mark and Seltzer 2003; Baraer et al. 2009). Spatial variability of glacier melt water was evaluated based on samples collected in the case study watersheds, spread along the Cordillera Blanca, during the 2008 dry season. A simplified two-component hydrograph separation using hydrochemical and isotopic signatures was used to deconvolve watershed discharge into melt water and groundwater. The interannual changes in water isotopes from tributary streams to the Rio Santa were also analyzed to trace the relative changes in glacier melt contribution to the larger Callejón de Huaylas watershed (after Mark and McKenzie 2007).

We formulated a mixed set of social science methods directed toward maximizing levels of objective confidence and minimizing potential biases in our findings. We purposively selected case study communities in each watershed to generate the largest set of observable livelihood activities and possible impacts of recent glacier recession. We developed a stratified random sampling frame for individual household selection based on preliminary participatory community mapping activities. Individual case study households were then selected in the field based on their proximity to randomly generated coordinates using portable Global Positioning System field units and 1 m resolution satellite imagery (Figure 2). We conducted intensive semistructured household surveys and a diverse array of unstructured key interviews in each community. Overall, the findings reported in this article are based on seventy-two semistructured household surveys, twenty-one unstructured household surveys, and thirty-seven formal key interviews in the Querococha (QUER) and Quilcay (QUIL) case study watersheds (Table 2).

Table 2. Demographic variables listed for both the Querococha (QUER) and Quilcay (QUIL) case study watersheds

Case study sample demographic summary	QUER	QUIL
Total case study area population	3,249	1,200
Total households in sample population	40	32
Total sample population	181	124
Household sample percentage of total community population	5.5	10.3
Average age of respondents	47	51
Gender percentage of respondents		
Male	35	53
Female	65	47

Results

Shifts in Regional Water Resources

Glacier-fed stream discharge from the Cordillera Blanca correlates strongly with climate changes, but the magnitude of glacier melt influence is scale dependent. Streams draining the two watersheds with greatest amounts of glacier coverage (glacial > 20% of watershed by area, $n = 2$) experienced a significant increase ($p = 0.023$) in average annual discharge over the forty-three-year period of historical records (Figure 3). However, there is no significant trend in annual discharge averaged for all tributaries with glaciers ($n = 7$), implying that glacier melt enhancement is not discernable on an annual basis below a threshold amount of glacier coverage. A significant ($p = 0.004$) correlation between annual discharge and regional mean air temperature over the same time suggests that streamflow responds rapidly to regional-scale climatic forcing.

In the same glaciated watersheds, dry season (May–October) stream discharge increased significantly until the early 1980s but since 1983 has declined considerably (Figure 4). This change in trend indicates that glacier melt water buffers discharge only temporarily in local watersheds, and future dry season streamflow is likely to decline further. On the Santa River scale, discharge draining the entire Callejón de Huaylas watershed (< 10 percent glacier coverage) at La Balsa has declined

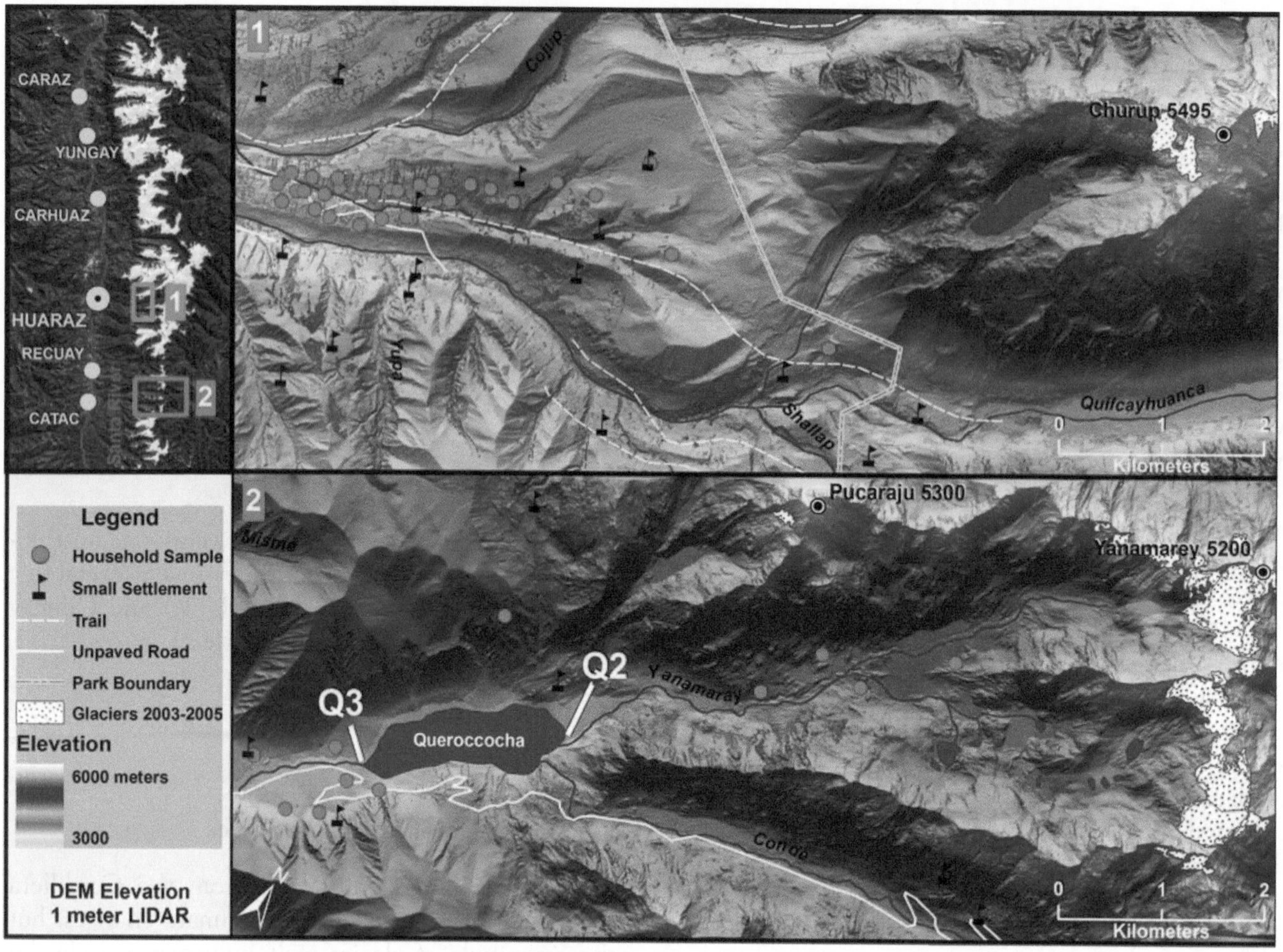

Figure 2. Shaded relief maps of portions of the populated sections of the Quilcay (1) and Querococha (2) case study watersheds, illustrating the location and distribution of case study locations and glaciers. Digital elevation data obtained by airborne light distance and ranging (LIDAR) survey flown in 2008 (by Horizons South America, S.A.C.).

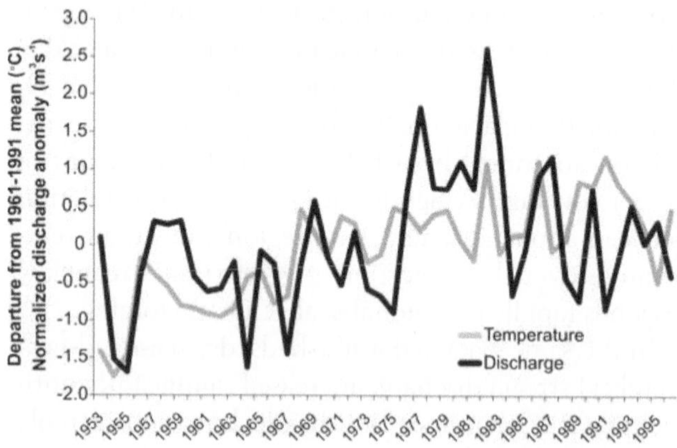

Figure 3. Average normalized anomalies of annual discharge for those tributary watersheds of the Santa River with > 20 percent glacier coverage ($n = 2$; see Table 1) and annual deviation of temperature from the 1961–1991 average from twenty-nine Peruvian meteorological stations between 9–11°S (from Mark and Seltzer 2005).

very significantly ($p = 0.004$), equaling a 17 percent decline from 1954 to 1997 (Figure 5).

Our analysis of stable isotope values from Callejón de Huaylas subcatchments with and without glaciers confirms a relative increase in dry season discharge due to recent glacier melt (Figure 6). Annual dry season samples (2004–2008) from Cordillera Blanca tributaries with glaciers describe a progressive depletion trend in $\delta^{18}O$ (i.e., more negative), whereas nonglacierized Cordillera Negra waters show no systematic change. The total isotopic change correlates to watershed glacier coverage (area percentage), translating to an increase in area-averaged discharge of 1.6 (± 1.1) percent (Mark and McKenzie 2007).

Relative groundwater contributions during the dry season were first evaluated using HBCM at Q2, the inflow to Querococha Lake (Figure 2), to explore interannual variability. This location drains a 28 km^2

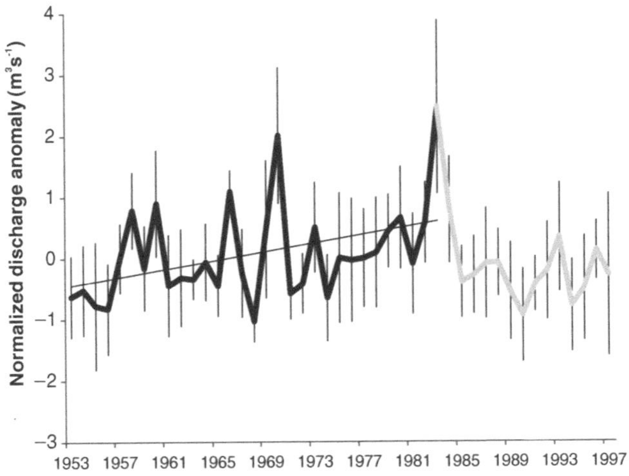

Figure 4. Normalized discharge anomalies for glacierized tributaries ($n = 7$) of the Santa River from 1953 to 1997. Prior to 1983, there is a significant ($p = 0.025$) positive trend shown with dark line; after 1983 there is a downward trend that is not quite significant ($p = 0.057$) shown in lighter gray. Vertical bars show range of variability at 2σ level ($\pm$ 1 *SD*).

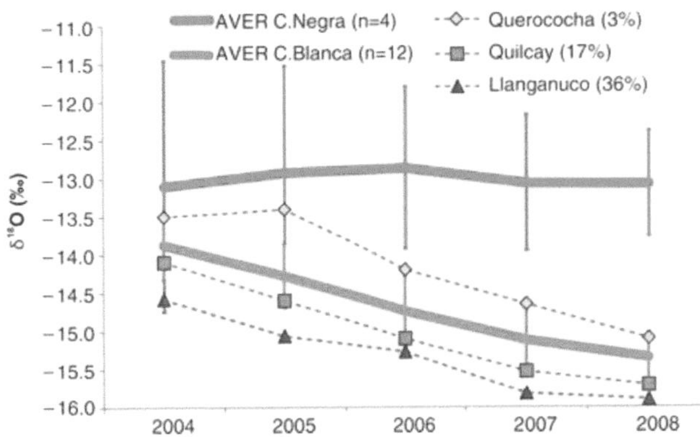

Figure 6. Oxygen isotope values of annual dry season water sampled from three watersheds with varying amounts of glacier coverage (dotted lines with symbols) plotted with average isotopic values for the Cordillera Blanca (with glaciers) and the Cordillera Negra (no glaciers) from 2004 to 2008, shown in thick lines with 2σ variability in vertical lines ($\pm$ 1 *SD*).

subwatershed with 7.2 percent glacier coverage. Results indicate a median 59 percent dry season groundwater contribution to the Q2 discharge (Figure 7). The largest calculated contribution was 74 percent in 2007 and the minimum contribution was 18 percent in 1998. With such high variability and dominant overall contribution, groundwater is likely as essential as melt water in Q2 discharge. Second, HBCM applied to different tributary watersheds simultaneously indicates that groundwater aquifers having significant yield likely exist in each case study watershed. A simplified two-component model of the 2008 dry season estimates groundwater contributes 0.19 m^3s^{-1} (77 percent) of discharge at Q2 and 0.15 m^3s^{-1} (26 percent) of Quilcay discharge. The difference between these estimated groundwater contributions is partly explained by the relative amount of glacier coverage, equal to 3 and 17 percent in the Querococha and Quilcay watersheds, respectively.

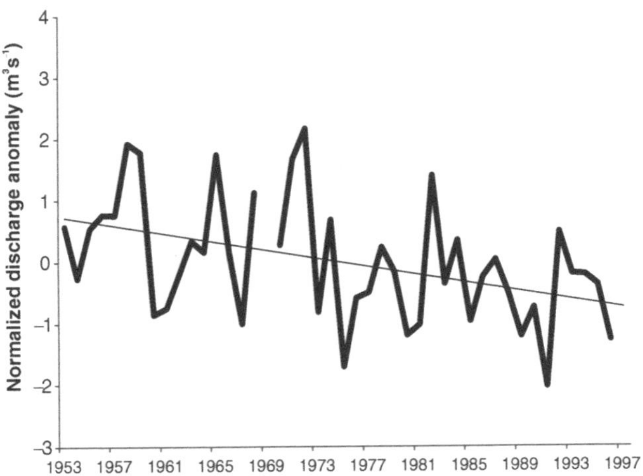

Figure 5. Normalized anomaly of annual discharge for La Balsa station on the Santa River (1953–1997).

Livelihoods and Emerging Vectors of Vulnerability

Communities in the Querococha and Quilcay watersheds are embedded within diverse communal, private, and public governance institutions that affect household access to key livelihood resources such as land and water. Land tenure and land use vary across watersheds and among a variety of institutions beginning with privately titled land parcels at the bottoms of the watersheds, ranging upward across communally managed lands and terminating at high lakes and the glaciers that remain the sole property of the state and are within Huascaran National Park. Access to the upper watersheds is thus governed by a complex and often conflictual nexus of state–community regimes (see Figure 2). For example, grazing access to the upper Querococha watershed is controlled by the 2,200 members of the Campesino Community of Catac and access to the upper Quilcay watershed is controlled by the 240 associates of the Quilcayhuanca Users Group. In addition, a number of land management practices across both watersheds are officially proscribed by National Park authorities.

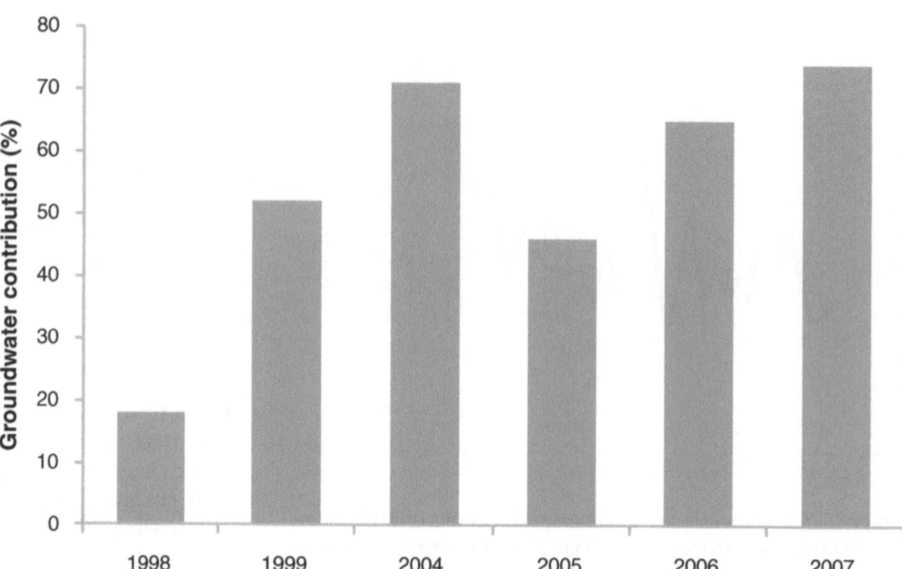

Figure 7. Interannual variation of the groundwater relative contribution modeled by hydrochemical basin characterization method at the inflow to Querococha Lake, Q2, in the Querococha watershed.

Households engage in a variety of livelihood activities across both watersheds, but the principal sources of household subsistence are agriculture and livestock. Table 3 presents a comparative summary of household livelihoods across both case study communities. Households in Catac are more dependent on livestock production largely due to the fact that they have access to much more communal land (~66,000 ha) and that the lowest vertical range of the community (3,500 m above sea level) limits the production of key staple crops such as corn. Conversely, households in the Quilcay watershed engage in more agricultural, commercial, and manual labor activities because part of the community is located at a lower elevation (3,300 m above sea level) and due to the proximity of Huaraz, where many household members travel on a frequent basis to generate income.

Table 3. Major livelihood activities in the Querococha (QUER) and Quilcay (QUIL) case study watersheds

		Percentage of households	
Type of activity	Description and variety	QUER	QUIL
Livestock	Cattle, sheep, pigs, horses, guinea pigs, burros	85	78
Agriculture	Potatoes, olluco, oca, mashua, corn, wheat, barley, oats, quinua, beans, herbs	68	100
Agroforestry	Eucalyptus, pine, quenual (*Polylepis*), capuli (*Prunus serotina*), colle (*Buddleja coriacea*)	68	72
Commercial	Livestock, agricultural products, market, prepared food	45	78
Manual labor	Temporary agricultural labor, carpentry, manufacturing	38	40
Tourism	Guiding, animal caretaker, burro rental, cook, boat rental	25	27
Dairy products	Milk, cheese, eggs	15	9
Artisanal	Hand-spun fabrics, clothing, ceramics	15	3

Our evaluation of the ways in which household livelihood vulnerability is being affected by climate change and glacier recession demonstrates the combined ways in which new vectors of environmental and social change have begun to affect household activities and access to resources. Table 4 presents a summary of findings on household perceptions of these changes. Nearly all of the households in the case study communities are acutely aware of the pace of glacier recession that is taking place in nearby watersheds and frequently noted in great detail the accelerating rate and magnitude of recession that has been taking place over the past several decades. In addition, older respondents often presented detailed accounts of the extent of glacial coverage in the upper watersheds long before instrumental records were begun. Overall, households noted uniformly that recent climate change is already negatively affecting their lives. The key factors identified by respondents that are currently affecting household livelihood activities and pose significant challenges to their future trajectory are shifting water availability, increasing weather extremes, and threats to tourism.

Table 4. Climate change perception and new vectors of household vulnerability (95 percent confidence interval) in the Querococha (QUER) and Quilcay (QUIL) case study watersheds

	Percentage	
Perceptions and vectors	QUER	QUIL
Household perceptions of recent climate change		
Households reporting significant recession of nearby glaciers	100	94 ± 3.96
Households responding that they are very preoccupied by recent climate change taking place in the region	91 ± 4.3	84 ± 6.11
Households reporting that they have noted negative changes in their lives due to recent climate change	94 ± 3.66	88 ± 5.42
New vectors of vulnerability affecting households		
Households reporting that water supplies during the dry season have been decreasing	93 ± 3.61	81 ± 6.54
Households reporting significant changes in recent weather patterns that negatively affected crops or livestock	95 ± 3.61	100
Households reporting recent freezing or extreme precipitation events that negatively affected human health, crops, or livestock	91 ± 3.61	100
Respondents reporting that glacier recession already is or will negatively significantly affect tourism	100 (*n* = 27)	100 (*n* = 26)

Water constitutes one of the most important resources for households in the region for human consumption as well as agro-pastoral activities. Regular glacial discharge has historically provided perennial supplies of water to households with access to land resources in the case study communities, but according to case study respondents this temporal regularity is shifting due to increasing local and regional demands on water supplies and because the magnitude of water resources during the dry season is declining. Across both case study communities, more than 90 percent of respondents (QUER, 93 percent; QUIL, 94 percent) indicated that water supplies in the watersheds have been decreasing during the dry season. Although less than 5 percent of households in both communities argued that they did not have enough water for human consumption, roughly one quarter of households (QUER, 26 percent; QUIL, 22 percent) indicated that they did not have enough water for irrigation. Households reported that these changes have had significant impacts on agricultural and livestock productivity. Respondents in both communities also noted that many of the perennial and intermittent springs in the watersheds have begun to disappear. Households noted that diminishing spring flows in the upper reaches of the Querococha watershed are negatively affecting livestock productivity in terms of health and growth rates, and that the decline or disappearance of springs in the lower reaches of the Quilcay watershed would have grave consequences, as 90 percent of case study households noted that they depend on fewer than five springs across the entire area for all of their potable water resources. Interviewees also uniformly indicated that one major spring above the community (located at ~4,000 m) disappeared in the past several years and that this has already affected dry season water availability.

Another key vector of vulnerability affecting livelihood activities in the case study communities is increasing short-term weather variability. As Table 4 illustrates, households across both communities report that intense precipitation events, freezing events, strong winds, shifting rainfall patterns, and intense heat spells have all negatively affected household health, agricultural productivity, and livestock health. Households also noted increasing interannual or seasonal variability in weather events and temperatures such as longer freezing periods, more damaging freezing events, and shifts in crop planting or harvest periods. Respondents indicated that the cumulative effects of these increasing climatic extremes pose new risks for basic household food security, are the source of greater uncertainty about agricultural cycles, and are often responsible for substantial declines in crop productivity.

Finally, another important set of vectors affecting household livelihoods are the effects of new climate-induced shocks on tourism activities in the region. More than one quarter of households in both case study communities engage in the provision of tourism services. Because the high glaciated peaks are one of the primary reasons tourists visit the region, glacier recession constitutes a significant threat to tourism-related income for households. In fact, glacier recession has already had significant consequences for households in the vicinity of the Querococha watershed as the Pastoruri glacier, which was visited by more than 40,000 people in 2006, became the first glacier in the Cordillera

Blanca to be closed due to "adverse climatic conditions" and is likely to completely disappear by 2015 (INEI 2007).

Discussion

Glaciers are an integral component of the coupled natural–human systems of the tropical Central Andes and their rapid recession is transforming downstream hydrology. Our integrated research has innovatively quantified these changes and initiated new understanding of groundwater processes. Our isotope analyses show that the relative amount of glacier melt water is increasing in glacierized tributaries of the Santa River, reflecting an increased rate of glacier recession in all the case study watersheds. Historically, a significant increase in discharge was observed until 1983, after which time the trend is weakly negative. This inflection point in the discharge trend is suggestive of a systemic threshold in glacier response to climate forcing, whereby glacierized watersheds initially provide more discharge but then diminish in influence over time. In addition, temperature is closely correlated to discharge, reflecting a regional climate control. It is important to note that 1983 was a major El Niño year, and the role of El Niño southern oscillation has been identified as controlling discharge and glacier mass balance on multiyear intervals (Francou et al. 1995; Vuille, Kaser, and Juen 2008). Nevertheless, the overall decrease in discharge at La Balsa is not altered by these episodic influences. Our results demonstrate that groundwater is proportionately at least as important as glacier melt with respect to total current dry season streamflows, and as glaciers recede, the influence of groundwater, and its role as a seasonal buffer, will become increasingly important.

Livelihood vulnerability in our case study watersheds is also being significantly affected by recent glacier recession. Our results clearly show that households are acutely aware of these changes and that new vectors of vulnerability, including shifting water availability, increasing weather extremes, and threats to tourism, are affecting household access to resources. Respondents across both watersheds uniformly indicated that these factors are significantly affecting their current livelihood activities. Because our methodology was designed to limit biases through the use of a random sampling frame, and our total household sample constitutes a statistically significant representative population for both watersheds, our findings are highly suggestive that livelihood vulnerability has already increased significantly and that there is a compelling need to address these concerns.

Our transdisciplinary findings suggest that there is an intriguing scale-dependent discontinuity between household perceptions, on one hand, and what our physical measurements and models demonstrate on the other. Ongoing glacier melt in the Cordillera Blanca tributaries is accompanied by increasing discharge and relatively more glacier melt water contribution in streamflows from highly glacierized watersheds. However, households from communities situated in both of our glacierized case study watersheds indicate very strongly that water supplies are declining, a trend that is only clearly measured in the Santa River discharge from the entire Callejón de Huaylas. Although we are considering a number of possible alternative explanations, given the minimal hydrologic influence of glaciers on the entire Santa River basin (i.e., $<$ 10 percent glacier cover), we hypothesize that increasing basin-wide water withdrawals from human activities might be an important explanatory factor.

With this report we present an initial synthesis of results from two of three case study watersheds and establish the context for our ongoing research project. Exploring the way in which these observed patterns and hydrologic processes are currently interacting provides an important rationale for continued observations, model development, and testing and more intensive social research so that we might better integrate our transdisciplinary research with the project's social and scientific components and provide useful analyses to broader scientific and policy audiences. Finally, although this research focuses rather exclusively on human vulnerability, our larger research goals are intended to address household resilience and adaptive capacity as well as larger governance questions affecting the management of water resources throughout the Santa River watershed.

Acknowledgments

This research was funded by the National Science Foundation (NSF No. 0752175, BCS—Geography and Regional Science) and included a Research Experience for Undergraduates (REU) Supplement. Additional funding for the LIDAR flight came from the National Aeronautics and Space Administration (NASA No. NNX06AF11G), The National Geographic Society Committee for Research and Exploration, and the Ohio State University Climate, Water & Carbon

Program, and the Faculty Senate of the University of California, Santa Cruz. We acknowledge the cooperative assistance of Peruvian colleagues Ing. Ricardo J. Gomez, Ing. Marco Zapata, and others at the Unidad de Glaciología y Recursos Hídricos in Huaraz, Peru, and the following interview research assistants: Carlos Torres Beraun, Oscar Lazo Ita, Erlinda Marilu Pacpac, Jesus Yovana Castillo, and Gladys Jimenez. We recognize REU undergraduates Laurel Hunt, Sarah Knox, Galen Licht, Sara Reid, Michael Shoenfelt, Patrick Burns, Alyssa Singer, and Shawn Stone for fieldwork assistance. We also thank Kyung In Huh for assisting with LIDAR data display. This is Byrd Polar Research Center contribution number 1396.

References

Ames, A., S. Dolores, A. Valverde, P. Evangelista, D. Javier, W. Gavnini, J. Zuniga, and V. Gómez. 1989. *Glacier inventory of Peru, Part 1, 105*. Huaraz, Peru: Hidrandina.

Baraer, M., J. M. McKenzie, B. G. Mark, J. Bury, and S. Knox. 2009. Characterizing contributions of glacier melt and groundwater during the dry season in the Cordillera Blanca, Peru. *Advances in Geosciences* 22:41–49.

Barry, R. G., and A. Seimon. 2000. Research for mountain area development: Climatic fluctuations in the mountains of the Americas and their significance. *AMBIO: A Journal of the Human Environment* 29 (7): 364–70.

Bebbington, A. 2000. Reencountering development: Livelihood transitions and place transformations in the Andes. *Annals of the Association of American Geographers* 90 (3): 495–520.

Bebbington, A. J., and J. T. Bury. 2009. Institutional challenges for mining and sustainability in Peru. *Proceedings of the National Academy of Sciences* 106 (41): 17296–301.

Bradley, R. S., F. T. Keimig, H. F. Diaz, and D. R. Hardy. 2009. Recent changes in freezing level heights in the Tropics with implications for the deglacierization of high mountain regions. *Geophysical Research Letters* 36.

Burton, I., G. White, and R. Kates. 1978. *The environment as hazard*. New York: Oxford University Press.

Bury, J. 2004. Livelihoods in transition: Transnational gold mining operations and local change in Cajamarca, Peru. *The Geographical Journal* 170 (1): 78–91.

———. 2005. Mining mountains: Neoliberalism, land tenure, livelihoods, and the new Peruvian mining industry in Cajamarca. *Environment and Planning A* 37 (2): 221–39.

Bury, J., B. Mark, J. M. McKenzie, A. French, and M. Baraer. Forthcoming. Glacier recession and human vulnerability in the Yanamarey watershed of the Cordillera Blanca, Peru. *Climatic Change*.

Byers, A. C. 2000. Contemporary landscape change in the Huascarán National Park and buffer zone, Cordillera Blanca, Peru. *Mountain Research and Development* 20 (1): 52–63.

Carey, M. 2010. *In the shadow of melting glaciers: Climate change and Andean society*. New York: Oxford University Press.

Cutter, S. 2003. The vulnerability of science and the science of vulnerability. *Annals of the Association of American Geographers* 93:1–12.

Denevan, W. M. 1992. The pristine myth: The landscape of the Americas in 1492. *Annals of the Association of American Geographers* 82 (3): 369–85.

Eakin, H. C. 2006. *Weathering risk in rural Mexico: Climatic, institutional, and economic change*. Tucson: University of Arizona Press.

Francou, B., P. Ribstein, H. Semiond, C. Portocarrero, and A. Rodriguez. 1995. Balances de glaciares y clima en Bolivia y Peru: Impacto de los eventos ENSO [Climate and glacier balances in Bolivia and Peru: Impact of the ENSO event]. *Bulletin de l'Institut francais d'études Andines* 24 (3): 661–70.

Hewitt, K. 1983. *Interpretations of calamity from the viewpoint of human ecology*. Winchester, MA: Allen & Unwin.

INEI. 2007. *The 2007 national census: XI of population and VI of houses*. Lima, Peru: Institute of National Statistics and Information.

Intergovernmental Panel on Climate Change. 2007. Climate change 2007: Impacts, adaptation and vulnerability. In *Working Group II Contribution to the Intergovernmental Panel on Climate Change fourth assessment report*, ed. M. L. Parry, O. F. Canziani, J. P. Palutikof, P. J. v. d. Linden, and C. E. Hanson. Cambridge, UK: Cambridge University Press.

Juen, I., G. Kaser, and C. Georges. 2007. Modelling observed and future runoff from a glacierized tropical catchment (Cordillera Blanca, Perú). *Global and Planetary Change* 59 (1–4): 37–48.

Kaser, G., I. Juen, C. Georges, J. Gómez, and W. Tamayo. 2003. The impact of glaciers on the runoff and the reconstruction of mass balance history from hydrological data in the tropical Cordillera Blanca, Peru. *Journal of Hydrology* 282 (1–4): 130–44.

Knapp, G. W. 1991. *Andean ecology: Adaptive dynamics in Ecuador*. Boulder, CO: Westview Press.

Leichenko, R. M., and K. L. O'Brien. 2008. *Double exposure: Global environmental change in an era of globalization*. New York: Oxford University Press.

Liverman, D. 1990. Drought impacts in Mexico: Climate, agriculture, technology, and land tenure in Sonora and Puebla. *Annals of the Association of American Geographers* 80:49–72.

Mark, B. G. 2008. Tracing tropical Andean glaciers over space and time: Some lessons and transdisciplinary implications. *Global and Planetary Change* 60 (1–2): 101–14.

Mark, B. G., and J. M. McKenzie. 2007. Tracing increasing tropical Andean glacier melt with stable isotopes in water. *Environmental Science and Technology* 41 (20): 6955–60.

Mark, B. G., J. M. McKenzie, and J. Gómez. 2005. Hydrochemical evaluation of changing glacier meltwater contribution to stream discharge: Callejon de Huaylas, Peru. *Hydrological Sciences Journal/Journal des Sciences Hydrologiques* 50 (6): 975–87.

Mark, B. G., and G. O. Seltzer. 2003. Tropical glacier meltwater contribution to stream discharge: A case study in the Cordillera Blanca, Peru. *Journal of Glaciology* 49 (165): 271–81.

———. 2005. Evaluation of recent glacial recession in the Cordillera Blanca, Peru (AD 1962–1999): Spatial distribution of mass loss and climatic forcing. *Quaternary Science Reviews* 24:2265–80.

Ministry of Energy and Mines. 2008. *Electrical statistics 2008*. Lima, Peru: Ministry of Energy and Mines.

Polsky, C. 2004. Putting space and time in Ricardian climate change impact studies: Agriculture in the US Great Plains, 1969–1992. *Annals of the Association of American Geographers* 94 (3): 549–64.

Postigo, J. C., K. R. Young, and K. A. Crews. 2008. Change and continuity in a pastoralist community in the high Peruvian Andes. *Human Ecology* 36 (4): 535–51.

Racoviteanu, A., Y. Arnaud, and M. Williams. 2008. Decadal changes in glacier parameters in Cordillera Blanca, Peru derived from remote sensing. *Journal of Glaciology* 54 (186): 499–510.

Sauer, C. O. 1941. Foreword to historical geography. *Annals of the Association of American Geographers* 31 (1): 1–24.

Turner, B., R. Kasperson, P. Matson, J. McCarthy, R. Corell, L. Christensen, N. Eckley, J. Kasperson, A. Luers, and M. Martello. 2003. A framework for vulnerability analysis in sustainability science. *Proceedings of the National Academy of Sciences* 100 (14): 8074–79.

Viviroli, D., H. H. Durr, B. Messerli, M. Meybeck, and R. Weingartner. 2007. Mountains of the world, water towers for humanity: Typology, mapping, and global significance. *Water Resources Research* 43 (7): 13.

Vuille, M., B. Francou, P. Wagnon, I. Juen, G. Kaser, B. G. Mark, and R. S. Bradley. 2008. Climate change and tropical Andean glaciers: Past, present and future. *Earth Science Reviews* 89 (3–4): 79–96.

Vuille, M., G. Kaser, and I. Juen. 2008. Glacier mass balance variability in the Cordillera Blanca, Peru and its relationship with climate and the large-scale circulation. *Global and Planetary Change* 62 (1–2): 14–28.

Watts, M. J., and H. G. Bohle. 1993. The space of vulnerability: The causal structure of hunger and famine. *Progress in Human Geography* 17 (1): 43.

White, G. F. 1942. *Human adjustment to floods*. Chicago: University of Chicago Press.

Young, K., and J. Lipton. 2006. Adaptive governance and climate change in the tropical highlands of western South America. *Climatic Change* 78 (1): 63–102.

Zimmerer, K. S. 1991. Wetland production and smallholder persistence—Agricultural change in a highland Peruvian Region. *Annals of the Association of American Geographers* 81 (3): 443–63.

Climate–Streamflow Linkages in the North-Central Rocky Mountains: Implications for a Changing Climate

Erika K. Wise

Water, already a scarce resource in the semiarid Western United States, has become increasingly threatened due to population growth pressures, natural climate variability, and the prospect of future climate change. Water managers are challenged by uncertainties concerning how climate variability and change interact with water supply at the regional level. This study aims to increase predictive capacity in one of the largest river systems in the Western United States, the Snake River, by identifying key atmosphere–ocean controls on streamflow and assessing how projected climate changes will translate into future water supply in the river. In contrast to previous research suggesting the Snake headwaters are in an area of weak teleconnection influence, the results of this study indicate that the region exhibits a response to El Niño-Southern Oscillation (ENSO) that is similar to the Pacific Northwest and strongly modulated by northern Pacific conditions. Although current projections of how dynamic features like ENSO might respond to climate change are highly uncertain, the lagged nature of the relationship between Pacific Ocean conditions and streamflow will be useful for near-term planning. The region was found to be sensitive to shifts in the winter westerly storm track system, with extreme low-flow years associated with a north-shifted storm track. Precipitation increases have been projected for the region, but other predicted consequences of climate change—including a poleward shift in storm track position, changing seasonality, and reduced snowpack—suggest an increased likelihood of future drought conditions. *Key Words: climate change, drought, Snake River, teleconnections, water resources.*

在半干旱的美国西部，水已经成为一种稀缺的资源，由于人口增长的压力，自然气候变化，以及未来气候变化的前景，水资源的匮乏已经构成日趋严重的威胁。水资源管理人员的挑战是如何面临在区域一级处理气候变异与变化和供水之间的相互作用所造成的不确定性。蛇河是美国西部最大的河流系统之一，通过确定那些大气海洋对径流的关键控制因素，并且分析所预测的气候变化是如何影响该河流未来的水供给，本研究的目的是增加对该河流的预测能力。以前的研究表明蛇河的源头处于远距联系影响较弱的地区，与此相对，本项研究的结果表明，该地区呈现出对厄尔尼诺一南方振荡现象（ENSO）的对应反应，与西北太平洋地区相似，受到北太平洋状况的强烈制约。至于动态的因素，例如 ENSO，是如何应对气候变化的，虽然目前的预测还处于高度不确定，太平洋状况和径流流量之间的滞后关系特性对于近期规划是有益的。我们发现该地区对冬季西向风暴的轨迹系统的变化比较敏感，当风暴轨迹北向变动时，伴随有极端的低流量年份。我们预测了该地区降水量的增加，但是，气候变化的其它预测后果，包括风暴轨迹位置向极地的转变，季节性变化，以及积雪的减少，都表明未来的干旱情况有增加的可能性。*关键词：气候变化，干旱，蛇河，远距联系，水资源。*

El agua, un recurso de por sí escaso en los semiáridos Estados Unidos occidentales, ha llegado a estar crecientemente amenazada por las presiones del crecimiento demográfico, variabilidad natural del clima y por la perspectiva de cambio climático futuro. Quienes administran este recurso enfrentan el reto de incertidumbres sobre la manera como interactúan la variabilidad climática y el cambio con el suministro de agua a nivel regional. Este estudio está orientado a mejorar la capacidad predictiva en uno de los sistemas fluviales más grandes del oeste de los Estados Unidos, el Río Snake, por medio de la identificación de controles críticos de atmósfera-océano sobre el flujo de las corrientes y la valoración de cómo los cambios climáticos proyectados se traducirán en la provisión futura de agua del río. En contraste con investigaciones anteriores que sugieren que las fuentes del Snake están en un área de débil influencia de teleconexión, los resultados del presente estudio indican que la región muestra una respuesta a El Niño-Oscilación del Sur (ENSO) que es similar a la experimentada en el Pacífico del noroeste y fuertemente modulada por las condiciones del Pacífico Norte. Aunque las proyecciones actuales sobre el modo como podrían reaccionar al cambio climático rasgos dinámicos por el estilo de ENSO son altamente inciertas, el retraso natural como se presentan las relaciones entre las condiciones del Océano Pacífico y el caudal de los ríos será útil para la planificación a plazo corto. Se encontró que la región es sensible a los cambios que se presentan en el sistema de tormentas invernales oestes, con años extremos de flujo bajo asociados con un desplazamiento hacia el norte de la ruta de las tormentas. Los incrementos de las precipitaciones se han proyectado para la región, pero otras consecuencias pronosticadas del cambio climático—incluyendo el cambio de posición de la ruta de tormentas en dirección polar, que cambia estacionalmente, y disminución de la cubierta de nieve—sugieren un incremento en la probabilidad de futuras condiciones de sequía. *Palabras clave: cambio climático, sequía, Río Snake, teleconexiones, recursos hídricos.*

The Snake River, like many river systems in the Western United States (the West), is dependent on snowmelt, perturbed by natural climate variability, and threatened by population growth pressures and the prospect of future climate change. With twenty-five dams along its 1,600-km length, the Snake River is one of the largest and most intensively managed rivers in the West. Mostly contained within Idaho (Figure 1), the majority of Snake River discharge has, historically, been used for agriculture (Marston et al. 2005). Like many Western rivers, though, the Snake has been facing increased demand from a variety of municipal, hydropower, and instream (recreational and wildlife) users (Miller et al. 2003; McGuire et al. 2006). As the largest tributary of the Columbia River, which is challenged by growing demands from many of the same sectors, the Snake River is also critical for Northwest cities and other users further downstream.

Water planning for the Snake River's water resources is hampered by the region's low hydroclimatic predictive capacity. In the West, decadal-scale fluctuations in atmospheric and oceanic conditions are important controls on water supply, accounting for 20 to 45 percent of annual precipitation variance (Cayan et al. 1998). A major feature of both interannual and decadal hydroclimatic variability in the West is a north–south pattern of precipitation variance (Dettinger et al. 1998), referred to as a *dipole*. The dipole pattern is centered on the Pacific Northwest and the Desert Southwest (Figure 1), which tend to behave in opposition to each other (i.e., when one is wet, the other is dry) and have opposite reactions to broad-scale atmosphere–ocean controls such as the El Niño-Southern Oscillation (ENSO) system (Redmond and Koch 1991). The Snake River Basin, located between these regions, is affected by several different winter storm patterns and tends to exhibit weaker and more complex connections to seasonal mean atmospheric circulation than other parts of the West.

There is a strong, lagged relationship between summer–fall ENSO conditions and winter precipitation in the West that has allowed for long lead-time forecasts in areas near the dipole centers of opposite association (Redmond and Koch 1991). Although the effects of Pacific Ocean conditions are well delineated in these centers, impacts are not as well understood in the western Wyoming region that contains the Snake headwaters, which earlier work indicated is at a latitudinal and longitudinal transition zone of precipitation patterns and teleconnection influence (Mock 1996; Cayan et al. 1998). This has led to low predictive capacity for water resources in this region.

Climate change projections add another layer of complexity to water management in the Snake River. Although scientists agree that the Earth will warm, precipitation forecasts have been less certain (Intergovernmental Panel on Climate Change [IPCC] 2007). Rising temperatures are expected to increase atmospheric moisture transport and convergence, resulting in an increase in annual precipitation in most parts of the United States north of approximately 40°N (IPCC 2007), which includes the Snake River Basin. However, it is not clear how this will translate to water resources when combined with other projected changes. This is particularly true for rivers such as the Snake that are fed by snowmelt (Figure 2), as snow variability is dependent on both temperature and precipitation conditions and influenced by seasonality changes. Temperature and precipitation are also both factors in the development of severe and sustained droughts, which are of major concern in this region. Available paleoclimate data suggest that there have been more extreme and long-lasting droughts in the preinstrumental period than what has been experienced in the past century (Wise 2009), and projections from global circulation models indicate that midlatitude continental interiors will experience increased summer dryness and probability of drought under global warming scenarios (IPCC 2007; Seager et al. 2007).

Interannual to multidecadal anomaly patterns in precipitation, temperature, and snowpack have their roots in circulation pattern changes, such as the interruption of zonal flow by strong and persistent meridional flow, leading to anomalous storm track positions (Carrera, Higgins, and Kousky 2004). These synoptic-scale circulation patterns and storm track positions are in turn modified by larger scale features such as ENSO (Cayan 1996; Cayan, Redmond, and Riddle 1999). The goal of this study is to improve near- and long-term water supply forecasting for the Snake River by addressing the following research questions:

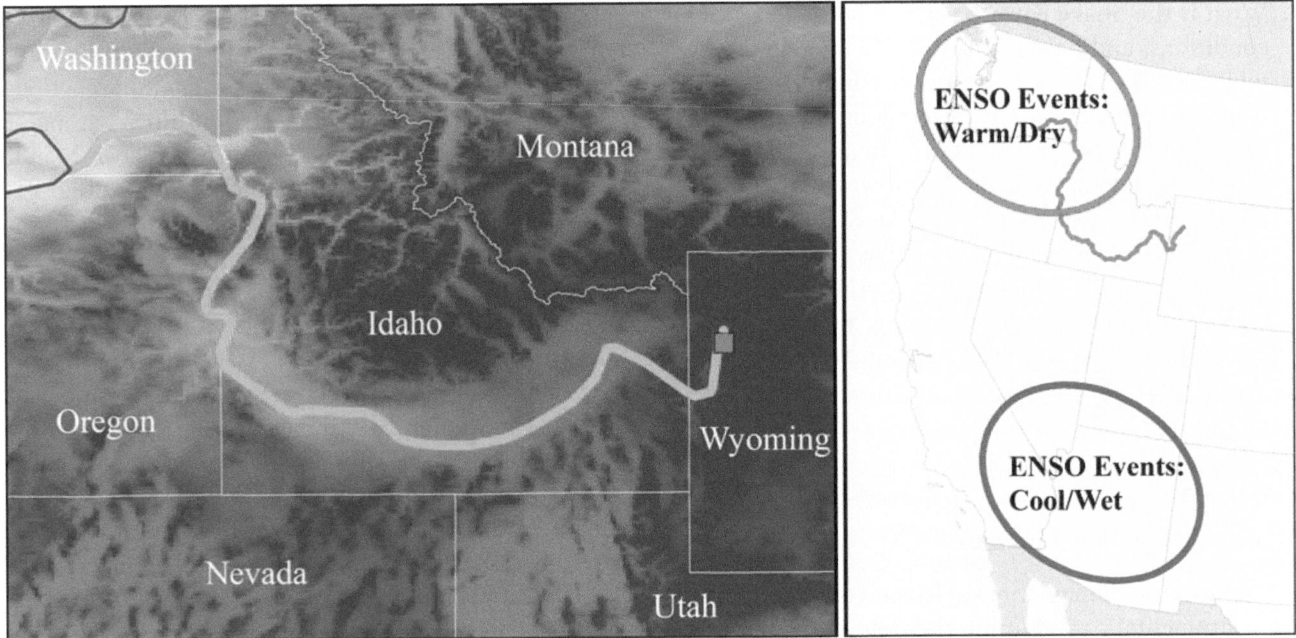

Figure 1. Climate overview map at right shows Snake River (cyan line) in the context of the West's climatic centers of opposite association. Regional map at left displays the Snake River (thick cyan line) and Jackson Lake Dam gauge (red square). Snake River flow is to the west, feeding into the Columbia River (thin blue line). ENSO = El Niño-Southern Oscillation.

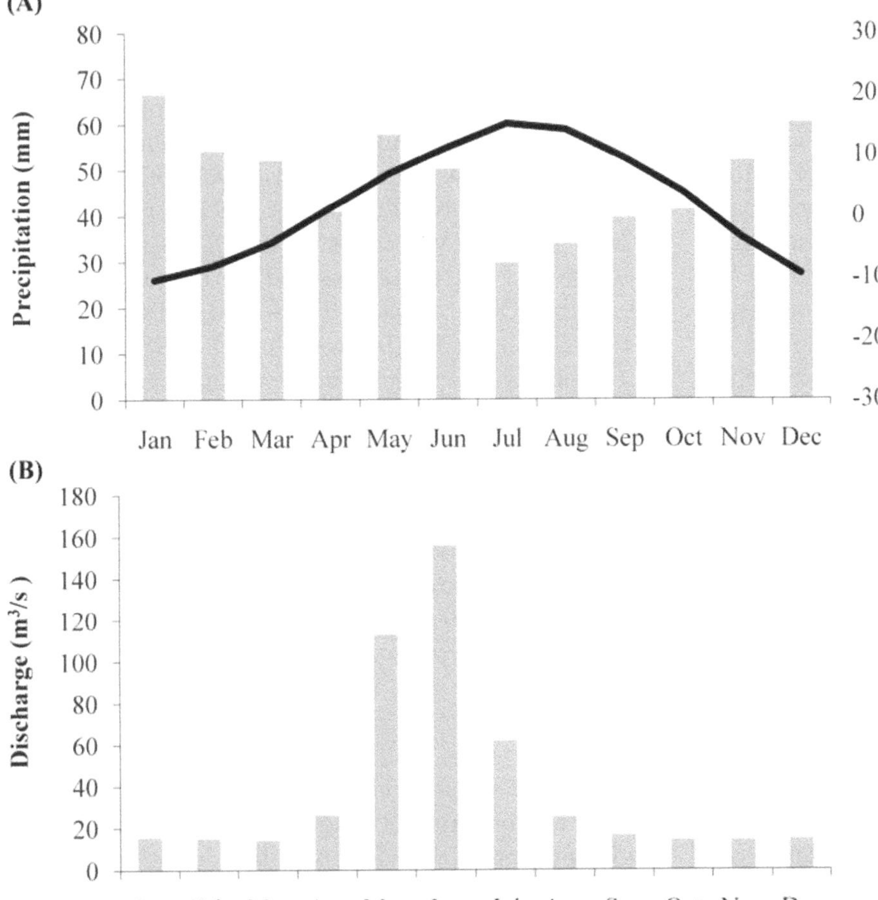

Figure 2. (A) Streamflow in the Snake River is dependent on snowmelt occurring in the headwaters region, located within Wyoming climate division 2. As seen in this climograph for that climate division, the majority (56 percent) of precipitation falls during the cool season (October–March). Bars show monthly average precipitation and line shows monthly average temperature, for 1896 through 2006. (B) The importance of snowmelt to Snake River flow can be seen in the average monthly discharge (1910–2008) recorded at the Jackson Lake Dam streamflow gauge in the Snake River headwaters region, where a snowmelt pulse contributes approximately 68 percent of yearly streamflow in May through July.

1. What is the Snake River Basin's response to ENSO conditions, and how is this response modulated by multidecadal oscillations in the northern Pacific?
2. Can the lagged relationship between summer–fall ENSO and winter precipitation, often used as a predictive tool in the Pacific Northwest and Desert Southwest, be applied in the Snake River region?
3. Is Snake River streamflow sensitive to storm track trajectories of the midlatitude westerlies?
4. Given the results from the preceding questions, what are the likely effects of projected climate changes on Snake River streamflow?

Methods

Climate–Streamflow Linkages

Analyses of climate–Snake River connections in the instrumental period utilize the following data sets: (1) monthly values of the Southern Oscillation Index (SOI) and the Pacific Decadal Oscillation (PDO) from the University of East Anglia's Climatic Research Unit and the University of Washington's Joint Institute for the Study of the Atmosphere and Ocean, respectively; (2) gridded climate data from the National Center for Environmental Prediction (NCEP)/National Center for Atmospheric Research (NCAR) Reanalysis data set (http://www.esrl.noaa.gov/psd) and the Parameter-elevation Regressions on Independent Slopes Model (PRISM) data set (http://prism.oregonstate.edu); and (3) naturalized streamflow records for the Snake River at Jackson Lake Dam (http://www.usbr.gov/pn/hydromet). Naturalized streamflow records are those that have been numerically adjusted to remove the effects of reservoirs, dams, diversions, and other human impacts to estimate the unimpaired flow hydrology (Wurbs 2006). Figure 3 displays smoothed time series of Snake River streamflow, SOI, and PDO.

Tropical Pacific ENSO conditions are considered one of the primary controls on global interannual climate variability (Mantua and Hare 2002). SOI, which is based on sea-level pressure differences between Tahiti and Australia and represents the atmospheric component of the ENSO cycle, is strongly linked to both precipitation and streamflow variability in the West (Redmond and Koch 1991; Cayan, Redmond, and Riddle 1999). Negative SOI values represent relatively low air pressure in the eastern Pacific and typically correspond to El Niño conditions. Conversely, positive SOI values represent above-average pressure in the eastern Pacific and are normally associated with La Niña episodes. The Northwest is generally positively correlated with SOI: Conditions tend to be drier than average in winters following summer and autumn periods when SOI is negative. This relationship is strongest between June–November SOI and October–March precipitation in the West (Redmond and Koch 1991; Wise 2010), and analyses in this article utilize those sets of months. I divided the SOI data into SOI+, SOI neutral, and SOI − categories based on the upper, middle, and lower terciles, respectively, of SOI values. When comparing to Snake River flow, I evaluate the streamflow in water-year y (which encompasses October of year $y - 1$ through September of year y) with June–November SOI values in year $y - 1$. For example, Snake River streamflow from October 1998 through September 1999 (referred to as water-year 1999) would be evaluated against SOI values from June to November 1998.

ENSO forecasts have not been highly utilized in the Snake River region, in part due to their seemingly weak predictive skill in this area (Clark, Serreze, and McCabe 2001). Northern Pacific Ocean conditions can modify the ENSO–precipitation relationship (Gershunov and Barnett 1998) and might have a larger impact than ENSO on Western snowpack variability (and consequently water resources) due to impacts on temperature conditions and winter storm and precipitation regimes (McCabe and Dettinger 2002). Northern Pacific conditions are analyzed in this study using the PDO, a multidecadal cycle of sea-surface temperature variability in the Pacific poleward of 20°N latitude (Mantua et al. 1997). There are three complete PDO phases covered by the instrumental record: PDO+ (warm) 1926–1943, PDO− (cool) 1944–1976, and PDO+ (warm) 1977–1998.

Monthly PRISM precipitation data, available at a 4 km grid from 1895 through 2007, were used to examine spatially the impacts of SOI and PDO on hydroclimatic conditions in the study area. Data preparation included reducing the resolution of the grid to 8 km by removing every other grid point and restricting the data to the 1926 through 1998 time period of complete PDO cycles. I converted these data to monthly standardized values using

$$Z = (X - \mu)/\sigma \qquad (1)$$

where X is the individual month's total precipitation (e.g., for January 1995) and μ and σ are the mean total precipitation and the standard deviation, respectively, for that month over the entire time period (e.g., for every January from 1926–1998). This procedure was

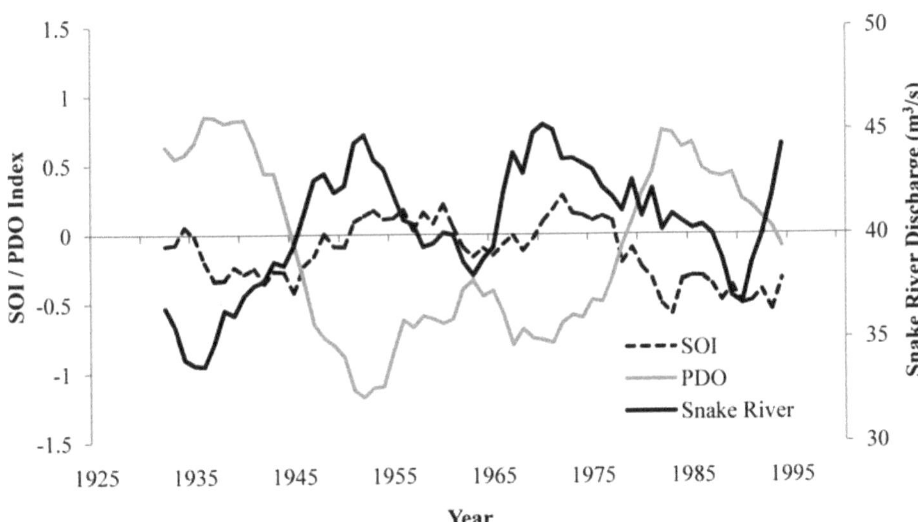

Figure 3. Smoothed time series (eleven-year moving averages) comparing Snake River streamflow (black line; annual water-year average, m^3/s), SOI (dashed line; June–November average, unitless index), and PDO (gray line; December–February average, unitless index). SOI = Southern Oscillation Index; PDO = Pacific Decadal Oscillation.

implemented on each grid cell to calculate precipitation anomalies for SOI phases and SOI/PDO phase combinations. I also calculated average Snake River discharge at the Jackson Lake Dam gauge for each of these phase combinations and conducted an analysis of variance to determine the significance of differences between groups (Table 1).

Analysis of pressure patterns and storm track position during high- and low-flow years allows a better understanding of synoptic climate controls on Snake River flow. Gridded daily values of geopotential heights at 850 hPa, obtained from the NCEP/NCAR Reanalysis data set, are used here to reflect surface conditions. These gridded data (2.5° × 2.5°) span the time period from 1948 to the present. I averaged geopotential height data for 1948 through 2008 over sets of months to create seasonal means. The analyses presented here are based on cool-season (October–March) averages, as this is the period of snow accumulation that is a vital component of total annual streamflow. I determined the highest and lowest flow years in the Snake River during the time period from 1948 through 2007, based on the 75 percent and 25 percent quartiles, respectively, and calculated separate geopotential height averages for these data sets.

Teleconnections and synoptic-scale atmospheric circulation patterns impact streamflow through their influence on storm track position, which delivers cool-season moisture to the West. The importance of storm track position necessitated its calculation and comparison with Snake River streamflow in this study. Storm track position can be calculated using a number of different methods that measure maxima in eddy quantities (Chang and Fu 2002). For this study, following the method outlined by Quadrelli and Wallace (2002) and McAfee and Russell (2008), I calculated storm track position based on the seasonal maximum variance in the meridional component of wind following the application of a first-difference filter on the daily data. I based my analyses on cool-season (October–March) maximum variance in gridded daily wind data from the NCEP/NCAR Reanalysis data set at 300 hPa, which represent conditions in the upper atmosphere free from topographic effects. After determining the sets of years comprising the lowest and highest quartiles of Snake River flow over the 1948 through 2008 time period, I calculated the average storm track position for those years to determine deviation from the mean position.

Table 1. Snake River discharge by SOI phase and SOI/PDO phase combinations

Phase	*N*	Discharge	*F*	*P*
SOI+	24	45.68	4.92	0.01
SOI neutral	25	40.18		
SOI–	24	36.95		
SOI+/PDO–	13	47.49	3.55	0.02
SOI+/PDO+	11	43.55		
SOI–/PDO–	11	38.06		
SOI–/PDO+	13	36.02		

Note: N is the number of observations in each category. SOI = Southern Oscillation Index; PDO = Pacific Decadal Oscillation. Discharge is the mean water-year discharge of the Snake River at Jackson Lake Dam in m^3/s, averaged by phase. The *F* statistic was calculated through an analysis of variance between the three SOI phases and the four SOI/PDO phase combinations, and the *p* value represents the significance of the difference between groups based on the *F* statistic.

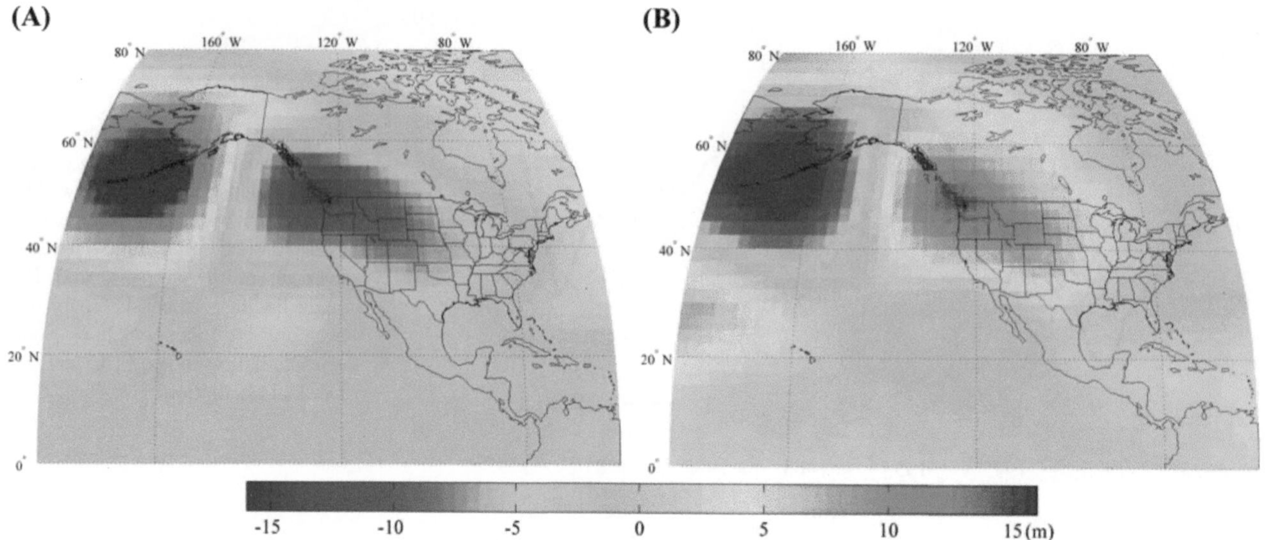

Figure 4. Composite anomalies of cool season (October–March) 850 hPa geopotential heights during (A) the upper quartile (highest flow) years and (B) the bottom quartile (lowest flow) years in the Snake River at Jackson Lake Dam, 1948–2007.

Future Climate Projections

The climate projections of precipitation rate (mm/day) and surface air temperature (°C) used here are from ten climate models for emissions scenario A1B (a "middle" emissions path) for two time periods (1950–1999 and 2050–2099). These data, based on climate model projections from the World Climate Research Programme's Coupled Model Intercomparison Project phase 3 (CMIP3) multimodel data set, were downscaled to a 0.5° grid through a joint Lawrence Livermore National Laboratory/Reclamation/Santa Clara University project (gdo-dcp.ucllnl.org). For this study, I averaged the output from all ten models to create a mean ensemble projection for each time period. I then calculated the difference between 2050–2099 projections and the 1950–1999 baseline.

Results and Discussion

Climate–Streamflow Linkages

Understanding the synoptic climate conditions and controlling processes that result in above- or below-average streamflow can improve the predictive capacity of a water supply system and help plan for the impacts of future climate change. Results of this research indicate that Snake River streamflow is strongly linked to changes in synoptic circulation patterns. In high streamflow years, the northern half of the Western United States, as well as western Canada, lies under an anomalously low pressure center (Figure 4A). The lowest streamflow years are characterized by a strong Aleutian low off the Alaskan coast and abnormally high pressure over the Northwest (Figure 4B). The geopotential height pattern seen in Figure 4B is characteristic of a positive Pacific North American (PNA) pattern, when a blocking high forces meridional flow across the United States (Leathers, Yarnal, and Palecki 1991). The resulting shift in storm systems causes incoming moist flow to miss the interior West, leading to decreased precipitation, snowpack, and streamflow.

The positive PNA pressure pattern, characterized by below-average geopotential heights near the Aleutian Islands and above-average heights over the Intermountain West, often occurs during SOI− and PDO+ phases (Figure 5), and these phases are associated with the lowest flow years of the Snake River record (Figure 3, Figures 6–7, and Table 1). The broader Snake River basin borders the transition between the north and south sides of the West's climatic dipole, but the headwaters region in western Wyoming has a strong Northwest-type response, with below-average precipitation in SOI− years (Figure 6). Based on the analysis of variance, the streamflow differences between SOI phases are significant at the 0.01 level (Table 1).

The SOI response has been modulated by northern Pacific conditions associated with the PDO over the past century (Figure 7). The division of years into SOI/PDO phase combinations results in low sample sizes (eleven to thirteen years per group), although the difference in streamflow between groups is still

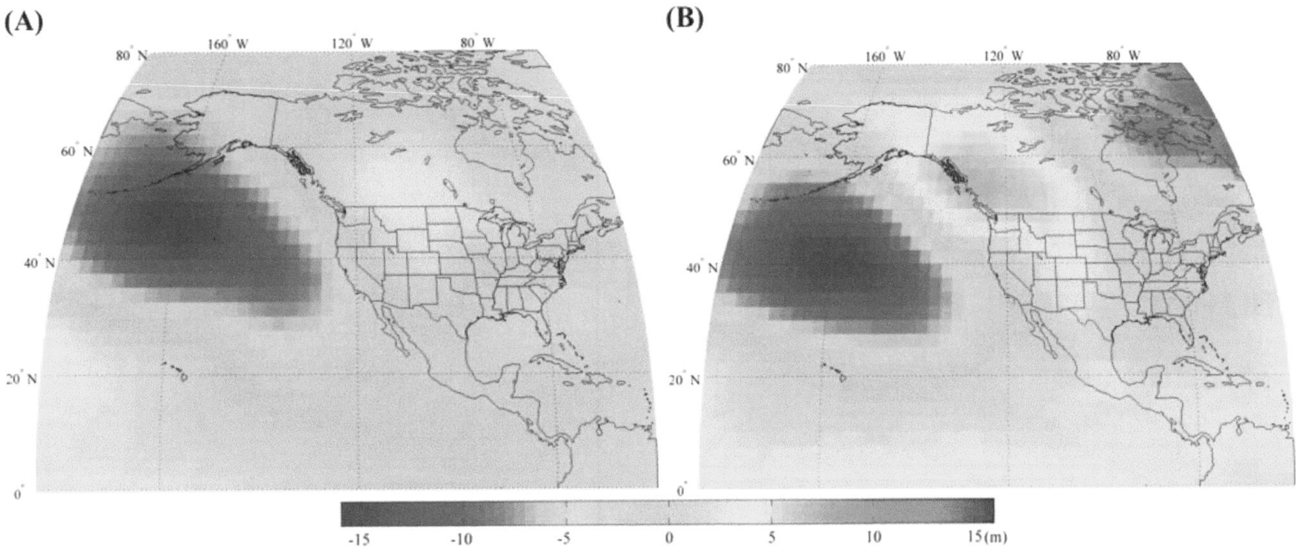

Figure 5. Average anomalies in cool season (October–March) 850 hPa geopotential heights following (A) June–November SOI– years, 1948–2007, and (B) during the most recent PDO+ phase, 1977–1998. SOI = Southern Oscillation Index; PDO = Pacific Decadal Oscillation.

significant ($p < 0.05$; Table 1). Streamflow was below average (i.e., water-year average < 40.92 m^3/s) in SOI– years during both PDO+ and PDO– phases. Consistent with previous research (e.g., Gershunov and Barnett 1998), the dipole response is stronger during SOI–/PDO+ and SOI+/PDO– phases. Streamflow was below average in just slightly over half (55 percent) of the years during SOI–/PDO– (compared to nearly 70 percent when SOI–/PDO+), and average discharge was higher in SOI–/PDO– than SOI–/PDO+ (38.06 m^3/s versus 36.02 m^3/s). Similarly, flow was above average in SOI+ years in both PDO phases but was higher (47.49 m^3/s versus 43.55 m^3/s) when SOI+/PDO–. It is particularly noteworthy that there was above-average flow in 77 percent of the SOI+/PDO– years but only in 36 percent of SOI+/PDO+ years. In the other three phase combinations, mean and median discharge values are similar, but median flow in SOI+/PDO+ years is actually below average (38.74 m^3/s). There were several very high-flow years, including an extreme outlier in the late 1990s of 73.14 m^3/s, which led to elevated flow when values were averaged over all SOI+/PDO+ years.

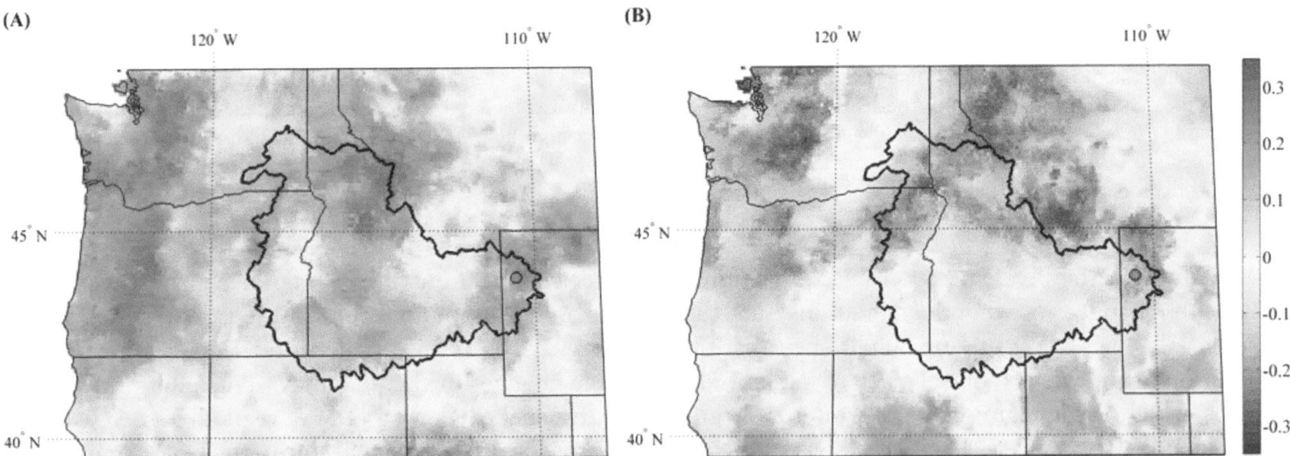

Figure 6. Average standardized October–March precipitation anomalies following June–November seasons with (A) SOI+ and (B) SOI– over the 1926–1998 time period. The Snake River Basin is outlined in black, and the location of the Jackson Lake Dam gauge is indicated by a red circle. SOI = Southern Oscillation Index.

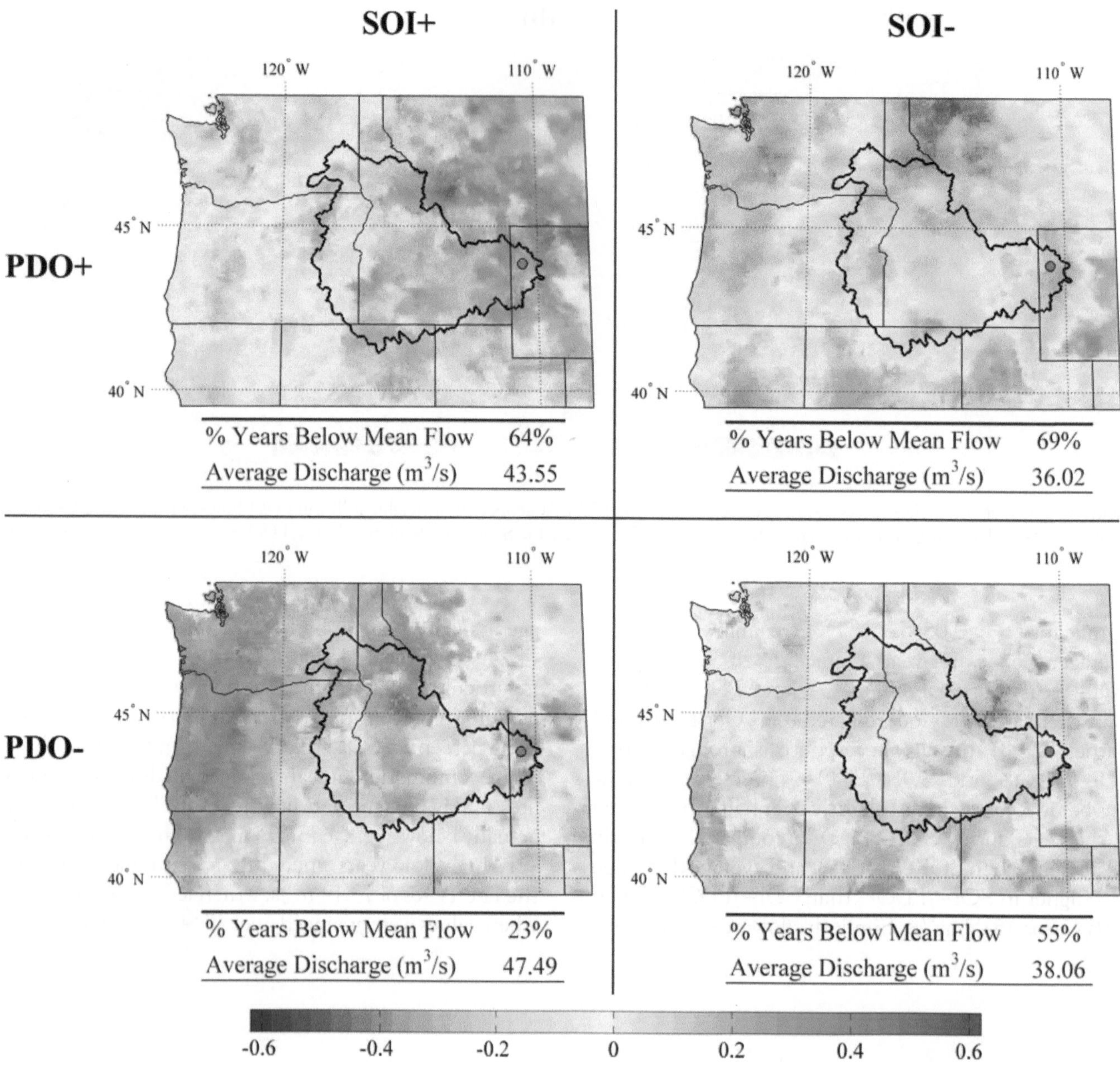

Figure 7. Standardized October–March precipitation anomalies by SOI/PDO phase over the 1926–1998 time period. Snake River Basin is outlined in black, and the location of the Jackson Lake Dam gauge is indicated by a red circle. Each corresponding table indicates the percentage of years in that category with below-mean Snake River streamflow and the average water-year mean discharge for years in that category. Mean water-year discharge for the entire 1926–1998 time period is 40.92 m^3/s. SOI = Southern Oscillation Index; PDO = Pacific Decadal Oscillation.

Oscillations in the PDO and other teleconnections, along with changes in synoptic conditions such as geopotential height, affect streamflow by altering the position of the westerly storm track of midlatitude cyclones, which is the main source of winter moisture in areas west of the Rocky Mountains. Maps showing the average spatial positioning of the winter storm track during years corresponding to the highest and lowest quartiles of Snake River discharge demonstrate this direct connection (Figure 8). The Snake River Basin sits on the latitudinal edge of the typical winter storm track. A slight shift in storm track position can cause large changes in the amount of precipitation received in the Snake headwaters. Wet years are associated with a more southerly, zonal storm track position across the United States (Figure 8A). The average storm track shifted northward over the northwestern United States during the lowest streamflow years, causing the

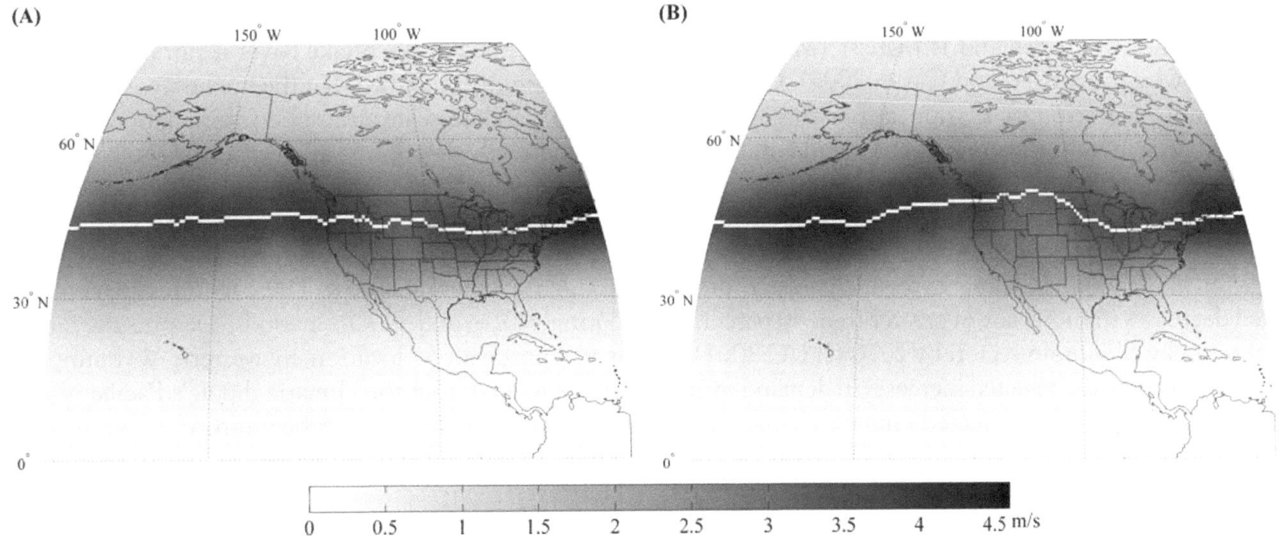

Figure 8. Cool season (October–March) storm track position during (A) the upper quartile (highest flow) years and (B) the bottom quartile (lowest flow) years measured in the Snake River at Jackson Lake Dam. White line signifies the line of maximum variance in 300 hPa meridional wind.

moisture delivery system to effectively miss the Snake River Basin and diminishing the mountain snowpack needed to feed the headwaters (Figure 8B).

Future Climate Projections

Models project warming temperatures across the Western United States for the upcoming century, with greater warming during the cool season and at higher elevations (IPCC 2007). There are many uncertainties concerning moisture-related changes, largely related to current models' limited abilities to represent the large-scale atmospheric and oceanic controls that significantly impact Snake River hydroclimatic variability. Projections have indicated that northern areas of the United States will experience increased annual precipitation (IPCC 2007). The downscaled climate model results show no change or slightly decreased precipitation in the warm season in the Snake River Basin (Figure 9). The models project increased precipitation in the cool season, particularly at higher elevations. Annually, the net effect of these seasonal changes is a small precipitation increase at higher elevations and little to no change elsewhere.

These modest high-elevation precipitation increases seem unlikely to offset other projected climate changes in the basin. Shifts in seasonality (e.g., earlier spring onset) are expected to impact the form and timing of

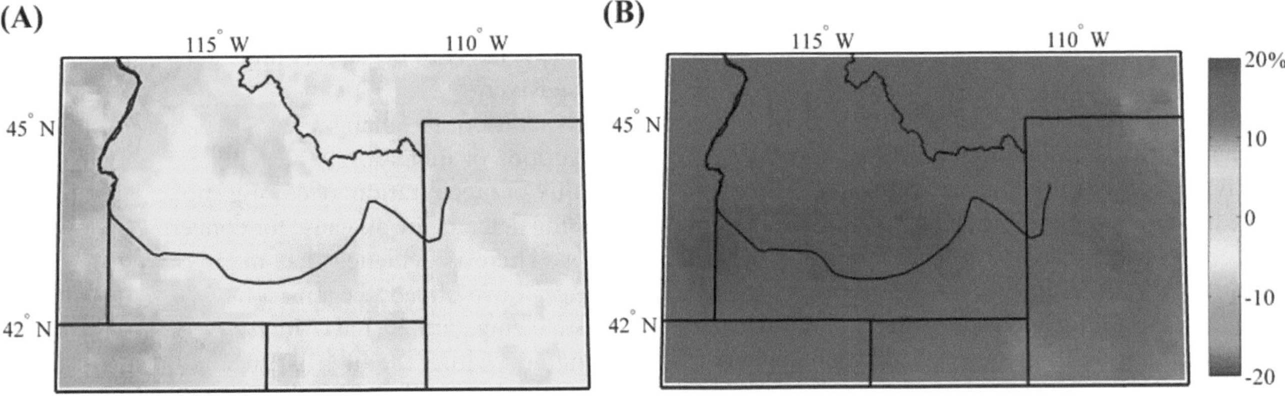

Figure 9. Downscaled climate projections of percentage difference in 2050–2099 precipitation versus a 1950–1999 baseline for the Snake River region during (A) the warm season (April–September) and (B) the cool season (October–March). Snake River shown with thick, black line.

moisture received in the region, reducing water supply in the summer when demand is highest (Miles et al. 2000; Knowles, Dettinger, and Cayan 2006). Changes in the form of precipitation (rain rather than snow) and increases in rain-on-snow events could increase episodes of heavy runoff and contribute to flooding in the river basin (IPCC 2007); increases in extreme precipitation events have also been projected (Diffenbaugh et al. 2005). Modeling experiments have projected decreases in snowpack of greater than 50 percent for the Rocky Mountain region by 2070 (IPCC 2007). Taken together with expected increases in demand over the next century, these projected climate changes are likely to exacerbate the strain on water resources.

Climate change is also expected to impact larger scale circulation patterns. Climate models have consistently projected a poleward shift in the storm track, due primarily to an increase in the height of the tropopause (Yin 2005; Lorenz and DeWeaver 2007). A northward shift in storm track position has already been detected in the West in the late winter and early spring (McAfee and Russell 2008). Other projected changes over the next century include an intensification of the storm track and a strengthening of the Aleutian low (Yin 2005; Salathé 2006). These projected changes are similar to the atmospheric conditions that have negatively impacted Snake River streamflow in the past: As shown here, a strong Aleutian low with meridional flow over the West and a northward shift in the storm track affects the Snake River watershed by shifting the moisture delivery system away from the headwaters region. A quasi-permanent poleward shift seems likely to result in decreased Snake River streamflow.

Conclusions

This study aimed to improve predictive capacity in the Snake River basin through better understanding of climate–streamflow linkages. Increased knowledge of the driving forces behind the variability in a system can enable better water planning and help mitigate drought and flood impacts. Despite year-to-year variability in response, data on basin-specific connections between streamflow variability and interannual to multidecadal Pacific Ocean conditions are valuable for improving probabilistic near-term forecasts. The lagged relationship between tropical Pacific conditions and winter precipitation is currently one of the best predictive tools available for regions in the West with a strong ENSO response. The results of this study indicate that this tool can be applied in the Snake River region.

This study has shown that the Snake River is strongly linked to broad-scale atmospheric circulation patterns. Distinctive high-pressure patterns over the Northwest United States and Canada and a strongly developed Aleutian low are features that have been present during years of low Snake River streamflow over the period of the instrumental record. These features are common during SOI− and PDO+ periods. Despite the Snake River headwaters' location in western Wyoming, an area removed from the climatic dipole's Pacific Northwest center of opposite association, the region has a strong response to ENSO signals similar to that of the Northwest.

Recognition of this connection between tropical Pacific conditions and streamflow can improve long lead-time water supply forecasting in the Snake River basin. However, this response is modulated by northern Pacific Ocean conditions. Below-average flow typically occurred when SOI−/PDO+, and the region was more reliably wet when SOI+/PDO−. Several extreme high-flow events occurred when SOI+ and PDO+, but the majority of years in this phase combination had below-average streamflow. Although highly persistent, shifts in northern Pacific conditions are currently not predictable. The ability to project these shifts would greatly improve multidecadal water supply forecasting in the Snake River Basin.

Results of this study indicate that Snake River streamflow is sensitive to winter storm track position. The Snake River Basin lies on the edge of the average storm track location, and slight shifts can significantly impact the amount of moisture delivery to the system. Low-flow years during the instrumental period were associated with a north-shifted storm track. The projected poleward shift in the storm track over the upcoming century is likely to negatively impact streamflow in the Snake River.

In addition to changes in storm track position, projections of increasing temperatures, changing seasonality of precipitation, declining snowpacks, and increasing demand are all cause for concern in the Snake River. There is evidence that many of the predicted changes have already begun to have an effect in the West. Temperatures have increased, particularly in spring and winter, at high latitudes, and in the continental interior (IPCC 2007). The fraction of precipitation falling as rain rather than snow has increased (Knowles, Dettinger, and Cayan 2006), snowpack has decreased (IPCC 2007), and streamflow peaks are earlier (Stewart, Cayan, and Dettinger 2005). One of the primary uncertainties in modeling regional climate changes, both in the Snake River Basin and elsewhere, is the incapacity of current models to estimate how dynamic features like ENSO and PDO might respond to climate change (IPCC 2007). Future changes in ENSO conditions—both in mean state and amplitude—are still quite unclear (Vecchi, Clement, and Soden 2008). Given the strong impact of ENSO and PDO conditions on the Snake headwaters region, resolving these uncertainties will be of key importance to long-term water supply planning for the Snake River.

Acknowledgments

I would like to thank Andrew Comrie, Katie Hirschboeck, Stephanie McAfee, Connie Woodhouse, and two anonymous reviewers for their helpful suggestions on this article. This project was supported by a Pruitt National Fellowship from the Society of Women Geographers and by the United States Environmental Protection Agency (EPA) under the Science to Achieve Results (STAR) Graduate Fellowship Program. EPA has not officially endorsed this publication and the views expressed herein might not reflect the views of the EPA.

References

Carrera, M. L., R. W. Higgins, and V. E. Kousky. 2004. Downstream weather impacts associated with atmospheric blocking over the northeast Pacific. *Journal of Climate* 17:4823–39.

Cayan, D. R. 1996. Interannual climate variability and snowpack in the Western United States. *Journal of Climate* 9:928–48.

Cayan, D. R., M. D. Dettinger, H. F. Diaz, and N. E. Graham. 1998. Decadal variability of precipitation over Western North America. *Journal of Climate* 11 (12): 3148–66.

Cayan, D. R., K. T. Redmond, and L. G. Riddle. 1999. ENSO and hydrologic extremes in the Western United States. *Journal of Climate* 12:2881–93.

Chang, E. K. M., and Y. Fu. 2002. Interdecadal variations in Northern Hemisphere winter storm track intensity. *Journal of Climate* 15:642–58.

Clark, M. P., M. C. Serreze, and G. J. McCabe. 2001. Historical effects of El Nino and La Nina events on the seasonal evolution of the montane snowpack in the Columbia and Colorado River basins. *Water Resources Research* 37 (3): 741–57.

Dettinger, M. D., D. R. Cayan, H. F. Diaz, and D. M. Meko. 1998. North–south precipitation patterns in Western North America on interannual-to-decadal timescales. *Journal of Climate* 11 (12): 3095–111.

Diffenbaugh, N. A., J. S. Pal, R. J. Trapp, and F. Giorgi. 2005. Fine-scale processes regulate the response of extreme events to global climate change. *Proceedings of the National Academy of Sciences* 102 (44): 15774–778.

Gershunov, A., and T. P. Barnett. 1998. Interdecadal modulation of ENSO teleconnections. *Bulletin of the American Meteorological Society* 79 (12): 2715–25.

Intergovernmental Panel on Climate Change (IPCC). 2007. *Climate change 2007: The physical science basis. Contribution of Working Group I to the Fourth Assessment Report of the Intergovernmental Panel on Climate Change*, ed. S. Solomon, D. Qin, M. Manning, Z. Chen, M. Marquis, K. B. Averyt, M. Tignor, and H. L. Miller. New York: Cambridge University Press.

Knowles, N., M. D. Dettinger, and D. R. Cayan. 2006. Trends in snowfall versus rainfall in the Western United States. *Journal of Climate* 19:4545–59.

Leathers, D. J., B. Yarnal, and M. A. Palecki. 1991. The Pacific/North American teleconnection pattern and United States climate: Part I. Regional temperature and precipitation associations. *Journal of Climate* 4:517–28.

Lorenz, D. J., and E. T. DeWeaver. 2007. Tropopause height and zonal wind response to global warming in the IPCC scenario integrations. *Journal of Geophysical Research* 112: D10119. http://www.agu.org/pubs/crossref/2007/2006JD008087.shtml (last accessed 12 July 2010).

Mantua, N. J., and S. R. Hare. 2002. The Pacific Decadal Oscillation. *Journal of Oceanography* 58:35–44.

Mantua, N. J., S. R. Hare, Y. Zhang, J. M. Wallace, and R. C. Francis. 1997. A Pacific interdecadal climate oscillation with impacts on salmon production. *Bulletin of the American Meteorological Society* 78 (6): 1069–79.

Marston, R. A., J. D. Mills, D. R. Wrazien, B. Bassett, and D. K. Splinter. 2005. Effects of Jackson Lake Dam on the Snake River and its floodplain, Grand Teton National Park, Wyoming, USA. *Geomorphology* 71:79–98.

McAfee, S. A., and J. L. Russell. 2008. Northern Annular Mode impact on spring climate in the Western United States. *Geophysical Research Letters* 35: L17701. http://www.agu.org/pubs/crossref/2008/2008GL034828.shtml (last accessed 12 July 2010).

McCabe, G. J., and M. D. Dettinger. 2002. Primary modes and predictability of year-to-year snowpack variations in the Western United States from teleconnections with Pacific Ocean climate. *Journal of Hydrometeorology* 3: 13–25.

McGuire, M., A. W. Wood, A. F. Hamlet, and D. P. Lettenmaier. 2006. Use of satellite data for streamflow and reservoir storage forecasts in the Snake River Basin. *Journal of Water Resources Planning and Management* 132 (2): 97–111.

Miles, E. L., A. K. Snover, A. F. Hamlet, B. Callahan, and D. Fluharty. 2000. Pacific Northwest Regional Assessment: The impacts of climate variability and climate change on the water resources of the Columbia River Basin. *Journal of the American Water Resources Association* 36 (2): 399–420.

Miller, S. A., G. S. Johnson, D. M. Cosgrove, and R. Larson. 2003. Regional scale modeling of surface and ground water interaction in the Snake River Basin. *Journal of the American Water Resources Association* 39:517–28.

Mock, C. J. 1996. Climate controls and spatial variations of precipitation in the Western United States. *Journal of Climate* 9:1111–25.

Quadrelli, R., and J. M. Wallace. 2002. Dependence of the structure of the Northern Hemisphere annular mode on the polarity of ENSO. *Geophysical Research Letters* 29 (23): 2132. http://www.agu.org/pubs/crossref/2002/2002GL015807.shtml (last accessed 12 July 2010).

Redmond, K. T., and R. W. Koch. 1991. Surface climate and streamflow variability in the Western United States and their relationship to large-scale circulation indices. *Water Resources Research* 27 (9): 2381–99.

Salathé, E. P., Jr. 2006. Influences of a shift in North Pacific storm tracks on Western North American precipitation under global warming. *Geophysical Research Letters* 33: L19820. http://www.agu.org/pubs/crossref/2006/2006GL026882.shtml (last accessed 12 July 2010).

Seager, R., M. Ting, I. Held, Y. Kushnir, J. Lu, G. Vecchi, H. P. Huang et al. 2007. Model projections of an imminent transition to a more arid climate in Southwestern North America. *Science* 316 (5828): 1181–84.

Stewart, I. T., D. R. Cayan, and M. D. Dettinger. 2005. Changes toward earlier streamflow timing across Western North America. *Journal of Climate* 18:1136–55.

Vecchi, G. A., A. Clement, and B. J. Soden. 2008. Examining the tropical Pacific's response to global warming. *Eos: Transactions of the American Geophysical Union* 89 (9): 81, 83.

Wise, E. K. 2009. Streamflow and the climate transition zone in the western United States. Ph.D. disseration, University of Arizona, Tucson.

———. 2010. Spatiotemporal variability of the precipitation dipole transition zone in the Western United States. *Geophysical Research Letters* 37: L07706. http://www.agu.org/pubs/crossref/2010/2009GL042193.shtml (last accessed 12 July 2010).

Wurbs, R. A. 2006. Methods for developing naturalized monthly flows at gaged and ungaged sites. *Journal of Hydrologic Engineering* 11 (1): 55–64.

Yin, J. H. 2005. A consistent poleward shift of the storm tracks in simulations of 21st century climate. *Geophysical Research Letters* 32: L18701. http://www.agu.org/pubs/crossref/2005/2005GL023684.shtml (last accessed 12 July 2010).

Adapting to Climate Change in Andean Ecosystems: Landscapes, Capitals, and Perceptions Shaping Rural Livelihood Strategies and Linking Knowledge Systems

Corinne Valdivia, Anji Seth, Jere L. Gilles, Magali García, Elizabeth Jiménez, Jorge Cusicanqui, Fredy Navia, and Edwin Yucra

In the Bolivian Altiplano, indigenous systems for dealing with weather and climate risk are failing or being lost as a result of migration, climate change, and market integration. Andean rural communities are particularly vulnerable to changing social and environmental conditions. Changing climate over the past forty years and current forecast models point to increasing temperatures and later onset of rains during the growing season. Current meteorological models are coarse grained and not well suited to the complex topology of the Andes—so local-scale information is required for decisions. This article outlines a process for developing new local knowledge that can be used to enhance adaptive processes. This is a three-step process that includes assessment of local knowledge, the development of future scenarios, and the use of participatory research methods to identify alternative adaptation strategies. Initial analyses based on the survey of 330 households in nine communities indicate that northern Alitplano communities are more vulnerable than central Altiplano ones. In both areas, losses from climate shocks are high, but the types of hazards vary by location. The use of local knowledge indicators of climate is declining, and downscaling of climate forecasts is unlikely to occur due to the lack of data points and the large number of microclimates. Participatory mapping and research, where knowledge is shared, are processes that enhance adaptive capacity and are critical to building resilience. This article outlines a strategy for linking science-based and indigenous methods to develop early warning systems that are an important part of coping strategies. This approach combines science and indigenous knowledge to enhance adaptive capacity. *Key Words: adaptation, climate change, knowledge systems, livelihoods, mapping.*

在玻利维亚高原，处理天气和气候风险的原生系统正趋于失败，或作为移民，气候变化，和市场一体化的结果正在丧失。安第斯农村社区特别容易受到变化的社会和环境条件的影响。过去40年变化的气候和目前的气候预测模型显示升高的温度和生长季节降雨的推后。目前的气象模型比较粗糙，不适合安第斯山脉复杂的拓扑结构一因此需要局地尺度的信息来作决策。本文概述了开发新的，用于增强适应过程的地方知识的过程。这是一个三步程序，其中包括对当地知识的评估，未来的情形发展，以及参与性研究方法的使用，以确定替代性适应战略。基于在9个社区330户调查的初步分析表明，北部高原社区比中央高原社区更脆弱。在这两个区域，来自气候的冲击损失高，但危害的种类所在位置有所不同。当地的气候知识指标的使用在下降，由于缺乏数据点和大量的微气候，气候预报的降尺度是不可能的。知识共享的参与性测绘和研究是提高适应能力和建立应变能力的至关重要的流程。本文概述了连接科学基础和当地方法以制定应对战略的重要组成部分即早期预警系统的一种战略。这种方法结合了科学和当地知识以增强适应能力。*关键词：适应，气候变化，知识系统，生计，绘图。*

Los sistemas indígenas del Altiplano boliviano para lidiar con los riesgos del tiempo y el clima están fallando, o se están perdiendo como resultado de la migración, el cambio climático y la integración de mercado. Las comunidades rurales andinas son particularmente vulnerables a las cambiantes condiciones sociales y amb ientales. Los cambios del clima durante los pasados cuarenta años y los modelos de predicción actuales apuntan a un aumento de las temperaturas y llegada tardía de las lluvias durante la estación de crecimiento vegetativo. Los actuales modelos meteorológicos son burdos y no muy apropiados para la compleja topografía de los Andes—de tal suerte que se requiere información a escala local para las decisiones. Este artículo delinea un proceso para desarrollar nuevo conocimiento local que pueda usarse para fortalecer los procesos adaptativos. Se trata de un proceso en tres etapas, que incluye la consideración de conocimiento local, el desarrollo de escenarios futuros y la utilización de métodos de investigación participativa para identificar estrategias alternativas de adaptación. Los análisis iniciales basados en un estudio de 330 hogares en nueve comunidades indican que las de la parte norte del Altiplano son más vulnerables que las comunidades de la parte central. En ambas áreas las pérdidas por calamidad climática son grandes, pero los tipos de catástrofes varían según la localización. El uso de indicadores de conocimiento local sobre el clima está declinando y la reducción de la escala para los pronósticos del clima es difícil que ocurra dada la falta de puntos proveedores de datos y al gran número de microclimas. La cartografía e investigación participativas, donde el conocimiento se comparte, son procesos que fortalecen la capacidad adaptativa y son cruciales para construir resiliencia. Este artículo esquematiza una estrategia para ligar los métodos de origen científico con los de los indígenas para desarrollar sistemas de alertas tempranas, una parte importante para la incorporación de estrategias varias. Este enfoque combina el conocimiento de la ciencia con el de los indígenas para mejorar la capacidad adaptativa. *Palabras clave: adaptación, cambio climático, sistemas de conocimiento, medios de vida, mapeo.*

Sustainable systems have the capacity to adapt to changing circumstances without undermining their long-term survival. They are resilient to shocks and are able to adapt to changes by reorganizing themselves (Folke et al. 2002). Vulnerable human and biological systems, on the other hand, either lack or are losing the ability to handle these shocks. Technological change could increase the adaptability and resiliency of social and ecological systems faced by the prospect of climate change. Resilience, stated Stadel (2008, 17), can be considered an "antipode to vulnerability," a pillar to ecological and social sustainability. This article focuses on issues that arise in adapting to climate change in regions where social, economic, political, and environmental factors threaten this capacity (Bebbington 1999; Rhoades 2008; Stadel 2008). We use examples from the Andean Alitplano to describe a strategy for increasing the adaptive capacity of vulnerable indigenous populations by developing linkages between local, indigenous knowledge systems and scientific ones (United Nations Development Programme 2004; Jetté 2005; Meinke et al. 2006; Young and Lipton 2006; Marx et al. 2007).

In this region, climate change is expected to increase temperatures and the frequency of extreme events. Production strategies that used to buffer against climate variability are being lost because of economic, social, and market conditions (Zimmerer 1993). These changes have led to an increase in food insecurity, a reduction in available protein, and increased losses due to drought, frost, disease, and pests.

Diversity and resilience are closely linked in human and in biological systems. The more options available to human populations, the less vulnerable they are to natural disasters or market shifts (Valdivia et al. 2003; Howden et al. 2007). The same is true for natural systems: Diversity enhances the ability of ecosystems to adapt to change (Peterson, Allen, and Holling 1998). "Increased knowledge is the best way to improve the effectiveness of response" (Committee on Abrupt Climate Change 2002, vi); that is, to reduce vulnerabilities and increase adaptive capabilities.

We address four questions in the context of the northern and central Altiplano of Bolivia:

1. What is the diversity of livelihoods?
2. What is the diversity of perceptions related to climate?
3. What do we know about Andean climate trends and change and its implications?
4. How do knowledge and access to new climate information improve adaptive response?

Participatory processes are implemented to link knowledge systems and to build human, social, and political capitals to strengthen adaptive capacities. Downscaling, by statistical or dynamical methods (Wilby and Wigley 1997), of global climate projections to the spatial detail of this region will require more high-quality climate observations. There is a declining use of traditional knowledge, even where scientific research has validated it. Participatory research allows farmers and scientists to develop a common set of expectations and vocabulary to discuss alternative strategies. By participating in research, farmers can make their own observations and can derive lessons from research beyond those conclusions presented by the researcher. This approach informs development of early warning systems to identify anticipating actions and address climate risks in four rural communities of a municipality in the northern Altiplano.

Climate Change, Sustainable Development, and Agriculture

Climate change will increase food insecurity in tropical regions (Brown and Funk 2008; Lobell et al. 2008). Adaptation requires an integrated approach that addresses sources of risk and uses a sustainable development approach (Howden et al. 2007). Adaptation requires an interdisciplinary approach that integrates the knowledge systems of decision makers and the contribution of science.

Agriculture in the Andes is highly sensitive to climate variability and change, especially when climate changes challenge traditional relationships between humans and land (Young and Lipton 2006; Rhoades 2008). Global climate change challenges the adaptive capacity of traditional knowledge that has evolved over centuries (Rhoades 2008; Sperling et al. 2008). Therefore, understanding and managing climate risk is key to developing adaptive capabilities (Howden et al. 2007). Adaptation requires adjusting practices, processes, and capitals in response to current or future climate change (Howden et al. 2007). Interdisciplinary approaches that involve the decision makers directly affected (Meinke et al. 2006; Marx et al. 2007; Rhoades 2008) are needed in situations where the changes represent major disruptions. Participatory research on adaptation options can help decision makers understand that taking action now is likely to be to their advantage (Howden et al. 2007). These approaches are especially needed where there are market failures and a lack of safety nets. Strategies that build knowledge to anticipate, plan, and strengthen the capacity of rural communities to effect change are key to building resilience and adaptive capacity (O'Brien et al. 2008).

Frameworks and Methods

Ostrom (2007) provided a diagnostic framework for understanding the drivers of systems that have interactions across multiple scales. Our focus is on how the relationship among livelihoods, strategies, capitals, and knowledge affects responses to the uncertainties of climate change at a local scale (Liverman, Yarnal, and Turner 2002; Young and Lipton 2006). Our research operates at multiple scales and uses multiple methods. The livelihood strategies of small farmers in peasant communities in the northern and central Altiplano, who operate in complex systems and require flexibility in management of their weather and markets, are analyzed (Collinson 2001). Climate trends for the past thirty to forty years in the entire Bolivian Altiplano are also analyzed. An analysis of the implications of regional climate change projections for current knowledge systems provides a context for future decisions. Participatory research processes are employed as a tool for integrating knowledge systems by mapping perceptions of hazards, and change, at a watershed and landscape level and are used to document changes in production systems and in local knowledge. Participatory research also concentrates on developing adaptive capacities.

Sustainable Livelihoods and Livelihood Strategies

Ecosystem resilience describes the capacity to adjust to disturbances, without shifting to a qualitatively different state, by absorbing shocks while maintaining function and adapting to change (Folke et al. 2002). Livelihood resilience depends on the capabilities and capacities of people to adapt to internal and external shocks and stresses (Chambers and Conway 1991). A livelihood encompasses the income-generating activities pursued by a household and its individuals that result from commanding human, natural, financial, social, cultural, and physical capitals (Bebbington 1999; Valdivia and Gilles 2001), which are claimed, accessed, or controlled through networks and institutions (de Haan 2000, 345–46; Pretty and Ward 2001). Capabilities (Chambers and Conway 1991) directly impact the ability to cope. Sustainable livelihoods (SLs) connect livelihood and ecosystem resiliency (Chambers and Conway 1991; Bebbington 1999), through the capability of individuals to build their capitals (Valdivia 2004). The strategies developed might be agriculturally based or could be rural strategies (Bebbington 1999), where farmers articulate with markets (Mayer 2002). Resilient livelihoods depend on the outcome of the "hinge" among individuals, their families, and communities and the structures in which they negotiate (de Haan 2000; Valdivia and Gilles 2001). Building human, social, and political capitals in rural communities are key elements of this hinge. Research has documented a great diversity in the ability of individuals and communities in Latin America to access resources, to control them, and to act on information (Bebbington 2000; Eakin 2000; Podestá et al. 2000; Valdivia et al. 2003; Rhoades 2008; Sperling et al. 2008; Yager, Resnikowski, and Halloy 2008).

Knowledge Systems

Scientists and agricultural producers rely on different knowledge systems to guide their decisions. Although both knowledge systems are empirical, they differ significantly in their focus and orientation. Networks for communicating forecast information in the Altiplano are local and do not utilize scientific forecasts. Networks that communicate scientific forecasts do not include farmers (Gilles and Valdivia 2009). Farmers base their production decisions on local knowledge systems, developed from years of observations, experiences, and experiments, which provide the basis for making forecasts and a set of behavioral rules followed in response to these forecasts (Gilles and Valdivia 2009).

Scientific knowledge focuses on relationships and phenomena that do not vary across time and space, whereas local knowledge is very context specific. Scientific knowledge is reductionist and looks at relationships among individual variables, whereas local knowledge is holistic and focuses on cases rather than on variables (Kloppenburg 1991). Because the knowledge systems of agricultural practitioners and scientists have different foci, teaching farmers scientific principles will not lead to technological change. If agricultural research is to contribute to the well-being of agricultural producers and their communities, bridges must be created between these two knowledge systems. These bridges assure that scientific research is conducted in priority areas for producers and ensure that research results can be used to improve their livelihoods. The traditional model of technological transfer assumes that once farmers understand the science, they can build these bridges. This model has not had great success with agricultural producers; for example, merely reporting climate data to farmers will not lead to adaptation (Ziervogel 2004; Patt and Schroter 2008). One of the principal responses to the weaknesses of the technology transfer model has been the development of participatory research approaches where farmers and researchers work together to bridge knowledge systems (Chambers and Ghildyal 1985; Collinson 2001). Here participation is viewed both as a means for achieving a purpose and as an end in itself (Hayward, Simpson, and Wood 2004). Participatory research involves scientists and farmers conducting research together but evaluating the results using their own criteria. The colearning that takes place can help the researcher develop technologies that are both appropriate to the needs of farmers and attractive to them. In addition, the process also helps farmers understand the nature of agronomic research and makes them better consumers of these products.

Participatory research can take a number of forms and stages (Hayward, Simpson, and Wood 2004). A rudimentary form is where farmers present their knowledge of climate change and its impacts to scientists and technicians and they in turn provide community members with summaries of results from existing weather stations and climate models. A second form of participatory research is where farmers and scientists conduct experiments together. Research on the ability of soil amendments to increase water retention was conducted in this manner in the region. A third type of research is scientists trying to validate existing local knowledge of climate prediction methods and then incorporating these techniques into an early warning system for producers. The participatory research maps of hazards and their change developed with farmers and the findings on climate trends are shared using the first approach.

Risk Perceptions and Decisions

For knowledge on forecasts to be useful for coping strategies, it must be skillful, timely, and relevant to potential users; expressed in the language of the potential user; and from a trusted source (Stern and Easterling 1999; Finan and Nelson 2001). Even then, economic and structural constraints might preclude producers from using these forecasts (Eakin 2000; Agrawala and Broad 2002).

Because forecasts are not inherently deterministic, the way people process uncertainty should be taken into account to facilitate communication systems that contribute to adaptive capacity. Marx et al. (2007) argued that sharing experiences is important because people process information relying on experiential systems.

Two-way participatory communication can enhance trust between knowledge systems (Wilkins 2001). People assess risks using rules-based systems and association-based systems (Slovic and Weber 2002). When the results of associational and rules-based systems conflict, it is more likely that people would assess risk based on the former. In the case of the Altiplano, this will probably mean that if the results of traditional and expert forecasts conflict, farmers will use the traditional assessment model, unless expert forecasts can be incorporated into local knowledge systems (Slovic and Weber 2002).

Altiplano farmers perceive that the climate is changing (Gonzales, Cusicanqui, and Aparicio 2005) and we use farmers' perceptions of climate change as a starting hypothesis for this multilevel, multidisciplinary study.

Findings

Improving the capacity of Andean farmers to adapt to climate change is a three-stage activity. First, farmers and researchers need to develop a set of understandings about changes in climate and production systems over recent decades. Three research activities were conducted to accomplish this. A survey of 330 households was undertaken in the northern and central Altiplano to identify livelihood strategies and constraints facing local households. At the same time, meteorological data from existing weather stations in the Altiplano were gathered and analyzed to estimate trends over the past three to four decades. Finally, participatory research was carried out in our study communities to identify perceptions of climate change, to map weather-related hazards, and to identify indicators that might be used in early warning systems.

Second, we need to learn how climate change might impact agriculture in the future. A research activity focused on the projections of future climate change for this region using existing global climate change models. Finally, this information must be returned to the community to identify mitigation and adaptation strategies.

The Livelihood Strategies, Capitals, and Perceptions of Risks in the Northern and Central Altiplano

A sustainable livelihoods framework of capitals and capabilities (Chambers and Conway 1991; Bebbington 1999; Valdivia and Gilles 2001) and a household economic portfolio approach (Valdivia, Dunn, and Jetté 1996; Valdivia 2004) inform our household survey. Diversification of activities might reduce the impact of climate variability with activities and technologies that reduce covariant risk. Farmers manage risk ex ante by diversifying the portfolio of economic activities that smooth income through the year (Valdivia 2004). Coping strategies occur after a shock and include liquidating assets, temporarily migrating, relying on remittances, pulling children out of school, or reducing consumption of food (Valdivia 2004). Markets (seed, credit, food) are unreliable, and nonmarket relations are important in accessing labor, land, food, and animals. Interviews in Spanish and Aymara were conducted in 2006 with 330 households. The capitals and activities (Table 1), shocks experienced and coping strategies (Table 2), and perceptions of the impact of climate hazards on well-being (Table 3) of rural communities in the northern and central Altiplano are presented by community. Analysis of variance is performed to determine differences in the mean values among communities, and significant differences are indicated.

Human capital (Table 1) in communities of the central Altiplano is higher than in the northern Altiplano. These communities also plow more land (natural capital), have more fallow area, and manage a larger number of potato varieties. In the northern Altiplano, land size is smaller, and a higher diversity of crops is planted. Differences in social capital, such as networks to access credit (local vs. institutions), exist. San Jose (a canton, or political unit, including five communities) and San Juan have access to institutions for credit through their engagement in dairy markets. Other communities rely on local networks, with higher probability of facing similar climate risks. Local knowledge about climate forecasts is the main source of information (Gilles and Valdivia 2009) in all communities. In the northern Altiplano communities, the radio is a common source of information. Economic capital, stored in livestock and sheep, is greater in the central Altiplano communities, which implies a greater level of accumulation and ability to cope with shocks. Central Altiplano communities possess larger amounts of buffer stocks, *chuño* (freeze-dried potatoes that depend on cold winter nights to be processed), which is essential for food security. Income from agriculture and total income are higher in the central Altiplano communities studied. Northern Altiplano communities have a higher proportion from off-farm activities, related to smaller average land size. Farmers in this region in recent years gradually introduced new cash crops, such as onions and peas, and more recently potato varieties that can be grown with warmer temperatures, farming intensively.

Perceptions of threats to family well-being, elicited through the same household interviews (Table 3), used rankings from 1 (*not a threat*) to 5 (*extreme threat*). Threats included climate hazards and change, market risks, and environmental stressors (pests and a changing climate). Impacts of frost, a changing climate, and pests are a very strong threat in northern Altiplano communities, whereas floods and frost are very strong threats in the central Altiplano. The communities with lower total income have greater concerns about adult unemployment and low prices of livestock, both important sources of income in this region. Ancoraimes communities have a higher vulnerability to shocks. With farmers

Table 1. Capitals, coping strategies, and income sources in rural communities of the central and northern Altiplano of Bolivia in 2006

	Ancoraimes municipality, northern Altiplano					Umala municipality, central Altiplano				
Variables	Chichaya	Karkapa-ta	Cohani	Calahuan-cane	Chojna-pata	San Jose	San Juan	Vinto Coopani	Kellhuiri	Significance
Number of households	57	15	27	23	27	96	31	29	25	
Human capital										
Education, head of household (years)	7.07	4.33	3.44	4.82	4.50	7.13	5.32	4.38	5.20	*
Age, head of household (years)	49.26	48.07	44.48	45.64	50.04	49.33	48.61	50.41	52.44	*ns*
Total household labor (adult equivalent)	3.53	3.68	3.09	4.43	2.86	3.86	3.32	3.84	3.16	$p < 0.07$
Natural capital										
Plowed fields (Has) 2006–2007	0.390	0.147	0.091	0.171	0.286	1.835	2.550	0.853	1.172	*
Fallow field (Has)	1.636	1.000	0.222	0.391	0.703	3.955	7.370	3.960	4.766	*
Alfalfa (Has)	0.182	0.047	0.002	0.106	0	2.105	2.112	0.796	0.474	*
Grasslands (Has)	0.644	3.470	0.647	0.224	0.250	2.272	3.812	0.612	2.167	*
Diversity potato (# of varieties)	1.88	1.87	1.59	2.52	2.67	3.79	3.16	3.83	4.36	*
Crops diversity (# of crops)	5.84	4.53	3.96	4.43	3.74	2.76	2.74	2.68	2.8	*
Social capital										
Access to credit last five years (% of households)	16	60	26	17	30	29	16	24	16	$p < 0.057$
Credit from institutions (%)	18	0	29	50	37	79	100	29	0	*
Credit from family, friends (%)	82	100	71	50	63	21	0	71	100	*
Access to information radio (% households)	35	15	15	35	7	7	0	7	0	*
Cultural capital										
Local knowledge of biophysical indicators for agriculture (% households)	31.58	26.67	62.96	65.22	55.56	59.46	51.61 16/31	34.48 10/29	52.00 13/25	$p < 0.018$
Economic capital										
Cattle (head)	3.7	2.01	1.27	2.00	3.47	8.16	8.70	3.66	4.23	*
Sheep (head)	14.69	16.71	11.15	19.36	42.79	34.14	35.20	27.16	51.71	*
Coping strategies										
Food buffer *Chuño* (*arroba* = 11 kg)	5.054	3.222	1.883	2.894	4.720	23.898	19.814	15.423	29.416	*
Livestock sale crop failure (% households)	82	87	63	74	73	9	0	4	0	*
Borrow from friends when crops fail (% households)	21	67	37	26	19	17	11	20	6	
Income sources										
Total income: Cash and in kind (Bs)	10,171	3,760	2,416	6,768	5,910	24,029	22,880	10,582	15,624	*
Total cash income (Bs)	8,700	2,659	1,774	5,305	4,077	14,310	13,449	5,937	8,637	*
Income ag cash (Bs)	6,770	998	497	2,248	2,649	12,020	11,742	2,978	5,846	*
Remittances (Bs)	379	148	121	467	808	724	421	485	890	0.037
Income from migrant labor (Bs)	4,016	2,515	1,633	3,517	1,692	3,372	4,645	4,447	3,960	*ns*

Note: Exchange rate: 1 U.S. dollar = 6.95 *Bolivianos* (Bs). Significance is the analysis of variance comparison of means across communities.
*$p < 0.000$.
Source: Household Survey of Capitals, Practices and Perceptions; 330 household interviews.

Table 2. Perceptions of risks to household well-being of various climate and market hazards in rural communities of the central and northern Altiplano of Bolivia (2006)

	Ancoraimes municipality, northern Altiplano					Umala municipality, central Altiplano				
Type of threat	Chichaya	Karkapata	Cohani	Calahuancane	Chojnapata	San Jose	San Juan	Vinto Coopani	Kellhuiri	Significance
No. of households	56	15	27	23	27	94	31	29	25	
Hail impacting on crops and livestock	3.51	3.80	3.96	4.09	3.56	3.88	3.77	4.24	4.32	*
Impact of floods	3.96	3.53	3.89	3.91	3.85	4.48	4.23	3.76	4.28	*
Impact of drought	2.41	2.93	2.96	3.00	2.67	2.98	2.90	3.00	3.00	*
Impact of frost on agriculture	3.89	4.00	3.93	4.26	3.59	4.29	4.52	4.66	4.32	*
Impact of changing climate	3.79	3.93	4.26	4.22	4.11	3.90	3.81	3.29	3.80	$p < 0.016$
Impact of pests	3.68	4.13	4.04	4.17	3.78	3.16	3.06	3.48	3.88	*
Impact of low livestock prices	3.84	4.20	3.78	3.96	3.78	3.69	3.80	4.13	4.07	$p < 0.072$
Impact of an adult becoming unemployed	3.70	4.20	4.17	4.30	4.04	2.29	2.46	2.76	3.24	*

Note: Scale values are as follows: 1 = it is not a threat; 2 = it is a minimal threat; 3 = it is a moderate threat; 4 = it is a very strong threat; 5 = it is an extreme threat.
$^{*}p < 0.000$.
Source: Household Survey of Capitals, Practices and Perceptions.

Table 3. Losses experienced in the last year in rural communities of the central and northern Altiplano of Bolivia (2006)

		Ancoraimes municipality, northern Altiplano					Umala municipality, central Altiplano				
Shocks	*n*	Chichaya	Karkapata	Cohani	Calahuancane	Chojnapata	San Jose	San Juan	Vinto Coopani	Kellhuiri	Significance
No. of households		56	15	27	23	27	94	31	29	25	
Important losses in agriculture % of households	160	0	0	0	0	0	97.33	70.97	89.66	92	*
% of losses due to drought	34	40	17.50	30	0	0	28.74	21.87	20.59	10	*
% of losses due to floods	50	21	0	0	0	28.33	30.13	12.43	0	11.50	*ns*
% of losses due to frosts	107	18.52	18.33	18.72	20.52	25.35	21.91	0	0	0	*ns*
% of losses due to hail	48	16.5	20	21.87	20	20.45	16.02	40	22.20	20	*ns*
% of losses due to crop pests	278	27.73	27.66	36.48	28.18	22.38	12.33	9.61	12.63	9.44	*
% of losses due to low prices crops	13	15	0	0	0	0	17.18	14.55	0	33	*ns*
% of losses frost on livestock	40	11.32	13.55	15	5	16.85	11.16	0	0	9.55	*ns*
% of losses livestock diseases	164	14.63	19.15	17.85	16.37	17.61	17.06	20.61	11.81	19.88	*ns*

Note: n = number of households experiencing the shock in 2006 agricultural year.

$^{*}p < 0.000$.

Source: Household Survey of Capitals, Practices and Perceptions.

we develop landscape analysis, participatory mapping of vulnerabilities analysis, and an example of a knowledge system for early warning in this region.

Climate Trends in the Altiplano

To analyze trends in temperature and precipitation we needed to obtain a long-term homogenous data set of monthly maximum and minimum temperatures (Tmax and Tmin) and daily precipitation (PP). To meet the criteria of homogeneity, only stations with less than 10 percent missing data could be included (García et al. 2007). Most data were not available in digital form and had to be transcribed from individual station records. Fourteen stations representing a subset of all Bolivian Altiplano stations had sufficient data to analyze trends in surface Tmax, Tmin, and PP. The stations had time series of varying lengths with the longest being 1950 through 2004 and the shortest being 1960 through 2004.

The temperature data were used to calculate reference evapotranspiration (ETo) with the Penman–Monteith method using the guidelines for estimating missing data (García et al. 2004). Finally the evapotranspiration deficit (ETo-PP) and its trend were calculated. Trends were analyzed using the Mann–Kendall test. Data from stations with statistical significance were plotted in a geographic information system (GIS) environment. Both ordinary kriging with and without the incorporation of anisotropy were used to interpolate point results. The lag-dependence (position independence) of the variances of the point results is represented in the semi-variogram (SV). To yield a sound interpolation for each climatic indicator, a suitable kriging model (ordinary or ordinary anisotropic), a good lag distance, and a good distribution for the semi-variogram (e.g., Gaussian, exponential) were selected using a trial-and-error procedure. Finally, the observed maps are linked to qualitative results coming from surveys conducted in several rural communities in the area.

Figure 1 presents the differences in the annual mean temperatures and precipitation before and after 1983. The results show that trends in Tmax and Tmin represent some warming over the last fifty years. However, the temporal and spatial structure of these trends varies. Significant cooling trends in Tmin are found in the southern Altiplano (Figure 1A), whereas significant warming is identified in the central and northern Altiplano. In contrast, Tmax trends show a more homogeneous pattern of warming, although with less significance (Figure 1B). Precipitation presents a homogeneous pattern of no change, with only one station (Copacabana) showing significant trends (Figure 1C).

The increasing Tmax is related to increases in reference evapotranspiration in the region (Figure 2). The impacts are larger in the southern and northern Altiplano and less significant in the central region. The cooling trends in Tmin in the southern Altiplano might be due to the reported advance of nonsustainable agricultural activities like quinoa production. These trends are not observed in the central and northern areas where land management has not changed. This might be reducing air humidity, thus permitting increased radiative cooling at night. This drying could be causing a local cooling trend that is not yet overcome by trends in global warming. The overall warming trend in Tmax might be related to both greenhouse warming and desertification (drying and cooling) effects, especially in the south. The results suggest regime changes in temperature and precipitation trends that could be due to a combination of large-scale forcing and local disturbances.

Projections of Climate Change for the Altiplano

Accurate climate projections for the Altiplano will require high-resolution scenarios because of the region's complex topography. At present the trade-off is between a sufficient range of model results to characterize uncertainties and high-resolution results. In this initial research we opt for a sufficient range of models to characterize future climate.

The global climate models employed in the Intergovernmental Panel on Climate Change (IPCC) Fourth Assessment Report (AR4; Solomon and Qin 2007) have been analyzed using the Coupled Model Intercomparison Project (CMIP3) data set (Meehl et al. 2007). The models are employing horizontal grid resolutions that vary from near 100 km to more than 400 km. The medium and higher resolutions of these models do represent an elevated plateau in this central Andean region and provide a rough starting point for this evaluation. Analysis of the present-day annual cycle of temperature and precipitation simulated by the models shows that the models capture the timing of the rainy season onset and termination and the dry season. They also capture the weaker annual cycle observed in monthly temperatures. However, compared with observations, the models have a small warm bias and a wet bias, with too much moisture transport from the Amazon (Thibeault, Seth, and García 2010).

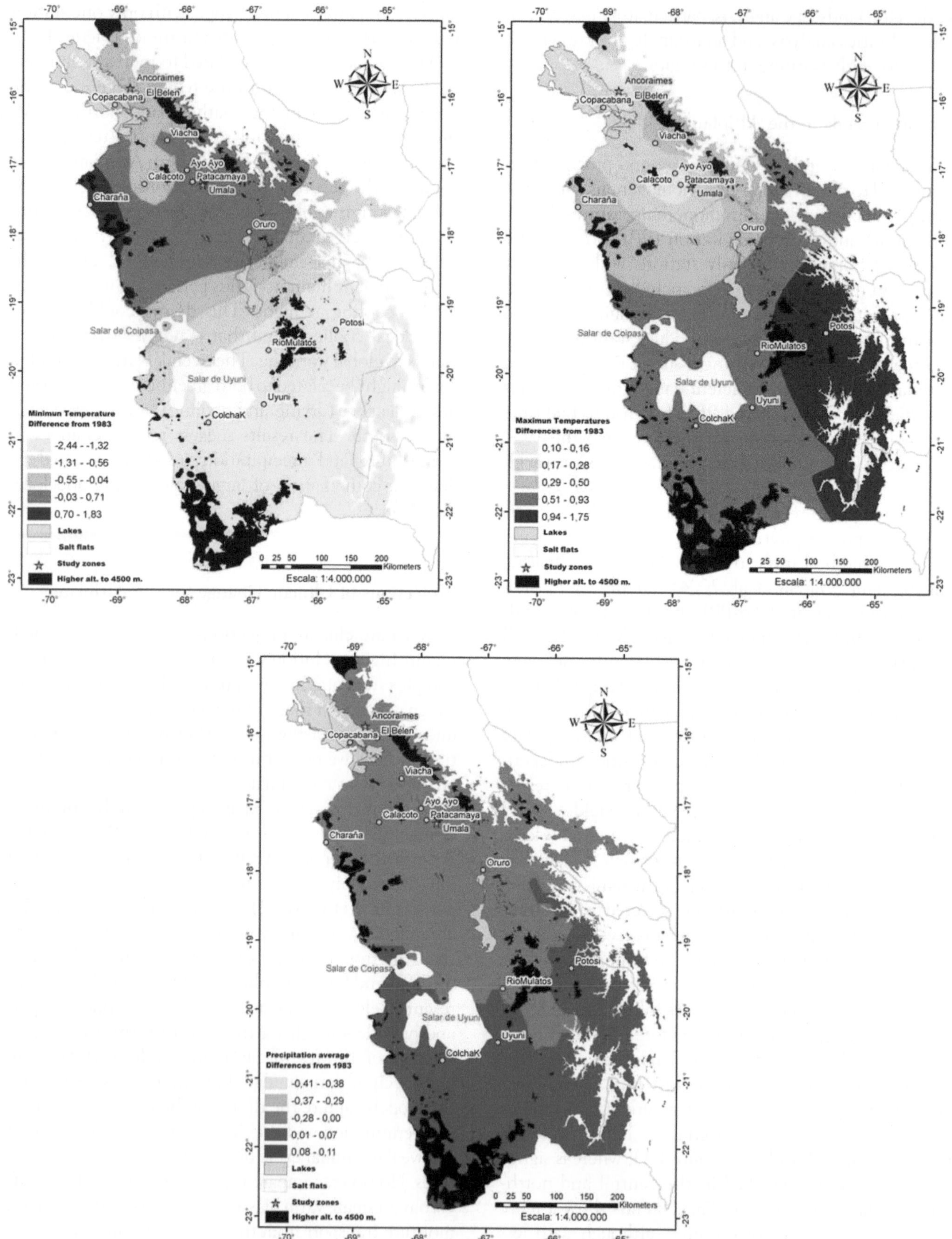

Figure 1. Differences in the mean of the records before and after 1983 in (A) Tmin, (B) Tmax, and (C) precipitation.

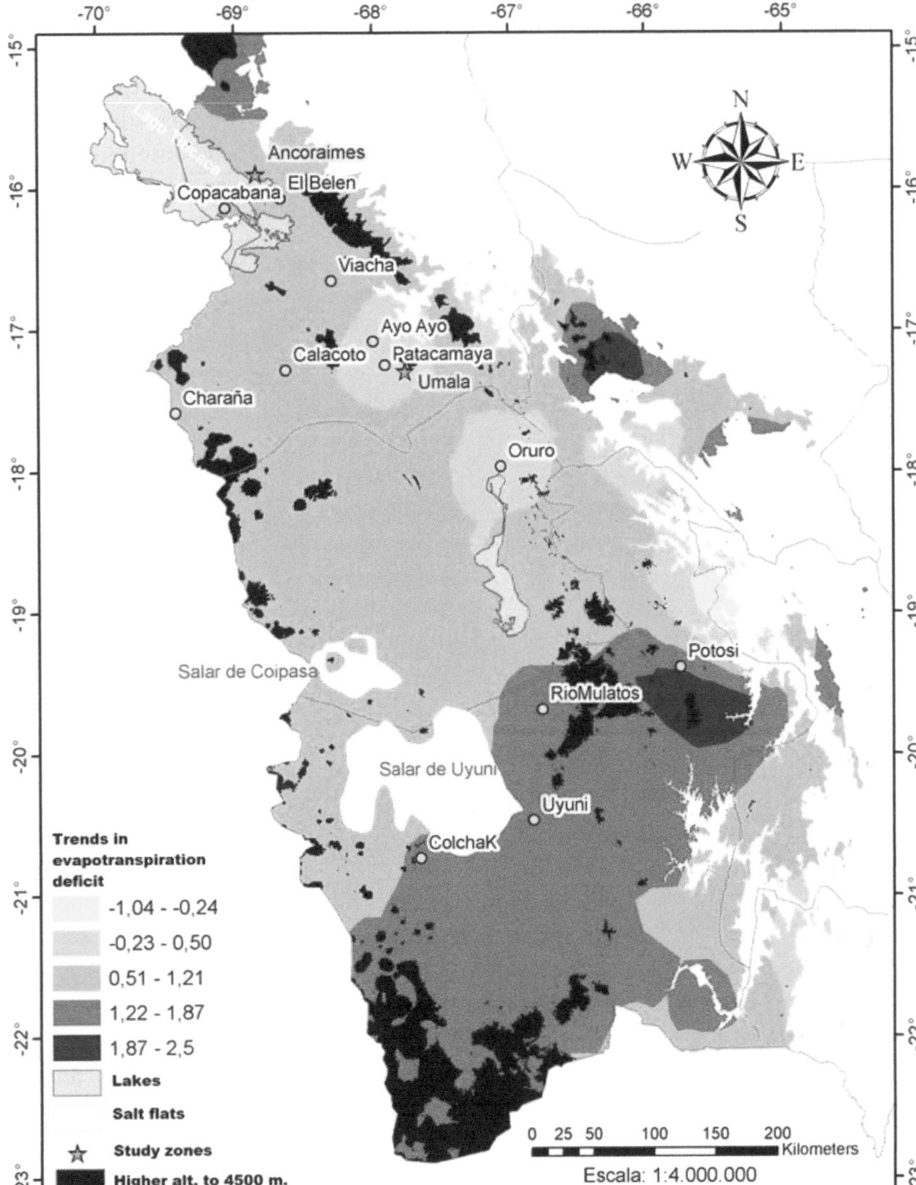

Figure 2. Trends in evapotranspiration deficit (ETo-PP) in the study area for the complete stations records.

Analysis of twenty-first-century projections shows agreement among the models that mean temperature increases of 1.5°C are likely by 2030 and greater than 4°C (5–6 *SDs*) through the annual cycle by the end of the century. Expected precipitation changes are negligible when averaged over all seasons. Our analysis suggests, however, that the early rainy season (September–November) is likely to be drier and the peak rainy season (January–March) is likely to be wetter, implying a shift toward a later and stronger rainy season (Thibeault, Seth, and García 2010), consistent with a larger scale analysis (Seth, Rojas, and Rauscher 2010).

Because changes in mean climate can have a large effect on the distribution of extremes, we also examine indexes of temperature and precipitation extremes in the scenarios. Temperature-related extremes show strong trends in increasing warm nights and heat wave duration, fewer frost days, and a larger extreme temperature variability (Thibeault, Seth, and García 2010). Precipitation extremes are projected to increase both in heavy rainfall events and in dry spells. These signals are seen in the observed records at Patacamaya, although the dry spell signal is not statistically significant (Thibeault, Seth, and García 2010).

These changes are different from present conditions, with increased variability and shifts in timing of the rains, in a context of rain-fed agriculture. Climate model projections also project that soil moisture is likely to be substantially reduced throughout the year, stressing the need for processes that build adaptive capacity of farmers, who will make decisions in very different conditions (Thibeault, Seth, and Wang forthcoming).

Participatory Research, Perceptions of Climate Change, and Hazard Mapping

Workshops were carried out in each study community to identify perceived changes in climate and in production systems and to develop local hazard maps. Farmer perceptions of temperature changes and frost frequencies were quite similar to the results obtained from the analysis of regional station data. Perceptions of changes in precipitation were quite different from the results of the analysis of station data. Farmers felt that there was a trend toward less rain, especially in the early part of the agricultural year. Further analysis and discussions with community members suggest that local perceptions might be related to the increasing atmospheric water demand due to higher temperatures and larger thermal amplitude. As a result, soils are drier and the first rains no longer provide enough water for early planting and consequently precipitation levels are perceived to be less. However, a spring drying trend (not yet significant) is seen in the station data (Seth, Rojas, and Rauscher 2010).

In addition to identifying perceptions, vulnerability maps were developed for all participating communities. The following example is from the northern Altiplano research site, the Huanquisco microbasin (Figure 3) located between 15°45′34.91″ and 15°56′15.29″ latitude and 68°54′00.66″ and 68°45′02.51″ longitude. The approximate area is 125.14 km^2. The communities studied represent an altitudinal gradient: Chojñapata is located at the highest level in this watershed, at 4,400 to 4,550 m above sea level; Cohani and Calahuancane are in the middle of this watershed, and Chinchaya is at 3,840 m above sea level, by Lake Titikaka. Community participatory workshops were conducted to learn about the changes in production systems and land use, to identify sources of risks, and to construct maps of the past and present situations. Vulnerability maps were developed through workshops organized with a community official from each participating rural community (Figure 3 and Figure 4), for each of the four communities. The borders of each community were geo-referenced. Participants identified on maps the zones where climate hazards are experienced. The hazards identified include hail, frost, snow, and droughts. Both perceptions of the past, defined by farmers as the time when things were different, and of the present were drawn and transferred to two maps (see Figure 3 and Figure 4). As the maps depict, at higher elevations there is a change from hail to snow, associated with more frost events in the present. The middle of the watershed also witnessed a shift from hail to snow and frosts, but there is also the appearance of drought events that especially impact crop production. Finally, the lowest part of the watershed, Chinchaya, saw a shift from drought events to damaging frosts near Lake Titicaca. These findings are consistent with those found through the household survey (Table 3).

Early Warning System: An Example of Linking Knowledge Systems to Develop Adaptive Capacity by Anticipating Weather Changes

The development of early warning systems is a priority for international and national disaster relief agencies, For example, early warning systems have been developed for floods, droughts, and tsunamis (Basher 2006). Most of these systems operate at the regional level and their focus is to provide information to government or corporate decision makers. The elements of the system include risk knowledge, monitoring and warning, dissemination and communication, and the capacity to take appropriate action (Basher 2006). The system described here operates at the community level and is based primarily on locally generated data complemented by forecast information where available.

We describe the methodology used to develop early warning systems in a situation where there is insufficient meteorological data and a multitude of microclimates created by the region's topography. The first step is to identify and validate local indicators to see what role they can play in a system incorporating local and scientific knowledge. After community discussions, groups of four individuals were chosen to work on the topic of weather risks. They were chosen because of their interest in climate prediction and mitigation rather than because of their expertise. These groups contained people with varying levels of knowledge about climate forecasting and included members from all four communities in the watershed.

The first task of each group was to interview community elders to get a list of local climate indicators

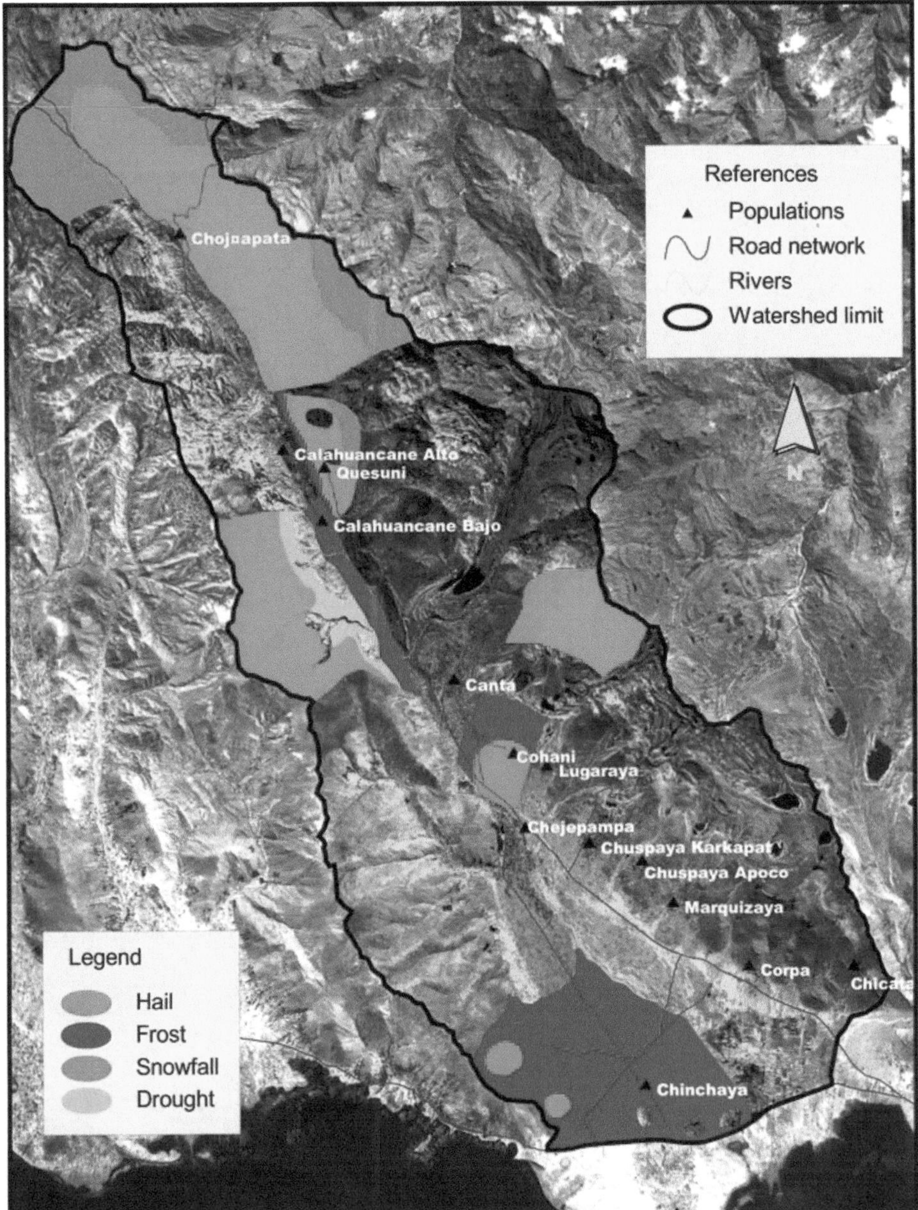

Figure 3. Map of the Ancoraimes watershed depicting present-day vulnerabilities in four rural communities. Red = hail; blue = frost; purple = snow; orange = drought.

and to understand how they were used. Once this list was assembled, group members went into the field to observe the indicators themselves. Based on their indicators, each of the group participants made his or her own forecasts for his or her farms and communities. As groups, they created matrices of the indicators, the observations, and their predictions. We now have two years of data and the results of this work will be presented to the communities. The data to date show that the indicators related to time of planting appear to have a high correlation with crop yields.

Group members, community members, and project agronomists discussed various mitigation and adaptation strategies to be tried. Some intended to reduce the effects of hazards, others to shift to alternative practices. Strategies currently under consideration include changing planting times and locations and the use of foliar fertilizers to enhance recovery from frost. This is the first component of the early warning system based on a seasonal forecast. The next phase of the early warning system will focus on strategies for reducing potential damage from frost and hail. This includes preparation for the use of smoke and other techniques for reducing frost damage and using linkages with national forecast agencies and small weather stations to indicate when these techniques should be used.

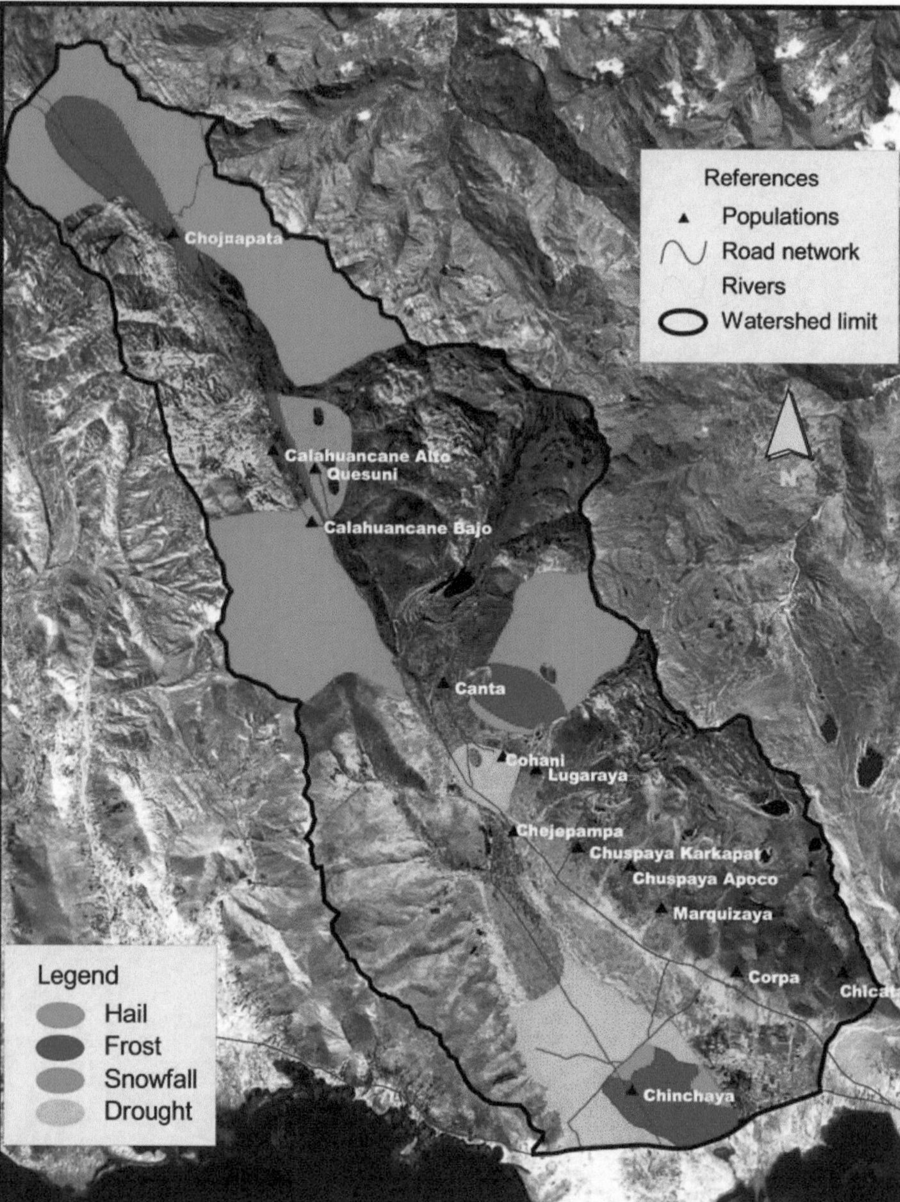

Figure 4. Map of the Ancoraimes watershed depicting past vulnerabilities in four rural communities. Red = hail; blue = frost; purple = snow; orange = drought.

Analysis and Conclusions

In the Altiplano of Bolivia, warming, later onset of the rainy season, and the presentation of more extreme weather undermine the production strategies that farmers have used to deal with weather-related risk. Local forecast techniques are falling into disuse and some are no longer working as they once did. In addition, the behavior of certain indicator species has changed due to climate and environmental changes. So as climate is changing, the ability to respond to climate-related risks is declining. The later onset of the rains is reducing food security by reducing the options that farmers have for planting dates and by threatening the production of two of the most important sources of dietary plant protein (quinoa and fava beans).

This project outlines a participatory methodology for linking the climate and weather forecasting communities with small farmers in indigenous communities. This increased understanding of the situation faced by producers permits the development of new production strategies. The interaction between researchers and local communities has generated a number of possible strategies that might mitigate the disruption caused by these changes. These opportunities include the introduction of new crops and the identification of varieties that will be more adapted to future conditions and require an iterative and long-term process of collaboration.

Can Information Improve Adaptive Capacity?

Researchers have found institutional, economic, and political constraints to the use of climate forecasts (Eakin 2000; Agrawala and Broad 2002). Better forecast information alone might not address vulnerability. Often those who benefit are those best placed to take advantage of knowledge and not necessarily the most vulnerable (Adger 2006). Information, with alternatives provided to vulnerable farmers, allows them to decide how to adjust (Valdivia et al. 2003; Patt, Suarez, and Gwata 2005). In this case, participatory approaches that incorporate the knowledge of trends, projections of climate change with implications for the Altiplano agriculture, and local knowledge of forecasts and the participatory research activities are all elements in developing adaptive capacity. Introducing an early warning system, developed through participatory processes, improves the ability to prepare for these events. Our focus on adaptive capacity to climate change does not negate our acknowledgment that economic, cultural, political, and social factors play key roles in livelihood strategies and outcomes, on well-being and the environment (Stadel 2008). The processes explored in the context of climate change are processes that in themselves provide opportunities to strengthen capabilities of a community to address its issues (Hayward, Simpson, and Wood 2004).

Acknowledgments

This publication was made possible by the U.S. Agency for International Development and the generous support of the American People for the Sustainable Agriculture and Natural Resources Management Collaborative Research Support Program (SANREM CRSP) under terms of Cooperative Agreement No. EPP-A-00-04-00013-00 to the Office of International Research and Development at Virginia Polytechnic Institute and State University. Adapting to Change in the Andean Highlands: Practices and Strategies to Address Climate and Market Risks in Vulnerable Agroecosystems is a research project of the SANREM CRSP award to the University of Missouri and collaborating institutions. We are thankful to the people of the eleven collaborating rural communities and the teams of researchers and students in Bolivia, Peru, and the United States who made this research possible.

References

Adger, W. N. 2006. Vulnerability. *Global Environmental Change* 16:268–81.

Agrawala, S., and K. Broad. 2002. Technology transfer perspectives on climate forecast applications. *Research in Science and Technology Studies* 13:45–69.

Basher, R. 2006. Global early warning systems for natural hazards: Systematic and people centred. *Philosophical Transactions of the Royal Society* A 364:2167–82.

Bebbington, A. 1999. Capitals and capabilities: A framework for analyzing peasant viability, rural livelihoods and poverty. *World Development* 27 (12): 2021–44.

———. 2000. Reencountering development: Livelihood transitions and place transformations in the Andes. *Annals of the Association of American Geographers* 90 (3): 495.

Brown, M. E., and C. C. Funk. 2008. Food security under climate change. *Science* 319:580–81.

Chambers, R., and G. R. Conway. 1991. Sustainable rural livelihoods: Practical concepts for the 21st century. Discussion Paper 296, Institute of Development Studies, London.

Chambers, R., and B. P. Ghildyal. 1985. Agricultural research for resource poor farmers: The farmer first-and last model. *Agricultural Research and Extension* 20:1–20.

Collinson, M. 2001. Institutional and professional obstacles to a more effective research process for small holder agriculture. *Agricultural Systems* 69 (1–2): 27–36.

Committee on Abrupt Climate Change. 2002. *Abrupt climate change: Inevitable surprises.* Washington, DC: National Academy Press.

de Haan, L. J. 2000. Globalization, localization and sustainable livelihood. *Sociologia Ruralis* 40 (3): 339–65.

Eakin, H. 2000. Smallholder maize production and climatic risk: A case study from Mexico. *Climatic Change* 45:19–36.

Finan, T., and D. Nelson. 2001. Making rain, making roads, making do: Public and private adaptations to drought in Ceará, Northeast Brazil. *Climate Research* 19:97–108.

Folke, C., S. Carpenter, T. Elmqvist, L. Gunderson, C. S. Holling, and B. Walker. 2002. Resilience and sustainable development: Building adaptive capacity in a world of transformations. *Ambio* 31 (5): 437–40.

García, M., D. Raes, R. Allen, and C. Herbas. 2004. Dynamics of reference evapotranspiration in the Bolivian highlands (Altiplano). *Agricultural and Forest Meteorology* 125:67–82.

García, M., D. Raes, S. E. Jacobsen, and T. Michel. 2007. Agroclimatic constraints for rainfed agriculture in the Bolivian Altiplano. *Journal of Arid Environments* 71:109–21.

Gilles, J., and C. Valdivia. 2009. Local forecast communication in the Altiplano. *Bulletin of the American Meteorological Society (BAMS)* 90:85–91.

Gonzales, J., J. Cusicanqui, and M. Aparicio. 2005. *Vulnerabilidad y adaptación al cambio climático en las regiones del lago Tititcaca y los Valles Cruceños de Bolivia* (Vulnerability and adaptation to climate change in the regions of Lake Titicaca and the Cruceño Valleys of Bolivia). La Paz, Bolivia: Ministry of Development Planning, Programa Nacional de Cambios Climáticos Bolivia (Bolivian National Program on Climate Change).

Hayward, C., L. Simpson, and L. Wood. 2004. Still left out in the cold: Problematizing participatory research and development. *Sociologia Ruralis* 44 (1): 95–108.

Howden, S. M., J. Soussana, F. N. Tubiello, N. Chhetri, and M. Dunlop. 2007. Adapting agriculture to climate change. *Proceedings of the National Academy of Sciences* 106 (50): 19691–696.

Jetté, C. 2005. *Democratic decentralization and poverty reduction: The Bolivian case*. Oslo, Norway: United Nations Development Programme, Oslo Governance Centre.

Kloppenburg, J. 1991. Social theory and the deconstruction of agricultural science: Local knowledge for an alternative agriculture. *Rural Sociology* 56 (4): 519–48.

Liverman, D., B. Yarnal, and B. L. Turner II. 2002. The human dimensions of global change. In *Geography in America at the dawn of the 21st century*, ed. G. L. Gaile and C. J. Willmott, 267–82. New York: Oxford University Press.

Lobell, D. B., M. B. Burke, C. Tebaldi, M. D. Mastrandrea, W. P. Falcon, and R. L. Naylor. 2008. Prioritizing climate change adaptation needs for food security in 2030. *Science* 319:607–10.

Marx, S. M., E. U. Weber, B. S. Orlove, A. Leiserowitz, D. H. Krantz, C. Rocoli, and J. Phillips. 2007. Communication and mental processes: Experiential and analytic processing of uncertain climate information. *Global Environmental Change* 17:47–58.

Mayer, E. 2002. *The articulated peasant: Household economies in the Andes*. Oxford, UK: Westview Press.

Meehl, G. A., C. Covey, T. Delworth, M. Latif, B. McAvaney, J. F. B. Mitchell, R. J. Stouffer, and K. E. Taylor. 2007. The WCRP CMIP3 multimodel dataset: A new era in climate change research. *Bulletin of the American Meteorological Society* 88:1383–94.

Meinke, H., R. Nelson, P. Kokic, R. Stone, R. Sevaraju, and W. Baethgen. 2006. Actionable climate knowledge: From analysis to synthesis. *Climate Research* 33 (December): 101–10.

O'Brien, G., P. O'Keefe, H. Meena, J. Rose, and L. Wilson. 2008. Climate adaptation from a poverty perspective. *Climate Policy* 8:194–201.

Ostrom, E. 2007. A diagnostic approach for going beyond panaceas. *PNAS* 104 (39): 15181–87.

Patt, A. P., and D. Schroter. 2008. Perceptions of climate risk in Mozambique: Implications for the success of adaptation strategies. *Global Environmental Change* 18:458–67.

Patt, A., P. Suarez, and C. Gwata. 2005. Effects of seasonal climate forecasts and participatory workshops among subsistence farmers in Zimbabwe. *Proceedings of the National Academy of the Sciences* 103 (35): 12623–28.

Peterson, G., C. R. Allen, and C. S. Holling. 1998. Ecological resilience, biodiversity and scale. *Ecosystems* 1 (1): 6–18.

Podestá, G., D. Letson, J. Jones, C. Mesian, F. Royce, A. Ferreyra, J. O'Brien, D. Legler, and J. Hansen. 2000. Experiences in application of ENSO-related climate information in the agricultural sector of Argentina. In *Proceedings of the International Forum on Climate Prediction, Agriculture and Development*, 217–21. Palisades, NY: International Research Institute for Climate Prediction.

Pretty, J., and H. Ward. 2001. Social capital and the environment. *World Development* 29 (2): 209–27.

Rhoades, R. 2008. Disappearance of the glacier on Mama Cotacachi: Ethnological research and climate change in the Ecuadorian Andes. *Pirineos* 163:37–50.

Seth, A., M. Rojas, and S. A. Rauscher. 2010. CMIP3 projected changes in the annual cycle of the South American Monsoon. *Climatic Change* 98:331–57.

Slovic, P., and E. U. Weber. 2002. Perception of risk posed by extreme events. Paper presented at Risk Management Strategies in an Uncertain World Conference, Palisades, NY, 12–13 April.

Solomon, S., and D. Qin. 2007. Climate change 2007: The physical science basis. Working Group I report to the Fourth Assessment of the Intergovernmental Panel on Climate Change. New York: Cambridge University Press.

Sperling, F., C. Valdivia, R. Quiroz, R. Valdivia, L. Angulo, A. Seimon, and I. Noble. 2008. Transitioning to climate resilient development—Perspectives from communities of Peru. Climate Change Series No. 115: The World Bank Environment Department Papers, Washington, DC.

Stadel, C. 2008. Resilience and adaptations of rural communities and agricultural land use in the tropical Andes: Coping with environmental and socioeconomic changes. *Pirineos* 163:15–36.

Stern, P. C., and W. E. Easterling, eds. 1999. *Making climate forecasts matter*. Washington, DC: National Academies Press.

Thibeault, J., A. Seth, and M. Garcia. 2010. Changing climate in the Bolivian Altiplano: CMIP3 projections for extremes of temperature and precipitation. *Journal of Geophysical Research—Atmospheres* 115: D08103. http://www.agu.org/pubs/crossref/2010/2009JD012718.shtml (last accessed 9 July 2010).

Thibeault, J., A. Seth, and G. Wang. Forthcoming. Increased surface drying in the 21st century Altiplano.

United Nations Development Programme. 2004. *Indice de desarrollo humano en los municipios de Bolivia* [*Human development index of the municipalities of Bolivia*]. La Paz, Bolivia: United Nations Development Programme.

Valdivia, C. 2004. Andean livelihoods and the livestock portfolio. *Culture and Agriculture* 26 (1–2): 19–29.

Valdivia, C., E. Dunn, and C. Jetté. 1996. Diversification as a risk management strategy in an Andean agropastoral community. *American Journal of Agricultural Economics* 78 (5): 1329–34.

Valdivia, C., and J. Gilles. 2001. Gender and resource management: Households and groups, strategies and transitions. *Agriculture and Human Values* 18 (1): 5–9.

Valdivia, C., J. L. Gilles, C. Jetté, R. Quiroz, and R. Espejo. 2003. Coping and adapting to climate variability: The role of assets, networks, knowledge and institutions. In *Insights and tools for adaptation: Learning from climate variability*, 189–99. Washington, DC: NOAA Office of Global Programs, Climate and Societal Interactions.

Wilby, R. L., and T. M. L. Wigley. 1997. Downscaling general circulation output: A review of methods and limitations. *Progress in Physical Geography* 21:530–48.

Wilkins, L. 2001. A primer on risk: An interdisciplinary approach to thinking about public understanding of agbiotech. *AgBioForum* 4 (3–4): 163–72.

Yager, K., H. Resnikowski, and S. Halloy. 2008. Grazing and climatic variability in the Sajama National Park, Bolivia. *Pirineos* 163:97–109.

Young, K. R., and J. K. Lipton. 2006. Adaptive governance and climate change in the tropical highlands of western South America. *Climatic Change* 78:63–102.

Ziervogel, G. 2004. Targeting regional forecasts for integration into household level decisions: The case of smallholder farmers in Lesotho. *The Geographical Journal* 170 (1): 6–21.

Zimmerer, K. S. 1993. Soil erosion and labor shortages in the Andes with special reference to Bolivia 1953–91: Implications for conservation with development. *World Development* 21 (10): 1659–75.

Making Sense of Twenty-First-Century Climate Change in the Altiplano: Observed Trends and CMIP3 Projections

Anji Seth, Jeanne Thibeault, Magali Garcia, and Corinne Valdivia

A synthesis is presented of the first phase in regionalizing climate projections for the Altiplano, an elevated central Andean plateau in Bolivia and Peru. A prerequisite to downscaling is analysis of the large-scale forcing provided by global, multimodel climate scenarios. Global climate models in the Coupled Model Intercomparison Project (CMIP3) archive are employed to qualitatively evaluate the direction of change in twenty-first-century projections of the annual cycle, indexes of extremes, and soil moisture. Analysis suggests the observed warming in the region is likely to accelerate in the coming decades under the high emissions scenario. Precipitation projections exhibit larger uncertainty but suggest an evolution toward a shorter, more intense wet season with weakened spring (September–November) precipitation and strengthened summer rainfall (January–March). These results are consistent with projections for the large-scale South American Monsoon, and station observations indicate trends similar to the projections. Extremes analysis suggests that precipitation may increasingly be experienced as intense storms, with more consecutive dry days. These results are consistent with soil moisture projections, which indicate drier conditions during the rainy season, despite the projected increase in precipitation. Our results suggest climatic changes in the Altiplano might have serious consequences for water management and indigenous agriculture. However, these climate model projections must be taken with caution, due to the relatively coarse grid scales employed and the model warm and wet biases. The results presented here will require further testing with improved, higher resolution climate models. *Key Words: Altiplano, climate change, climate extremes, climate models, South America.*

本文综述介绍了在区域化的高原，即玻利维亚和秘鲁的中央安第斯高原气候预测的第一阶段。降级分析的一个前提是分析由全球多模式气候情景所提供的大尺度的影响力。在耦合模式交互比较项目（CMIP3）存档中的全球气候模式被用来定性评价对二十一世纪的年度周期，极端指标，和土壤水分等预测的变化方向。分析表明该区域观测到的变暖在高排放的情况下很可能在未来几十年加快。降水预测虽显示更大的不确定性，但表面朝着一个更短，更激烈的，由削弱的春季（9月－11月）降水和加强的夏季降水（1月－3月）为特点的雨季演变。这些结果与大规模南美季风的预测相一致，站点观察也显示与预测类似的趋势。极值分析表明，降水可能越来越多表现为强风暴与更多的连续干旱天交错的形式。这些结果与土壤水分的预测相一致，即尽管预计降水量增加，雨季呈现更干旱的条件。我们的研究结果表明在高原地区的气候变化可能对水资源管理和本土农业造成严重后果。但是，由于使用的相对粗的网格尺度和模型的暖湿偏差，必须谨慎对待这些气候模型预测。在此显示的研究结果将需要用改进的，更高分辨率的气候模型进行进一步测试。*关键词：高原，气候变化，极端气候现象，气候模型，南美洲。*

Se presenta en este artículo una síntesis de la fase inicial en la regionalización de proyecciones climáticas para el Altiplano, la elevada meseta de los Andes centrales de Bolivia y Perú. Un prerrequisito para reducir las escalas es el análisis de compulsión a gran escala generado por escenarios climáticos globales multimodelados. Se utilizaron modelos climáticos globales que hacen parte del Proyecto de Comparación del Modelo Acoplado (Coupled Model Intercomparison Project, CMIP3) para evaluar cualitativamente la dirección del cambio del ciclo anual en las proyecciones para el siglo XXI, índices de extremos y humedad del suelo. El análisis sugiere que el calentamiento observado en la región probablemente se acelerará las próximas décadas bajo el escenario de emisiones altas. Las proyecciones de precipitación muestran mayor incertidumbre pero sugieren una evolución hacia una estación lluviosa más corta y más intensa, con debilitamiento de la precipitación en primavera (septiembre–noviembre) e intensificación de las lluvias en verano (enero–marzo). Estos resultados son consistentes con las proyecciones a gran escala para el monzón sudamericano y las observaciones de estaciones meteorológicas indican tendencias similares a las de las proyecciones. Los análisis de extremos sugieren que la precipitación puede estar cada vez más asociada con tormentas intensas, seguidas de un mayor número de días secos. Estos resultados son consistentes con las proyecciones de humedad del suelo, que indican condiciones más secas durante la estación lluviosa, a pesar del incremento proyectado para la precipitación. Nuestros resultados sugieren que los cambios climáticos del Altiplano podrían tener serias consecuencias para el manejo del agua y para la agricultura indígena. No obstante, estas proyecciones de modelo climático deben tomarse con cautela debido a las escalas de observación relativamente burdas, lo mismo que los sesgos que tiene el modelo hacia las condiciones cálida y húmeda. Los resultados que aquí se presentan requerirán cotejo adicional con modelos climáticos mejorados y de mayor resolución. *Palabras clave: Altiplano, cambio climático, extremos climáticos, modelos climáticos, Sudamérica.*

Information about future changes in precipitation and temperature in the Altiplano is essential for developing strategies to reduce climate-related risks to agriculture and water management. The study of regional climate change is a frontier in climate science and poses significant challenges. Current global coupled models (Coupled Model Intercomparison Project [CMIP3]; Meehl et al. 2007) employ grid resolutions ranging from 120 km to more than 300 km, which are insufficient to represent spatial detail related to complex topography, coastlines, and land use. In climate projections, however, it is essential to characterize uncertainties, which include those in emission scenarios, differences in coupled climate model CO_2 sensitivity (intermodel variability), the chaotic nature of the climate system (internal variability), and if downscaling is employed, differences among the regional models (e.g., Giorgi and Francisco 2000; Rowell 2006). Thus, a compromise is required at present to study climate projections for this region. The use of high-resolution models can provide spatial detail but an insufficient number of models and scenarios exist to characterize uncertainty. Alternatively, the use of intermediate-resolution models can characterize uncertainties but does not incorporate spatial detail. Pierce et al. (2009) have shown for regional climate projections that a large number of models is more critical than individual model skill in characterizing uncertainty. Thus, we apply the strategy of using as many intermediate-resolution models as possible, with a purpose to evaluate the ability of current models to indicate qualitative directions of change in projections of Altiplano climate with emphasis on the related uncertainties.

This research is conducted at the nexus of the geographic and atmospheric science approaches to climatology described by Carleton (1999), where the integration and societal impacts of local climate processes across scales meet the global transient drivers of change and their uncertainties.

We begin with evaluation of present-day and projected annual cycle of precipitation and temperature and provide new analysis of station observations to examine twentieth-century trends in the Altiplano. We then synthesize recent results, which suggest the changes in the annual cycle of precipitation are related to large-scale changes in the South American Monsoon (SAM; Seth, Rojas, and Rauscher 2010). Because large changes in the frequency of extreme climate events can result from even small changes in mean climate (Nicholls and Alexander 2007), we include analysis of precipitation and temperature-related extremes (Thibeault, Seth, and Garcia 2010) and soil moisture (Thibeault, Seth, and Wang forthcoming) in this synthesis of twenty-first-century climate projections for the Altiplano. Due to the coarse representation of the region in the current models and the limited observations available, this research cannot provide conclusive results. Rather, our intention is to link robust large-scale processes to a possible mechanism for understanding climate change in the region and to call attention to the needs for improved observations and models to enable study of climate change in this important region.

Previous Literature

Analyses of twenty-first-century climate projections for the Altiplano indicate that changes in mean temperature and precipitation are likely (Bradley et al. 2006; Urrutia and Vuille 2009; Thibeault, Seth, and Garcia 2010). Observed temperature increases in the Andes (0.11°C/decade) are larger than trends in global mean temperature (0.06°C/decade, between 1939 and 1998) and projected changes are expected to be greater than in the global mean (Bradley et al. 2006). Traditional methods of rain-fed agriculture are practiced by approximately 50 percent of the rural population (Garcia et al. 2007). More than 60 percent of annual precipitation occurs during the months of December through February and is associated with the southwest margin of the SAM (Garreaud and Aceituno 2001;

Garcia et al. 2007). Crop production is particularly vulnerable to climate-related shocks including drought, flooding, frost, and pests (Gilles and Valdivia 2009). Temperature increases are likely to have a substantial impact on agriculture in tropical regions. Battisti and Naylor (2009) point out that projected mean growing season temperatures through much of the tropics will likely be outside the range of extremes experienced in the twentieth century.

Analysis of the CMIP3 data set for the twenty-first century indicates an overall weakening of the tropical circulation. This weakening is seen predominantly in the Walker circulation (Held and Soden 2006; Vecchi and Soden 2007): The large-scale convective overturning must slow to compensate for the fact that atmospheric water vapor increases more than precipitation in a warmer world. Vecchi and Soden (2007) show enhanced upward motion in the eastern tropical Pacific and reduced upward motion in the western tropical Pacific. This mean atmospheric response is El Niño–like (see, e.g., Latif and Keenlyside 2009). Consistent with this result, over the South American continent enhanced subsidence is seen over the Amazon basin and enhanced vertical motion in the southeast. In addition, Seth, Rojas, and Rauscher (2010) show the increased subsidence in the Amazon extends into the South Atlantic Convergence Zone (SACZ) region, which is associated with a poleward shift in the South Atlantic Anti-Cyclone (SAAC, enhanced subsidence poleward and reduced subsidence equatorward compared with present day). These large-scale circulation changes appear to affect the annual cycle of the SAM, with weakened spring and strengthened summer season rainfall. These results are also consistent with the "upped ante" mechanism (Neelin, Chou, and Su 2003; Lintner and Neelin 2006); a warmer, more humid atmosphere is increasingly stable and requires additional moist static energy to initiate convection. Subtropical regions along the margins of precipitation, such as the seasonally dry Monsoon region, are especially disadvantaged according to this upped ante mechanism.

These model projections of large-scale changes provide important context for our study of future Altiplano climate. Although the global models cannot provide quantitative results for a small region, we investigate the ability of the models to indicate likely directions of change and ask these questions: Is the projected shift in the annual cycle of the SAM also seen in the Altiplano rainy season? Can changes in projected extremes of temperature and precipitation help to explain projected changes in the annual cycle?

Data and Methods

A broad, elevated plateau in the central Andes, the Altiplano spans 800 km north to south and 300 km east to west. We are interested in the northern Altiplano, defined here as 16°–19°S, 67°–70°W (see Figure 1), which experiences warm season precipitation associated with the SAM. This region spans 330 × 330 km and rises to an elevation of 4,000 m, posing a challenge for climate models. The medium and higher resolution climate models employed in this study represent the elevated plateau to varying degrees (between 2,000 and 3,800 m) and with varying numbers of grid points (from one to nine; Thibeault, Seth, and Garcia 2010). Thus, our purpose is to qualitatively evaluate the projections within the context of both observations and the robust, large-scale changes projected by the models.

Nine CMIP3 global coupled climate models (Meehl et al. 2007; detailed model descriptions are available from http://www-pcmdi.llnl.gov/ipcc/about_ipcc.php) are analyzed (Table 1) to evaluate for present-day and future scenarios, the annual cycles of monthly temperature and precipitation, and annual indexes of extremes computed from daily model output. To characterize the intermodel uncertainty, we examine all medium- and higher resolution models in the CMIP3 database having available the derived extreme indexes (e.g., Pierce et al. 2009). Twentieth-century simulations are initialized in 1860 and run with observed estimates of historical radiative forcing, which includes anthropogenic greenhouse gases and aerosols, and natural volcanic and solar variability (hereafter for the twentieth century, we analyze the period 1970–1999). The SRES A2 emissions scenario is initialized using the end of the twentieth century and assumes a continuously increasing global population with regionally oriented economic growth that results in higher overall emissions and atmospheric concentrations of CO_2 near 800 ppm by 2100 (hereafter A2). Recent observed emissions growth is larger than assumed in this high emission scenario ("Carbon Dioxide Emissions Rise to Record Levels" 2008).

Analysis presented here emphasizes the later twenty-first century (2070–2099) and more detail on mid-twenty-first-century projections is available in

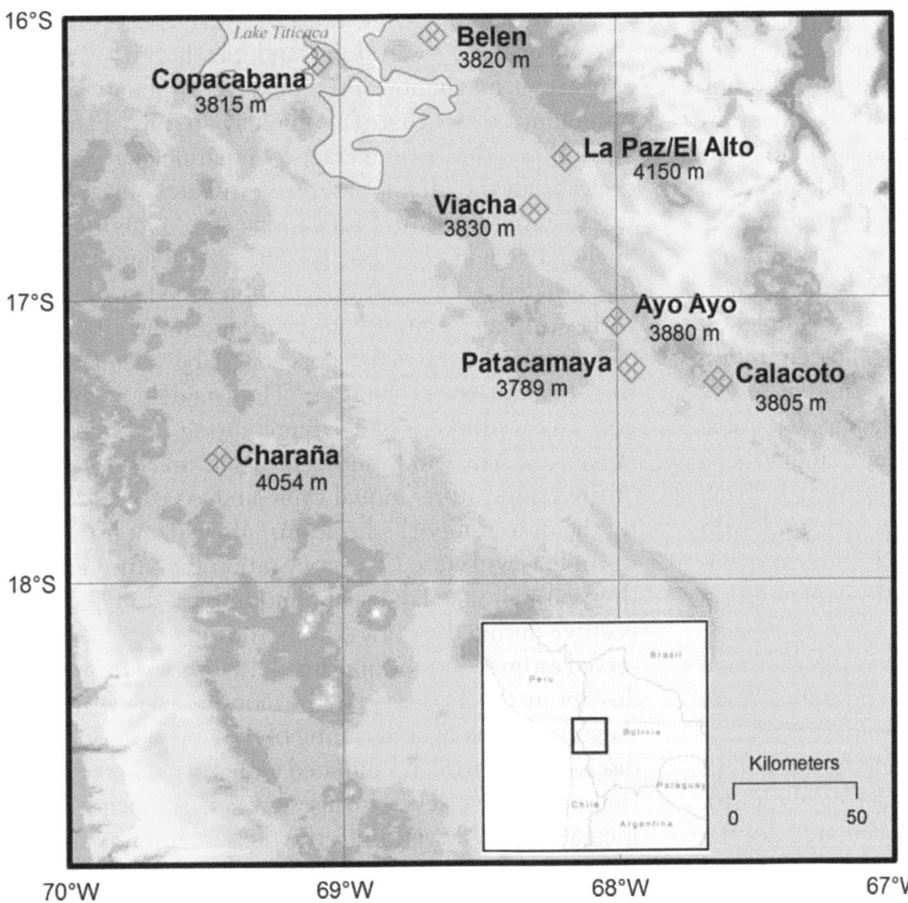

Figure 1. The northern Altiplano region (3° × 3°) defined for area-averaged analysis of gridded observations and coupled climate models in this study. Contours represent topographic elevation and stations are located (red) with elevation (m) specified for each. Inset locates region in the central Andes of South America.

Thibeault, Seth, and Garcia (2010), which shows less model uncertainty and smaller climate response in accordance with expectations (Cox and Stephenson 2007). The multimodel statistics for present-day and future climate are represented by box and whisker plots for temperature precipitation and soil moisture, to show the intermodel uncertainty. Differences between 2070–2099 and 1970–1999 multimodel mean monthly, seasonal (SON and JFMA), and annual precipitation are presented.

Table 1. Global coupled climate models employed in this study from the CMIP3 archive, with resolution of the atmosphere and ocean given in degrees longitude by latitude

Model	Sponsor	Atmosphere	Ocean	AnCy	Ext	SM
CCSM3	NCAR	1.4 × 1.4	0.3–1 × 1	X		X
ECHAM5	Max Planck Institute	1.9 × 1.9	1.5 × 1.5			X
GFDL CM2.1	GFDL	2.0 × 2.5	0.3–1 × 1	X	X	X
GFDL CM2.0	GFDL	2.0 × 2.5	0.3–1 × 1	X	X	X
MRI CGCM2.3	MRI	2.8 × 2.8	1 × 2	X	X	
MIROC3.2	JAMSTEC	2.8 × 2.8	0.5–1.4 × 1.4	X	X	X
MIROC3.2HR	JAMSTEC	1.1 × 1.1	1.1 × 1.1	X		
CNRM CM3	CNRM	2.8 × 2.8	0.8 × 1	X	X	
PCM	NCAR	2.8 × 2.8	0.5–0.7 × 1.1	X	X	X
IPSL CM4	IPSL	2.5 × 3.75	2 × 2	X	X	

Note: The data available for specific analyses, annual cycle (AnCy), extremes (Ext), and soil moisture (SM) are marked by X. CMIP3 = Coupled Model Intercomparison Project; CCSM3 = Community Climate System Model version 3; NCAR = National Center for Atmospheric Research; ECHAM5 = European Community Hamburg Model version 5; GFDL = Geophysical Fluid Dynamics Laboratory; MRI = Meteorological Research Institute (Japan); CGCM = Canadian Global Climate Model; MIROC = Model for Interdisciplinary Research on Climate; JAMSTEC = Japan Agency for Marine-Earth Science and Technology; CNRM = Centre National Researches Météorologiques (France); PCM = Parallel Climate Model; IPSL = Institut Pierre Simon Laplace (France).

The annual cycles of temperature and precipitation are evaluated using the University of Delaware (UDEL) monthly gridded (.5°) temperature and precipitation data sets (Willmott and Matsuura 2009). Monthly climatologies (1970–1999) are computed for the Altiplano region. Due to the inherent differences in the character of point and area-averaged precipitation, station observations cannot be directly compared with simulated grid averaged data (e.g., Legates and Willmott 1992). As an additional qualitative check, simple averages of monthly mean temperature and precipitation are computed for eight stations representative of the region including Ayo Ayo, Belen, Calacoto, Charaña, Copacabana, LaPaz/El Alto, Patacamaya, and Viacha (Figure 1, station locations and elevations). Each station has a minimum of thirty years of record with a common period between 1970 and 1999 and has been quality controlled and statistically validated on monthly and annual bases.

Analysis of daily observations involves four annual indexes: two derived from daily minimum temperature (frost days and warm nights) and two from daily precipitation (consecutive dry days and precipitation >95th percentile; Frich et al. 2002; Alexander et al. 2006). Temperature extreme indexes are calculated for La Paz/El Alto (hereafter, La Paz) using the U.S. National Climatic Data Center's Global Surface Summary of the Day, covering the period from 1973 to 2007. Precipitation extreme indexes for Patacamaya (1951–1999) were provided by the CCl/CLIVAR/JCOMM Expert Team on Climate Change Detection and Indices (ETCCDI) Climate Extreme Indices data set available at their Web site (http://cccma.seos.uvic.ca/ETCCDI/data.shtml). Although direct, quantitative comparison between observed extreme indexes and the same indexes computed from daily model output is not possible due to differences in their definitions and the character of area versus point measurements (Alexander et al. 2006; Chen and Knutson 2008), a check on the consistency between observed and projected trends and direction of change is useful.

Results

Annual Cycle

The Altiplano experiences a distinct dry season during the cold (May–August) months and wet season (September–April) with peak rainfall during Austral summer. A simple average of the station observations (blue lines in Figure 2) shows the monthly mean annual cycle of temperature varies from 10°C in December to 4°C in July, and precipitation varies between a maximum of 4 mm/day in January and a minimum of less than 0.5 mm/day in July. Figure 2 also shows the area-averaged climatological annual cycles from the UDEL observed estimates (black) and the twentieth-century simulations (box plots) for the analyzed region. For the specified Altiplano region, the gridded UDEL estimates show good agreement with Altiplano station observations, indicating that our simple area average appears to be adequate for this qualitative evaluation. The models (standard box and whisker plots representing interquartile range, median, and outliers) agree in the timing of the temperature annual cycle and, as expected due to the lower model elevations, they demonstrate a warm bias with respect to the station data throughout the year. The models' precipitation bias are large and positive during the rainy season, showing twice the observed estimates in the multimodel average. The models do capture the dry season in the months of May through August and the overall timing of the rainy season. The discrepancy between simulated and observed precipitation could have several causes. Whereas precipitation increases with elevation in the high mountains and in the rain shadow of the Andes to the west and south (in the Atacama Desert; Houston and Hartley 2003), on the windward side the largest amount of precipitation is in lower elevations and the eastern slope leading up to the Altiplano, where moisture is abundant. Because the models' elevation are too low, surplus simulated moisture is transported to the Altiplano from the Amazon basin. In addition, some of the disagreement between observed and simulated precipitation may be due to inaccuracy in the measurements, such as rain gauge undercatch (e.g., Legates and Willmott 1990).

The projected changes in the annual cycle, given the assumptions of the A2 scenario for the period from 2070 through 2099, are shown in Figure 3 for standardized temperature and precipitation differences between twenty-first- and twentieth-century climatologies in the multimodel ensemble statistics. The model projections are in agreement that temperatures in the Altiplano region are likely to increase (median differences are greater than five standard deviations, σ, or ~4°C) through the annual cycle. These temperature projections are consistent with previous analyses for the Andes (Bradley et al. 2006; Urrutia and Vuille 2009). Figure 3B shows that precipitation increases during the rainy season (January–April, +8 percent) and the dry season intensifies, as seen in the negative median

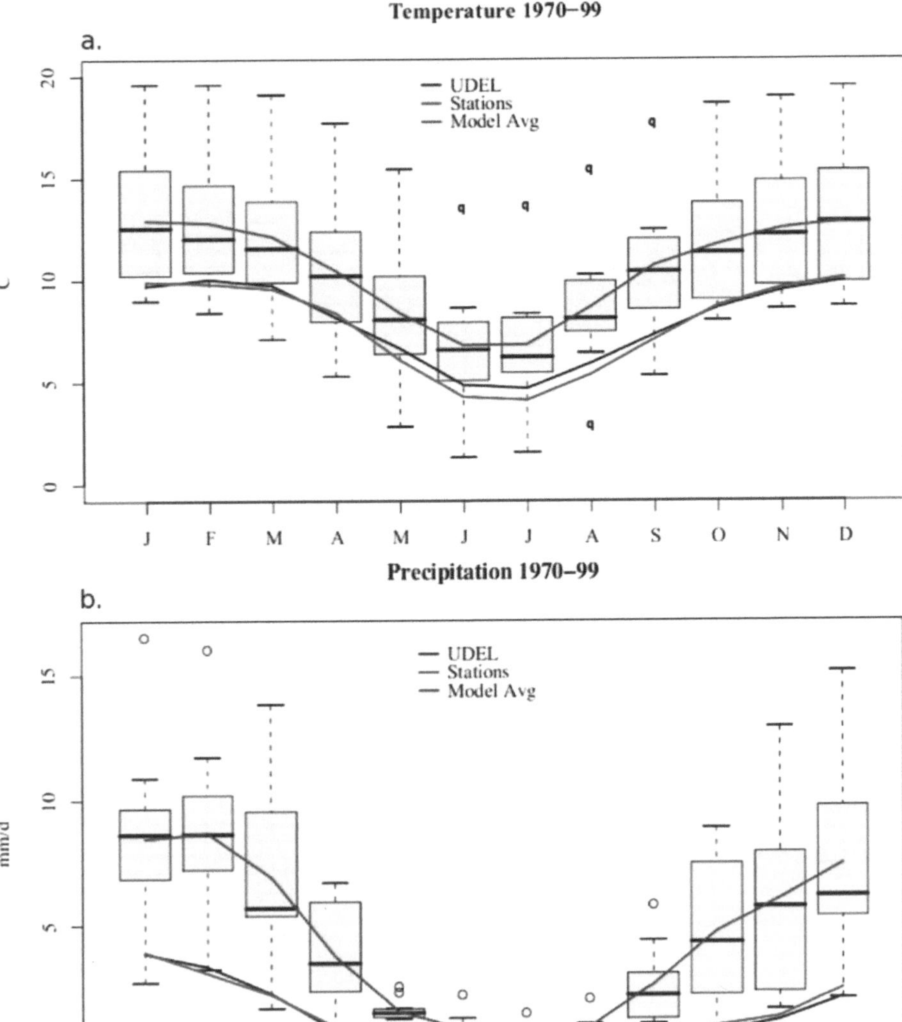

Figure 2. (A) Annual cycle of temperature (°C) for the Altiplano region (67°–70°W, 16°–19°S) from stations (blue), UDEL (black), multimodel average (red), and twentieth-century multimodel distribution (box and whiskers). (B) Same, but for precipitation (mm/day). UDEL = University of Delaware.

differences during winter (May–July) with decreases persisting during the early rainy season (September–November, -5 percent). The precipitation results are consistent with the projections of the large-scale SAM (Seth, Rojas, and Rauscher 2010), however, it must be noted that the spread among the models is larger than the projected changes, which suggests large uncertainty.

We then ask if there is observational evidence of such changes in the precipitation annual cycle that might support the model projections, which show drying in spring and increased rainfall during summer. We examine the Patacamaya station precipitation records and show in Figure 4 time series of September through November and January through March precipitation since 1960, with small linear trends (red) suggesting drying in spring and increases in summer. Time series for Viacha and Belen (not shown) indicate similar trends. Although these trends are not definitive, the directions of change support the model projections of drier early and wetter main rainy seasons. Further, these results are consistent with the changes projected for the large-scale SAM (Seth, Rojas, and Rauscher 2010).

Extremes

We begin our discussion of extreme indexes with analysis of observations. Probability distribution functions (PDFs) of precipitation extreme indexes at Patacamaya are shown for two periods (1951–1979 and

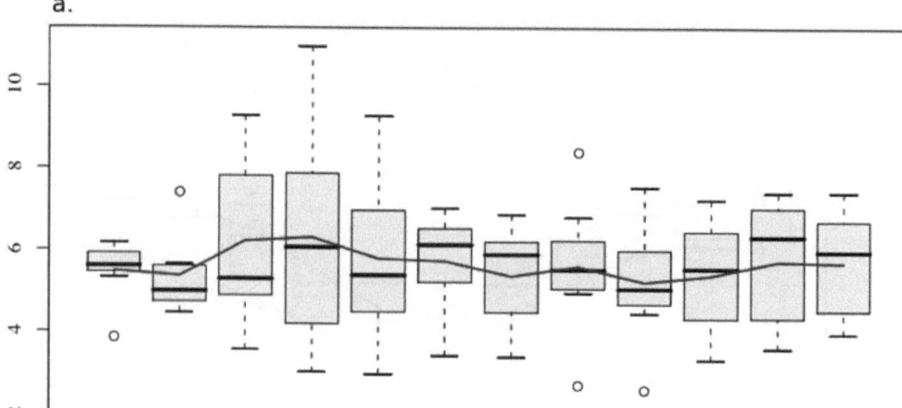

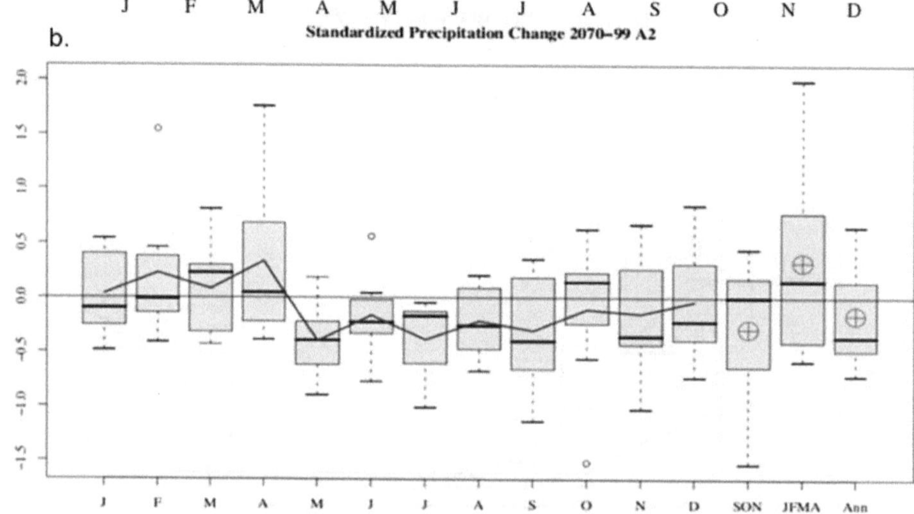

Figure 3. (A) Altiplano region monthly standardized temperature differences A2–twentieth century from multimodel ensemble (boxes and whiskers) and multimodel average (line). (B) Same, but for precipitation. Averages for September, October, and November (SON), January, February, March, and April (JFMA), and annual are also shown with stars representing model means. *Source:* Adapted from Thibeault, Seth, and Garcia (2010).

1980–1999) and indicate positive shifts in both dry days (Figure 5A) and precip >95th (Figure 5B) from the middle to late twentieth century, confirming earlier results (Haylock et al. 2006). La Paz temperature indexes are shown in Figures 5C and 5D (1980–1990) for qualitative evaluation of the multimodel ensemble for the recent period. Positive trends have been found in both frost days and warm nights (Thibeault, Seth, and Garcia 2010). Although quantitative comparisons between the simulated and observed indexes are problematic (Chen and Knutson 2008), a comparison of the directions of these observed trends with those found for the twenty-first-century projections provides a check on the consistency of the model results.

The models do not show much change in the distribution of dry days during the recent period (Figure 6A). However, consistent with the observed positive shift, multimodel projections indicate a shift toward longer periods of dry days by 2020 through 2049, which increases further by the end of the century. Similarly, the simulated distributions of precip > 95th for the twentieth century also show little change (Figure 6B). By 2020 through 2049, precip > 95th shows a shift toward higher values, with increasing variability, and continues into 2070 through 2099. Compared with the observed values of the precipitation indexes, the multimodel indexes have values that are consistent with lower elevation and excess moisture in the models; dry days are too few and precip > 95th has minimum and median values that are too high. Nevertheless, increases in both dry days and precip > 95th in the simulated precipitation indexes agree with the direction of trends identified in the observations at Patacamaya.

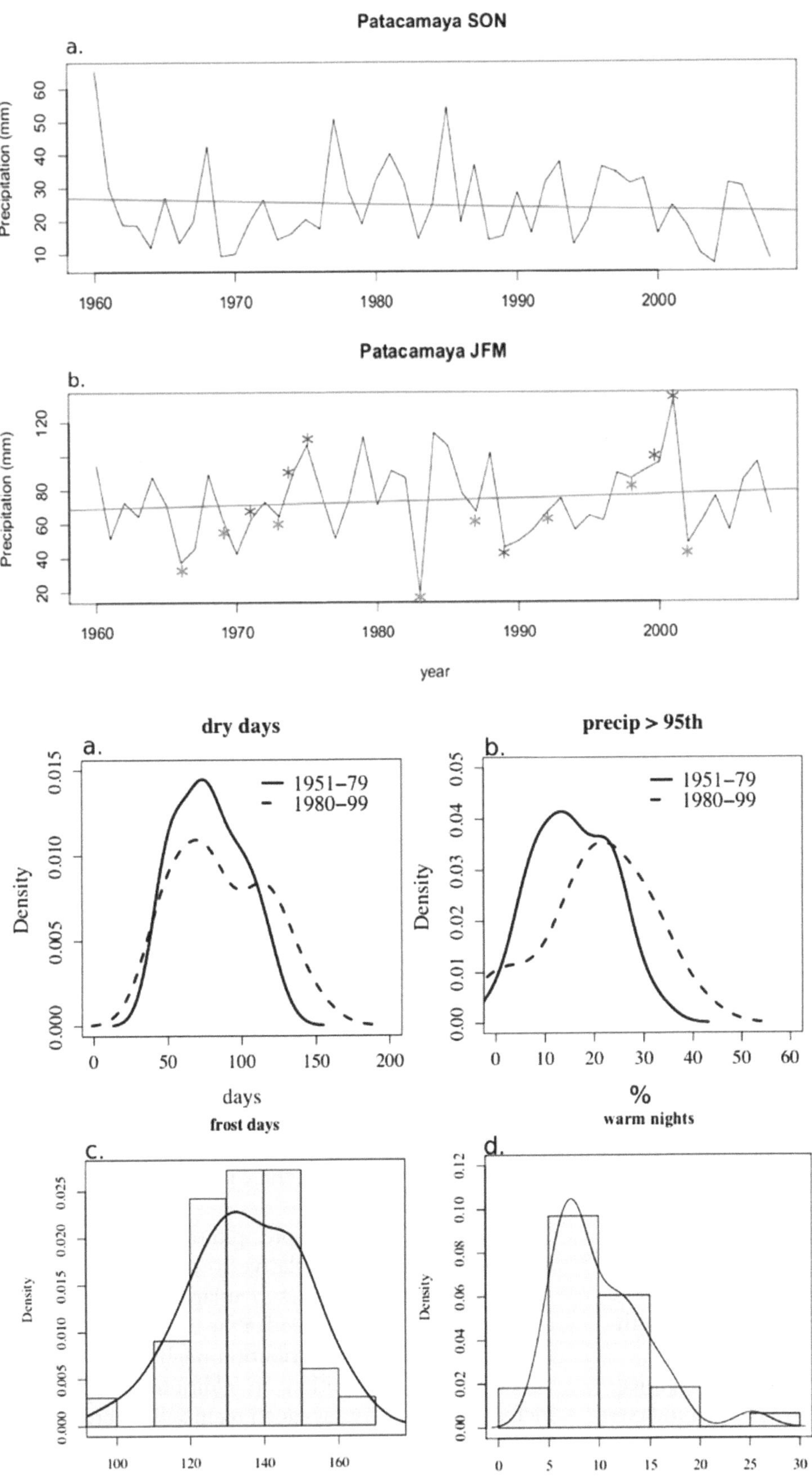

Figure 4. Observed 1960 through 2008 time series from Patacamaya for (A) September, October, and November (SON) and (B) January, February, March (JFM) precipitation (mm) shown with linear trends (red line) and for JFM, years defined by central equatorial Pacific (Niño 3.4) sea surface temperature anomalies above 1°C (El Niño, red stars) and below –1°C (La Niña, blue stars).

Figure 5. Probability distribution functions of extreme indexes computed from Patacamaya daily precipitation distribution for 1951 through 1979 (solid) and 1980 through 1999 (dashed) (A) dry days and (B) precip > 95th, and from La Paz/El Alto daily minimum temperature for 1973 through 2007, (C) frost days and (D) warm nights. *Source:* Adapted from Thibeault, Seth, and Garcia (2010).

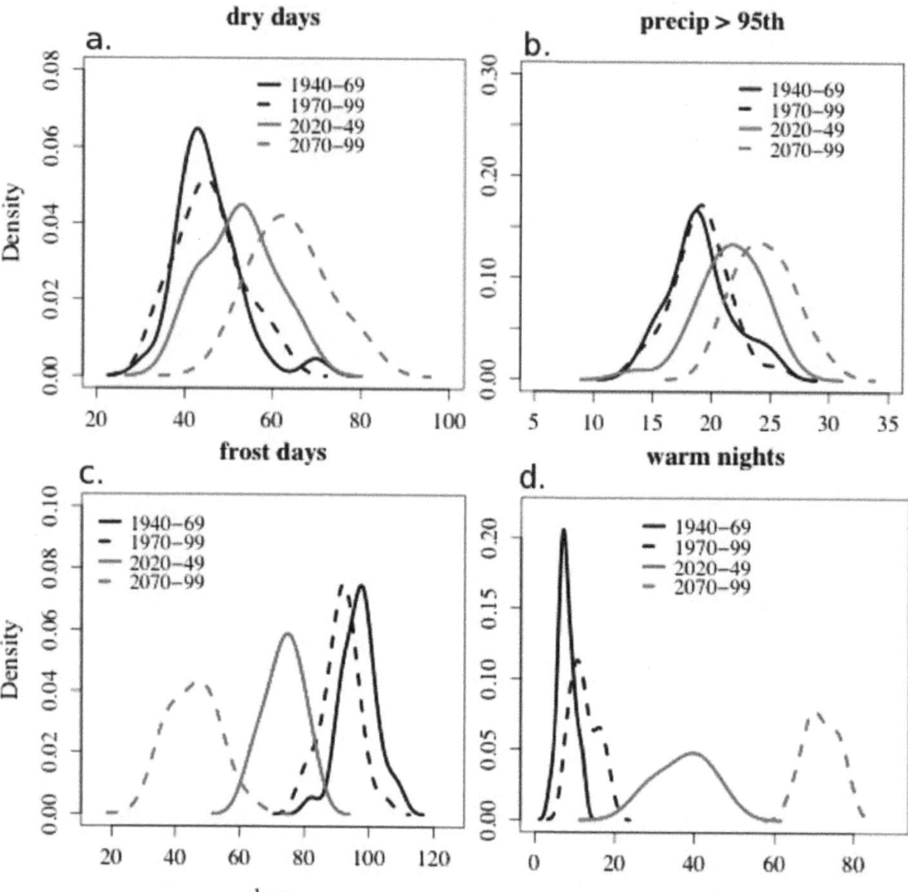

Figure 6. Extreme indexes computed from the multimodel ensemble for the periods shown with gray representing the A2 scenario for (A) dry days and (B) precip >95th, (C) frost days, and (D) warm nights. *Source:* Adapted from Thibeault, Seth, and Garcia (2010).

Distributions of frost days (Figure 6C) show a steady shift toward lower values from the middle twentieth century to the late twenty-first century, by which time the distribution of frost days do not share any values with those from the twentieth century, suggesting that the maximum number of annual frost days in the late twenty-first century will be fewer than the minimum number of annual frost days in the twentieth century. PDFs of the multimodel averages for the twentieth century suggest a shift toward more warm nights and higher variability in 1970 through 1999 relative to 1940 through 1969 (Figure 6D). By the middle century (2020–2049), there is a shift toward more warm nights with increased variability. By the late century (2070–2049), the PDFs show larger shifts toward more warm nights, not sharing any values with the twentieth-century distributions.

Both the simulation of recent trends and projections of warm nights agree well with observed indexes at La Paz. The models underpredict frost days at La Paz, likely due to warm and wet biases in the models. The decrease in simulated frost days is not in agreement with the increasing trend observed at La Paz. However, the increase at La Paz can be understood within the context of observed precipitation changes: In the early stages of warming, an increase in clear nights might lead to an increase in radiation frosts. The projections of both frost days and warm nights are consistent with warming and show shifts in PDFs that are completely outside the range of twentieth-century values by the end of the century. However, because the models at present are not able to simulate the recent increase in frost days, these projections have added uncertainty.

Projected PDF shifts of precipitation indexes suggest increases will be small relative to temperature; many twenty-first-century values overlap with twentieth-century values, yet by the end of the twenty-first century the mean of the new distribution might be in the tail of the present distribution. In addition, the expected increases in both dry days and precip > 95th are consistent with and provide further support for the projected changes in the annual cycle, with an extension of the dry season and intensification of the rainy season.

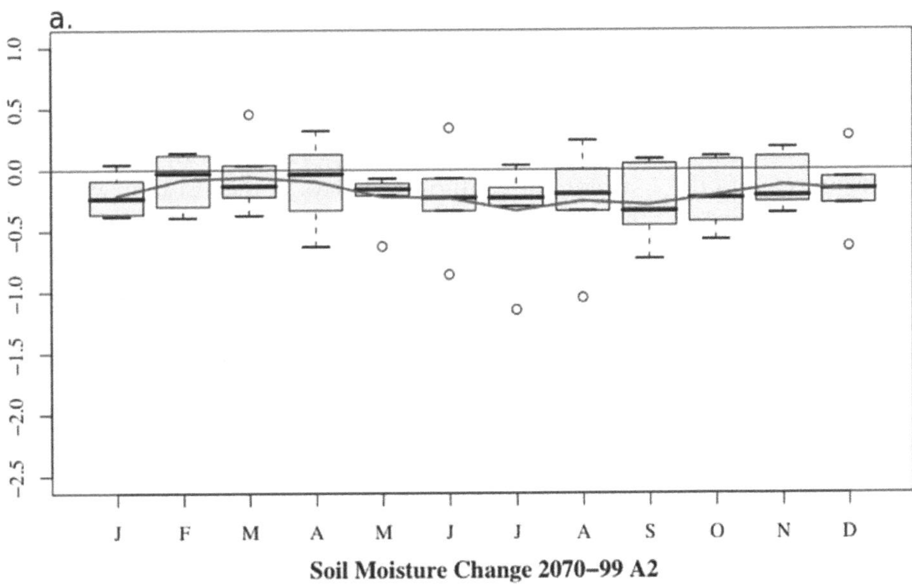

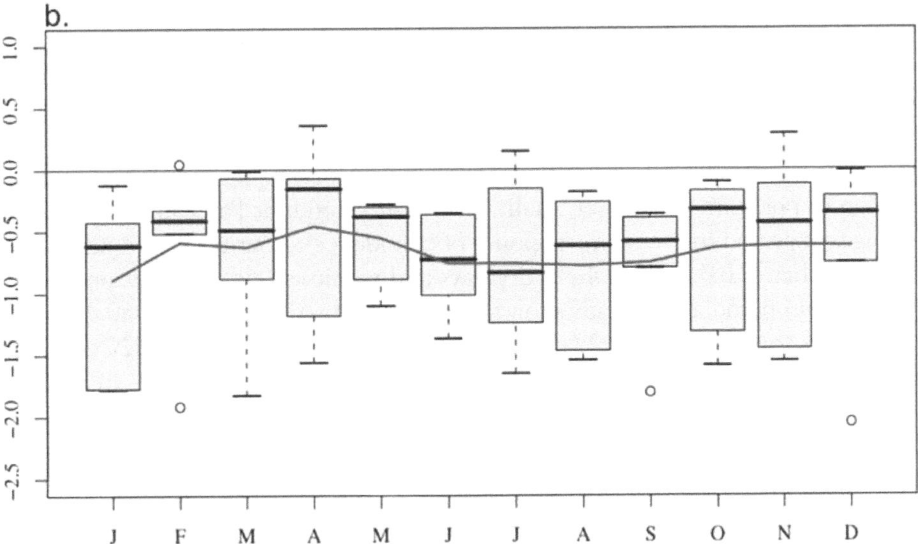

Figure 7. Monthly standardized soil moisture differences A2–twentieth century from multimodel ensemble (boxes and whiskers) and multimodel average (line) for (A) 2020–2049 and (B) 2070–2099.

Soil Moisture

The semi-arid Altiplano is dotted with wetland areas, peat bogs called *bofedales*, which provide critical water and feed (pastures) for highland farm animals (South American Camelids; e.g., alpaca, llama) and also sequester carbon. The *bofedales* are fragile and changes in the water regime can result in rapid and irreversible loss of habitat (Alzérreca et al. 2006). Land surface parameterizations incorporated by the global models vary in number of soil layers and soil water capacity and the absolute value of soil water (in centimeters) is not meaningful. For this reason, we focus on the standardized soil moisture changes (see Wang 2005) to enable intermodel comparison. Standardized differences in soil moisture between the twenty-first and twentieth centuries from the multimodel ensemble are shown in Figure 7. By the mid-twentieth century, decreases of up to $.5\sigma$ are seen in the dry and early rainy seasons, and during the late rainy season there is little or no drying in soil moisture due to the increased precipitation in the projections. By the end of the twenty-first century a larger decrease ($\sim.5$–1σ) is projected despite the expected increase in rainfall during the rainy season. This decrease is due to the large increases in temperature, which result in further losses of soil moisture due to increased evaporative demand. There are questions as to whether this mechanism is likely to dominate under realistic warming conditions and clearly further study is needed.

Discussion and Conclusions

Climate model projections for regions with complex topography must be taken with caution (e.g., Biasutti, Battisti, and Sarachick 2003), and the results presented here will require further testing with improved, high-resolution climate models. The intermediate-resolution global models represent a plateau with elevations ranging from 2,000 to 3,800 m. Although quantitative comparisons between station and grid-average (simulated) data are not emphasized, observed extreme indexes computed from daily precipitation at Patacamaya and daily minimum temperature at La Paz have been used to interpret the projections and suggest that the models might be able to provide reliable qualitative information about the direction of future changes in temperature and precipitation for the Altiplano.

Our results suggest the timing of the annual cycle is well simulated by the multimodel ensemble for the twentieth century, although the models exhibit a warm and wet bias in the Altiplano region. In a warmer world, the multimodel ensemble indicates little change in the annual mean precipitation. However, the results do suggest a change in the seasonal distribution of precipitation, which is consistent with the changes expected in the large-scale SAM (Seth, Rojas, and Rauscher 2010). The models indicate a more intense rainy season and a dry season, which extends into and weakens the early season rains. Weakening spring and strengthening summer season rainfall trends are also seen in the station observations for the period from 1960 through 2008. We note that the projected changes for the twenty-first century are small compared with the intermodel variability, suggesting substantial uncertainty in the projections for precipitation.

However, the consistency with large-scale projected changes is intriguing. The mechanism adapted from Seth, Rojas, and Rauscher (2010) for this change involves increased drying associated with the subsiding branch of the Hadley circulation during winter; as solar heating increases in spring, additional time is needed to build the moist static energy required to initiate convection (Lintner and Neelin 2006); the weakening of the tropical circulation (increased stability) further inhibits the development of convection (Vecchi and Soden 2007); and once the threshold of moist static energy is achieved, the atmospheric moisture content is greater, creating the opportunity for more intense precipitation. This consistency (between the Altiplano and the large-scale SAM) implies that in these seasonally dry areas the transition from dry to wet season might require more time.

Analysis of extreme indexes shows qualitative agreement between observed recent trends and changes projected by the multimodel ensemble for dry days, precip > 95th, and warm nights. The decrease in simulated twentieth-century frost days is inconsistent with the observed increasing trends in frost days at La Paz. Observed increases in frost days can be understood within the context of observed precipitation indexes; reductions in precipitation frequency might lead to an increase in the frequency of clear nights and radiation frosts. That models are not able to simulate this effect highlights the need for further study with improved tools.

Projections of precipitation extremes provide insight into the projected shift in the annual cycle for the Altiplano (Thibeault, Seth, and Garcia 2010). Increasing dry days are likely to occur during the dry season, which is expected to lengthen and weaken the early rainy season. Positive trends in precip > 95th are consistent with annual cycle projections indicating a shorter rainy season characterized by less frequent but more intense rainfall events. That projected precipitation extremes agree qualitatively with trends at Patacamaya provides some measure of confidence to our results. In addition, PDFs of dry days and precip > 95th at Patacamaya already show shifts in their distributions for 1980 through 1999 that are not seen in the models until 2020 through 2049, suggesting the possibility that changes in precipitation might occur sooner than the models project. Because the warming in the tropical Andes might be larger and occur at a faster rate than what the CMIP3 models project (Bradley et al. 2006; Vuille et al. 2008), the changes in extremes might also be larger and felt sooner than our results indicate.

Battisti and Naylor (2009) pointed out that projected mean growing season temperatures through much of the tropics will likely be outside the range of extremes experienced in the twentieth century and emphasized that present knowledge systems might be inadequate to deal with such changes. Our results indicate similar large shifts in temperature indexes and although shifts in precipitation indexes are more moderate, the projected means are beyond present-day extremes. This synthesis of observations and current global model projections suggests, given a business-as-usual scenario, that changes in the Altiplano are likely to have serious consequences for water management and indigenous agriculture. However, a suite of improved, high-resolution model scenarios and additional high-quality observations are essential to the further evaluation of climate projections for this region.

Acknowledgments

The authors thank two anonymous reviewers for insightful and constructive comments, which have improved the presentation of this research. We are grateful to Edwin Yucra for painstaking assembly of station observations used here and to Richard Mrozinski for lending expertise in map-making. The authors thank the international modeling groups for providing their data for analysis, the Program for Climate Model Diagnosis and Intercomparison for collecting and archiving the model data, and the JSC/CLIVAR Working Group on Coupled Modeling and their Coupled Model Intercomparison Project for organizing the model data analysis activity. The IPCC Data Archive at Lawrence Livermore National Laboratory is supported by the Office of Science, U.S. Department of Energy. This research was supported by a Long-Term Research award (LTR-4) in the Sustainable Agriculture and Natural Resource Management Collaborative Research Program with funding from USAID.

References

Alexander, L. V., X. Zhang, T. C. Peterson, J. Caesar, B. Gleason, A. M. G. Klein Tank, M. Haylock et al. 2006. Global observed changes in daily climate extremes of temperature and precipitation. *Journal of Geophysical Research-Atmospheres* 111:D05109. http://www.agu.org/pubs/crossref/2006/2005JD006290.shtml (last accessed 9 July 2010).

Alzérreca, H., J. Laura, F. Loza, D. Luna, and J. Ortega. 2006. The importance of carrying capacity in sustainable management of key high-Andean Puna rangelands (Bofedales) in Ulla Ulla, Bolivia. In *Land use change and mountain biodiversity*, ed. E. Spehn, M. Lieberman, and C. Kömer, 167–86. Boca Raton, FL: CRC Press.

Battisti, D. S., and R. L. Naylor. 2009. Historical warnings of future food insecurity with unprecedented seasonal heat. *Science* 323:240–44.

Biasutti, M., D. S. Battisti, and E. S. Sarachick. 2003. The annual cycle over the Tropical Atlantic, South America, and Africa. *Journal of Climate* 16:2491–508.

Bradley, R. S., M. Vuille, H. F. Diaz, and W. Vergara. 2006. Threats to water supplies in the tropical Andes. *Science* 312:1755–56.

Carbon dioxide emissions rise to record levels. 2008. *Nature* 455:581.

Carleton, A. M. 1999. Methodology in climatology. *Annals of the American Association of Geographers* 89:713–35.

Chen, C.-T., and T. Knutson. 2008. On the verification and comparison of extreme rainfall indices from climate models. *Journal of Climate* 21:1605–21.

Cox, P., and D. Stephenson. 2007. Climate change—A changing climate for prediction. *Science* 317:207–08.

Frich, P., L. V. Alexander, P. Della-Marta, B. Gleason, M. Haylock, A. M. G. K. Tank, and T. Peterson. 2002. Observed coherent changes in climatic extremes during the second half of the twentieth century. *Climate Research* 19:193–212.

Garcia, M., D. Raes, S. E. Jacobsen, and T. Michel. 2007. Agroclimatic constraints for rainfed agriculture in the Bolivian Altiplano. *Journal of Arid Environments* 71: 109–21.

Garreaud, R. D., and P. Aceituno. 2001. Interannual rainfall variability over the South American Altiplano. *Journal of Climate* 14:2779–89.

Gilles, J. L., and C. Valdivia. 2009. Local forecast communication in the Altiplano. *Bulletin of the American Meteorological Society* 90:85–91.

Giorgi, F., and R. Francisco. 2000. Uncertainties in regional climate change prediction: A regional analysis of ensemble simulations with the HADCM2 coupled AOGCM. *Climate Dynamics* 16:169–82.

Haylock, M. R., T. C. Peterson, L. M. Alves, T. Ambrizzi, Y. M. T. Anunciação, J. Baez, V. R. Barros et al. 2006. Trends in total and extreme South American rainfall in 1960–2000 and links with sea surface temperature. *Journal of Climate* 19:1490–512.

Held, I. M., and B. J. Soden. 2006. Robust responses of the hydrological cycle to global warming. *Journal of Climate* 19:5686–99.

Houston, J., and A. J. Hartley. 2003. The central Andean west-slope rainshadow and its potential contribution to the origin of hyper-aridity in the Atacama Desert. *International Journal of Climatology* 23: 1454–64.

Latif, M., and N. S. Keenlyside. 2009. El Niño/Southern Oscillation response to global warming. *Proceedings of the National Academy of Sciences* 106:20578–83.

Legates, D. R., and C. J. Willmott. 1990. Mean seasonal and spatial variability in gauge-corrected, global precipitation. *International Journal of Climatology* 10: 111–27.

———. 1992. A comparison of GCM-simulated and observed mean January and July precipitation. *Global and Planetary Change* 97:345–63.

Lintner, B. R., and J. D. Neelin. 2006. A prototype for convective margin shifts. *Geophysical Research Letters* 34:L05812. http://www.agu.org/pubs/crossref/2007/2006GL027305.shtml (last accessed 9 July 2010).

Meehl, G. A., C. Covey, T. Delworth, M. Latif, B. McAvaney, J. F. B. Mitchell, R. J. Stouffer, and K. E. Taylor. 2007. The WCRP CMIP3 multimodel dataset—A new era in climate change research. *Bulletin of the American Meteorological Society* 88:1383–94.

Neelin J. D., C. Chou, and H. Su. 2003. Tropical drought regions in global warming and El Niño teleconnections. *Geophysical Research Letters* 30(24): 2275. http://www.agu.org/pubs/crossref/2003/2003GL018625.shtml (last accessed 9 July 2010).

Nicholls, N., and L. Alexander. 2007. Has the climate become more variable or extreme? Progress 1992–2006. *Progress in Physical Geography* 31:77–87.

Pierce, D. W., T. P. Barnett, B. D. Santer, and P. J. Gleckler. 2009. Selecting global climate models for regional climate change studies. *Proceedings of the National Academy of Sciences* 106:8441–46.

Rowell, D. 2006. A demonstration of the uncertainty in projections of UK climate change resulting from regional model formulation. *Climatic Change* 79:243–57.

Seth, A., M. Rojas, and S. A. Rauscher. 2010. CMIP3 projected changes in the annual cycle of the South American Monsoon. *Climatic Change* 98:331–57.

Thibeault, J., A. Seth, and M. Garcia. 2010. Changing climate in the Bolivian Altiplano: CMIP3 projections for extremes of temperature and precipitation. *Journal of Geophysical Research—Atmospheres* 115: D08103. http://www.agu.org/pubs/crossref/2010/2009JD012718.shtml (last accessed 9 July 2010).

Thibeault, J., A. Seth, and G. Wang. Forthcoming. Precipitation mechanisms of precipitation variability in the Bolivian Altiplano: Present and future.

Urrutia, R., and M. Vuille. 2009. Climate change projections for the tropical Andes using a regional climate model: Temperature and precipitation simulations for the end of the 21st century. *Journal of Geophysical Research—Atmospheres* 114: D02108. http://www.agu.org/pubs/crossref/2009/2008JD011021.shtml (last accessed 9 July 2010).

Vecchi, G. A., and B. J. Soden. 2007. Global warming and the weakening of the tropical circulation. *Journal of Climate* 20:4316–40.

Vuille, M., B. Francou, P. Wagnon, I. Juen, G. Kaser, B. G. Mark, and R. S. Bradley. 2008. Climate change and tropical Andean glaciers: Past, present and future. *Earth Science Reviews* 89:79–96.

Wang, G. L. 2005. Agricultural drought in a future climate: Results from 15 global climate models participating in the IPCC 4th assessment. *Climate Dynamics* 25:739–53.

Willmott, C. J., and K. Matsuura. 2009. Terrestrial air temperature and precipitation: Gridded monthly time series V2.01 (1900—2008). http://climate.geog.udel.edu/~climate/html_pages/Global2_Ts_2009/README.global_t_ts_2009.html (last accessed 3 February 2010).

Parameterization of Urban Characteristics for Global Climate Modeling

Trisha L. Jackson, Johannes J. Feddema, Keith W. Oleson, Gordon B. Bonan, and John T. Bauer

To help understand potential effects of urbanization on climates of varying scales and effects of climate change on urban populations, urbanization must be included in global climate models (GCMs). To properly capture the spatial variability in urban areas, GCMs require global databases of urban extent and characteristics. This article describes methods and characteristics used to create a data set that can be utilized to simulate urban systems on a global scale within GCMs. The data set represents three main categories of urban properties: spatial extent, urban morphology, and thermal and radiative properties of building materials. Spatial extent of urban areas is derived from a population density data set and calibrated within thirty-three regions of similar physical and social characteristics. For each region, four classes of urbanization are identified and linked to a set of typical building morphology, thermal, and radiative characteristics. In addition, urban extent is simulated back in time to 1750 based on national historical population and urbanization trends. A sample set of simulations shows that the urban characteristics do change urban heat island outcomes. In general the simulations show greater urban heat islands with increasing latitude, in agreement with observations. [Supplemental material is available for this article. Go to the publisher's online edition of *Annals of the Association of American Geographers* for the following free supplemental resource: (1) a table of the Global Data Set of Urban and Building Properties © 2007–2009.]
Key Words: climate simulation, global climate change, urban climate, urban properties.

为了有助于理解不同尺度下城市化对气候的潜在影响，以及气候变化对城市人口的影响，全球气候模型（GCMs）必须包括城市化因子。要想正确把握城市空间的变异性，GCMs 需要包含城市范围和特性的全球数据库。本文介绍了一些方法和特性，它们被用来建立一个数据集，该数据集被用于在 GCMs 里模拟全球范围内的城市系统。数据集表现了城市特性的三个主要类别：空间范围，城市形态，建筑材料的热和辐射特性。城区的空间范围来自于人口密度数据集，并在 33 个具有相似的物质和社会特征的地区内进行了校准。对于每个地区，城市化的四个类别被确定并与一整套的典型建筑形态，热和辐射特征相链接。此外，根据国家人口和城市化的历史趋势，对城市空间范围的模拟一直回溯到 1750 年。模拟样本表明，城市特性确实影响城市热岛的结果。总的来说，模拟研究显示纬度增加往往伴随更大的城市热岛，次结果与实际观测相一致，。[本文的辅助材料可以通过下述途径获取。出版者的网络版本：美国地理学家协会年鉴网络版：（1）城市和建筑特性的全球数据集表格 c_ 2007–2009。] *关键词：气候模拟，全球气候变化，城市气候，城市特性。*

Para ayudar a comprender los efectos potenciales de la urbanización sobre los climas a escalas variables y los efectos del cambio climático sobre las poblaciones urbanas, tenemos que incorporar la urbanización en los modelos del clima global (GCMs, sigla en inglés). Para que la variabilidad espacial en áreas urbanas se capte adecuadamente, los GCMs deben contar con bases de datos globales sobre la extensión y características urbanas. Este artículo describe los métodos y características usados para crear un conjunto de datos susceptible de utilizarse en los GCMs para la simulación de sistemas urbanos a escala global. El conjunto de datos representa las tres principales categorías de las propiedades de lo urbano: extensión espacial, morfología urbana y propiedades térmicas y radiantes de los materiales con los que está construida la ciudad. La extensión espacial de las áreas urbanas se derivó desde un conjunto de datos de densidad de población y se calibró al interior de treinta y tres regiones de de características físicas y sociales similares. Para cada región se identificaron cuatro clases de urbanización y éstas fueron relacionadas con un conjunto de típicas características termales, radiantes y morfológicas de las edificaciones. Por otra parte, se hizo una proyección de la extensión urbana hacia el pasado para simular las condiciones de 1750, con base en las tendencias históricas nacionales de población y urbanización. Un conjunto de muestras de las simulaciones indica que las características urbanas en verdad cambian la magnitud de las islas de calor urbano. En general, a mayor latitud las simulaciones muestran mayores islas de calor urbano, lo cual concuerda con las observaciones. [Hay disponible material suplementario para este artículo. Ir a la edición electrónica del publicista de *Annals of the Association of American Geographers* por el siguiente recurso suplementario gratuito: (1) una tabla del Conjunto de Datos Globales de las Propiedades Urbanas y de las Edificaciones © 2007–2009.] *Palabras clave: simulación climática, cambio climático global, clima urbano, propiedades de lo urbano.*

Beginning with observations of the London heat island in the late nineteenth century (Howard 1833) and subsequent observational evidence (Landsberg 1981), there is clear evidence that urban systems represent the greatest human impact on local climates. Further analysis of these systems shows that urban systems affect the radiation and energy balance in predictable ways that allow these impacts to be effectively simulated (Oke 1974; Arnfield 2000, 2003). In addition to modifying local climates, urban systems are the primary interface between human activities and other human impacts on climate. For example, most anthropogenic emissions originate from urban areas and their peripheral regions (e.g., Decker et al. 2000; Mills 2007).

With the rapid growth of urban areas in modern society, it is likely that these urban systems will play an important role in human climate interactions in the future; thus, they should be included as a component of the future global climate models (GCMs; e.g., Oleson, Bonan et al. 2008a). Incorporating urban systems within a GCM framework has two main purposes: first, to understand how urban areas might be contributing to climate change and the spatial distribution of emissions and, second, to learn how urban populations will be affected by climate change (Oleson, Bonan et al. 2008a, 2008b) and to improve our understanding of urban climates on a global scale (Oleson et al. forthcoming). In addition, as governments work to address global warming, policymakers will first look at areas where most people reside, our cities. Using GCMs as a learning tool will help scientists test ideas and make recommendations about how to alter urban characteristics to lessen their impact on climate and energy consumption (Oleson, Bonan, and Feddema 2010). Additionally, if GCM projections show that urban responses to climate change differ compared to the response in natural environments, it is important that this differential response be included in climate impact assessment studies.

Realizing a need for and justifying the need for global urban modeling is a first step. However, any global scale modeling effort requires that the model work for all places and in all seasons. In addition, spatial characteristics and physical properties of urban systems vary widely because of cultural factors and the differing sources of primary building materials (Grimmond and Souch 1994; Grimmond and Oke 1999; Grimmond 2007). This work addresses the need for a global urban data set that provides urban spatial and physical properties that fulfill the parameter needs for modeling urban systems in GCMs. The three core parameterizations are (1) spatial extent of urban areas; (2) urban morphology characteristics including building heights, urban canyon height-to-width ratios, plan area of buildings, impervious (e.g., sidewalks and roads) and pervious covers (e.g., parks and lawns); and (3) thermal and radiative properties of building and road materials (Oleson, Bonan et al. 2008a). To illustrate the importance of these parameters we present simulations showing differences in the global distribution of urban heat islands with different assumptions about urban system properties.

Background

Urban systems affect climate through a number of mechanisms that affect the absorption and disposition of energy because of (1) urban canyon geometry, (2) heat storage in building materials, (3) reduced evapotranspiration due to the increase in impervious cover and subsequent decrease in vegetation, and (4) the trapping of atmosphere-warming pollutants via temperature inversion (Oke 1982, 1987; Oke et al. 1991).

Theoretical and satellite studies highlight the urban characteristics that result in urban climate modification, most often expressed as urban heat islands. First, the size, shape, and orientation of physical components, or urban morphology, explain in part the temperature contrast with adjoining rural areas. Additional alteration of the urban surface response depends on physical properties of these factors. These physical properties are subdivided into thermal properties (i.e., heat capacity, thermal conductivity) and radiative properties (i.e., emissivity, albedo), which combine to account for temperature differences. The combined morphology and physical properties of urban areas explain the differential radiation balance at the surface compared to surrounding rural areas. Moreover, these factors account for storage of heat in urban materials and the partitioning of energy into latent and sensible heat fluxes

(Landsberg 1981; Oke 1982, 1987, 1988, 1995). Because of the complexity of these interactions, urban climate modelers face many challenges, particularly when working at the global scale.

Several advances of modeling theory of urban climate interactions have occurred since Oke (1974) published an initial review of urban climatology. From that point forward, modelers began considering individual buildings, including wall and roof facets, each experiencing varying time exposure of solar radiation, net long-wave radiation exchange, and ventilation (Arnfield 1984, 2000; Paterson and Apelt 1989; Verseghy and Munro 1989a, 1989b). The horizontal aspect of urban systems contains impervious (e.g., paved roadways) and pervious surfaces (e.g., lawns) with different properties and interactions including radiative, thermal, aerodynamic, and hydrological elements (Oke 1979, 1989; Suckling 1980; Doll, Ching, and Kaneshiro 1985; Asaeda, Ca, and Wake 1996; Anandakumar 1999; Grimmond, Souch, and Hubble 1996; Kjelgren and Montague 1998). These horizontal surfaces along with walls of buildings define the urban canyon (Figure 1). This fundamental morphological urban unit can be scaled up to encompass the area needed for a particular model, from a city block to an entire city.

Previous modeling approaches include detailed mapping of urban morphology (Ellefsen 1990–1991; Grimmond and Souch 1994; Cionco and Ellefsen 1998) or direct observation of urban systems at aggregate scales, such as the objective hysteresis model by Grimmond, Cleugh, and Oke (1991) and mesoscale modeling to study interactions between urban systems and local weather patterns (Changnon 1992; Shepherd 2005). Other models have used the urban canyon concept to incorporate joint interactions of the combined budgets of component facets (Terjung and O'Rourke 1980a, 1980b; Mills 1993; Arnfield 2000; Masson 2000). Development of the parameter set in this study is intended for use with such urban canyon type models but can be modified to meet the needs of other modeling strategies.

A number of factors must be considered when designing a model that operates on a global scale, including computational efficiency and its ability to simulate urban systems in a wide variety of climates (Oleson, Bonan et al. 2008a). However, the response of urban systems will differ geographically based on the properties related to urban design and building materials used. These characteristics must be documented for a global model to effectively simulate urban responses to climate change. In addition, these characteristics must be able to change over time to better assess the impact of climate change on urban populations, which includes more than half of the total world population today and is projected to include about 65 to 70 percent of the global population by 2050 (Center for International Earth Science Information Network et al. 2004; Shepherd 2005).

The data set described here was developed to correspond with requirements of the National Center for Atmospheric Research's (NCAR) urban submodel (CLMU; Oleson, Bonan et al. 2008a, 2008b). CLMU is

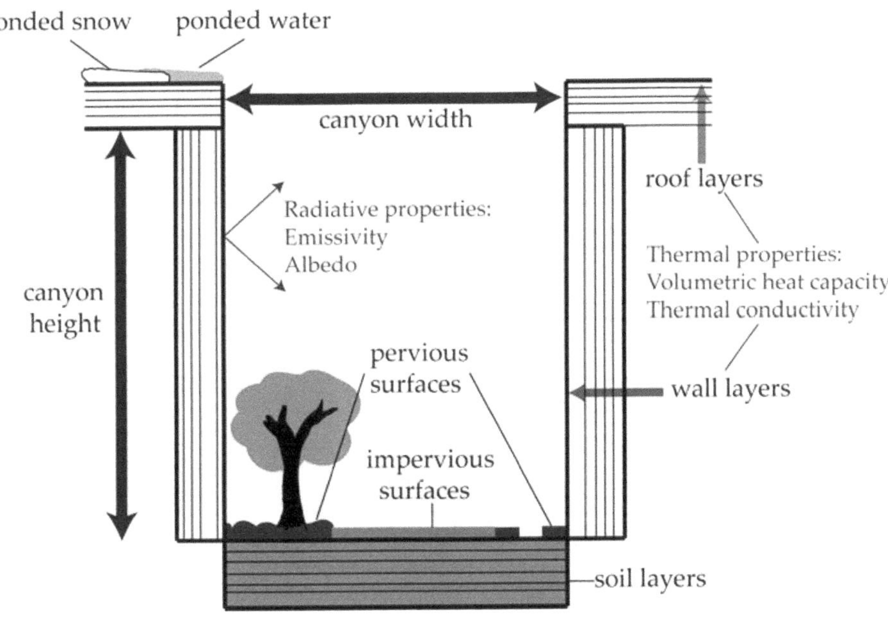

Figure 1. National Center for Atmospheric Research's (NCAR) urban submodel parameter requirements as discerned by the urban canyon concept.

a submodel of the Community Land Model (CLM; Oleson, Niu et al. 2008) of the Community Climate System Model (CCSM; Collins et al. 2006). In the newly released CCSM 4.0, the Earth's surface is divided into grid cells approximately at one degree resolution. To characterize the varied landscapes that occur within a grid cell at this resolution, each grid cell has the capacity to be unique. A CLM grid cell consists of a nested subgrid hierarchy including land units, plant functional types (PFTs), and snow/soil columns (Oleson et al. 2004). Presently, the CLM contains five land unit types of glacier, wetland, lake, vegetation, and, for version 4.0, urban. Of the five land unit types, only vegetation and urban classes can be further subdivided. Vegetation is divided into PFTs and the urban area is subdivided into five urban facets: sunlit and shaded walls, roofs, impervious surfaces (roads), and pervious areas in the canyon floor (Figure 2). The full description of the urban model's design and sensitivity tests can be found in Oleson, Bonan et al. (2008a, 2008b).

CLMU simulates six main energy balance processes: (1) absorption and reflection of solar radiation; (2) absorption, reflection, and emission of long-wave radiation; (3) momentum, storage, sensible, and latent heat fluxes; (4) anthropogenic heat fluxes due to waste heat from building heating and air conditioning; (5) heat transfer in roofs, building walls, and roads; and (6) hydrology (Oleson, Bonan et al. 2008a). The urban model calls for morphologic, thermal, and radiative input parameters to efficiently simulate these processes. Urban canyon morphology is captured by urban canyon height-to-width ratio, average building height, roof fraction (as a percentage of total area), pervious and impervious canyon floor fractions, roof and wall thicknesses, and impervious canyon floor thickness. Thermal parameters include thermal conductivity and volumetric heat capacity for roofs, walls, impervious and pervious canyon floor materials, minimum and maximum interior building temperatures, and soils of differing textures. Radiative properties are emissivity and albedo of roofs, walls, and impervious and pervious canyon floor materials.

This detailed information provided to the model aids in capturing the character of an urban area. Considering

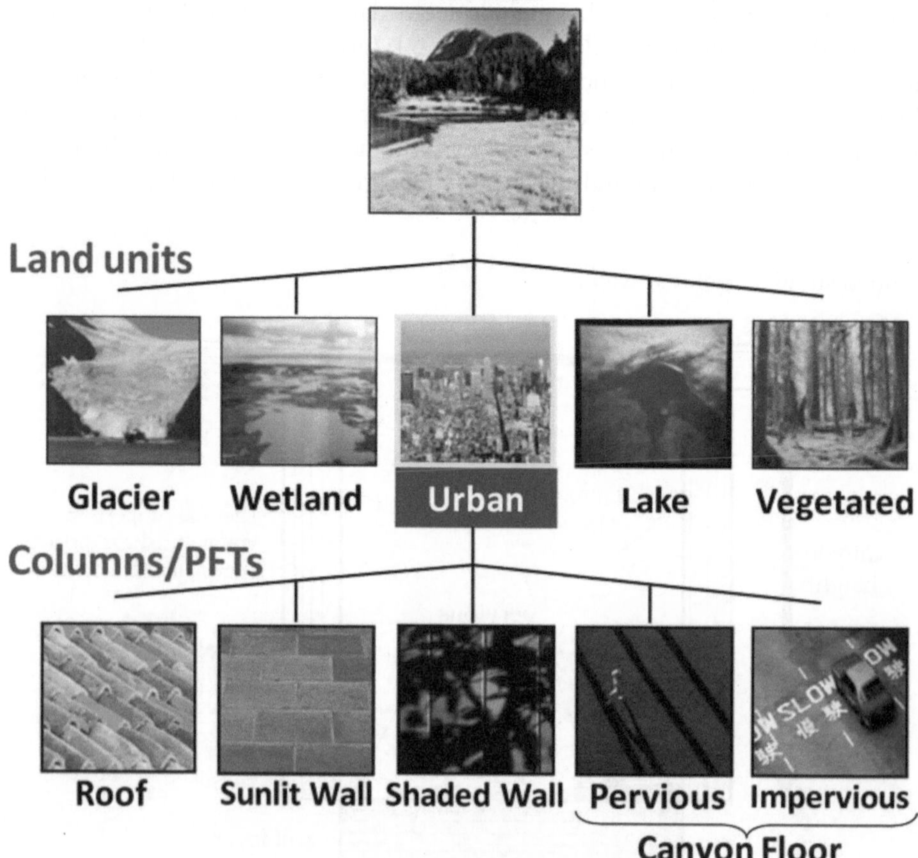

Figure 2. Schematic representation of the makeup of a typical Community Land Model grid cell (Oleson, Bonan, and Feddema 2006). The urban component is broken up into five facets to calculate the exchange of energy between the surfaces and the urban canyon. PFT = plant functional types.

Figure 3. Thirty-three regions with similar urban character. Delineating factors include climate and housing characteristics.

that cities across the world vary widely in terms of structure (i.e., different canyon widths, different building heights) and building materials, these parameters successfully portray the dissimilarities from region to region and between urban categories (e.g., tall building districts vs. low-density residential areas).

Methodology

A three-step process was employed to complete the urban data set. Initially, the Earth's land surfaces were divided into manageable regions with similar urban character. Next, spatial extent of four categories of urban intensity was determined and, finally, a database of building and road properties for each regional category was compiled.

To better manage the information at a global scale we divided the world into regions. With generalization of regions, an initial data set can be completed in a timely manner to provide a framework that allows for future improvements in the detail of the data. Therefore, the division of the globe into regions with similar urban character was our first step (Figure 3). In making these generalizations, physical and cultural geography played a significant role, as well as the realization that for many regions this information is difficult to obtain, limiting our ability to further subdivide areas that we realize are not necessarily homogenous in political or even building characteristics. Thus, the overriding criteria are based on the fact that buildings are designed to withstand weather conditions of an area, so climate was one determinant, as well as available resources for building materials (Givoni 1976; Olgyay 1992) in discerning regional boundaries. Culture also plays a role in the structure of buildings and cities (Clarke 1989).

For this study we evaluated each of the thirty-three regions (Figure 3) to determine unique urban–rural boundaries and then to delineate among four levels of urban intensity based on population density. Population density proved to be the most reliable proxy for urban intensity after several satellite-derived landcover products were evaluated for defining urban extent (e.g., MODIS, GLC2000, and DISCover). Not only do these products show significant disagreement on the location and size of urban areas, but they also do not provide levels of urban intensity, which is a required parameter of CLMU. Population density as provided by LandScan already considers slope, landcover, nighttime lights, proximity to roadways, and census data to determine where people are most likely living (Dobson et al. 2000). Population density is not a perfect determinant of urban concentration, but it provides an easy transition to temporal usage, so it offers the ability to answer more questions than a satellite-based snapshot of urban areas. With emergence of new population data sets, the time series of population densities lengthens, assisting in the study of urban evolution over time.

Using LandScan 2004 population densities (Oak Ridge National Laboratories 2005), a lower limit of population density for each region was selected to define the urban–rural boundary. Once urban areas were defined, they were further delineated into four levels of urban intensity, including tall building district (TBD), high density (HD), medium density (MD), and low density (LD). Population density boundaries for each of these urban levels differ widely to account for different types of urban communities. For example, East Africa contains many small farm plots housing large families. This causes the population density to be relatively high for an agriculture-based community, and it has many characteristics of, for example, U.S. suburbs, such as low buildings and a high percentage of green space.

Intraurban boundaries (e.g., between low and medium density) were based on population density and observations of satellite imagery in at least ten sample cities per region (i.e., validation cities). Validation cities were chosen based on size (i.e., large urban population), location (i.e., to geographically represent the entire region), and best available resolution of satellite imagery in Google Earth. The images in Google Earth come from a variety of sources (e.g., TerraMetrics, NASA, DigitalGlobe), and the available resolutions vary as well, from 15 m in most cases (EarthSat 2007) to less than 1 m (DigitalGlobe 2007). During comparisons of LandScan population density (Dobson et al. 2000) to satellite images, natural boundaries typically presented themselves in the form of a relatively large disparity of neighboring LandScan pixels corresponding to a change in urban land surface properties (e.g., vegetated fraction, building density, etc.) as seen in Google Earth.

For each region, boundaries were assigned based on an initial validation city and then adjusted for each subsequent validation city. The initial cities were then rechecked to ensure that the adjustments still properly represented their intraurban boundaries.

To maintain consistency from region to region during this process, definitions for the urban categories were established from the start. TBDs here are defined as an area of at least 1 km^2 with buildings greater than or equal to ten stories tall, with a small fraction of vegetation (i.e., 5–15 percent of plan area). Many cities that might appear to have a TBD were not included in this data set because the aerial extent was too small (i.e., less than 1 km^2) or because the population density of the TBD was too sparse. HD areas can encompass commercial, residential, or industrial areas and are characterized by buildings three to ten stories tall with a vegetated or pervious fraction typically in the range of 5 to 25 percent. MD areas are usually characterized by row houses or apartment complexes one to three stories tall with a vegetated or pervious fraction of 20 to 60 percent. Finally, the LD category covers areas with one- to two-story buildings and a vegetated or pervious fraction of 50 to 85 percent. LD includes a variety of urban types, from the suburbs of the United States to urban agricultural parts of East Africa.

As a final step in defining urban extent, an inter-region comparison looked at disparities on the global scale such as population percentile rank of each boundary. This ensured that regions with similar qualities were consistent in their assignment of urban categories. In this process, minor adjustments were made to the urban–rural and intraurban boundaries to maintain consistency in urban levels from region to region.

On further evaluation we concluded that only TBD, HD, and MD urban types can be adequately simulated in the CLMU. This decision is based on the fact that in suburban or high-population-density agricultural areas the assumptions about radiation trapping and aerodynamic processes in the urban canyon model do not hold true when vegetation is a dominant component of the landscape. We propose to develop a separate suburban model in the future to properly handle such landscapes. Thus, for the purpose of global models only, the TBD, HD, and MD classes should generally be used to simulate urban systems in an urban canyon type model. The intraurban categories (TBD, HD, MD, and LD) must be adaptable to any location. For example, LD encompasses suburbs in the United States and urban agriculture in East Africa, and similar differences exist for the other urban classes. For the model to delineate between places that fall within the same category, additional information is required for the buildings and surrounding area at each locale. Therefore, we define urban morphology characteristics for each of the 132 regional categories (i.e., 33 regions × 4 categories) including average building heights (H), urban canyon height-to-width ratios (H:W), and fraction of pervious surface (e.g., vegetation), roof area, and impervious surfaces (e.g., roads and sidewalks). In this article we summarize our methods, but precise values and references for each measurement are presented in the database provided as supplemental material to this article.

To estimate average building heights for the validation cities, a global database of tallest buildings by cities was used for TBDs within each region (Emporis Corporation 2007). Where available, the twenty-five tallest buildings were averaged for five cities within a region, then the average building heights of these cities were calculated to determine an average regional tall building value. This average was then decreased proportionately to account for the remaining buildings in the TBD, resulting in an average TBD building height representative of each regional category. In regions with no TBD (e.g., Greenland), or those with few cities with a TBD (e.g., West Africa), the maximum number of buildings in the buildings database was used in the calculation.

For the HD category, Emporis Corporation (2007) data were used where available and were supplemented by imagery of verification cities. By studying imagery of city skylines, building heights could be estimated by comparing TBD heights to HD buildings. To further aid the estimation process, the number of stories of HD buildings was counted to approximate heights. For the remaining categories of MD and LD, building heights were determined based primarily on dominant housing types. For instance, a typical two-story frame home in the United States is approximately 8 m tall. References to local building codes and estimations from imagery supplement this method. Further, because part of the data set presented here includes typical building types for each category, the gathering of this information also helped resolve typical building heights.

Local building codes and other municipal documentation were also used to ascertain road widths and other required information (e.g., lawn area) for calculation of H:W ratios of representative urban canyons. Where official documentation was absent, road widths were found on the Internet, usually as part of information necessary for emergency planners and other government agencies. Once road widths were known and additional canyon floor lengths (e.g., lawns, sidewalks) were estimated based on imagery or municipal documentation, H:W could be calculated with building heights for each regional category. To illustrate, consider a suburban area with 8-m-tall homes with a 24-m canyon floor (including roads and lawns); the H:W equals 0.33 (i.e., 8 m/24 m = 0.33).

Urban morphology also includes fraction of pervious area (e.g., vegetation), roof area, and impervious area (e.g., roads and sidewalks). Imagery from Google Earth for each regional category's validation cities provided the primary tool for estimating these fractions. In essence, the spatial extent data set was used to identify a cluster of pixels within a validation city for the desired urban category (i.e., TBD, HD, MD, LD). Then the location was identified in the satellite imagery and studied (which often involved using the distance measuring tool to calculate actual plan area) before an estimation of these fractions was assigned. Percentage pervious and roof area were initially estimated, and then percentage impervious could be deduced based on this information. Because the majority of impervious surfaces in urban areas are roadways, it is considered to be the road fraction in the model input.

Urban morphology considers the three-dimensional form of the urban canyon, but additional information on thermal and radiative properties of building materials is required to effectively capture urban interaction with climate at many scales (Bonan 2002). The primary reason for distinguishing building types is to promote the model's ability to simulate thermal and radiative exchanges within the urban canyon. Buildings are the main component of the urban canyon, especially in terms of how their structure and materials influence the canyon's thermal storage and radiative balance. Therefore, a database of typical building types and their associated properties was compiled.

By studying imagery, construction data by country, and other published documentation on building types and housing for every region (e.g., Canada's National Research Council Institute for Research in Construction), a list of the most common building types in the world was assembled. For the 132 regional categories, the three most common building types and roof types were determined and their relative abundance was reported in percentage terms (e.g., walls: 50 percent wood frame home, 40 percent brick home, 10 percent stone home; roofs: 70 percent asphalt shingle, 20 percent ceramic tile, and 10 percent wood shingle). Learning what constitutes common building and roof types involved background research in building trade literature as well as studying documentation describing historical and cultural determinants of typical buildings in a region today and in the recent past (Givoni 1976; Olgyay 1992; Straube and Burnett 2005). However, there were many cases in which imagery and geographical knowledge played a large role in approximating the most common building and roof types as well as establishing the primary road type found in each category.

Establishing common wall, roof, and road types laid the groundwork for ascertaining their individual makeup as represented by portioning walls and roof into discrete layers. With each building type there was a typical method of construction, or a way of assembling walls. Once the order of the construction materials was ascertained, each respective wall or roof type was divided into ten layers with information on each layer's thickness (Oleson, Bonan et al. 2008a). The road types were limited to the number of actual layers present; subsequent layers are occupied by soil.

In a wall or roof with fewer than ten actual layers of materials, each material can occupy more than one layer in a row to satisfy model requirements. For instance, one building type is "mud or adobe." Because mud is the only wall material, mud was stated as the material for every layer and the sum of individual layer thicknesses equals the standard thickness of a mud or adobe home. Thus, by determining common construction techniques for different types, a list of typical building materials divided into ten layers could be established for each wall and roof type (see the supplemental Excel file).

Next, thermal and radiative properties were collected for each construction material including density, specific heat capacity, and thermal conductivity. Only for surface materials were albedo and emissivity values collected. CLMU calls for volumetric heat capacity (Oleson, Bonan et al. 2008a), which was calculated using density and specific heat capacity where necessary. The compiled lookup table of material thermal and radiative values was used; the literature sources of the values are presented in the supplemental materials. A final calculation was the summation of total wall thickness. The entire process was repeated for roofs and roads. The result was a database of typical wall, roof, and road types with associated thermal properties by layer and radiative properties of surface layers. Based on this background, a table of three wall types, three roof types, their relative abundances, and one road type was constructed for each regional category.

At present, CLMU can accommodate only one urban class, and only one set of representative thermal and radiative properties within the TBD, HD, and MD categories can be used. The class selected is based on which is most abundant in a GCM grid cell, and the values are made up of weighted averages of the three typical wall and roof types within that class. To arrive at these representative values, a first step is to calculate whole-wall thermal conductivity, which requires that the values must be converted to thermal resistance (i.e., R-value), an additive property. Thermal conductivity is conceptually and mathematically related to thermal resistance by dividing the wall layer width by its thermal conductivity. R-values of the ten layers are then summed to arrive at a whole-wall (or whole-roof) R-value for each type. This value is divided by its respective wall or roof width to arrive at the whole-wall or whole-roof thermal conductivity. Calculating R-values for the whole wall or roof allowed comparison of these values to published estimates of these values. Because the published values show that walls in reality are much less thermally efficient compared to the theoretical values associated with the layers of materials that make up a wall, we must consider sources of heat transfer. For instance, we calculated R-values in series for some walls (i.e., mud or adobe) because the wall is consistent throughout (i.e., the layers are in series) and calculated R-values in parallel for other walls (i.e., wood frame walls) to help account for thermal differences in layers that occur in parallel (i.e., in the same lateral space; for example, windows in a wall).

Another important consideration in developing a data set of thermal building properties that represents the real world is to consider thermal bridging and air leakage. Thermal bridging initiates heat transfer through materials where there is a discontinuity in the insulation or construction material (e.g., staples or nails used to hold insulation in place, wood or steel columns in a wall, internal corners of walls, floors, and ceilings, etc.). Thermal bridging along with air leakage (e.g., from chimneys, windows, doors, porous building materials such as brick, joints between materials, etc.) cause buildings to be less thermally efficient.

To account for the majority of thermal inefficiencies of buildings, which are poorly documented and known to vary from region to region and within regions (Rashkin et al. 2008), total wall resistance was multiplied by a factor of 0.2 for buildings with windows based on published estimates of typical resistance losses and estimates that suggest that windows account for 22 to 37 percent of total heat loss and occupy 15 to 40 percent of wall area (Rashkin et al. 2008). Roofs and glass curtain and corrugated iron walls were excluded from this adjustment because roofs have no windows and glass curtain walls are made up almost wholly of windows, so their thermal properties already account for this, and corrugated iron walls typically have no windows.

Using these whole-wall or whole-roof thermal conductivities, along with the other thermal and radiative values (i.e., volumetric heat capacity, emissivity, and albedo), weighted average thermal values were calculated for TBD, HD, and MD regional categories to arrive at representative values considering the three main building types. For example, if three wall and roof types in a given category are weighted at 40, 30, and 30 percent, then their respective properties are also weighted in this way to arrive at one representative value. This value represents the weighted average properties of the typical wall or roof in a given category. Only one road type was listed for each category, eliminating the need for calculating averages.

A final step in preparing the urban properties database for CLMU was to provide estimates of temperature ranges for building interiors for the 132 urban categories. These general estimates are based on seasonal temperature averages and extremes, and a subjective evaluation of availability of heating and cooling methods (i.e., prominence of air conditioning).

The final product includes a geographic information system (GIS)-based global data set of urban extent divided into four urban categories. The 1-km resolution data set is in geographical grid format. For use in the CCSM, data are also aggregated to a half-degree grid (i.e., the 0.05 degree resolution data is summed by class to 0.5 degree aggregates) and, based on historical population estimates and histories of urbanization by region, we have extrapolated urban extent back to 1750. Associated data tables can be linked to the urban extent grid via region names. Tables included are wall types, roof types, road types, material properties, percentage of pervious and roof areas, thermal bridging and insulation calculations, a master table of three common building and roof types with their relative frequency as well as road types and internal temperature ranges per regional category.

Evaluation of the Data Set

To explore the importance of spatially explicit urban characteristics we performed two uncoupled global CLM simulations at T42 resolution (~2.8° longitude by 2.8° latitude). The simulations were driven by a fifty-seven-year (1948–2004) atmospheric forcing data set from Qian et al. (2006). In other words, the climate conditions (radiation, precipitation, etc.) were held constant for both experiments. In the control experiment we summed global regional areas of TBD, HD, and MD to create an urban extent and applied the properties described by Voogt and Grimmond (2000)

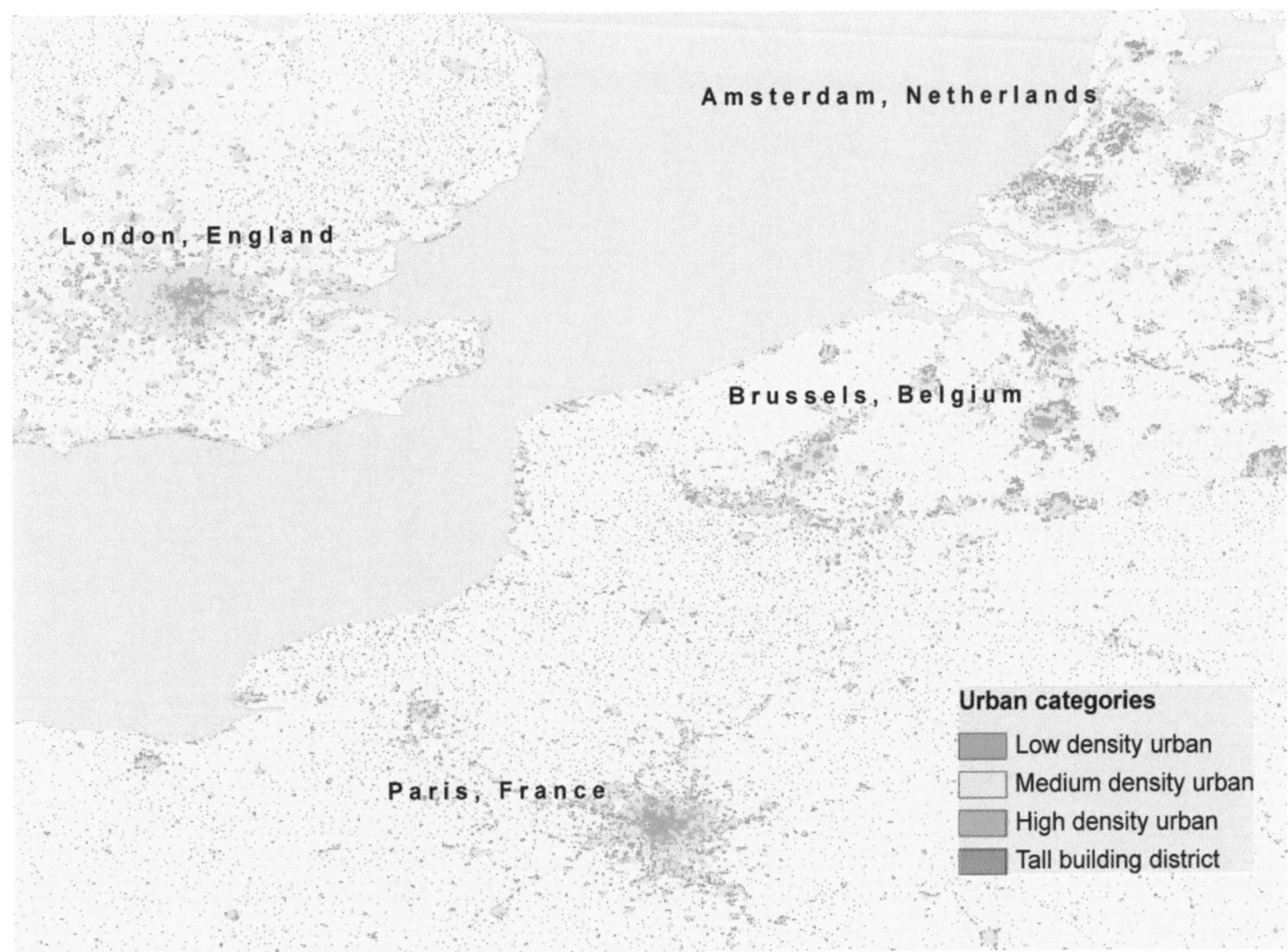

Figure 4. Urban areas, by category, of Western Europe. One pixel represents about 1 km.

for Vancouver, British Columbia, to all urban areas. Building interior temperatures were maintained between 288.72 and 299.83 K; however, we did not implement the model function that includes expelling waste heat from heating and air conditioning back into the urban canyon (Oleson, Bonan et al. 2008a). In a second experimental simulation we assumed the same urban extent but applied the regional urban properties associated with the urban class (TBD/HD/MD) that represented the largest urban area in the grid cell. Because unintended interactions can occur when mixing properties of different urban classes, only one class's properties should be used in an analysis. Analysis of the simulations is based on an evaluation of the heat island as measured by the difference between 1984 through 2004 climatologies of the urban reference height temperatures in a grid cell minus the average rural (the vegetation and soil fraction areas in the grid cell) reference height temperature. For those interested in how other variables are simulated by the model, please refer to a more detailed analysis of model responses in Oleson et al. (forthcoming). In our analysis, observations are only shown for grid cells with urban extent.

Results and Discussion

Most striking about the final data set presented here is the stark differences in urban characteristics from place to place, whether looking at the form of cities, the materials they are built from, or the density of the populations living there. Spatial extent shows variations in the form of cities, such as the dense, concentric ring form of older cities such as Paris, France, and London, England (Figure 4), contrasted against most U.S. cities, which sprawl to cover large areas (Figure 5). Urban morphology displays another aspect of this contrast, with modern cities teeming with skyscrapers resulting in very high urban canyon height-to-width ratios versus the shorter, denser TBDs in places like India.

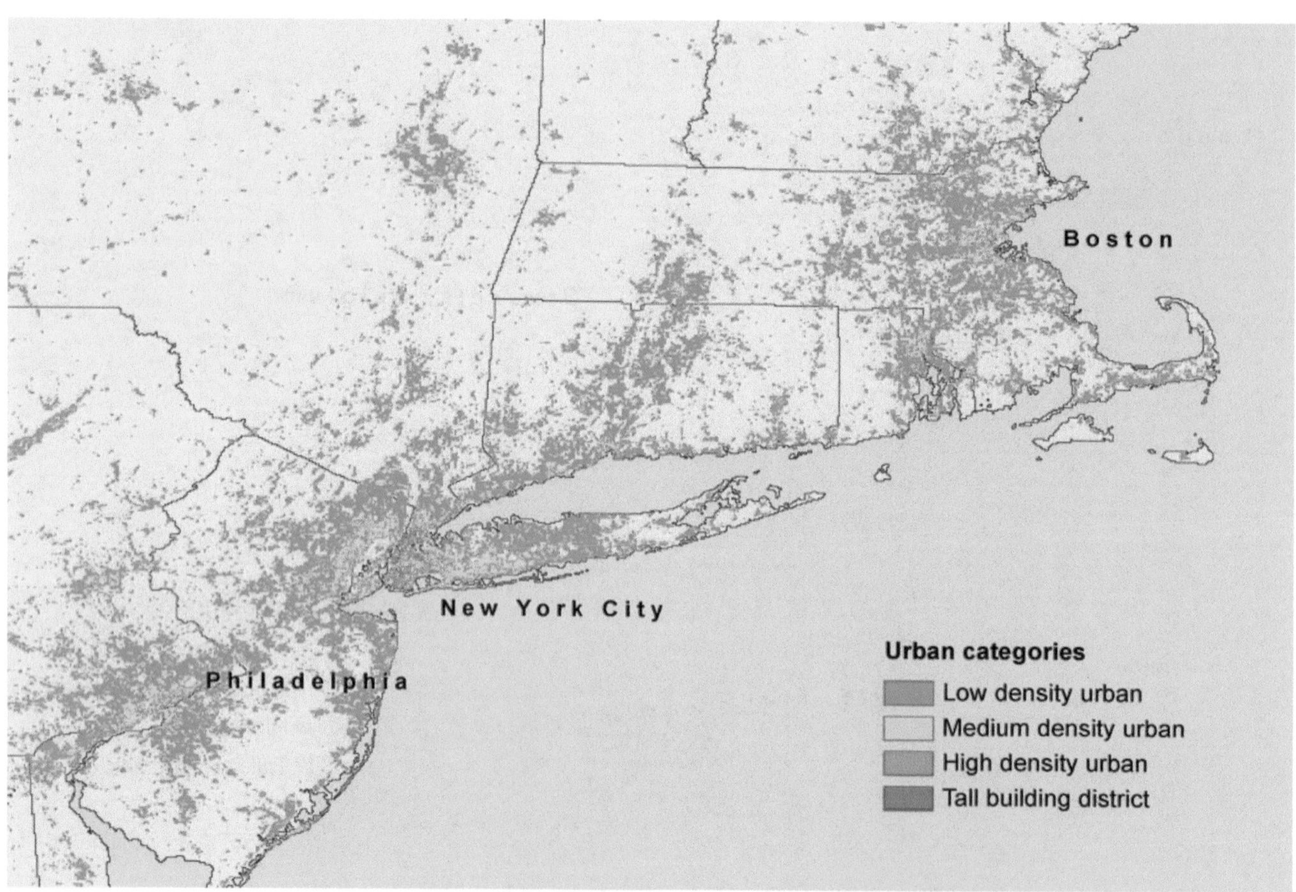

Figure 5. Urban areas, by category, of the Northeast United States. One pixel represents about 1 km.

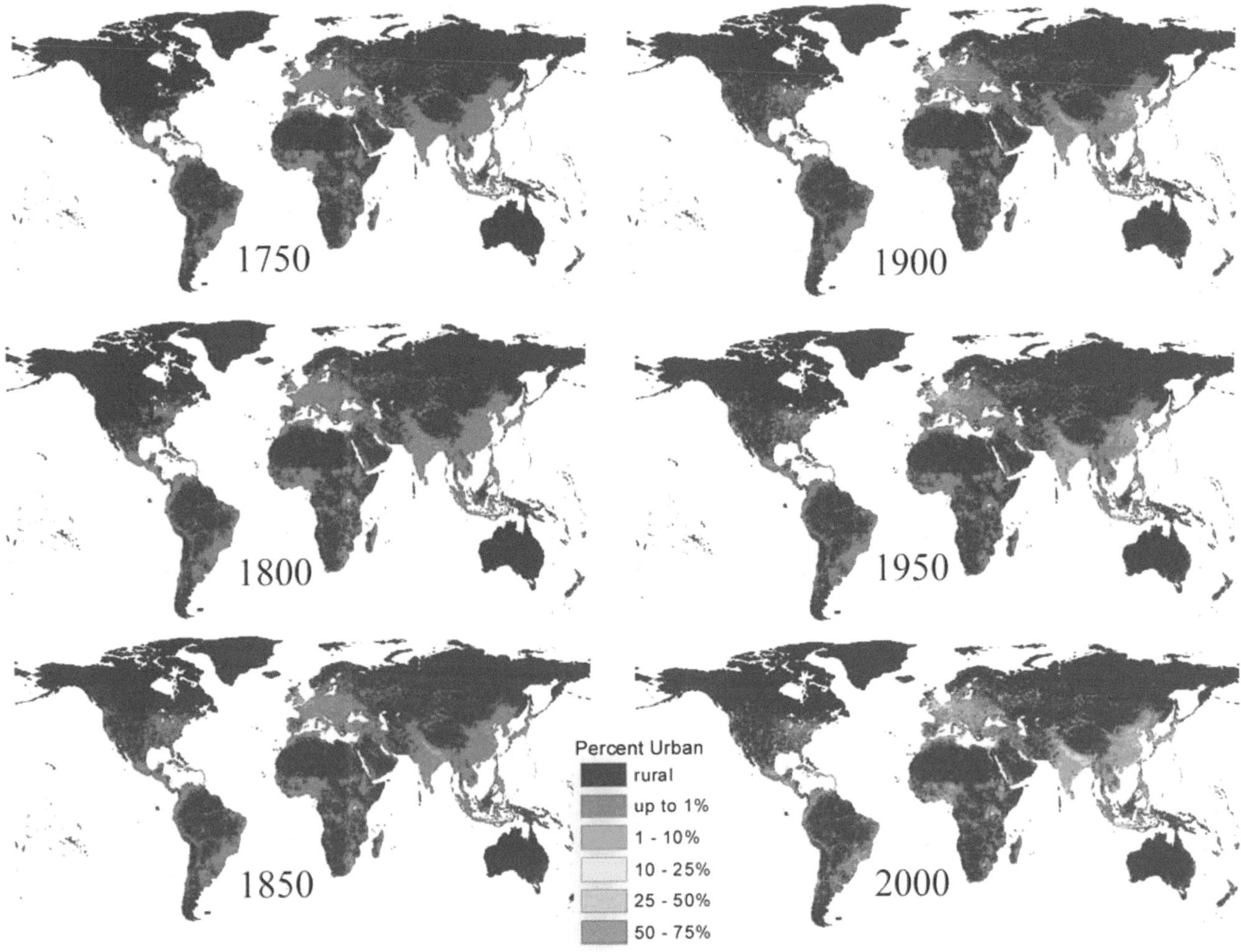

Figure 6. Global estimates of urban extent from 1750 to present.

For the CLM simulation only the TBD, HD, and MD urban extent were used. We extrapolated these areal extents backward in time to 1750 using historical national population estimates and our own estimates of urbanization (Figure 6). It is also possible to compare our estimate of present-day urban area for these three classes against existing data from the HYDE 3.0 (Klein Goldewijk and van Drecht 2006) data set (Figure 7). The HYDE 3.0 urban areas are largely based in United Nations–defined urban areas, which vary from country to country. Hence, we see large differences in the data sets, with the United States showing less urban area in our data set because we omitted the suburban class, and regions like India and China being more urban in our data because our data set aggregates relatively small but high-frequency urban centers like villages that show high population densities in the LandScan 2004 data set.

Urban spatial extent and morphological parameters effectively capture visible differences in urban areas, whereas thermal and radiative properties capture less apparent yet sharp contrasts between regions. A summary of estimated median, quartile, and ranges for each parameter are summarized over the thirty-three regions (Figure 8) including published values for two test sites used to validate the model (see Oleson, Bonan et al. 2008a). Take, for instance, average thermal conductivity for typical wall materials for each region in the middle density category (Figure 9). The figure is presented in such a way to display a wall's insulating capability. Note that many low-latitude regions contain walls that function as poor to extremely poor insulators,

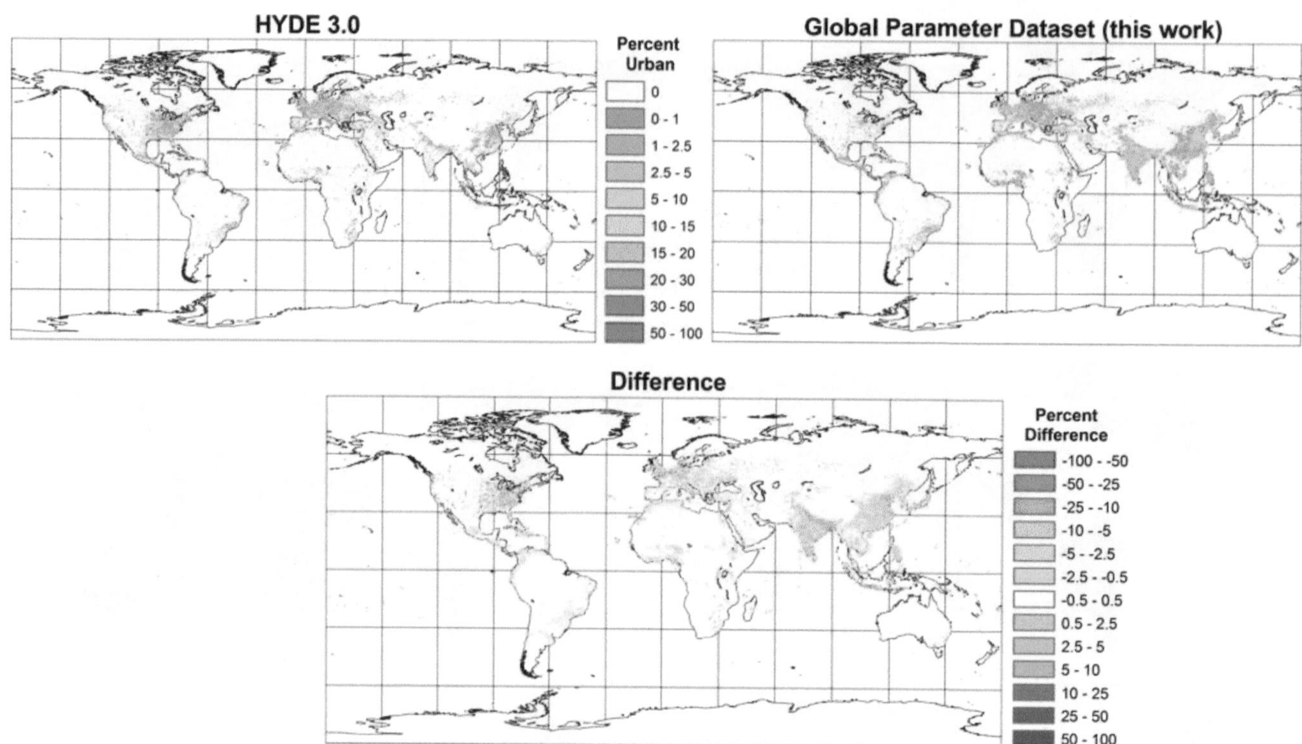

Figure 7. Comparison with the summed tall building district, high density, and medium density urban extent based on this work and the HYDE 3.0 (Klein Goldewijk and van Drecht 2006) urban areas of the world.

an effective adaptation for warm climates. On the contrary, there is no apparent latitudinal pattern associated with albedo of roof surfaces in low-density areas (Figure 10A). However, Australia's effort to increase usage of roof materials with high albedo is apparent (Salhani 2007). Using highly reflective roofs or other less conventional materials for roof cover (e.g., grass) is a topic receiving a great amount of attention in the United States and other countries as a way of mitigating the urban heat island effect (Rashkin et al. 2008). CLMU has the capability to measure the effects of these types of changes on the atmosphere, which will be useful to policymakers, builders, and other stakeholders with the common goal of creating more efficient cities (e.g., Oleson, Bonan, and Feddema 2010).

The final CLM input data set differs slightly from the data presented in the supplementary materials because in each grid cell the CLM selected the properties associated with the most abundant urban type. For comparison we show the albedo for a single LD urban class (Figure 10A), which can be compared to the input for roof albedo as implemented in the model (Figure 10B).

Our model simulations show that whether using the Vancouver (Figure 11) or data set parameterization (Figure 11) we simulate effective heat island trends across the globe, with tropical areas typically having lower heat island impacts compared to higher latitude regions where more energy is added to the system to maintain building temperatures (for more details on the complete model response, please refer to Oleson et al. forthcoming). Yet, there are also significant differences between the Vancouver simulation and the parameter data set simulation. Differences in wall and roof thermal and radiative properties and interior temperature settings result in tropical heat islands being significantly lower, typically about 0.5 K, when using the new parameters compared to a typical Vancouver building structure in the same location. Heat islands increase by a maximum of about 1 K in parts of Eurasia primarily because of different wall and roof properties and use of home heating.

As with any first effort of building a global database, there are issues and limitations associated with the process and with the proper use of the data. The primary concern in constructing a database of this scope was

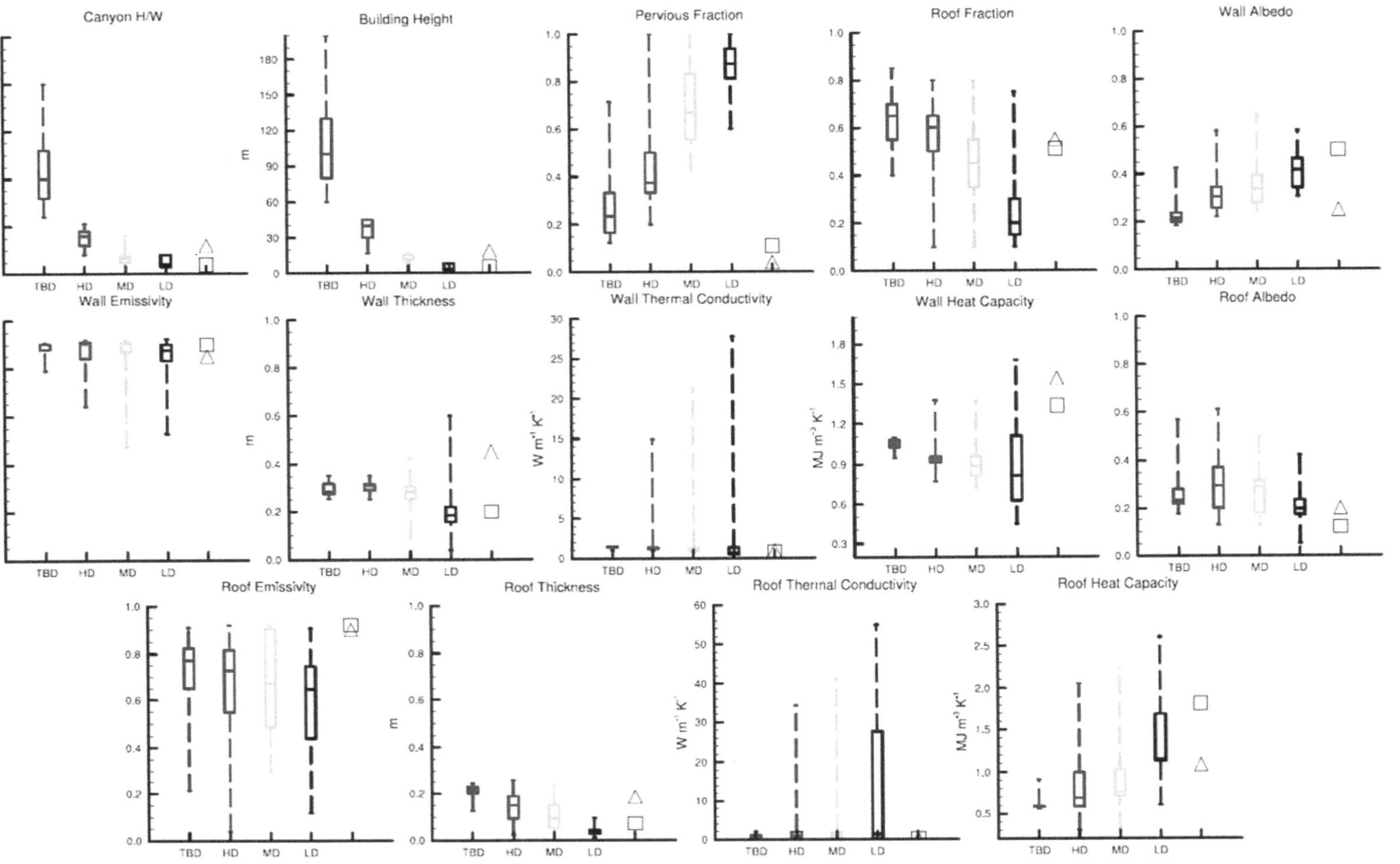

Figure 8. Box plots of parameter values estimated for the thirty-three regions. For reference, values for Vancouver (open triangles) used in the control experiment (Voogt and Grimmond 2000) and Mexico City (open squares) are also included (Jauregui 1986). H/W = height/width; TBD = tall building district; HD = high density; MD = medium density; LD = low density.

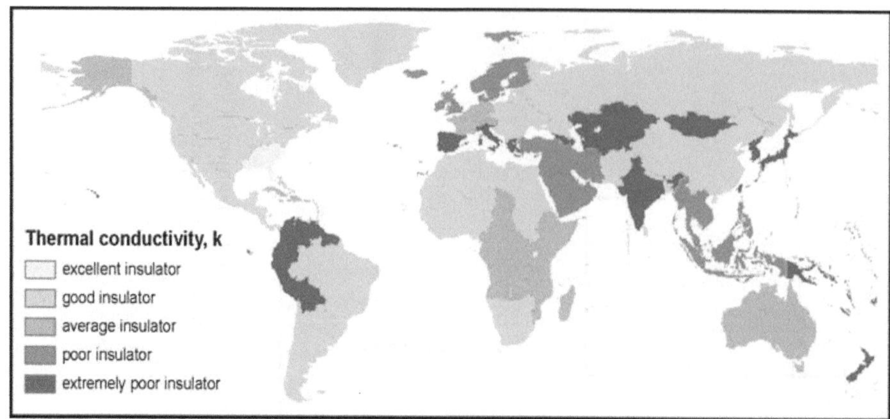

Figure 9. Thermal efficiency of typical walls in medium-density category by region. Insulator rankings are based on thermal conductivity values ranging from 0.7 to 3.1 $Wm^{-1}K^{-1}$.

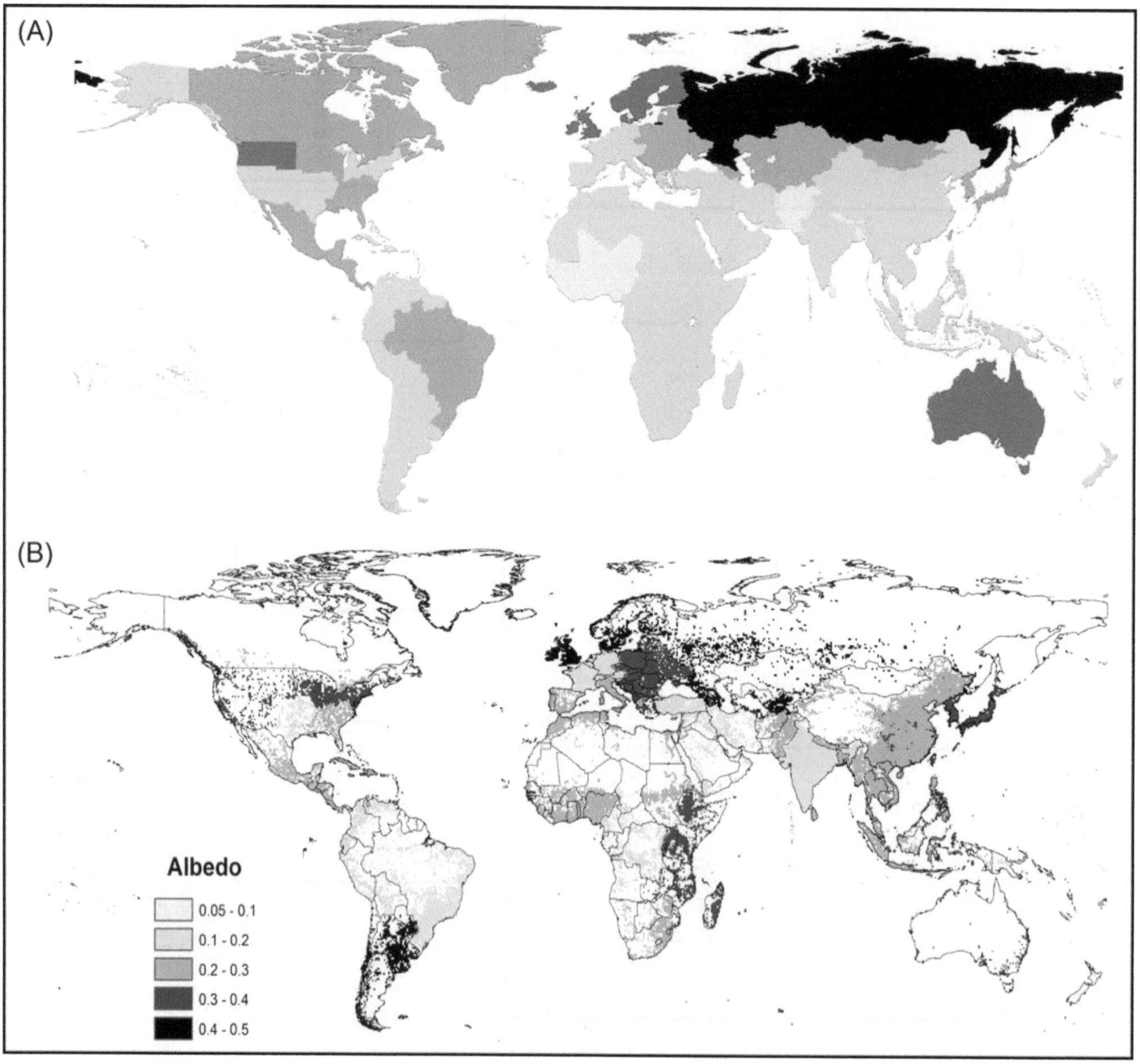

(A)

(B)

Albedo
0.05 - 0.1
0.1 - 0.2
0.2 - 0.3
0.3 - 0.4
0.4 - 0.5

Figure 10. (A) Albedo of typical roof materials used in low-density category by region; (B) Roof albedo values as actually used in the Community Land Model experimental simulation based on the dominant urban type in a global climate model grid cell.

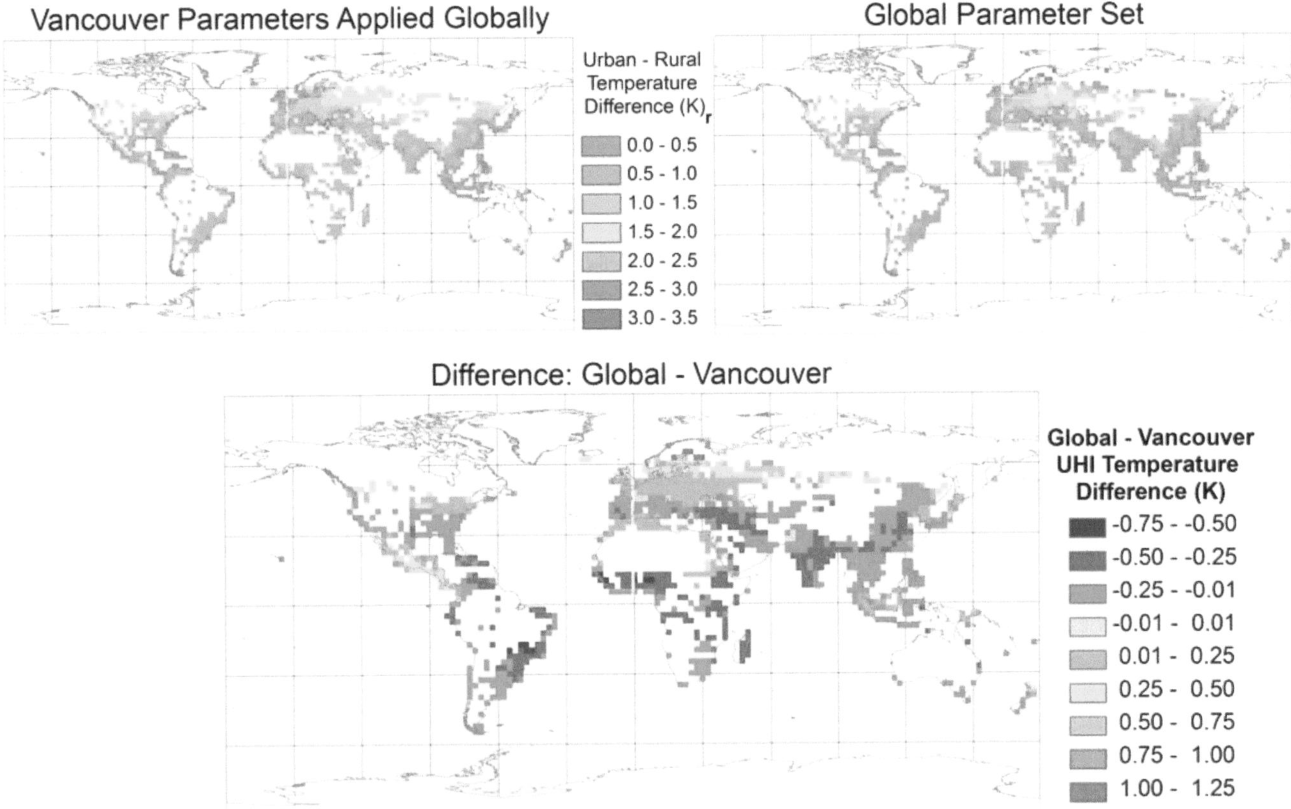

Figure 11. Comparison of simulations assuming all urban areas have the properties of the Voogt and Grimmond (2000) Vancouver location versus simulations using the new parameter data set.

to identify methods resulting in a meaningful outcome. Although some data exist on urban extent and building materials, none have uniform global coverage. The methods employed here are, to an extent, arbitrary. At the same time, we must consider that this is a first attempt at characterizing a complex system on a global scale.

Using population density as a proxy to determine urban boundaries also presents issues when attempting to include urban areas such as industrial complexes and airports. With sparse population in these pixels, they often are either classified as rural, or the population of surrounding areas captures the district as low density. Although low density might be a suitable classification for airports, it does not properly describe the properties of industrial complexes. The urban spatial extent requires further development to enhance coverage of these sparsely populated sites.

Determining spatial extent was likely the most straightforward process, but it was certainly not the simplest. After initially having only three classes of urban (Jackson 2007), there was too large a disparity between the qualities of low-density urban across regional boundaries. Thus, the medium-density class was added, and low density was reserved to describe suburban settings and areas employing urban agriculture, which have a relatively dense population but with a large fraction of vegetation. Because of the large fraction of vegetation in this class, it will be excluded from use in CLMU, although it will likely be used in future suburban models and useful for improving global landcover data sets.

Another challenge involved the classification of TBDs. In the United States, for example, the densest population areas are usually found adjacent to actual TBDs. Therefore, the location of TBDs in the data set might not be precise, as most effort was put toward resolving the proper extent of the TBD for the region. We erred on the conservative side here, as for a given region top population densities varied from city to city, creating challenges in finding the middle ground. But because the data set was to be scaled up to a half-degree resolution, accurate representation of the actual location for any given TBD is not a vital concern. Such errors can hopefully be corrected as additional information on urban areas becomes available in the future (e.g., Schneider, Friedl, and Potere 2009).

The urban spatial extent data set has a resolution of 1 km, which, considering the scale of the project, allows a reasonable level of generalization. LandScan also has a resolution of 1 km, so the method outlined here can be repeated for later versions of LandScan. Furthermore, because the intended use of the data set is to represent urban systems in GCMs with a grid resolution on the order of at least 50 km, the 1 km resolution is sufficient to serve this purpose. Yet if a user of the data set wished to use the data set for higher resolution studies, users must first consider if the data set is accurate enough for meaningful representations at smaller scales.

The scale limitation also exists for the building characteristics data. Three building types were described for each regional category, which reasonably describes the overall character of a place. Even more importantly, three building types are sufficient to point out contrasting elements between regions. Again, the database is suitable for global studies, but it is unlikely to be appropriate at smaller scales without addition of more detail. In addition to providing reasonable descriptions of building characteristics, the results show a realistic simulation of heat islands that qualitatively agrees well with observed heat island characteristics on a global scale (i.e., lower heat islands in the tropics and higher in high latitudes; also see Oleson et al. forthcoming).

Future Work

As described throughout this article, urban areas across the globe vary greatly. TBDs in Japan and Australia tower over those found in India and South Africa. Housing varies from sophisticated, wood frame walls with solar panel roofs to simple mud huts with thatch roofs. In some places, extensive, manicured lawns are the norm, whereas elsewhere, grounds surrounding homes are cultivated for food. This data set effectively captures these variances and, at the same time, invites improvement. As the first of its kind, the urban data set presented here is far from a final product. It is the hope of the authors that the data set will be viewed and used as a living document. This data set was created with the idea that it can be added to over time and to invite wide-ranging participation from other researchers to fill in areas presently lacking in detail. Using this project as a starting point, it would be fairly simple to further segregate regions or to add to or improve the precision of the data that are already present. The format of the database allows for progressive expansion of the data, and its design allows for scale modification to make it suitable for local or regional studies. The data sets described here are available from NCAR's Web site or from the authors.

Acknowledgments

This research was partly supported by the National Center for Atmospheric Research Weather and Climate Impact Assessment Program and the Water System Program, National Science Foundation Grants ATM-0107404 and ATM-0413540, and the University of Kansas, Center for Research. The National Center for Atmospheric Research is sponsored by the National Science Foundation.

References

Anandakumar, K. 1999. A study on the partition of net radiation into heat fluxes on a dry asphalt surface. *Atmospheric Environment* 33:3911–18.

Arnfield, A. J. 1984. Simulating radiative energy budgets within the urban canopy layer. *Modeling and Simulation* 15:227–33.

———. 2000. A simple model of urban canyon energy budget and its validation. *Physical Geography* 11:220–39.

———. 2003. Two decades of urban climate research: A review of turbulence, exchanges of energy and water, and the urban heat island. *International Journal of Climatology* 23:1–26.

Asaeda, T., V. T. Ca, and A. Wake. 1996. Heat storage of pavement and its effect on the lower atmosphere. *Atmospheric Environment* 30:413–27.

Bonan, G. 2002. *Ecological climatology: Concepts and applications*. Cambridge, MA: Cambridge University Press.

Center for International Earth Science Information Network, Columbia University; International Food Policy Research Institute, the World Bank; and Centro Internacional de Agricultura Tropical. 2004. *Global rural-urban mapping project (GRUMP): Urban extents*. Palisades, NJ: CIESIN, Columbia University. http://beta.sedac.ciesin.columbia.edu/gpw (last accessed 17 July 2009).

Changnon, S. A. 1992. Inadvertent weather modification in urban areas: Lessons for global climate change. *Bulletin of the American Meteorological Society* 73:619–27.

Cionco, R. M., and R. Ellefsen. 1998. High resolution urban morphology data for urban wind flow modeling. *Atmospheric Environment* 32:7–17.

Clarke, D., ed. 1989. *Approaches to the study of traditional dwellings and settlements*. Berkeley: Center for Environmental Design Research, University of California at Berkeley.

Collins, W. D., C. M. Bitz, M. L. Blackmon, G. B. Bonan, C. S. Bretherton, J. A. Carton, P. Chang et al. 2006. The

Community Climate System Model: CCSM3. *Journal of Climate* 19:2122–43.

Decker, E. H., S. Elliott, F. A. Smith, D. R. Blake, and F. S. Rowland. 2000. Energy and material flow through the urban ecosystem. *Annual Review of Energy and the Environment* 25:685–740.

DigitalGlobe. 2007. FAQ. http://www.digitalglobe.com/index.php/16/About+Us (last accessed 17 July 2009).

Dobson, J. E., E. A. Bright, P. R. Coleman, R. C. Durfee, and B. A. Worley. 2000. LandScan: A global population database for estimating populations at risk. *Photogrammetric Engineering and Remote Sensing* 66 (7): 849–57.

Doll, D., J. K. S. Ching, and J. Kaneshiro. 1985. Parameterization of subsurface heating for soil and concrete using net radiation data. *Boundary-LayerMeteorology* 32: 351–72.

EarthSat. 2007. MDA EarthSat satellite imagery. http://www.earthsat.com/ (last accessed 17 July 2009).

Ellefsen, R. 1990–1991. Mapping and measuring buildings in the canopy boundary layer in ten U.S. cities. *Energy and Buildings* 15–16:1025–49.

Emporis Corporation. 2007. Emporis buildings—The building industry platform. http://www.emporis.com/en/ (last accessed 17 July 2009).

Givoni, B. 1976. *Man, climate, and architecture*. 2nd ed. London: Applied Science.

Grimmond, C. S. B. 2007. Urbanization and global environmental change: Local effects of urban warming. *The Geographical Journal* 173:83–88.

Grimmond, C. S. B., H. A. Cleugh, and T. R. Oke. 1991. An objective urban heat storage model and its comparison with other schemes. *Atmospheric Environment* 25B:311–26.

Grimmond, C. S. B., and T. R. Oke. 1999. Heat storage in urban areas: Local-scale observations and evaluation of a simple model. *Journal of Applied Meteorology* 4 (7): 922–40.

Grimmond, C. S. B., and C. Souch. 1994. Surface description for urban climate studies: A GIS based methodology. *Geocarto International* 1:47–59.

Grimmond, C. S. B., C. Souch, and M. D. Hubble. 1996. Influence of tree cover on summertime surface energy balance fluxes, San Gabriel Valley, Los Angeles. *Climate Research* 6:45–57.

Howard, L. 1833. *Climate of London deduced from meteorological observations*. Vol. 1–3. London: Harvey and Darton.

Jackson, T. 2007. Developing a dataset for simulating urban climate impacts on a global scale. Master's thesis, Department of Geography, University of Kansas, Lawrence.

Jauregui, E. 1986. The urban climate of Mexico City. In Proceedings WMO Technology Conference (Tech Note 652), ed. T. R. Oke, 63–86. Geneva, Switzerland: World Meteorological Organization.

Kjelgren, R., and T. Montague. 1998. Urban tree transpiration over turf and asphalt surfaces. *Atmospheric Environment* 32:35–41.

Klein Goldewijk, K., and G. van Drecht. 2006. HYDE 3: Current and historical population and land cover. In *Integrated modelling of global environmental change: An overview of IMAGE 2.4*, ed. A. F. Bouwman, T. Kram, and K. Klein Goldewijk, 93–111. Bilthoven, The Netherlands: Netherlands Environmental Assessment Agency (MNP).

Landsberg, H. E. 1981. *The urban climate*. New York: Academic.

Masson, V. 2000. A physically-based scheme for the urban energy budget in atmospheric models. *Boundary-Layer Meteorology* 94:357–97.

Mills, G. M. 1993. Simulation of the energy budget of an urban canyon—I. Model structure and sensitivity test. *Atmospheric Environment B* 27:157–70.

———. 2007. Cities as agents of global change. *International Journal of Climatology* 27:1849–57.

Oak Ridge National Laboratories. 2005. *LandScan™ 2004 global population database*. Oak Ridge, TN: Oak Ridge National Laboratory. http://www.ornl.gov/landscan/ (last accessed 17 July 2009).

Oke, T. R. 1974. *Review of urban climatology, 1968–1973*. Technical Note 134 (Publication No. 383). Geneva, Switzerland: World Meteorological Organization.

———. 1979. *Review of urban climatology, 1973–1976*. Technical Note 169 (Publication No. 539). Geneva, Switzerland: World Meteorological Organization.

———. 1982. The energetic basis of the urban heat island. *Quarterly Journal of the Royal Meteorology Society* 108:1–24.

———. 1987. *Boundary layer climates*. 2nd ed. New York: Methuen.

———. 1988. The urban energy balance. *Progress in Physical Geography* 12:471–508.

———. 1989. The micrometeorology of the urban forest. *Philosophical Transactions of the Royal Society of London, Series B* 324:335–49.

———. 1995. The heat island characteristics of the urban boundary layer: Characteristics, causes and effects. In *Wind climate in cities*, ed. J. E. Cermak, A. G. Davenport, E. J. Plate, and D. X. Viegas, 81–107. Amsterdam: Kluwer Academic.

Oke, T. R., G. T. Johnson, D. G. Steyn, and I. D. Watson. 1991. Simulation of surface urban heat islands under "ideal" conditions at night, part 2: Diagnosis of causation. *Boundary-Layer Meteorology* 56:339–58.

Oleson, K., G. Bonan, and J. Feddema. 2006. Development of an urban parameterization for a global climate model. Paper presented at the annual meeting of the American Meteorological Society, Atlanta, GA.

———. 2010. The effects of white roofs on the global urban heat island. *Geophysical Research Letters* 37:L03701.

Oleson, K. W., G. Bonan, J. Feddema, and T. Jackson. Forthcoming. An examination of urban heat island characteristics in a global climate model. *International Journal of Climatology*.

Oleson, K. W., G. B. Bonan, J. Feddema, M. Vertenstein, and C. S. B. Grimmond. 2008a. An urban parameterization for a global climate model: 1. Formulation and evaluation for two cities. *Journal of Applied Meteorology and Climatology* 47 (4): 1038–60.

———. 2008b. An urban parameterization for a global climate model: 2. Sensitivity to input parameters and the simulated urban heat island in offline simulations. *Journal of Applied Meteorology and Climatology* 47 (4): 1061–76.

Oleson, K.W., Y. Dai, G. Bonan, M. Bosilovich, R. Dickinson, P. Dirmeyer, F. Hoffman et al. 2004.

Technical description of the Community Land Model (CLM). NCAR Technical Note NCAR/TN-461+STR. Boulder, CO: National Center for Atmospheric Research.

Oleson, K. W., G.-Y. Niu, Z.-L. Yang, D. M. Lawrence, P. E. Thornton, P. J. Lawrence, R. Stockli, et al. 2008c. Improvements to the Community Land Model and their impact on the hydrological cycle. *Journal of Geophysical Research* 113:G01021–B01035.

Olgyay, V. 1992. *Design with climate*. New York: Van Nostrand Reinhold.

Paterson, D. A., and C. J. Apelt. 1989. Simulation of wind flow around three-dimensional buildings. *Building and Environment* 24:39–50.

Qian, T., A. Dai, K. E. Trenberth, and K. W. Oleson. 2006. Simulation of global land surface conditions from 1948 to 2004: Part I. Forcing data and evaluations. *Journal of Hydrometerology* 7:953–75.

Rashkin, S., R. A. G. Chinery, D. Melsegeler, and D. Gamble. 2008. *Technology adoption plan: Advanced new home construction*. http://www.epa.gov/cppd/climatechoice/Adv%20New%20Home%20Constr%20Adopt%20Plan3.pdf (last accessed 17 July 2009).

Salhani, P. 2007. Go green with style. *The Sydney Morning Herald* 31 March. http://www.smh.com.au/news/domain/australian-capital-territory/go-green-with-style/2007/04/03/1175366226679.html (last accessed 27 July 2010).

Schneider, A., M. A. Friedl, and D. Potere. 2009. A new map of global urban extent from MODIS satellite data. *Environmental Research Letters* 4:1–11.

Shepherd, J. M. 2005. A review of current investigations of urban-induced rainfall and recommendations for the future. *Earth Interactions* 9:1–27.

Straube, J. F., and E. F. P. Burnett. 2005. *Building science for building enclosures*. Westford, MA: Building Science Press.

Suckling, P. W. 1980. The energy balance microclimate of a suburban lawn. *Journal of Applied Meteorology* 19: 606–08.

Terjung, W. H., and P. A. O'Rourke. 1980a. Influences of physical structures on urban energy budgets. *Boundary-Layer Meteorology* 19:421–39.

———. 1980b. Simulating the causal elements of urban heat islands. *Boundary-Layer Meteorology* 19:93–118.

Verseghy, D. L., and D. S. Munro. 1989a. Sensitivity studies on the calculation of the radiation balance of urban surfaces: I. Shortwave radiation. *Boundary-Layer Meteorology* 46:309–31.

———. 1989b. Sensitivity studies on the calculation of the radiation balance of urban surfaces: II. Longwave radiation. *Boundary-Layer Meteorology* 48:1–18.

Voogt, J. A., and C. S. B. Grimmond. 2000. Modeling surface sensible heat flux using surface radiative temperatures in a simple urban area. *Journal of Applied Meteorology* 39:1679–99.

Climatic Shifts in the Availability of Contested Waters: A Long-Term Perspective from the Headwaters of the North Platte River

Jacqueline J. Shinker, Bryan N. Shuman, Thomas A. Minckley, and Anna K. Henderson

Early summer snowmelt from mountains in northern Colorado and southeastern Wyoming supplies the North Platte River, supporting nationally important agriculture, energy production, and urban development. Repeated decisions from the U.S. Supreme Court have fully apportioned Platte River waters among Colorado, Wyoming, and Nebraska, underscoring societal strains on this system. Now, climate change threatens the regional allocation of water. Tree-ring records indicate that past centuries contained multidecadal "megadroughts" far more severe than those of the historic period. However, the potential for even more persistent droughts, as the result of climate change, is poorly known. We document and evaluate the severity of recent and prehistoric droughts via a combination of data sources: modern temperature, precipitation, and stream gauge data; evidence of low lake-level stands; and related estimates of past hydroclimate change. Modern climate and stream data show an increase in spring temperatures of 2.21°C since 1916, an increase in the frequency of peak spring runoff before 1 May, and a reduction in winter precipitation. Lakes, however, that have only experienced minor hydrologic changes historically were desiccated during prehistoric dry periods during the past 12,000 years. Prehistoric lake shorelines indicate that water supplies were substantially reduced over centuries and millennia, such as from >8,000 to <5,000 years before present. The magnitude of these droughts likely also resulted in ephemeral river flows and thus indicates the potential for persistent shifts in regional hydrology. Such shifts should, therefore, be considered as part of long-term economic and legal planning. *Key Words: climate change, drought, lake levels, North Platte River, water supply.*

来自科罗拉多州北部和怀俄明州东南部山区的夏季融雪为北普拉特河提供了水源，支援了全国性的重点农业，能源生产和城市发展。美国最高法院多次的裁决已经完全地分配了普拉特河在科罗拉多州，怀俄明州和内布拉斯加州的水资源，凸现了该系统所经受的社会压力。现在，气候变化威胁着水的区域分配。树轮记录表明，在过去的几百年中，历时几十年的“大干旱”的严重性远远超过了其它历史时期。但是，更加持续的干旱是否就是气候变化的结果，对此可能性，我们所知甚少。通过集成大量的数据来源：现代温度，降水，和径流量数据；低湖泊水位的证据；有关过去水文气候变化的估算，我们记录和评估了近期和史前旱灾的严重程度。现代气候和径流数据显示，自1916年以来，春季的温度升高了2.21°C，5月1日前出现春季径流高峰期的频率有所增加，而冬季的降水量呈现减少。但是，在过去12000年的干旱时期里，湖泊仅仅经历了轻微的水文变化而略有实际上的干涸。史前湖泊岸线表明，在过去的几百年和几千年里，例如距今8000年以前到距今5000年以内，水的供应已大幅削减。干旱的严重程度有可能会影响到短暂的河流流量，从而表明了区域性水文特征持续变化的可能性。因此，这种变化应被视为长期的经济和法律规划的一部分。*关键词：气候变化，干旱，湖水水位，北普拉特河，水供应。*

El derretimiento de la nieve al comienzo del verano en las montañas del norte de Colorado y sudeste de Wyoming alimenta el Río North Platte, dando así soporte a una agricultura de importancia nacional, la producción de energía y desarrollo urbano. Las aguas del Río Platte han sido repartidas enteramente entre los estados de Colorado, Wyoming y Nebraska por medio de repetidas decisiones de la Corte Suprema de los EE.UU., lo cual subraya las tensiones sociales que afectan este sistema fluvial. Esta asignación regional de agua está ahora amenazada por el cambio climático. Los registros de los anillos de los troncos de árboles indican que en siglos pasados se presentaban allí "megasequías" multidecadales mucho más severas que las registradas en tiempos históricos. Sin embargo, muy poco se conoce del potencial para que puedan ocurrir sequías de persistencia aun mayor como resultado del cambio climático. Documentamos y evaluamos la severidad de sequías prehistóricas recientes por medio de una combinación de fuentes de información: datos modernos de temperatura, precipitación y caudales de las corrientes; evidencia de menor nivel de la superficie de lagos; y estimativos relacionados con cambios hidroclimáticos pasados. Los datos modernos sobre clima y corrientes muestran un incremento de las temperaturas de primavera de 2.21°C desde 1916, un incremento en la frecuencia del tope de escorrentía antes del 1 de mayo y una reducción de la precipitación invernal. Sin embargo, los lagos, que habían experimentado solo cambios hidrológicos insignificantes en tiempos históricos, se secaron en períodos secos prehistóricos durante los pasados 12.000 años.Las líneas de costa de lagos prehistóricos indican que el suministro de agua fue sustancialmente reducido durante siglos y milenios, por ejemplo de >8.000 a < 5.000 años antes del presente. Es probable que la magnitud de estas sequías resultara en flujos efímeros de ríos, indicando así el potencial de cambios persistentes en la hidrología regional. Tales cambios deben ser consecuentemente considerados como parte de una planeación económica y legal a largo plazo. *Palabras clave: cambio climático, sequía, niveles lacustres, Río North Platte, abasto de agua.*

Water in the Western United States has long been a source of conflict within the region, with lessons from the past century (e.g., overallocation of the Colorado River) revealing the value of a long-term perspective on the variability of water resources (MacDonnell, Getches, and Hugenberg 1995). Climate change is likely to exacerbate uncertainties in water supplies (Bates et al. 2008; Milly et al. 2008), including the potential for hydroclimatic changes to persist beyond reasonable resource planning horizons (Seager et al. 2007; Barnett et al. 2008). Most Western waters derive from early summer snowmelt in the mountains, and much of the region is currently experiencing a shift to earlier snowmelt as a result of anthropogenic-forced warming (Stewart, Cayan, and Dettinger 2004). Superimposed on this trend, a nearly decade-long drought has contributed to declines in riverflows (Rood et al. 2005) and is the most recent manifestation of changes in precipitation detected over the last fifty years (Dai et al. 2009). At longer time scales, prolonged multidecadal "megadroughts" have been inferred from tree-ring records (Cook et al. 2004), but even these records might be unable to provide a full range of drought variability necessary for long-term planning, especially in the context of future climate change. The past 12,000 years provide an opportunity to examine how water resources respond to global climate forcing different from the past few centuries and thus might provide a wider range of contingencies to consider for the future (Shuman et al. 2009; Shuman et al. 2010).

We focus here on the hydrologic history of the North Platte River in Colorado, Wyoming, and Nebraska as an illustrative example of legally contested waters affected by climate change. The river has been a source of multiple interstate lawsuits arbitrated by the U.S. Supreme Court between 1911 and 2001 (see later). The 12,000-year history of water supplies in the river's headwaters is used to demonstrate the potential for climate change to radically affect the current legal consensus. The headwaters' region covers approximately 72,520 km^2 in the mountains of northern Colorado and southeastern Wyoming and is an important source of water for irrigation, energy production, and urban development (Figure 1). The smallest portion of the watershed is in northern Colorado, with the majority in southeastern Wyoming. However, primary state water rights on the North Platte are held by Nebraska through prior appropriations.

The goal of this article is to examine recent trends in temperature, moisture balance, and streamflow in the

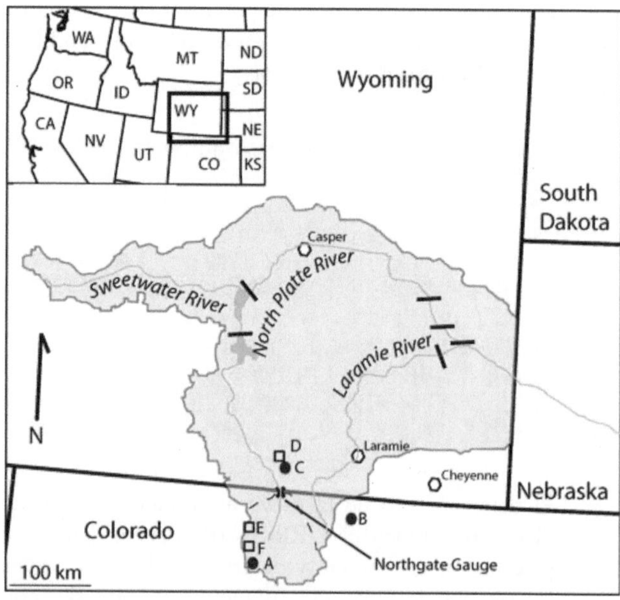

Figure 1. Map of the North Platte River Basin in Colorado and Wyoming. The Northgate stream gauge is shown on the Colorado–Wyoming border and a dashed line denotes the watershed above the gauge. Filled circles represent AMS-dated lake-level study sites: (A) Hidden Lake, (B) Creedmore Lake, (C) Little Windy Hill Lake. Open squares represent additional sites with ground-penetrating radar data: (D) Long Lake, (E) Teal Lake, (F) Tiago Lake. Black bars indicate location of dams.

North Platte drainage in a 12,000-year context. We pursue three main questions:

1. How has recent climate change affected water supply in the region?
2. How do recent hydrologic trends compare to those related to past climate change?
3. What are the implications of these past changes in water supply?

In answering these questions, we place the current drought in a context that reveals that the historical range of variability in water supplies can be woefully inadequate for considering possible futures. We focus on past aridity, but our evidence also indicates prolonged wet periods outside the range of historic experience.

The North Platte River Basin: The Regional–Legal Context

The North Platte River and its main tributaries, the Laramie and Sweetwater rivers, originate in high elevations of northern Colorado and Wyoming, turning southeast toward Nebraska with several impoundments and diversions, mostly in Wyoming (Figure 1). Currently, the river water is fully appropriated by legal agreement with water rights used primarily for agriculture, followed by industrial and energy production, municipal and domestic use, and recreation. The allocation has been continuously contested, however.

The first interstate lawsuit challenging allocation of North Platte waters was filed in 1911 by Wyoming against Colorado related to usage of the Laramie River. The U.S. Supreme Court ruled in the 1922 Laramie River Decree (*Wyoming v. Colorado*, 259 U.S. 419 [1922], amended in 1957) to set a specific limit (15,500 acre-feet/year) on Colorado diversions with the rest allocated to Wyoming. Such a ruling assumes that 15,500 acre-feet/year has been and will be continuously available from the Laramie River, but the claim has never been validated. Historic climate variability, however, challenges this assumption. During the 1930s, severe regional drought led Nebraska to sue, claiming that diversions in Wyoming were preventing priority water rights downriver. The U.S. Supreme Court decision in 1945 (*Nebraska v. Wyoming*, 108 (Orig.), 325 U.S. 589, 665) then set limitations on the appropriations of North Platte water in Wyoming, although litigation continued in 1953, 1993, and 1995.

The most recent modification to this ruling, the Modified North Platte Decree of 2001 (*Nebraska v. Wyoming & Colorado*, 534 U.S. 40 [2001]), declared that sections of the North Platte River originating in Wyoming are divided for consumptive use during irrigation season (May–September) between Nebraska (75 percent) and Wyoming (25 percent). The hydrologic assumptions for this allocation arise from the documented availability of water over any ten consecutive years during the period from 1952 to 1999 as determined by agricultural consumptive use and climate data (see Exhibit 6 of *Nebraska v. Wyoming & Colorado*, 534 U.S. 40 [2001]). This decree also allows Colorado to claim 1.28 million acre-feet of water over any ten-year period. The proportional allocation of water is based on the recognition of variable flows, such as during drought years, but the assumption remains that natural declines in flow are likely to be short lived (Milly et al. 2008).

Climate change and other environmental issues (e.g., endangered species; local resource extraction) have raised new concerns about the allocation of water. The climate of the North Platte Basin has historically been semiarid, which limits water supplies throughout much of the watershed where annual evapotranspiration is greater than annual precipitation despite a slight peak in precipitation in May (Shinker and Bartlein 2010). The majority of surface water in the region, therefore, comes from a single source: winter snowpack that falls mostly above 3,000 m in a small fraction of the watershed along its southwestern and northwestern margins. Reduced average annual snowpack in these mountain ranges can have large downstream consequences, and an increase in springtime temperatures has advanced the timing of spring snowmelt and peak river discharge throughout the West (Cayan et al. 2001; Stewart, Cayan, and Dettinger 2004). Early spring snowmelt, in turn, threatens to indefinitely reduce runoff, leaving less water available later in the summer than in past decades (Barnett and Pierce 2008).

Data and Methods

Modern Climate and Streamflow Data

Continuous climate and stream gauge data from 1916 through 2007 were used to characterize temperature, precipitation, and streamflow trends. Temperature and precipitation data from Wyoming Climate Division 10 represent the Upper Platte region (National Climatic Data Center 1994). Trends were examined in time series of mean annual and spring (March, April, and

May [MAM]) temperatures with spring temperatures used specifically to assess rates of change during peak snowmelt and maximum riverflows. Annual and winter (December, January, and February [DJF]) precipitation anomalies (1971–2000 base period) were also examined for comparison with riverflows. Winter precipitation anomalies were analyzed specifically because winter precipitation (via snowpack) is a strong indicator of variability in streamflow (Changnon, McKee, and Doesken 1991). Stream gauge data, including annual flow, magnitude of peak runoff, and timing of peak discharge, were examined for the North Platte River at the Northgate stream gauge near the Colorado–Wyoming border; data were obtained through the U.S. Geological Survey (USGS) National Streamflow Information Program (NSIP). Simple linear regressions were used to establish temporal trends, and residuals were statistically assessed for normality. Additionally, the frequency of events of certain magnitude (e.g., years with mean annual discharge of $<6\ m^3s^{-1}$, which is less than 50 percent of the mean historic flow) was calculated using a ten-year moving window (i.e., the number of events per decade centered on each year in the record). We use the term *mean historic flow* to refer to the average of the mean annual discharge over the period of record.

Holocene Lake-Level Reconstruction

In much the same way that data from local meteorological stations can be compared to assess weather patterns, arrays of lakes (via sediment deposits) can be studied to reconstruct past climate conditions. Submerged shorelines in lakes are indicative of periods of drier-than-present conditions (e.g., Dearing 1997; Shuman and Finney 2006). Stratigraphic analyses of small lakes, therefore, can provide accurate records of hydroclimatic change (Digerfeldt, Almendinger, and Bjorck 1992).

When lake levels are low, sandy inorganic substrates expand toward the center of the lake, and sediment accumulation becomes nonconstant (~10 cm/1,000 years or less) toward shore (Dearing 1997; Shuman 2003). As a result, sandy layers and hiatuses (erosion surfaces) extend out from shore, and deepwater sediments "pinch out" near shore. Our studies of such evidence from lakes in the North Platte watershed were conducted with (1) ground-penetrating radar (GPR) to identify past shifts in shoreline position and associated changes in sediment geometry and (2) coring and trenching to sample and date submerged and buried shoreline deposits. Plots of sediment type, age, and elevation provide constraints on the water-level history. GPR data were collected with a GSSI, Inc., SIR-3000 GPR with 200 and 400 MHz antenna towed in rafts on the lake surface. GPR surveys were conducted at more than ten lakes in the region and representative data are shown here. Sediment cores and samples from near-shore trenches were collected from Hidden Lake, Jackson County, Colorado (40.50°N, 106.61°W, 2,728 m elevation); Creedmore Lake, Larimer County, Colorado (40.86°N, 105.59°W, 2,515 m); and Little Windy Hill Lake Carbon County, Wyoming (41.43°N, 106.33°W, 2,980 m). Core analyses include loss-on-ignition (Shuman 2003) and radiocarbon dating, conducted at the W. M. Keck Carbon Cycle Accelerator Mass Spectrometry (AMS) Laboratory. AMS radiocarbon dates were calibrated to account for atmospheric radiocarbon variations using CALIB 5.0.2 (Reimer et al. 2004). Here we provide new data for Creedmore and Little Windy Hill lakes, and review published data from Hidden Lake (Shuman et al. 2009).

Calculation of Impacts on Water Resources

Differences between historic and Holocene riverflows at the Northgate gauge were calculated based on estimates of Holocene lake drawdown. The magnitude of drawdown was inferred from the difference in the past and modern elevations of sandy inorganic deposits (near shore). These elevational differences were multiplied by lake area to calculate a volume of water withdrawn from each lake in the past compared to today and then were divided by the area of each lake's watershed to determine the annual change in precipitation minus evapotranspiration (P – E; lake inputs versus lake outputs) across the watershed. Our P – E estimates, therefore, represent anomalies from the conditions associated with ca. 2008 lake levels. P – E values in mm/day were then calculated iteratively by dividing the annual total P – E for each lake by a range of 46 different lake-climate equilibration times from 0.5 to 5 years (182.5–1,825 days) in 0.1 year increments (see Shuman et al. 2010). Table 1 lists the means and standard deviations of these P – E estimates.

We cannot differentiate between changes in precipitation and evapotranspiration because each lake's water budget depends on the balance of both factors. Groundwater in each watershed is a product of the balance of P – E and is not an additional source of lake water (i.e., derived from outside the watershed) because each of the watersheds is rimmed by bedrock with limited permeability: Precambrian quartz monzonite and granite at Creedmore Lake; Cretaceous Dakota Sandstone

Table 1. Moisture balance and river flow estimates

Lake	Area (ha)	Watershed (ha)	Drawdown (mbmls at 8-5 ka)	Volume change (m^3)	$\Delta P - E$ mm/year/ watershed	$\Delta P - E^a$ mm/d/ watershed	Δflow at North Gate (m^3/s)	Δflow (10^5 acre-ft/year)	Δflow (% of mean)
Hidden Lake	6.50	94.00	3.5	227,500	−242	−0.35 ± 0.27	−14.9 ± 6.9	−7.27	−124 ± 58
Creedmore Lake	3.62	17.04	1.2	36,234	−255	−0.37 ± 0.28	−15.8 ± 7.3	−7.67	−131 ± 61
Little Windy Hill Lake	2.21	10.64	1.4	30,940	−291	−0.42 ± 0.32	−18.0 ± 8.2	−8.74	−149 ± 69
Top 5 low flow years since 1948									
2002							−9.5	−2.41	−78
1977							−8.7	−2.23	−72
1934							−8.5	−2.19	−71
1954							−7.9	−2.03	−66
1981							−7.6	−1.94	−63

Note: Mbmls = meters below modern lake surface; ka = thousands of years before 1950; $\Delta P - E$ = change in the balance of precipitation minus evapotranspiration.

[a] Standard deviations are based on forty-six estimates of $\Delta P - E$ using lake-equilibrium times in 0.1 year steps from 0.5 to 5.0 years.

underlain by Proterozoic granitic rocks and pelitic schist at Hidden Lake; and Proterozoic Deep Lake Group quartzite and other metasediments, and Archean mafic intrusive rocks at Little Windy Hill Lake (USGS 2005). Low lake levels would also be associated with reduced hydrostatic head and groundwater outflow from each watershed. Thus, by assuming no change in groundwater loss over time, we are making conservative estimates of the changes in P – E.

The annual change in P – E was multiplied by the area of the North Platte drainage (3,706 km^2) above the Northgate stream gauge to heuristically estimate an annual volume of water removed from the river system in the past. We present these values as flow rates (as m^{3}s^{-1} by dividing by the number of seconds per year; and as acre-feet per year, by converting the volume to the commonly and legally used hydrologic unit of acre-feet) and as a percentage deviation from the mean historic flow rate (12.0 m^{3}s^{-1}). Our calculated values are compared with values for the five lowest flow years since 1916 based on USGS NSIP data.

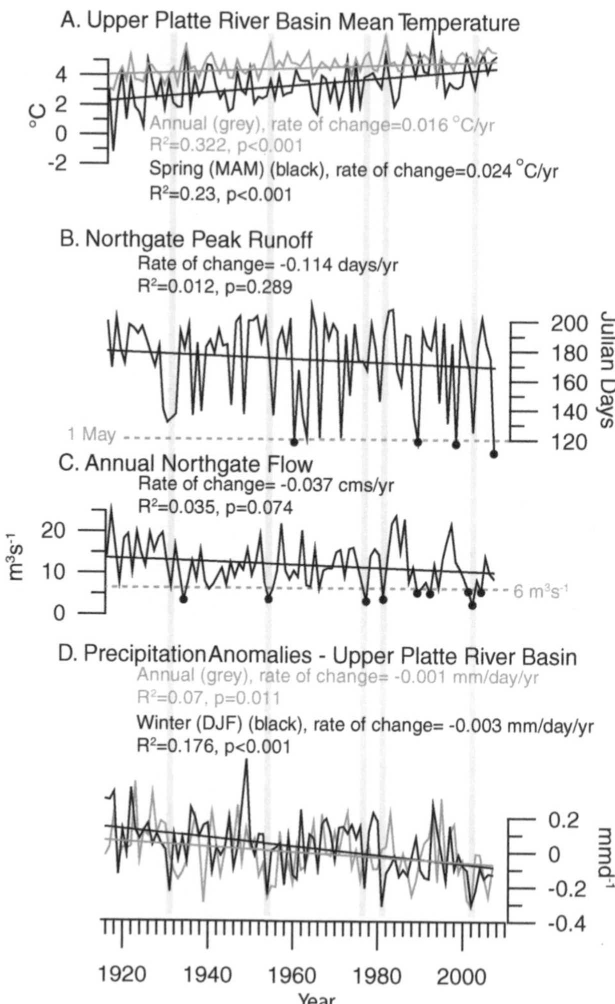

Figure 2. Trends in modern temperature, river flow, and precipitation data. Top panels show annual (gray) and spring (March, April, and May [MAM], black) temperature (A) and the timing of peak runoff (B). Lower panels show annual flow (C) and annual (gray) and winter (December, January, and February [DJF], black) precipitation anomalies (1971–2000 base period) (D). Gray dashed lines denote thresholds used to evaluate the frequency of (B) early peak runoff events and (C) low annual flow episodes; black circles mark years that cross the thresholds. Temperature and precipitation data represent Wyoming climate division 10, in the upper Platte River Basin (National Climatic Data Center 1994). Peak runoff and annual flow data from Northgate stream gauge obtained through the USGS National Streamflow Information Program.

Results

Recent Trends in Climate and Streamflow

The modern (1916–2007) climate time series illustrates trends in temperature (Figure 2A) consistent with global and regional patterns (Intergovernmental Panel on Climate Change 2007) with an increase in both mean annual and MAM temperatures. The rate of change in spring temperatures exceeds that for any other time of the year. Based on the linear regression (Figure 2A), mean annual temperatures in the North Platte River Basin have increased approximately 1.44°C over the ninety-two-year record ($R^2 = 0.322$, $p < 0.001$). Mean MAM temperatures in the region increased 1.5 times as fast as the mean annual temperatures at approximately 2.21°C over the same time ($R^2 = 0.230$, $p < 0.001$).

The increase in mean temperatures corresponds with an increase in the frequency of early spring peaks in discharge at Northgate (Figure 2B). For example, peak runoff before 1 May (Julian Day 120) has only taken place since 1960, and three of the four occurrences have taken place since 1989 (Figure 2B). A linear regression explains 45.0 percent of the variance in the frequency of early flow years per moving ten-year window. Linear regression of peak discharge dates, however, is not significant ($R^2 = 0.012$, $p = 0.289$) but qualitatively confirms the change in frequency of early runoff events, with peak runoff occurring now 0.114 days per year earlier than at the beginning of the record. The trends equal an advance in the timing of peak runoff by approximately 10.5 days over the record or an advance of 4.75 days per 1°C increase in MAM temperatures. Peak timing has not advanced smoothly but experienced both interannual and interdecadal variation (Figure 2C). The rate of change tending toward early runoff is consistent with other regional studies (e.g., Stewart, Cayan, and Dettinger 2004; Mote et al. 2005).

In addition to the shift in timing of peak runoff (Figure 2B), annual riverflow from the North Platte headwaters has decreased approximately 3.4 m^3s^{-1} (25 percent of the mean; Figure 2C). Annual river discharge dropped to only 21.5 percent of the mean discharge in 2002 and averaged 75.6 percent for the decade (1999–2008). Although the apparent linear trend ($R^2 = 0.035$, $p = 0.074$) in mean annual discharge depends on the time period analyzed and is driven by late runoff and high flows before 1930, the residuals of the trend have a near-normal distribution. A linear trend in the frequency of low flow years (with mean annual discharge <6 m^3s^{-1}; $R^2 = 0.433$) is also not dependent on the time period considered and indicates a significant change in hydrologic regimes ($p < 0.03$).

Peak discharge has similarly declined by approximately 58 m^3s^{-1} (52 percent) from 122 m^3s^{-1} for the decade centered around 1925 to 64 m^3s^{-1} for the decade centered around 2004. The frequency of both low mean annual and low peak flow years, therefore, is greater since 1977 compared to the earlier portion of the historic record. For example, low mean annual flow years were recorded in 1934 and 1954 and seven years since 1977 (including three years since 2001; Figure 2C). Linear regression explains 69.3 percent of the variance in the frequency of low peak flow years (with peak annual flow <40 m^3s^{-1}). Of the low peak flow years, 62.5 percent have taken place since 1977. Years with anomalously low DJF precipitation (e.g., 2002, 1981, 1977, 1954, and 1931) correspond to the lowest annual streamflow, and also increase in frequency after 1980 (Figure 2D). The rate of precipitation change is slightly faster in DJF than annually, equivalent to –0.28 mm/day over the period (based on our linear regression, $R^2 = 0.176$, $p < 0.001$).

Based on the changes in the frequency of low mean and peak runoff, and runoff timing, we conclude that meaningful hydrologic change has affected the North Platte headwaters—even if such changes are only qualitatively captured by our linear regressions (Figure 2). Importantly, the changes in frequency are not dependent on the inclusion of the pre-1930 period. All regressions (Figure 2) also have normally distributed residuals except for those calculated for peak runoff timing (Figure 2B) and the annual flow at Northgate (Figure 2C). Additionally, the trends of all climatological and hydrological variables examined in this study are consistent with similar hydroclimatic trends identified in other studies from Western North America (see Mote et al. 2005; Rood et al. 2005; Barnett et al. 2008). The last decade was the warmest and driest on record, with 2002 as the driest year. These results indicate a trend toward less water in the system.

Lake-Level Variations

The past decade of drought has caused low lake levels at Creedmore Lake (Figure 3A) and other lakes in the region. GPR and sediment core data, however, reveal the presence of submerged paleoshorelines and thus evidence of drier episodes than the recent drought. GPR profiles show dense near-shore deposits and erosion/nondeposition surfaces, which represent paleoshorelines (Figure 3B). A sediment core from 120 cm of water in Creedmore Lake terminated in 30 cm of sandy sediment (12–87 percent sand) comprising one of the paleoshorelines. A calibrated radiocarbon age on sedimentary charcoal from the base of 9 cm of organic-rich silt (>15 percent loss-on-ignition) above the sands (Figure 4A) indicates that the sands accumulated before 158 to 303 calibrated years before 1950 (cal year BP). The lake level before 1792–1647, therefore, was >120 cm lower than in 2008, consistent with dendroclimatic evidence of regional "megadroughts" (Cook and Krusic 2004). Profiles of sediments exposed near the beach of Creedmore Lake contain fine-grained intervals (Figure 3B; <75 percent sand and >5 percent loss-on-ignition), which date to 503–471, 934–834, 2,055–1,991, 3,878–3,778, and before 7,965–7,938 cal year BP as though the water level were at least as high as modern during those times (Figure 4A). Note that little evidence exists for higher than modern levels between ca. 7,900 and 3,900 cal year BP and that evidence for high levels is most frequently from recent millennia.

Examination of a regional network of lakes shows extensive evidence of prolonged low-water intervals including inorganic near-shore layers and associated nondeposition or erosion surfaces at lakes in the North Platte River headwaters (Figure 5). The inferred positions of ancient shorelines indicate significant reduction in the surface area of these lakes (Figure 5, right column). For example, from ca. 8,400 to 4,400 cal year BP, Hidden Lake was approximately 3.5 m below the modern lake surface with a change in volume of approximately 227,500 m^3 (Table 1; Shuman et al. 2009).

Sediment cores from Creedmore, Hidden, and Little Windy Hill lakes (Figure 4) contain inorganic sediment layers, which appear to represent repeated dry periods since ca. 8,000 cal year BP. The layers interrupt periods

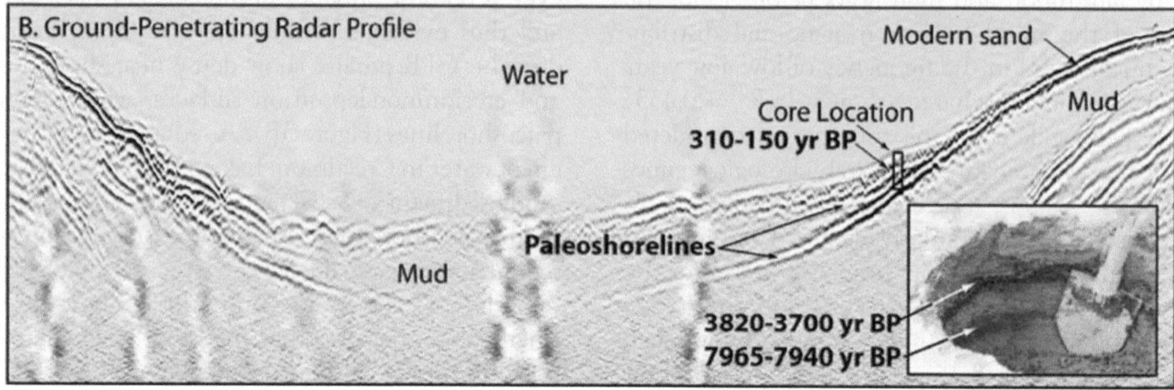

Figure 3. Recent low water levels at Creedmore Lake in May 2008 (A) illustrate the hydrological impacts of the past decade of drought. Subsurface geology (B), however, captured by ground-penetrating radar (GPR) and sediment core data reveal that the lake has been much lower in the past. Paleoshorelines submerged within the lake date from before ca. 200 years before present (BP), and shovel pits (inset photo) show no evidence of higher than modern lake levels from ca. 7,950–3,750 years BP. Bold numbers in B indicate ages and positions of radiocarbon ages on sedimentary features.

of organic sediment deposition near shore and are associated with reductions in net sediment accumulation consistent with periods of shallow water (see Shuman et al. 2009). At Hidden Lake, cores collected near shore in five and six meters of water (cores E and C respectively; Figure 4B) contain inorganic paleoshoreline deposits (<15 percent loss-on-ignition) with very low net sedimentation rates (~3.8 cm/1,000 years in core E) between dates of 2,043–1,993 and 1,262–1,179 cal year BP when tree-ring drought reconstructions reveal repeated periods with more than fifty drought years per century (Cook and Krusic 2004). These core locations also accumulated no sediment and were likely desiccated, before ca. 4,400 years ago. Similarly, at Little Windy Hill Lake, net sediment accumulation even in the center of the lake (core 1A; Figure 4C) was more inorganic than recently and was reduced to 9.0 cm/1,000 years between dates of 2,341–2,210 and 1,694–1,613 cal year BP as though it were also shallow at that time. Earlier periods with similar sediment characteristics (e.g., net sedimentation rates below ~10 cm/1,000 years) at Creedmore, Hidden, and Little Windy Hill lakes indicate other periods of prolonged drought or frequent extreme droughts, particularly from ca. 3,700–2,900 and 7,900–5,000 cal year BP. The low elevations of inorganic paleoshorelines deposits from these periods provide a constraint on changes in lake volumes (Table 1).

Estimated Holocene Water Supply Reductions

Overall, we find that all of the lakes we studied were commonly low from >7,900 to <5,000 years ago. To sustain these low levels, we estimate that precipitation minus evaporation (P − E) dropped by 0.35 to 0.42 mm/day (Table 1). These estimates are consistent with climate model simulations of the region for 6,000 cal year BP (Diffenbaugh et al. 2006) and indicate regional aridity more intense in magnitude than in 2002 (and yet far longer in duration).

The change of flow at Northgate, calculated by multiplying the estimated P − E anomalies by the

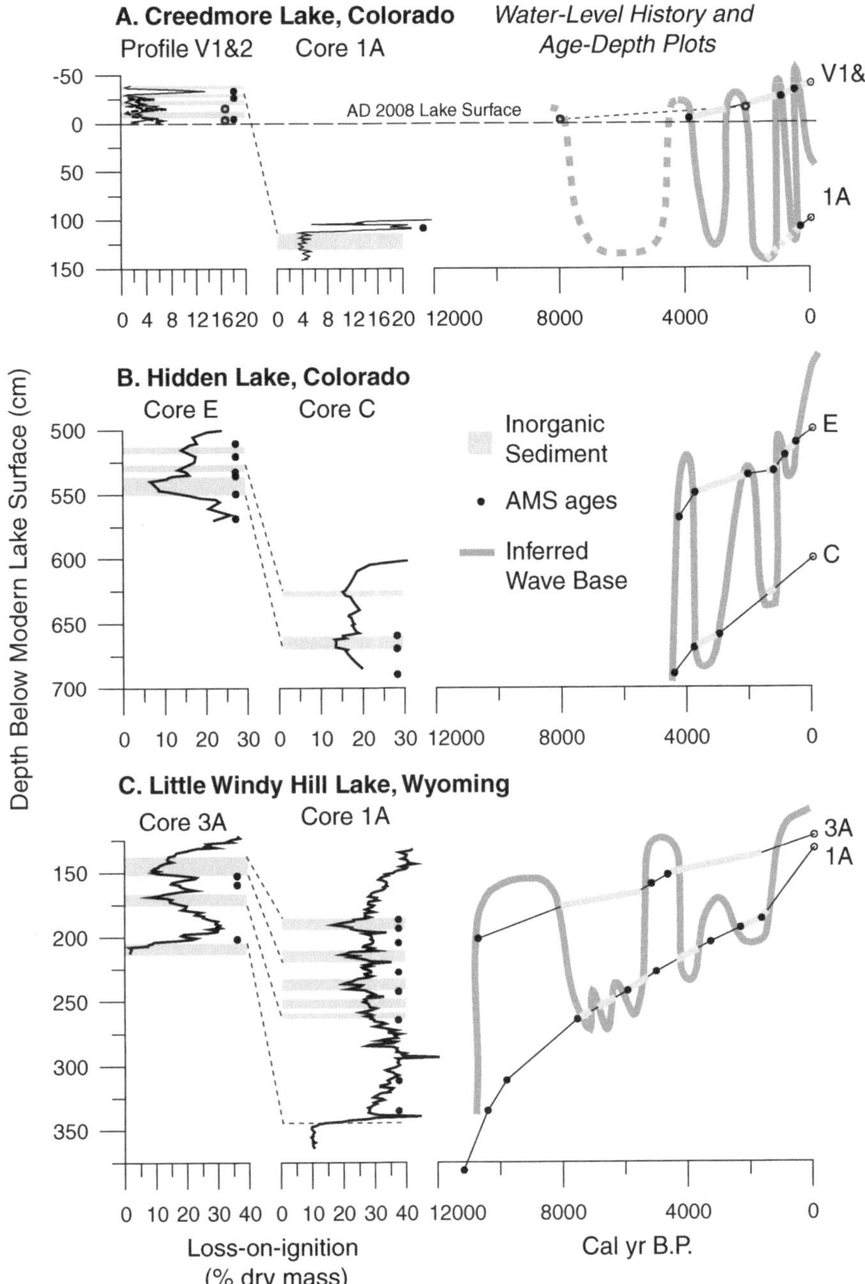

Figure 4. Sediment stratigraphies and inferred water-level histories from (A) Creedmore Lake, Larimer County, Colorado; (B) Hidden Lake, Jackson County, Colorado; and (C) Little Windy Hill Lake, Carbon County, Wyoming. Dark gray curves (at right) indicate the water-level history as inferred from the elevations of inorganic deposits with slow sediment accumulation rates. Left columns show percentage loss-on-ignition (LOI) data versus elevation relative to the modern water surface from two sediment cores or near-shore sediment profiles from each lake; the LOI indicates the amount of organic matter in the sediments, which is low in shallow, near-shore environments (Shuman 2003). Gray bars denote inorganic intervals thought to represent paleoshorelines. Dashed lines indicate points of correlation between cores. Right graphs show the age–depth relationships for the cores and profiles based on the elevation and ages of calibrated radiocarbon dates (closed circles; Table 2). Light gray lines indicate inorganic sediment intervals, and relatively horizontal lines indicate periods of low net sediment accumulation consistent with shallow water. Gray circles in (A) indicate calibrated radiocarbon ages from sediment profile V2, which provides replication of sediment profile V1. Open circles (right) denote the top of the sediment sequences.

watershed area above Northgate, ranges from −14.9 to −18.0 m^3s^{-1} compared to the flows associated with the historic hydrologic regime and lake volumes. The anomalies far exceed the historic mean surface flow of 12 m^3s^{-1} and thus would indicate insufficient water for even base river flow levels when the lakes occupied their paleoshorelines. Negative surface flow rates have specific legal implications (e.g., Exhibit 7 *Nebraska v. Wyoming & Colorado*, 534 U.S. 40 [2001]). Here, however, they numerically result from the estimated change in lake volumes (Table 1) and thus indicate that groundwater, as well as surface flows, would have been depleted. The river channel was likely dry—except perhaps during spring snowmelt. Therefore, if the Holocene reconstructions are accurate, the North Platte River experienced (repeated or prolonged) reductions in flow over several thousand years that were much more severe than the historic low flow years (Table 1).

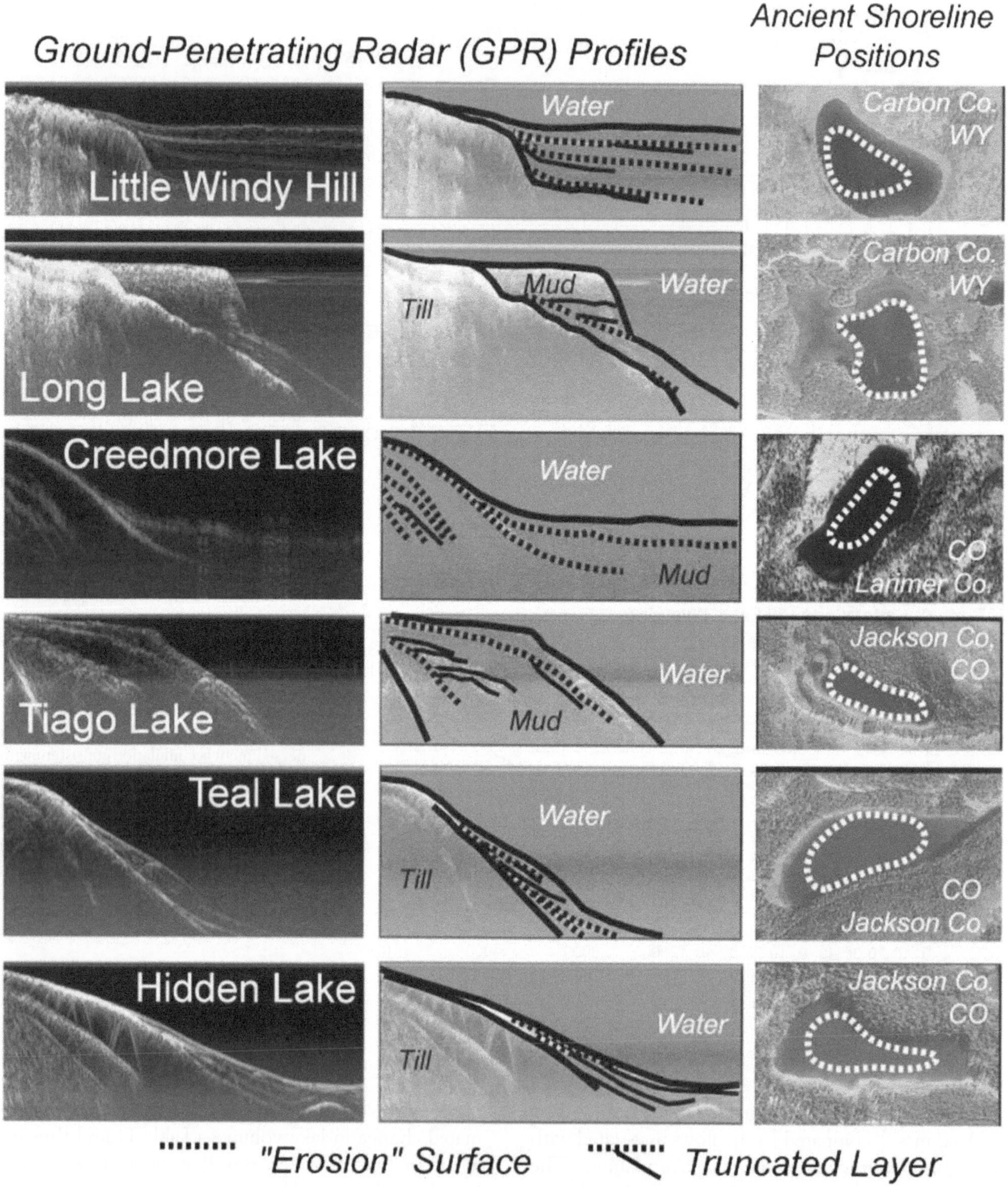

Ground-Penetrating Radar (GPR) Profiles

Ancient Shoreline Positions

Little Windy Hill
Water
Carbon Co. WY

Long Lake
Till
Mud
Water
Carbon Co. WY

Creedmore Lake
Water
Mud
CO Larimer Co.

Tiago Lake
Water
Mud
Jackson Co. CO

Teal Lake
Water
Till
CO Jackson Co.

Hidden Lake
Water
Till
Jackson Co. CO

"Erosion" Surface

Truncated Layer

Figure 5. Lakes throughout the region contain submerged erosion surfaces at their margins that indicate prolonged periods of low water levels. Representative ground-penetrating radar profiles from a geographically diverse selection of lakes are shown. White indicates areas of high-density contrast within the vertical radar profiles. The middle column shows the interpretation of the profile with black lines denoting the submerged lake sediments, dashed lines marking erosion surfaces, and thin black lines showing truncated layers. At the right, the inferred positions of the most prominent ancient shorelines (based on the erosion surfaces) are shown. Where dated at Little Windy Hill, Creedmore, and Hidden lakes, the inferred shoreline positions were occupied from 8,000 to 5,000 years ago.

Table 2. Ages of radiocarbon samples.

Lab	Sample #	Core	Depth (cm)	Material dated	Radiocarbon age (Years before AD 1950)	SD	Calibrated age Maximum	Median	Minimum
Little Windy Hill Lake									
UCI	63879	1A	53.5	>125 μm charcoal pieces	1735	20	1694	1651	1613
UCI	58941	1A	59.5	>125 μm charcoal (0.11 mgC)	2270	20	2341	2316	2210
UCI	63880	1A	69.5	>125 μm charcoal pieces	3040	20	3323	3267	3218
UCI	63881	1A	91.5	>125 μm charcoal pieces	4430	20	5045	5010	4972
UCI	58939	1A	105.5	>125 μm charcoal pieces	5160	20	5928	5921	5909
UCI	58940	1A	126.5	>125 μm charcoal pieces	6640	25	7566	7530	7506
UCI	63882	1A	172.5	>125 μm charcoal pieces	8785	25	9888	9799	9709
UCI	58938	1A	195	>125 μm charcoal (0.12 mgC)	9235	45	10494	10401	10297
UCI	63883	1A	240.5	>125 μm charcoal (0.14 mgC)	9700	35	11198	11154	11125
UCI	58935	3A	30.5	>125 μm charcoal (0.065 mgC)	4110	50	4806	4642	4529
UCI	58936	3A	37.5	>125 μm charcoal (0.092 mgC)	4480	40	5281	5160	5044
UCI	58937	3A	79.5	>125 μm charcoal pieces	9460	30	10740	10698	10609
Creedmore Lake									
UCI	61005	1A	8.5	>125 μm charcoal pieces	240	20	303	291	158
UCI	63886	V1	6.5	>125 μm charcoal pieces	400	20	503	485	471
UCI	63887	V1	13.5	>125 μm charcoal pieces	990	20	934	922	834
UCI	63888	V1*	21.5	>125 μm charcoal pieces	−30	20	n/a	n/a	n/a
UCI	63889	V1	35.5	>125 μm charcoal pieces	3545	20	3878	3843	3778
UCI	63890	V2	11.5	>125 μm charcoal pieces	2060	20	2055	2027	1991
UCI	61006	V2	23.5	>125 μm charcoal pieces	7120	20	7965	7953	7938

Note: UCI = University of California, Irvine (W. M. Keck Carbon Cycle Accelerator Mass Spectrometry [AMS] Laboratory).
* Out of order; not used.

Discussion

Historic climate data indicate that regional climate change is affecting regional hydrology and water supplies, but our lake-level reconstructions suggest that much more dramatic changes in water resources have taken place in the past. P − E and riverflow reconstructions indicate streamflow at the Northgate stream gauge could have been ephemeral (assuming some seasonal flow from snowmelt) from ca. 8,000 to 5,000 years before present. Similarly, dune activity in Nebraska indicates that extended drought periods in the past likely led to a "perennially dry" North Platte River Valley (Muhs et al. 2000, 214); dune activity in Nebraska spanned from 9,900 to 6,600 years before present (Miao et al. 2007). Such changes coincide with extensive regional evidence of severe aridity (e.g., evidence of low water at more than fourteen lakes across the Rocky Mountains [Stone and Fritz 2006; Shuman et al. 2009]; evidence of aridity from fossil pollen records [Thompson et al. 1993]) and probably relate to major changes in global climate controls during the Holocene (Bartlein et al. 1998; Diffenbaugh et al. 2006). Legal water allocation, therefore, could be severely strained by the effects of similar changes in global climate controls in the future (i.e., high greenhouse gas concentrations).

How reasonable are our riverflow inferences (i.e., seasonally dry riverbeds)? In 2002, the lowest flow year on record, flow at Northgate dropped to 21.5 percent of the mean annual flow and was fed by groundwater sources that had probably not equilibrated to the single year of severe drought (e.g., Winter, Rosenberry, and LaBaugh 2003). Despite the lack of hydrologic equilibrium, areas of the North Platte River channel were dry during ninety-four days in 2002 at the Wyoming–Nebraska border based on USGS NSIP data and were also commonly dry in the nineteenth century (Williams 1978; Muhs and Holliday 1995). Therefore, persistent shifts in regional moisture balance that lasted for centuries or millennia, such as before ca. 5,000 years BP (Figure 4), could be reasonably expected to reduce flows to ephemeral levels given that the low lake levels indicate substantial reductions in groundwater. Groundwater would have had time to reach equilibration with the persistently low precipitation, high evapotranspiration rates, or both, especially given sufficient time for paleoshoreline formation. Consistent with these ideas, the lakes were only minimally drawn down during recent droughts (Figure 3) but show evidence of multiple meters of drawdown during the mid-Holocene (Figures 3–5). Net erosion in uplands and small valleys produced alluvial terraces on the North Platte (Condon 2005), which is also consistent with large changes in riverflow regimes.

The current flow regime changes are both temperature related (e.g., increasing temperatures leading to early spring runoff) and precipitation related (e.g., decreasing winter precipitation reducing annual flow). The flow-regime trends are sensitive to the time period examined, but the regional temperature trend is consistent in sign and becomes stronger after 1948 and indicates that flow regimes could continue to shift (Stewart, Cayan, and Dettinger 2004). Even in the absence of any anthropogenic trends in the climate and river data, our Holocene results indicate that major shifts in hydrological regimes, although not frequent, have taken place in the North Platte watershed (Figure 4). The region, since settlement, therefore has not experienced the full range of possible drought severity, frequency, or duration that can result from variability intrinsic to the climate system. Our specific flow calculations are overly simplified, but Holocene flow reductions must have been greater than in 2002, when flows were reduced at Northgate by more than 75 percent (Table 1), because paleoshorelines were lower than historic levels (Figure 3) and associated with sufficient time for groundwater and base river flows to equilibrate to the climatic conditions (Figure 4).

The reductions have important implications for the allocation of river water. For example, the 1922 Laramie River Decree limited Colorado's usage of the Laramie River (which joins the Platte substantially downstream of the Northgate gauge) to 15,500 acre feet per year. However, if we apply the P – E anomalies estimated from the lake-level reconstructions to the 1,124 km^2 watershed of the Laramie River above Woods Landing, Wyoming (near the Colorado–Wyoming border and near Creedmore Lake) in the same way as for the Platte above Northgate (Table 1), we find that the Laramie River would also be reduced by much more than 15,500 acre-feet per year to ephemeral or zero flow. Such a reduction would enable Colorado to take all of the flow of the river (if any existed) and severely impact water use in municipalities, such as Laramie, Wyoming, where 50 percent of the water supply derives from the Laramie River.

The basis for the current legal allocations of the North Platte River, therefore, might not be applicable in the context of zero or near-zero flow, such as took place 5,000 to 8,000 years ago. The allocation of 1.28 million acre-feet to the state of Colorado over any consecutive ten-year period (per the Modified North Platte Decree of 2001) might account for annual variations within the system including short-term droughts, but our evidence shows that severe and persistent droughts have occurred over multiple decades to millennia. Such persistent aridity would prohibit the Colorado allocation over a ten-year time frame and similarly affect water delivery to the other states. Our results, therefore, confirm the importance of a ten-year reevaluation of the allocation of North Platte River water, as prescribed by the 2001 decree, because the hydrologic assumptions of the decree might not be met under climate conditions that differ from those of 1952 through 1999.

Ephemeral flows across the Wyoming portion of the North Platte River system would similarly have broad socioeconomic impact on cities and industries, such as Casper, Wyoming, where the regional fossil fuel industry is centered. Wyoming's ability to produce enough coal, oil, and natural gas to account for 17.9 percent of energy produced in the United States, which is approximately as much as Texas and more than any other individual state or nation (U.S. Energy Information Administration 2009), depends on historic levels of availability of water (U.S. Department of Energy 2006). For example, industrial users of water in the Platte River Basin in Wyoming include Chevron Environmental Services, Conoco-Phillips, and BP North America, which together are permitted to use 8,366 gallons per minute for oil exploration, refining, and reclamation (Wyoming Water Development Commission 2009). Ephemeral river flows and low groundwater could substantially impact such uses. Our results indicate that contingency plans for future energy production, municipal uses, agriculture, and other societal sectors should therefore consider extreme possibilities, including persistent absence of major surface water flows. Indeed, the current inability of many federal reservoirs on the Great Plains to store water (Brikowski 2008) demonstrates that Holocene-based scenarios are not unwarranted.

Acknowledgments

Thanks to two anonymous reviewers for their comments that improved this article. Victoria Perez, Paul Pribyl, and Robert Shriver assisted in data collection and analysis. Funding was provided by USGS and Wyoming Water Development Commission grants

(06HQGR0129 and WWDC25) to Bryan N. Shuman, Jacqueline J. Shinker, and Thomas A. Minckley; by the National Science Foundation (SBE-0623442) to Bryan N. Shuman and Anna K. Henderson; and (EPS-0447681) funded undergraduate fellowships to Paul Pribyle and Robert Shriver and a summer research apprenticeship program for Victoria Perez.

References

Barnett, T. P., and D. W. Pierce. 2008. When will Lake Mead go dry? *Water Resources Research* 44: W03201. http://www.agu.org/pubs/crossref/2008/2007WR006704.shtml (last accessed 9 July 2010).

Barnett, T. P., D. W. Pierce, H. G. Hidalgo, C. Bonfils, B. D. Santer, T. Das, G. Bala et al. 2008. Human-induced changes in the hydrology of the Western United States. *Science* 319:1080–83.

Bartlein, P. J., K. H. Anderson, P. M. Anderson, M. E. Edwards, C. J. Mock, R. S. Thompson, R. S. Webb, T. Webb III, and C. Whitlock. 1998. Paleoclimate simulations for North America over the past 21,000 years: Features of the simulated climate and comparisons with paleoenvironmental data. *Quaternary Science Reviews* 17:549–85.

Bates, B. D., Z. W. Kundzewicz, S. Wu, and J. P. Palutkof, eds. 2008. Climate change and water. Technical paper of the Intergovernmental Panel on Climate Change, IPCC Secretariate, Geneva, Switzerland.

Brikowski, T. H. 2008. Doomed reservoirs in Kansas, USA? Climate change and groundwater mining on the Great Plains lead to unsustainable surface water storage. *Journal of Hydrology* 354:90–101.

Cayan, D. R., S. Kammerdiener, M. D. Dettinger, J. M. Caprio, and D. H. Peterson. 2001. Changes in the onset of spring in the western United States. *Bulletin of the American Meteorological Society* 82:399–415.

Changnon, D., T. B. McKee, and N. J. Doesken. 1991. Hydroclimatic variability in the Rocky Mountains. *Water Resources Bulletin* 27:733–43.

Condon, S. M. 2005. Geologic studies of the Platte River, South-Central Nebraska and adjacent areas—Geologic maps, subsurface study, and geologic history. U.S. Geological Survey Professional Paper 1706, Washington, DC.

Cook, E. R., and P. J. Krusic. 2004. North American summer PDSI reconstructions. IGBP PAGES/World Data Center for Paleoclimatology Data Contribution Series #2004-045, NOAA/NGDC Paleoclimatology Program, Boulder, CO.

Cook, E. R., C. Woodhouse, C. M. Eakin, D. M. Meko, and D. W. Stahle. 2004. Long-term aridity changes in the western United States. *Science* 306:1015–18.

Dai, A., T. Qian, K. E. Trenberth, and J. D. Milliman. 2009. Changes in continental freshwater discharge from 1948–2004. *Journal of Climate* 22:2773–92.

Dearing, J. A. 1997. Sedimentary indicators of lake-level changes in the humid temperate zone: A critical review. *Journal of Paleolimnology* 18:1–14.

Diffenbaugh, N. S., M. Ashfaq, B. Shuman, J. W. Williams, and P. J. Bartlein. 2006. Summer precipitation in the United States: Response to Mid-Holocene changes in insolation, ocean mean-state, and ocean variability. *Geophysical Research Letters* 33: L22712. http://www.agu.org/pubs/crossref/2006/2006GL028012.shtml (last accessed 9 July 2010).

Digerfeldt, G., J. E. Almendinger, and S. Bjorck. 1992. Reconstruction of past lake levels and their relation to groundwater hydrology in the Parkers Prairie sandplain, west-central Minnesota. *Palaeoceonography, Palaeoclimatology, Palaeoecology* 94:99–118.

Intergovernmental Panel on Climate Change. 2007. Summary for policymakers. In *Climate change 2007: The physical science basis. Contribution of Working Group I to the Fourth Assessment Report of the Intergovernmental Panel on Climate Change*, ed. S. Solomon, D. Qin, M. Manning, Z. Chen, M. Marquis, K. B. Averyt, M. Tignor, and H. L. Miller, 1–18. New York: Cambridge University Press.

MacDonnell, L. J., D. H. Getches, and W. C. Hugenberg Jr. 1995. The law of the Colorado River: Coping with severe sustained drought. *Water Resources Bulletin* 31: 825–36.

Miao, X., J. A. Mason, J. B. Swinehart, D. B. Loope, P. R. Hanson, R. J. Goble, and X. Liu. 2007. A 10,000 year record of dune activity, dust storms, and severe drought in the central Great Plains. *Geology* 35:119–22.

Milly, P. C. D., J. Betancourt, M. Falkenmark, R. M. Hirsch, Z. W. Kundzewicz, D. P. Lettenmaier, and R. J. Stouffer. 2008. Stationarity is dead: Whither water management? *Science* 319:573–74.

Mote, P. W., A. F. Hamlet, M. P. Clark, and D. P. Lettenmaier. 2005. Declining mountain snowpack in Western North America. *Bulletin of the American Meteorological Society* 86:39–49.

Muhs, D. R., and V. T. Holliday. 1995. Evidence of active dune sand on the Great Plains in the 19th century from accounts of early explorers. *Quaternary Research* 43:198–208.

Muhs, D. R., J. B. Swinehart, D. B. Loope, J. Been, S. A. Mahan, and C. A. Bush. 2000. Geochemical evidence for an eolian sand dam across the North and South Platte Rivers in Nebraska. *Quaternary Research* 53:214–22.

National Climatic Data Center. 1994. Time bias corrected divisional temperature-precipitation-drought index: Documentation for dataset TD-9640. Available from DBMB, NCDC, NOAA, Federal Building, 37 Battery Park Ave., Asheville, NC28801–2733.

Nebraska v. Wyoming, 108 (Orig.), 325 U.S. 589, 665.

Nebraska v. Wyoming & Colorado, 534 U.S. 40 (2001).

Reimer, P. J., M. G. L. Baillie, E. Bard, A. Bayliss, J. W. Beck, C. J. H. Bertrand, P. G. Blackwell et al. 2004. IntCal04 terrestrial radiocarbon age calibration, 26–0 ka BP. *Radiocarbon* 46:1029–58.

Rood, S. B., G. M. Samuelson, J. K. Weber, and K. A. Wywrot. 2005. Twentieth-century decline in streamflows from the hydrographic apex of North America. *Journal of Hydrology* 306:215–33.

Seager, R., M. Ting, I. Held, Y. Kushnir, J. Lu, G. Vecchi, H.-P. Huang et al. 2007. Model projections of an imminent transition to a more arid climate in Southwestern North America. *Science* 316:1181–84.

Shinker, J. J., and P. J. Bartlein. 2010. Spatial variations of effective moisture in the Western United States. *Geophysical Research Letters* 37: L02701. http://www.agu.org/pubs/crossref/2010/2009GL041387.shtml (last accessed 9 July 2010).

Shuman, B. 2003. Controls on loss-on-ignition variation in cores from small New England lakes. *Journal of Paleolimnology* 30:371–85.

Shuman, B., and B. Finney. 2006. Late-Quaternary lake-level changes in North America. In *Encyclopedia of Quaternary Sciences*, ed. S. Elias, 1374–83. Amsterdam: Elsevier.

Shuman, B., A. K. Henderson, S. M. Colman, J. R. Stone, S. C. Fritz, L. R. Stevens, M. J. Power, and C. Whitlock. 2009. Holocene lake-level trends in the Rocky Mountains, USA. *Quaternary Science Reviews* 28:1861–79.

Shuman, B., P. Pribyl, T. A. Minckley, and J. J. Shinker. 2010. Rapid hydrologic shifts and prolonged droughts in Rocky Mountain headwaters during the Holocene. *Geophysical Research Letters* 37: L06701. http://www.agu.org/pubs/crossref/2010/2009GL042196.shtml (last accessed 9 July 2010).

Stewart, I. T., D. R. Cayan, and M. D. Dettinger. 2004. Changes in snowmelt runoff timing in Western North America under a "Business as usual" climate change scenario. *Climatic Change* 62:217–32.

Stone, J. R., and S. C. Fritz. 2006. Multidecadal drought and Holocene climate instability in the Rocky Mountains. *Geology* 34:409–12.

Thompson, R. S., C. Whitlock, P. J. Bartlein, S. P. Harrison, and W. G. Spaulding. 1993. Climatic changes in the Western United States since 18,000 Yr. B.P. In *Global climates since the last glacial maximum*, ed. H. E. Wright Jr., J. E. Kutzbach, T. Webb III, W. F. Ruddiman, F. A. Street-Perrot, and P. J. Bartlein, 468–513. Minneapolis: University of Minnesota Press.

U.S. Department of Energy. 2006. Energy demands on water resources: Report to Congress on the interdependency of energy and water. Congressional Report. U.S. Department of Energy, Washington, DC.

U.S. Energy Information Administration. 2009. State energy data system: Annual state energy production estimates through 2006. http://www.eia.doe.gov/emeu/states/_seds_production.html (last accessed 27 October 2009).

U.S. Geological Survey (USGS). 2005. Preliminary integrated geologic map databases for the United States: Central states: Montana, Wyoming, Colorado, New Mexico, North Dakota, South Dakota, Nebraska, Kansas, Oklahoma, Texas, Iowa, Missouri, Arkansas, and Louisiana. USGS Open-File Report 2005–1351. http://pubs.usgs.gov/of/2005/1351/ (last accessed 27 October 2009).

Williams, G. P. 1978. The case of the shrinking channels—The North Platte and Platte Rivers in Nebraska. U.S. Geological Survey Circular 781. U.S. Geological Survey, Washington, DC.

Winter, T. C., D. O. Rosenberry, and J. W. LaBaugh. 2003. Where does the ground water in small watersheds come from? *Ground Water* 41 (7): 989–1000.

Wyoming Water Development Commission. 2009. Wyoming state water plan: Platte River Basin water atlas. http://waterplan.state.wy.us/plan/platte/atlas/atlas.html (last accessed 27 October 2009).

Wyoming v. Colorado, 259 U.S. 419 (1922, 1957).

Climate Change, Drought, and Jamaican Agriculture: Local Knowledge and the Climate Record

Douglas W. Gamble, Donovan Campbell, Theodore L. Allen, David Barker, Scott Curtis, Duncan McGregor, and Jeff Popke

The purpose of this study is to reach a basic understanding of drought and climate change in southwestern Jamaica through an integration of local knowledge and perception of drought and its physical characteristics manifested in remotely sensed precipitation and vegetation data. Local knowledge and perception are investigated through a survey of sixty farmers in St. Elizabeth Parish and physical characteristics of drought are examined through statistical analysis of satellite precipitation and vegetation vigor time series. The survey indicates that most farmers are concerned about an increase in drought occurrence. Satellite estimates of rainfall and vegetation vigor for St. Elizabeth Parish support this perception and suggest that severe drought events are becoming more frequent. The satellite precipitation time series also suggest that the early growing season is becoming drier as compared to the primary growing season, especially since 1991. This recent divergence in growing season moisture conditions might add to farmers' observations that drought is becoming more prevalent. Consequently, Jamaican farmers perceptions of drought are not driven by magnitude and frequency of dry months alone but rather by the difference between growing seasons. Any development of drought adaption and mitigation plans for this area must not focus solely on drought; it must also compare moisture conditions between months and seasons to be effective. *Key Words: agriculture, climate change, drought, Jamaica, local knowledge.*

本项研究的目的是要通过结合当地知识和基于降水和植被的遥感数据而获得的对干旱和其物理特性的认识，以达到对牙买加西南地区干旱和气候变化的基本了解。我们对圣伊丽莎白教区的 60 个农户进行了统计调查，以获取当地的知识和看法；通过研究卫星降水和植被生机时间序列来检验干旱的物理特性。统计调查显示，大多数农民对旱灾频发感到关切。卫星估算的圣伊丽莎白教区的降水和植被生机结果支持这种看法，并且显示严重干旱正在日趋频繁。卫星降水时间序列还显示，相比主要生长季节，早期生长季节变得更加干燥，特别是自 1991 年以来。最近发生的水分条件在生长季节的离散状况，可能加深了农民对于干旱正在变得更加普遍的看法。因此，牙买加农民对干旱的看法不仅仅是依据干旱的规模和频率，而是基于生长季节之间的差异。如果想要更有效地设计针对该地区的干旱适应和缓解计划，就不能仅仅将重点放在干旱上，同时也必须比较各月之间和季节之间的水分条件。*关键词：农业，气候变化，干旱，牙买加，当地知识。*

El propósito de este estudio es lograr una comprensión básica de la sequía y el cambio climático en el sudoeste de Jamaica, mediante la integración del conocimiento local y la percepción de la sequía y sus características físicas, según se manifiestan en datos sobre precipitación y vegetación obtenidos por sensores remotos. El conocimiento y percepción locales se investigaron mediante el estudio de sesenta cultivadores de la Parroquia de St. Elizabeth, mientras las características físicas de la sequía se examinan por medio del análisis estadístico de series de tiempo existentes de precipitación y vegetación obtenidas con satélites. El estudio indica que la mayoría de los cultivadores están preocupados por el aumento en la ocurrencia de sequías. Los cálculos de satélite existentes sobre precipitaciones y vegetación para la Parroquia de St. Elizabeth apoyan esa percepción y sugieren que los eventos de sequía severa se están haciendo más frecuentes. Las series de tiempo de satélite sobre precipitación también sugieren que la estación de crecimiento temprana se está haciendo más seca, comparada con la estación de crecimiento primaria, especialmente desde 1991. Esta reciente divergencia en las condiciones de humedad durante la estación de crecimiento podría añadirse a las percepciones de los cultivadores de una sequía de prevalencia cada vez mayor. En consecuencia, las percepciones de los cultivadores jamaiquinos sobre sequía no están solo determinadas por la magnitud y frecuencia de los meses secos, sino principalmente por la diferencia entre las estaciones de crecimiento. Cualquier desarrollo de planes sobre adaptación a la sequía o mitigación de la misma en esta área debe centrarse no solo en la sequía misma; para que esos planes sean efectivos, deben compararse también las condiciones de humedad entre meses y estaciones. *Palabras clave: agricultura, cambio climático, sequía, Jamaica, conocimiento local.*

Global climate change is projected to increase the vulnerability of small island developing states (SIDS) to natural hazards. SIDS are vulnerable to the effects of global climate change and natural hazards because of their small size; location in regions prone to natural disasters; high population densities; poorly developed infrastructure; limited natural, human, and economic resources; and influence of external forces such as terms of trade, economic liberalization, and migration flows (Gable et al. 1990; Pelling and Uitto 2001; Lewsey, Cid, and Kruse 2004; Mimura et al. 2007). For the Caribbean region, the deterioration of water resources through climate change–induced drought and the consequent negative impact on agriculture are of particular concern (Watts 1995; Intergovernmental Panel on Climate Change [IPCC] 2001). Indeed, the IPCC Fourth Assessment Report indicates with high confidence that islands in the Caribbean will experience increased drought as a result of climate change (Mimura et al. 2007). Most agriculture in the region is dependent on rainfall because of a lack of sustainable groundwater resources (Gamble 2004). Recent trends and predictions suggest greater variability in rainfall and drier summers in the Caribbean (Singh 1997; Peterson et al. 2002; Angeles et al. 2007; Gamble 2009). Thus, effective drought adaptation and mitigation plans must be developed to increase the resilience of Caribbean SIDS in the face of climate change and drought.

There is no universally accepted definition of *drought* (Wilhite 2000; Keyantash and Dracup 2002). For example, one study identified more than 150 definitions of drought that used a variety of meteorological, hydrological, agricultural, and socioeconomic variables (Wilhite and Glantz 1985). Wilhite (2000, 6) therefore stated that: "realistically, definitions of drought must be region and application (or impact) specific." Given this, there is a need for measurements and assessments of drought at the local or regional scale.

Effective adaption to and mitigation of drought also requires an understanding of local perception and knowledge of a hazard. Such a location-specific understanding of drought is even more important given the broad, ambiguous definitions surrounding global climate change. Pettenger (2007, 5) pointed out: "Ask ten people how to define climate change, its causes and effects, and you will get ten different answers." Many scholars have noted that discussions and descriptions of climate change contain value-laden language that might have different meanings for different groups of people (Palutikof, Agnew, and Hoar 2004; Smith 2007). Thus, an understanding of local knowledge and perception of weather and climate is essential for the development of effective climate change policy at any given location, especially if it is to effectively address drought.

In regard to agriculture, farmers' perceptions of climate are inseparable from their accumulated local knowledge, through which they cognize, negotiate, and adapt to environmental and climate change (Osunade 1994; Ovuka and Lindqvist 2000; Roncoli, Ingram, and Kirshen 2002; Luseno et al. 2003; Meze-Hausken 2004; Vedwan 2006). Studies indicate that, although farmers might not understand the science behind climate change, their continual observation of and interaction with the surrounding environment heightens their awareness of even minor deviations from what is perceived to be normal weather and climate conditions. The majority of these studies are from Africa, Asia, and Latin America, with a paucity of studies for the Caribbean (Beckford, Barker, and Bailey 2007). The few studies that have been completed in Jamaica include analysis of environmental knowledge and decision making in small-farming systems (Davis-Morrison and Barker 1997) and the use of indigenous knowledge in yam production (Beckford 2002; Barker and Beckford 2006). No studies have been completed to date that have specifically assessed local knowledge and perceptions of drought and climate change and compared these to the climate and vegetation data for Jamaica.

Accordingly, the overall purpose of this study is to reach a basic understanding of drought hazard in southwestern Jamaica through an integration of local knowledge and perception of drought and its physical characteristics manifested in the climate record. Specifically, this study examines Jamaican farmers' knowledge and perceptions of drought in light of trends in local precipitation and remotely sensed vegetation data. The aim is to highlight the relationships between the climate record and the narrated practical experiences of farmers. As such, the analysis presented here seeks to integrate local and scientific knowledge and to increase our understanding of the nexus between the perception of weather and climate and data analysis within the Jamaican and Caribbean settings. Such a study provides an opportunity to examine the impacts of climate change from different epistemological perspectives, allowing for greater diversity in the climate change dialogue and avoiding an inappropriate application of a uniform definition of global climate change (Demeritt 2001; Smith 2007).

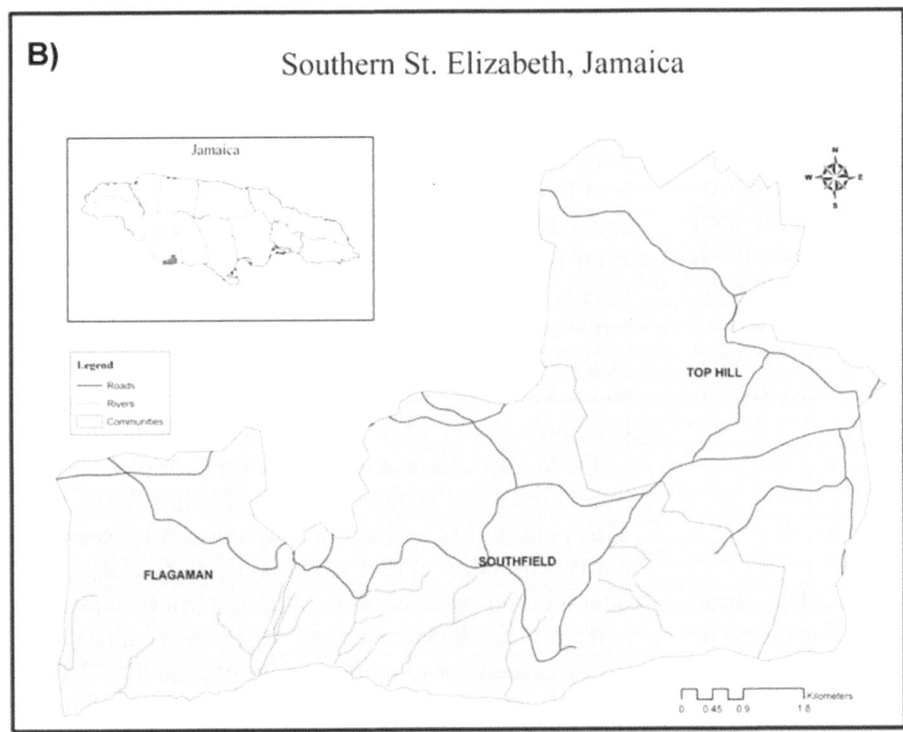

Figure 1. (A) A typical St. Elizabeth agricultural plot with scallions and guinea grass mulch (Photographer: E. J. Popke). (B) Map of St. Elizabeth Parish and towns included in survey.

Study Area

St. Elizabeth Parish is known as the "bread basket of Jamaica" and has been the top or second highest producer of domestic food crops in each of the last twenty years (Beckford, Barker, and Bailey 2007). Crops produced by local farmers include tomatoes, melons, cabbage, scallion, yam, beans, and cassava. The majority of the farms are small, individually owned plots, ranging from 0.1 to thirty acres (Figure 1A). The rainfall pattern for the parish as a whole is highly variable with marked differences in annual totals experienced over relatively short distances due to a complex interaction among topography, proximity to the coast, and the Jamaican rain shadow. Recent (2000–2008) average annual rainfall totals range from over 2,000 mm at stations in the north of the parish to around 1,000 mm in stations on the coast and below 800 mm in the dry coastal fringe.

This study focuses on the relatively drier southern portion of the parish, an area that serves as Jamaica's principal vegetable-producing region (Figure 1B). Farmers here have adapted to the dry environment through ingenious use of hand and drip irrigation combined with grass mulch techniques, offering a community for study that has a long tradition of using accumulated local knowledge to guide agricultural innovation (Beckford, Barker, and Bailey 2007; McGregor, Barker, and Campbell 2009). Recently, the agriculture system is perceived by farmers as being under stress through a series of recent early season droughts (2004, 2005, 2008) that were directly followed by stormy primary growing season conditions.

The primary growing season in St. Elizabeth Parish occurs in August through November due to relatively reliable rain. An early growing season also exists from April through June and has less reliable rain. The two growing seasons are separated by a midsummer dry spell that typically occurs in July (Chen and Brown 1994; Gamble and Curtis 2008; see Figure 2A). Many farmers in St. Elizabeth Parish view the early growing season as an opportunity to create capital that can be used to increase planting in the primary season, resulting in greater profit for the year.

Methodology

A survey of sixty farmers was conducted in three communities (Top Hill, Southfield, and Flagaman) in southern St. Elizabeth in June 2008, immediately after an early season drought event (Figure 1B). Twenty farmers were surveyed in each community to assess their perceptions of climate and climate change, with particular reference to drought. Findings from this survey are validated and complemented by two previous surveys (2006 and 2007), each of sixty farmers, and a random sample of 282 questionnaire-based interviews. The primary purpose of the survey reported here was to assess domestic food production and drought vulnerability in St. Elizabeth Parish. The general research questions addressed by the survey include the following: What have been the effects of droughts on farming over the last twenty years? What have been the adaptive strategies of farmers to these changes? What are the perceptions and attitudes of farmers to these events? For the purposes of this study, only results concerning farmers' knowledge, perceptions, and attitudes of drought are presented.

Precipitation data are available for multiple locations in St. Elizabeth Parish for the period from 1993 to 2006. This period of record is not long enough to assess climatic variability and it is necessary to find data that offer a better representation of long-term precipitation patterns. To develop a time series more appropriate for assessment of climatic variability, satellite-derived rainfall data can be used to represent rainfall in St.

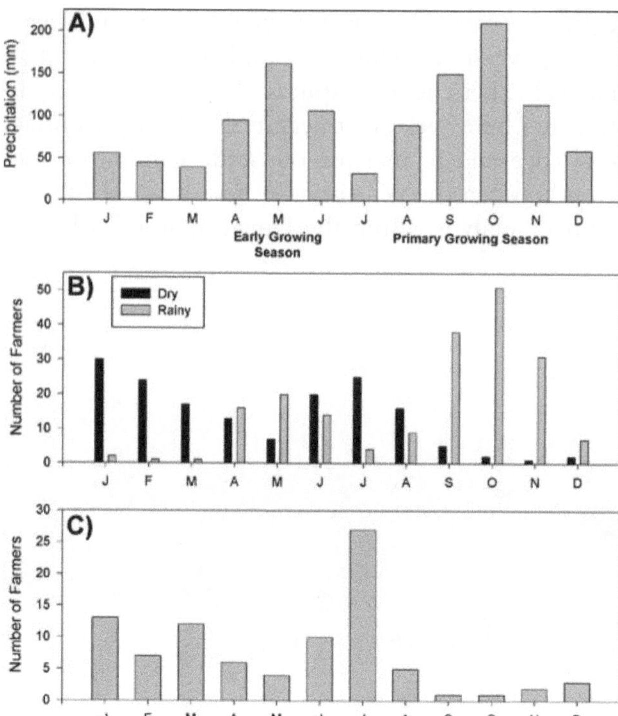

Figure 2. (A) Mean monthly rainfall (mm) for Southfield, St. Elizabeth Parish, Jamaica. (B) Frequency of farmers' perceptions of dry and rainy months. (C) Frequency of farmers' perceptions of the most severe dry month in terms of crop damage.

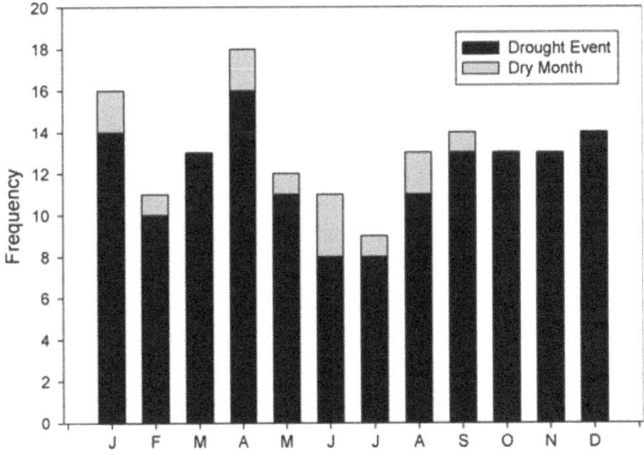

Figure 3. Monthly frequency of Standardized Precipitation Index defined drought events and dry months for the St. Elizabeth Parish Global Precipitation Climatology Project node, 1980–2007.

Elizabeth Parish. The best option for satellite precipitation data is the recently released Global Precipitation Climatology Project (GPCP) Version 2.1 monthly data set (Adler et al. 2003; Huffman et al. 2009). The GPCP data extend from 1979 to 2008 and are available at a 2.5° latitude/longitude grid. Analysis indicates that the GPCP data offer the best satellite estimator of rainfall over the Caribbean with the lowest errors and uncertainty over the western Caribbean and Jamaica, despite a potential dry bias (Jury 2009). The grid point closest to St. Elizabeth Parish (centered on 78.75°W, 18.75°N) was fairly well correlated ($r = 0.67$) with the average of available monthly St. Elizabeth Parish rainfall data from 1998 through 2006. However, it should be noted that these results are most applicable to the southern (coastal) portion of St. Elizabeth Parish, which has a better correlation ($r = 0.66$) with the GPCP data as compared to the northern portion ($r = 0.52$), which is at a higher elevation.

Drought events and dry months were defined in this study using the Standardized Precipitation Index (SPI; McKee, Doesken, and Kleist 1993). The SPI has been described as the best indicator of climatic drought in previous studies (Keyantash and Dracup 2002; Akhtari et al. 2008). It is calculated by first creating a three-month moving average from the raw precipitation data, then a gamma distribution is fit to the three-month moving average series and z scores are assigned to each observation based on the gamma distribution probability. The end result is a dimensionless unit similar to the Palmer Drought Index (Palmer 1965) that gives an indication of the magnitude of drought or absence of precipitation. A positive value indicates a wet month, whereas a negative value indicates a dry month. The SPI is a climatological drought index that only assesses frequency and magnitude of rain and does not take into consideration surface conditions or farmer water needs.

One modification of previous SPI methodology was implemented here. Typically, the SPI is calculated by fitting the gamma distribution to the entire time series, allowing for comparison of all months within a given time period. Here, the SPI is calculated by fitting the

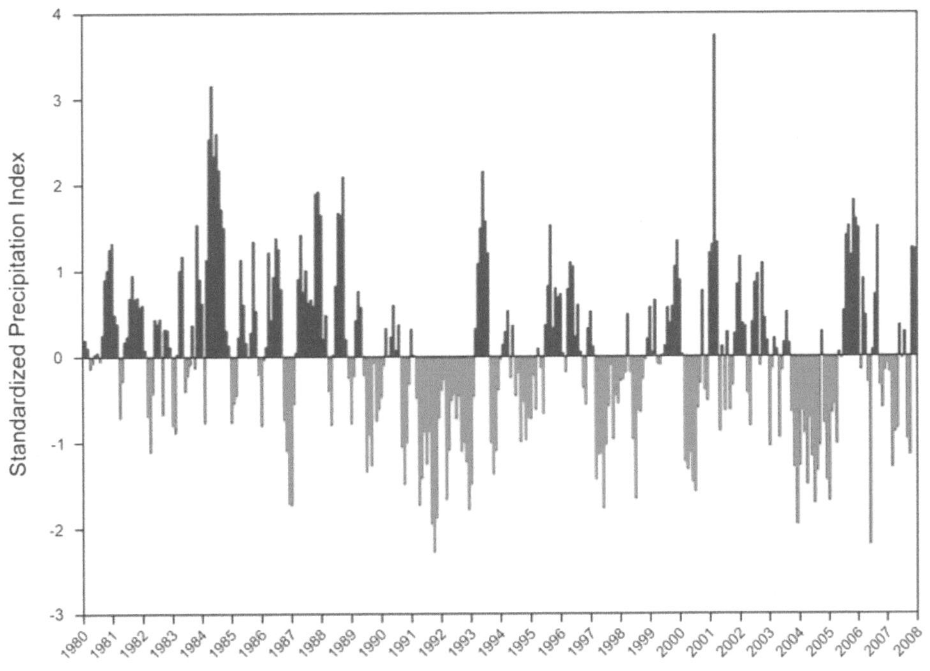

Figure 4. Calendar month Standardized Precipitation Index for St. Elizabeth Parish Global Precipitation Climatology Project node, 1980–2007.

gamma distribution to a time series for each calendar month. Thus, a total of twelve gamma distributions were fitted to twelve monthly time series. The rationale for this modification of standard SPI methodology is that farmers were most likely to compare rainfall of an individual month to that same month in previous years. For example, a farmer is more likely to compare this May's precipitation to the previous May's precipitation than to all months in the time series, as called for by traditional SPI methodology.

Once the SPI was calculated using the GPCP 1979–2008 time series for the St. Elizabeth Parish grid, both drought events and dry months were identified. The years 1979 and 2008 were dropped from classification of the monthly SPI values because calculation of three-month moving averages reduced the number of observations available in 1979 and 2008, creating an unequal number of observations for January and December as compared to the rest of the months. Consequently, all analyses of SPI data were completed for the years 1980 through 2007 to allow for consistent comparison. For this study, a drought event is defined as a minimum of two consecutive months with a negative SPI. A dry month is defined as a single month with a negative SPI value, preceded and followed by positive SPI values. After identification of drought event and dry months, basic descriptive statistics were calculated and linear trend analysis was completed. The linear trend analysis is not intended to produce statistically robust predictive models of the time series. Rather, the intention is to use the trend analysis to provide a descriptor (the slope) of the time series that can be used to compare with farmer perception.

Beyond analysis of an SPI monthly time series for the period of record from 1980 through 2007, two additional SPI time series were constructed, a primary season and an early season SPI time series. The reason for the inclusion of these two additional SPI time series in the analysis is that farmers frequently refer to moisture conditions in terms of seasons (Ovuka and Lindqvist 2000). Thus, data analysis will include the appropriate time frame for comparison to farmer perception.

Land surface and vegetation moisture conditions are represented in this study through the use of vegetation vigor estimates from satellite images. Specifically, vegetation vigor is estimated through the Global Inventory Modeling and Mapping Studies Normalized Difference Vegetation Index (NDVI; Pinzon, Brown, and Tucker 2005; Tucker et al. 2005). The NDVI is calculated by dividing the difference of the spectral reflectance between the visible channel (0.58–0.68 μm) and near infrared (NIR) channel (0.73–1.10 μm) by the sum of the two spectral ranges (Rouse et al. 1974):

$$NDVI = (NIR - visible)/(NIR + visible)$$

The result is a value ranging between -1 and $+1$, where higher values are more representative of greener or more vigorous vegetation. NDVI is calculated with the Advanced Very High Resolution Radiometer satellite, which provides 8-km, fifteen-day averages of NDVI

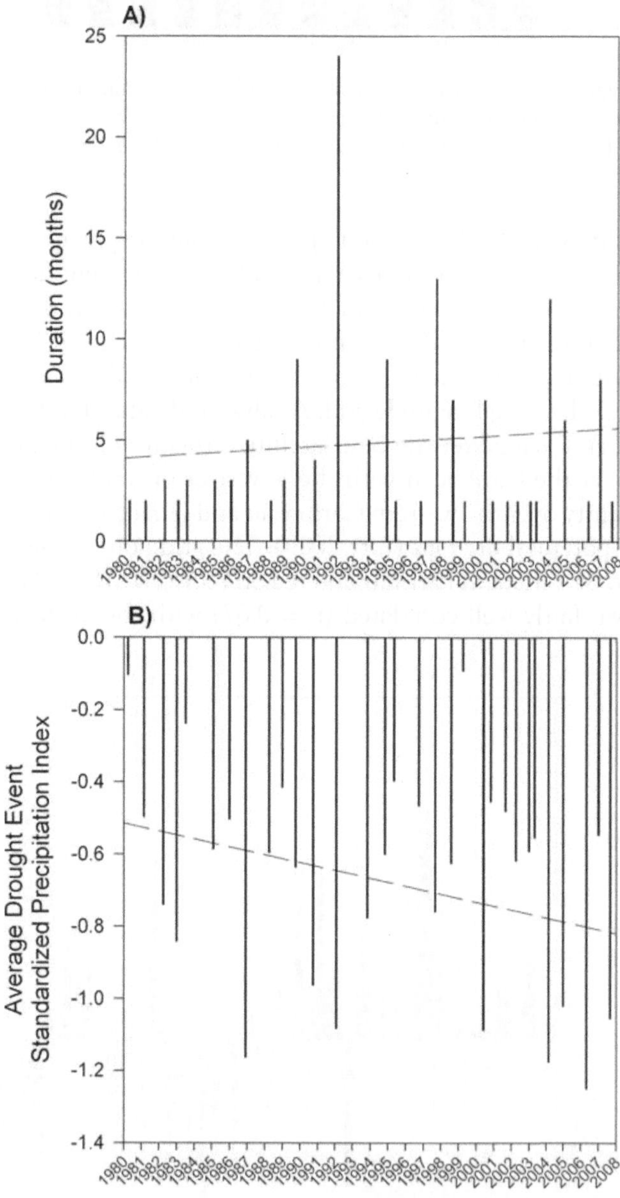

Figure 5. (A) Duration of individual Standardized Precipitation Index (SPI) defined drought events for the St. Elizabeth Parish Global Precipitation Climatology Project (GPCP) node, 1980–2007. (B) Average drought event SPI for St. Elizabeth Parish GPCP node, 1980–2007. Dashed lines represent linear regressions.

for cloud-free scenes. The NDVI period of record for the study, similar to GPCP period, is 1981 to 2003. Monthly averages of NDVI were computed for a representative grid box in the southeastern corner of St. Elizabeth Parish (77.6°W, 17.9°N). Then a monthly *z* score was calculated using the average NDVI values.

Results

Crop Calendars, Drought, and Seasonality

Analysis of the survey data and interview content reveals that farmers in southern St. Elizabeth possess appreciable knowledge of seasonal drought and climate variability. The farmers emphasize the timing of drought as opposed to the magnitude and duration of drought events as being the most harmful to crop production. Generally speaking, the farmers surveyed and interviewed were more interested in *when* below-normal levels of rainfall events are likely to occur as opposed to magnitude and duration of a drought event. Similar results of farmers being more concerned with the interaction of drought and cropping schedules have been reported for Kenya (Ovuka and Lindqvist 2000) and Burkina Faso (Roncoli, Ingram, and Kirshen 2002).

In St. Elizabeth Parish, the focus on timing of drought by farmers is not only a result of the prevailing rainfall regime but it is also a result of a cropping schedules being adapted to maximize production and profit. In recent years, farmers indicated that cropping schedules have changed in response to growth in tourism and the hotel industry. Farmers are utilizing more "quick crops" that mature in less than eight weeks and are popular with hotel guests (lettuce, zucchini, turnip, string bean, beetroot, cauliflower, broccoli, yellow squash, watermelon, honeydew melon, and cantaloupe). Such quick crops allow farmers to generate income during the dry season and "go out big" with primary crops in the rainy season. Consequently, if a drought occurs during a quick crop cycle, not only can it reduce income during the

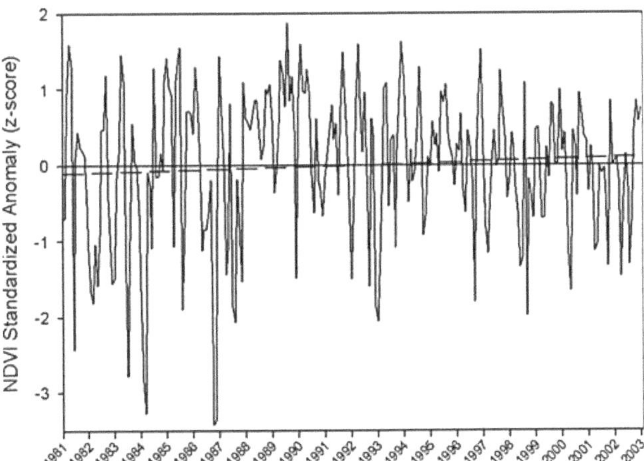

Figure 6. Calendar month Normalized Difference Vegetation Index (NDVI) *z* score for southern St. Elizabeth Parish, 1981–2003. Dashed line represents linear regression.

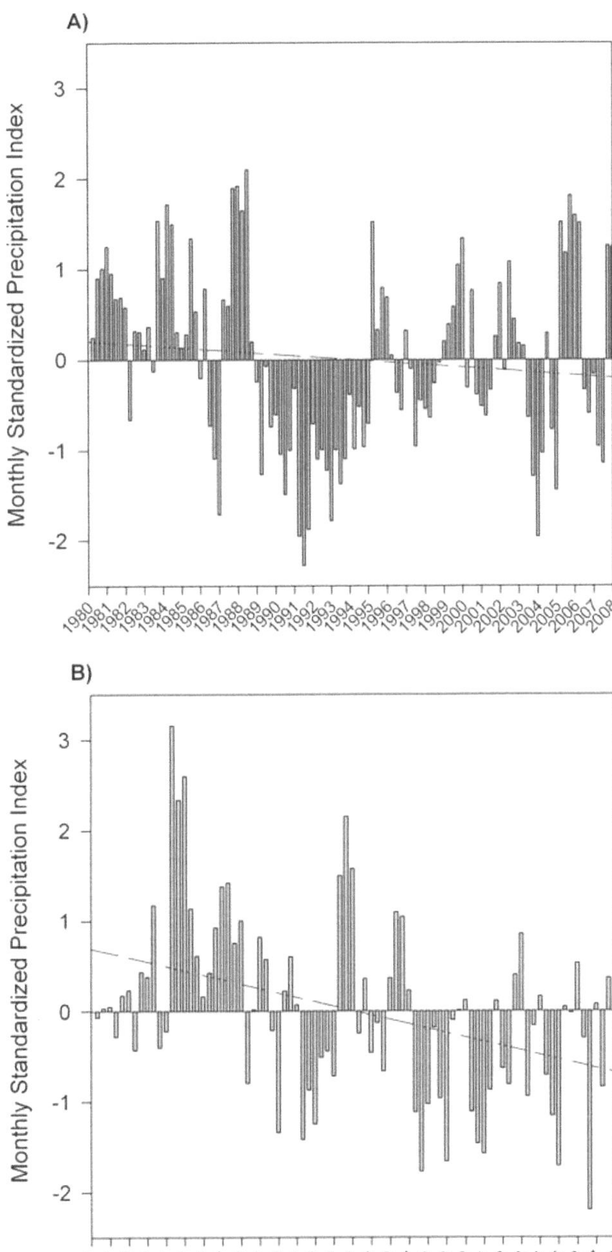

Figure 7. The (A) primary (August–November) and (B) early (April–June) growing season monthly Standardized Precipitation Index for the St. Elizabeth Parish Global Precipitation Climatology Project node, 1980–2007. Dashed lines represent linear regressions.

dry season, but it can diminish potential planting and eventual income during the rainy season.

As mentioned previously, the rainfall pattern of Jamaica is bimodal by nature (Figure 2A) with one initial peak that extends from April to June and the other from August to November (Chen and Brown 1994; Gamble and Curtis 2008). Both peaks are separated by a relatively low period of rainfall termed the midsummer drought or midsummer dry spell (Magaña, Amador, and Medina 1999; Gamble and Curtis 2008). This climatology is consistent with the farmers' perceptions described earlier (Figure 2B). Farmers are in agreement that in a typical year they experience four dry months (January–March and July) and four rainy months (May and September–November; see Figure 2B). The months of April, June, August, and December are "dry–wet"

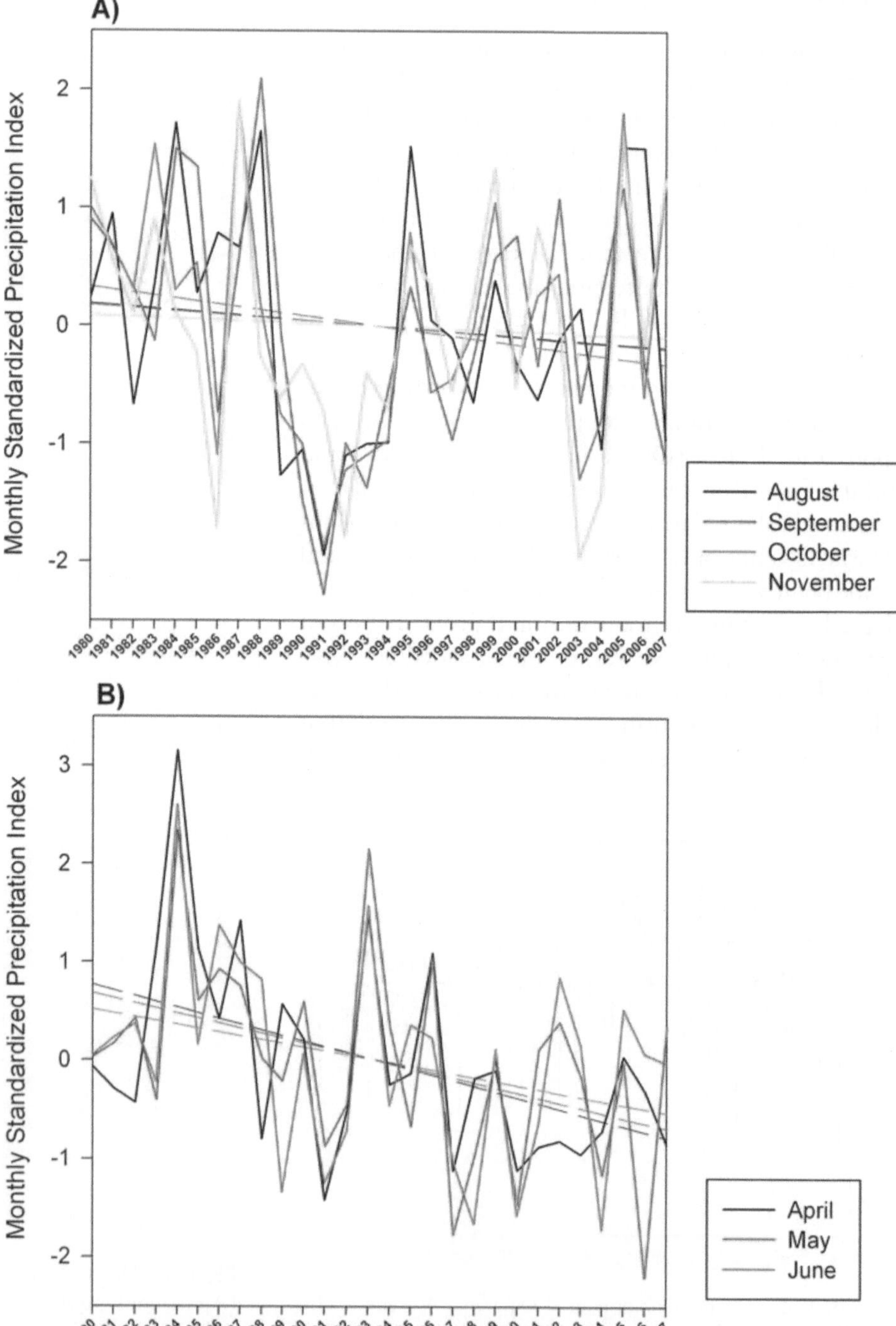

Figure 8. (A) August to November and (B) April to June monthly Standardized Precipitation Index for the St. Elizabeth Parish Global Precipitation Climatology Project node, 1980–2007. Dashed lines represent linear regressions.

months representing a transition between wet and dry conditions, as indicated by comparable frequencies of farmers identifying these months as either wet or dry.

For many decades farmers in southern St. Elizabeth have been fine-tuning farming practices to fit within this climatic pattern (McGregor, Barker, and Campbell 2009). These farmers know which crop to plant when, but extended or unusual periods of low rainfall remain a stress to most farmers, especially those with limited resources. Climate data analysis shows that such dry periods are not uncommon; there were a total of thirty-one drought events and thirteen dry months during the 1980 to 2007 study period. The range of drought event duration was two to twenty-four months with an average duration of five months. The range of total drought event magnitude (mean monthly SPI values in an event) was −0.91 to −0.25 and the average mean magnitude was −0.67. April and January have

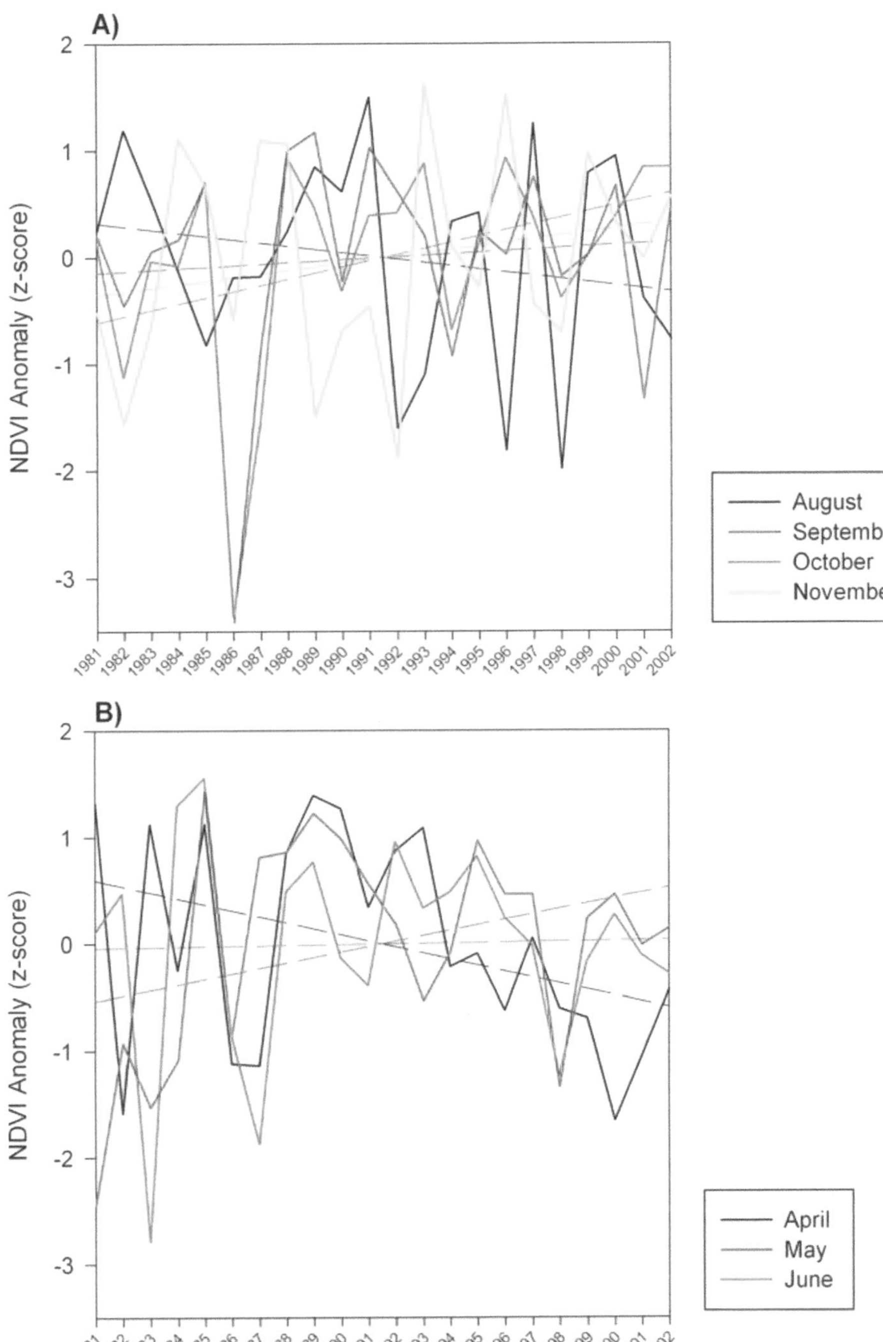

Figure 9. Growing season Normalized Difference Vegetation Index (NDVI) z score for southern St. Elizabeth Parish, 1981–2003. Dashed line represents linear regression.

the highest drought event frequency and July, June, and February have the lowest (Figure 3).

For farmers, however, drought during certain months might be much more damaging for agriculture. When asked to identify the most critical month in terms of drought, farmers displayed an understanding that July represents the midsummer dry spell and has the largest potential impact on agricultural yields (Figure 2C). Such a response is an indication of how important the quick crops for tourism have become in farmer production strategies. Because farmers are growing quick crops more frequently during the dry period of the year (December–March, July) a drought during this time can have a significant impact on farmer production. A low yield from a quick crop diminishes the amount of money that can be used on crops in the primary season. Thus, a month like July, which might have a low frequency of climatological drought as defined by the SPI, has an inordinate impact on agricultural production

Drought and Rainfall Variability

One of the key questions addressed in farmer surveys was their perception of long-term climate change. A majority of farmers (67 percent) indicated that there have been changes in weather patterns in the area. One of the most often-noted changes related to drought, with 65 percent of the surveyed farmers indicating that droughts are worse than in the past. Farmers reported that droughts have had specific impacts on agricultural practices. Sixty percent of the surveyed farmers, for example, reported having to buy water during recent drought, and 53 percent of farmers said they lost more than half of their crops, especially melon, tomatoes, and cucumbers, due to drought. Consequently, some farmers lament that the days of relying on the old mariner's poem to guide planting schedules (*June—too soon, July—stand by, August—look out you must, September—remember, October—all over*) has long gone.

Graphical analysis of the St. Elizabeth Parish monthly SPI time series from 1980 through 2007 indicates oscillation back and forth between wet and dry conditions throughout the time series (Figure 4). However, the frequency of longer drought events appears to change in the early 1990s. Before 1991, positive SPI months occur more frequently (86 out of 132 months or 58 percent) and the mean monthly SPI is above zero (0.31). After 1991, negative SPI months occur more frequently (119 out of 205 months or 76 percent), and the mean monthly SPI is less than zero (−0.17). An analysis of means test was performed at the 95 percent confidence level to test the null hypothesis that there is no statistically significant difference in pre-1991 mean SPI and post-1991 mean SPI. The test ($t = 4.62$, $p = 0.001$) results in the rejection of the null hypothesis, indicating a wetter pre-1991 (mean SPI = 0.306) and drier post-1991 (mean SPI = −0.17).

Analysis of a drought event duration time series indicates a slight linear increase in the duration of drought events ($y = 4.11 + 0.0045x$) and drought event

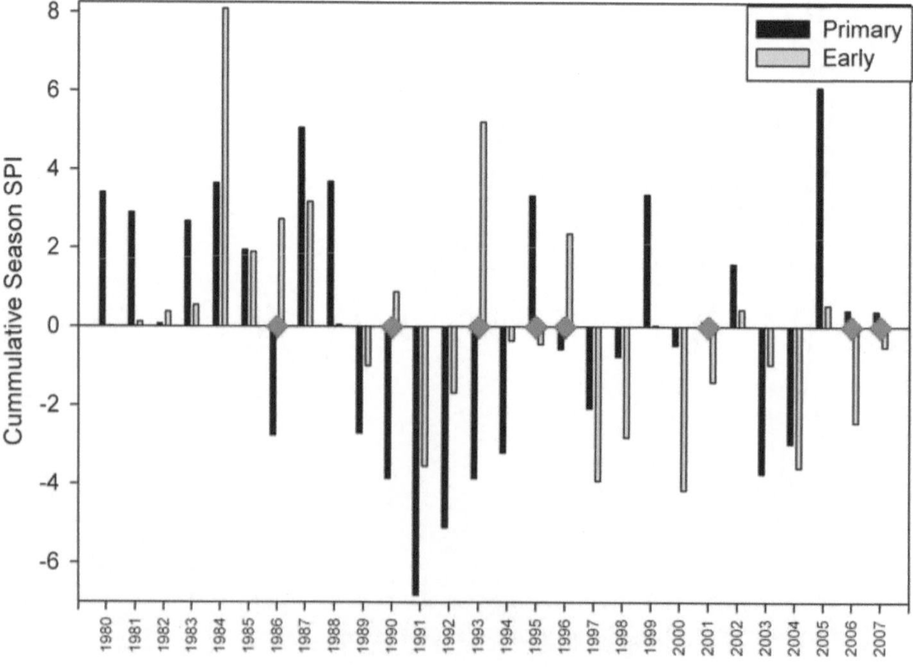

Figure 10. The primary (August–November) and early (April–June) growing season cumulative Standardized Precipitation Index (SPI) for the St. Elizabeth Parish Global Precipitation Climatology Project node, 1980–2007. Diamonds represent opposite signs between primary and early growing season.

magnitude ($y = -0.51 - 0.0009x$) in St. Elizabeth Parish between 1980 and 2007 (Figure 5). These trends are not statistically significant due to the considerable variability in the data. A clear or significant linear trend does not exist in the time series of the NDVI, with a linear trend close to zero (Figure 6).

Primary and Early Growing Season

Farmer perceptions of increasing drought might reflect relative changes in the early (April–June) and principal (August–November) growing seasons. Specifically, many farmers commented in interviews that drought is becoming more prevalent in the early growing season as compared to the primary growing season. In particular, the farmers contend that the dry season before the early growing season is getting longer and the midsummer dry spell is starting sooner, in effect reducing the early growing season to one month, May.

A time series graph of monthly SPI values for the primary and early growing season indicates a negative trend for both growing seasons from 1979 through 2007, although these trends are not statistically significant (Figure 7). Despite the lack of statistical significance the trend lines can be compared, suggesting greater change in the early season time series (-0.016 SPI_{month}^{-1}) as compared to the primary time series (-0.0036 SPI_{month}^{-1}), supporting farmers' perceptions that the early season is becoming drier as compared to the primary season. In terms of within growing season variability, each monthly SPI within each growing season (primary = August, September, October, November; early = April, May, June) was analyzed for the existence of a trend. Each of the primary season months exhibited a negative trend from 1980 through 2007 with similar slopes (-0.0067 to -0.24 $SPIyr^{-1}$; Figure 8). There is greater variability in trends for the early growing season with April, May, and June each having negative slopes of -0.057, -0.051, and -0.039 $SPIyr^{-1}$, respectively (Figure 8). Also, the slopes for the early season months are all steeper than the primary season months, further suggesting that the entire early season is drying faster than the primary season. April and May have the steepest slopes, supporting the farmer contention that the dry season comes before and shortens the early growing season. Vegetative vigor shows no consistent trend for either the primary or early growing season (Figure 9).

The aggregation of the monthly SPI values into a cumulative seasonal SPI further supports the contention that dry months are more frequent in the latter half of the early growing season time series in comparison with the primary growing season time series (Figure 10). The frequency of opposite sign primary and early season SPI has increased since 1991. Since 1991 the primary and early season have exhibited opposite sign

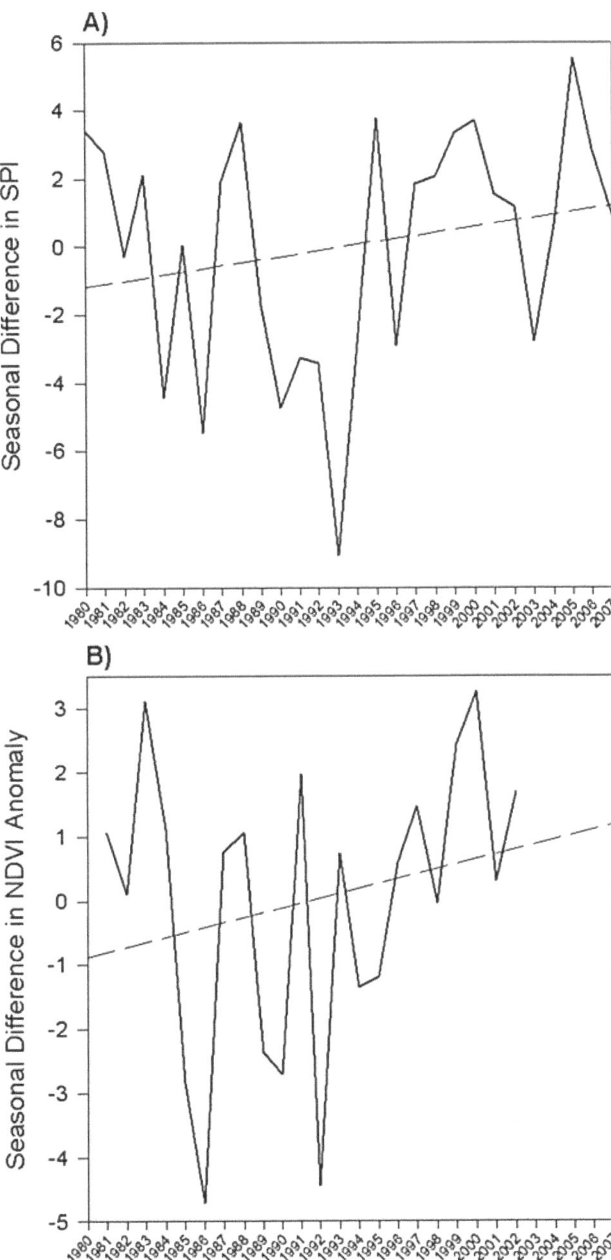

Figure 11. (A) The difference in the primary (August–November) and early (April–June) growing season cumulative Standardized Precipitation Index (SPI) for the St. Elizabeth Parish Global Precipitation Climatology Project node, 1980–2007. (B) The difference in the primary (August–November) and early (April–June) growing season Normalized Difference Vegetation Index (NDVI) for southern St. Elizabeth Parish, 1980–2007. Dashed lines represent linear regressions.

SPI values six times (35 percent), whereas this only occurred two times pre-1991 (18 percent). For the six incidents of opposite sign, the last three represent a drier early season and a wet primary season. Such recent occurrence of dry early and wet primary season might have a strong influence on current farmer perception. Overall, analysis of the difference between the primary SPI and early SPI (Figure 11) indicates a trend toward a positive difference or a wetter primary season as compared to a drier early season. This is also the case with the normalized NDVI (Figure 11).

Discussion and Conclusions

Farmers in St. Elizabeth Parish are keenly aware of annual rainfall patterns and have adapted their agricultural practices to take advantage of natural moisture surpluses. The somewhat marginal environment for agriculture in this area means that farmers are particularly sensitive to changes in weather and climate, with many expressing concerns about an increase in drought. In accordance with farmer perceptions of worsening drought conditions, satellite estimates of rainfall suggest that severe (high magnitude/long-duration) drought events are becoming more frequent in St. Elizabeth Parish. Such results, even though they are not statistically significant, are consistent with previous reports of a negative trend in Caribbean rainfall since the 1960s (Peterson et al. 2002) and a decrease in rainfall in Jamaica over the past century (McGregor, Barker, and Campbell 2009).

In terms of the primary versus the early growing season, the climate data suggest that the early growing season is drying faster than the primary growing season, particularly since 1991. Even though the trend for normalized NDVI is slightly positive, this same seasonal difference is depicted. Such a discrepancy between the data sets is due to the fact vegetation vigor is dependent on more than just precipitation, such as evapotranspiration and fires. This recent divergence in growing season moisture conditions might be reflected in farmers' observations that the early growing season is becoming drier than the primary, or drought is becoming more prevalent in the early season. Consequently, Jamaican farmer perception of drought may be driven not only by magnitude and frequency of dry months alone but also the difference between seasons. Specifically, farmers notice oscillation between a dry early season and wet primary season just as they do persistent dry conditions. This sensitivity to early season rainfall deficit might in part be due to regional changes in cultivation patterns, as farmers increase their production of dry season quick crops to take advantage of an expanding tourist market. Thus, any development of drought adaption and mitigation plans must not focus solely on the magnitude and frequency of drought; it must also compare moisture conditions between seasons in the context of changing livelihood strategies among local farmers.

At a more general level, the perception that droughts are worse now than in the past underscores farmers' feelings of increased vulnerability to changes in their local environmental setting. Beyond climate conditions, socioeconomic factors might serve to increase the sense of insecurity among farmers and thus amplify the impacts of environmental change. In recent years, Jamaican farmers have been hampered by increased production costs, declining state support, and enhanced competition from imports due to trade liberalization (Weis 2004). In such a context, farmers might be more sensitive to minor changes in weather and climate patterns. For example, the impacts of the increasingly unpredictability of rainfall in St. Elizabeth Parish can normally be offset by the increasing use of irrigation systems or the purchase of water from a vendor. In the context of declining rural fortunes and declining state support, however, this might no longer be possible, leaving many farmers at the mercy of climate variability and change as well as a dependence on family members and friends to survive. More than 90 percent of the farmers said that they did not receive any assistance from the government following the 2008 drought, a higher percentage than those who reported having received no assistance after Hurricane Ivan (McGregor, Barker, and Campbell 2009). This combination of physical and socioeconomic changes works to increase the stress on agricultural systems, as has been noted in Jamaica, the Caribbean, and other parts of the world (Glantz 1987; Liverman 1990; Barker 1993; Weis 2004).

It is clear from the record of interviews that smallholder farmers need help to adapt to changes in rainfall patterns. They are doing their best to cope with the changes by utilizing their accumulated local knowledge. However, the implementation of future policies and programs that complement existing local knowledge and perception, particularly by focusing on the range of moisture conditions as opposed to drought alone, will be essential in alleviating stresses, as well as protecting the local knowledge base that supports the existence of these farming communities. Without such efforts, adaption and mitigation plans risk the error of devaluing local expertise and cultural notions about the nature of knowledge (Roncoli, Ingram, and Kirshen 2002).

Ultimately, many researchers and policymakers desire to increase the capacity of vulnerable farmers to effectively adapt to climate change through utilization of a combination of traditional, local knowledge and scientific understanding of climate change (Roncoli, Ingram, and Kirshen 2002). Through such a balanced approach to combining local knowledge and scientific information, efforts can harness a rich history of adaptive capacity and resilience on small islands where individuals, communities, and governments have traditionally adapted to environmental change over a long period of time (Mimura et al. 2007). Specific future actions that can assist in developing balanced approaches to adapting to climate change and drought in Jamaica and the Caribbean include a refinement of downscaling techniques to appropriately assign climate databases to a specific location; a more in-depth analysis of farmer experience in forecasting and adaption to specific drought events; a better understanding of how government policy and socioeconomic forces intersect with environmental change; and development of end-user-focused drought management products. Through such efforts, Caribbean SIDS will be able to build greater adaptive capacity and resilience to an uncertain future brought on by global climate change.

Acknowledgments

This material is based on work supported in part by the National Science Foundation under Grant No. BCS-0718257. We gratefully acknowledge the willingness and enthusiasm of the farmers of St. Elizabeth Parish for participation in this project.

References

Adler, R. F., G. J. Huffman, A. Chang, R. Ferraro, P. Xie, J. Janowiak, B. Rudolf et al. 2003. The version-2 Global Precipitation Climatology Project (GPCP) monthly precipitation analysis (1979–present). *Journal of Hydrometeorology* 4:1147–67.

Akhtari, R., S. Morid, M. H. Mahdian, and V. Smakhtin. 2008. Assessment of areal interpolation methods for spatial analysis of SPI and EDI drought indices. *International Journal of Climatology* 29:135–45.

Angeles, M. E., J. E. Gonzalez, D. J. Ericson, and J. L. Hernandez. 2007. Predictions of change in the Caribbean region using global circulation models. *International Journal of Climatology* 27:555–69.

Barker, D. 1993. Dualism and disasters on a tropical island: Constraints on agricultural development in Jamaica. *Tijdschrift voor Economische en Sociale Geografie* 84 (5): 332–40.

Barker, D., and C. Beckford. 2006. Plastic yam and plastic yam sticks: Perspectives on indigenous technical knowledge among Jamaican farmers. *Tidjschrift voor Economische en Sociale Geographie* 97 (5): 535–47.

Beckford, C. 2002. Decision-making and innovation among small-scale yam farmers in central Jamaica: A dynamic, pragmatic, and adaptive process. *The Geographic Journal* 168 (3): 248–59.

Beckford, C., D. Barker, and S. Bailey. 2007. Adaptation, innovation and domestic food production in Jamaica: Some examples of survival strategies of small-scale farmers. *Singapore Journal of Tropical Geography* 28: 273–86.

Chen, A. A., and P. Brown. 1994. Characterization of April–May rainfall. *Jamaican Journal of Science and Technology* 5:1–11.

Davis-Morrison, V., and D. Barker. 1997. Resources management, environmental knowledge and decision-making in small farming systems in the Rio Grande Valley, Jamaica. *Caribbean Geography* 8 (2): 96–106.

Demeritt, D. 2001. The construction of global warming and the politics of science. *Annals of the Association of American Geographers* 91 (2): 307–37.

Gable, F. J., M. M. Affs, J. H. Gentile, and D. G. Aubrey. 1990. Global climate and coastal impacts in the Caribbean region. *Environmental Conservation* 17:51–60.

Gamble, D. W. 2004. Water resource development on small carbonate islands: Solutions offered by the hydrologic landscape concept. In *Worldminds: Geographic perspectives on 100 problems*, ed. D. J. B. Warf and L.K. Hansen, 503–8. Dordrecht, The Netherlands: Kluwer.

———. 2009. Caribbean vulnerability: An appropriate climatic framework. In *Global change and Caribbean vulnerability: Environment, economy and society at risk?*, ed. D. F. M. McGregor, D. Dodman, and D. Barker, 22–46. Kingston, Jamaica: University of West Indies Press.

Gamble, D. W., and S. Curtis. 2008. Caribbean precipitation: Review, model, and prospect. *Progress in Physical Geography* 23 (3): 265–76.

Glantz, M. H., ed. 1987. *Drought and hunger in Africa: Denying famine a future*. New York: Cambridge University Press.

Huffman, G. J., R. Adler, D. Bolvin, and G. Gu. 2009. Improving the global precipitation record: GPCP version 2.1. *Geophysical Research Letters* 36:L17808.

Intergovernmental Panel on Climate Change (IPCC), ed. 2001. *Climate change 2001: The scientific basis. Contribution of Working Group I to the Third Assessment Report of the Intergovernmental Panel on Climate Change*. Cambridge, UK: Cambridge University Press.

Jury, M. R. 2009. An intercomparison of observational, reanalysis, satellite, and coupled model data on mean rainfall in the Caribbean. *Journal of Hydrometeorology* 10:413–30.

Keyantash, J., and J. A. Dracup. 2002. The quantification of drought: An evaluation of drought indices. *Bulletin of the American Meteorological Society* August:1167–80.

Lewsey, C., G. Cid, and E. Kruse. 2004. Assessing climate change impacts on coastal infrastructure in the Eastern Caribbean. *Marine Policy* 28 (5): 393–409.

Liverman, D. M. 1990. Drought impacts in Mexico: Climate, agriculture, technology, and land tenure in Sonora and

Puebla. *Annals of the Association of American Georgaphers* 80 (1): 49–72.

Luseno, W. K., J. G. McPeak, C. B. Barrett, P. D. Little, and G. Gebru. 2003. Assessing the value of climate forecast information for pastoralists: Evidence from Southern Ethiopia and Northern Kenya. *World Development* 31:1477–94.

Magaña, V., J. A. Amador, and S. Medina. 1999. The midsummer drought over Mexico and Central America. *Journal of Climate* 12:1577–88.

McGregor, D. F. M., D. Barker, and D. Campbell. 2009. Environmental change and Caribbean food security: Recent hazard impacts and domestic food production in Jamaica. In *Global change and Caribbean vulnerability: Environment, economy and society at risk?*, ed. D. F. M. McGregor, D. Dodman, and D. Barker, 197–217. Kingston, Jamaica: University of West Indies Press.

McKee, T. B., N. J. Doesken, and J. Kleist. 1993. The relationship of drought frequency and duration to timescales. *Proceedings of the Eighth Conference on Applied Climatology*, 179–84. Boston, MA: American Meteorological Society.

Meze-Hausken, E. 2004. Contrasting climate variability and meterological drought with perceived drought and climate change in northern Ethiopia. *Climate Research* 27:19–31.

Mimura, N., L. Nurse, R. F. McLean, J. Agard, L. Brigulio, P. Lefale, R. Payet, and G. Sem. 2007. Small islands. In *Climate change 2007: Impacts, adaption and vulnerability. Contribution of working group II to the Fourth Assessment of Intergovernmental Panel on Climate Change*, ed. M. L. Parry, O. F. Canzaina, J. P. Paultikof, P. J. van der Linden, and C. E. Hanson, 687–716. Cambridge, UK: Cambridge University Press.

Osunade, A. 1994. Indigenous climate knowledge and agricultural practice in southwestern Nigeria. *Malaysian Journal of Tropical Geography* 25:21–28.

Ovuka, M., and S. Lindqvist. 2000. Rainfall variability in Murang'a District, Kenya: Meteorological data and farmers' perception. *Geografiska Annaler* 82A (1): 107–19.

Palmer, W. C. 1965. Meteorological drought. Research Paper No. 45. Washington, DC: U.S. Weather Bureau.

Palutikof, J. P., M. D. Agnew, and M. R. Hoar. 2004. Public perceptions of unusually warm weather in the UK: Impacts, responses and adaptions. *Climate Research* 26:43–59.

Pelling, M., and J. I. Uitto. 2001. Small island developing states: Natural disaster vulnerability and global change. *Environmental Hazards* 3:49–62.

Peterson, T. C., M. A. Taylor, R. Demeritte, D. Duncombe, S. Burton, F. Thompson, A. Porter, et al. 2002. Recent changes in climate extremes in the Caribbean region. *Journal of Geophysical Research* 107 (D21): 4601.

Pettenger, M. E. 2007. Introduction: Power, knowledge and the social construction of climate change. In *The social construction of climate change: Power, knowledge, norms, discourses*, ed. M. E. Pettenger, 1–22. Aldershot, UK: Ashgate.

Pinzon, J., M. Brown, and C. Tucker. 2005. Satellite time series correction of orbital drift artifacts using empirical mode decomposition. In *Interdisciplinary mathematical sciences: Vol. 5. Hilbert-Huang transform: Introduction and applications*, ed. N. Huang and S. S. P. Shen, 167–86. Hackensack, NJ: World Scientific.

Roncoli, C., K. Ingram, and P. Kirshen. 2002. Reading the rains: Local knowledge and rainfall forecasting in Burkina Faso. *Society and Natural Resources* 15:411–30.

Rouse, J. W., R. H. Haas, J. A. Schell, and D. W. Deering. 1974. Monitoring vegetation systems in the Great Plains with ERTS. *Goddard Flight Center 3rd ERTS-1 Symposium* 1 (A): 309–17.

Singh, B. 1997. Climate changes in the greater and southern Caribbean. *International Journal of Climatology* 17:1093–114.

Smith, H. A. 2007. Disrupting the global discourse of climate change: The case of indigenous voices. In *The social construction of climate change: Power, knowledge, norms, discourses*, ed. M. E. Pettenger, 197–216. Aldershot, UK: Ashgate.

Tucker, C., J. Pinzon, M. Brown, D. Slayback, E. Pak, R. Mahoney, E. Vermote, and N. El Saleous. 2005. An extended AVHRR 8km NDVI dataset compatible with MODIS and SPOT vegetation NDVI data. *International Journal of Remote Sensing* 26 (20): 4485–98.

Vedwan, N. 2006. Culture, climate and the environment: Local knowledge and perception of climate change among apple growers in northwestern India. *Journal of Ecological Anthropology* 10:4–18.

Watts, D. 1995. Environmental degradation, the water resource and sustainable development in the Eastern Caribbean. *Caribbean Geography* 6 (1): 2–15.

Weis, T. 2004. Restructuring and redundancy: The impacts and illogic of neoliberal agricultural reforms in Jamaica. *Journal of Agrarian Change* 4 (4): 461–91.

Wilhite, D. A. 2000. Drought as a natural hazard: Concepts and definitions. In *Drought: A global assessment*, ed. D. A. Wilhite, 3–18. London: Routledge.

Wilhite, D. A., and M. H. Glantz. 1985. Understanding drought phenomena: The role of definitions. *Water International* 10:111–20.

Correspondence: Department of Geography and Geology, University of North Carolina Wilmington, Wilmington, NC 28403, e-mail: gambled@uncw.edu (Gamble); Department of Geography and Geology, University of West Indies at Mona, Mona, Jamaica, e-mail: donovan.campbell@uwimona.edu.jm (Campbell); david.barker@uwimona.edu.jm (Barker); Rosenstiel School of Marine and Atmospheric Science, University of Miami, Miami, FL 33149, e-mail: tallen@rsmas.miami.edu (Allen); Department of Geography, East Carolina University, Greenville, NC 27858, e-mail: curtisw@ecu.edu (Curtis); popkee@ecu.edu (Popke); Department of Geography, Royal Holloway, University of London, Egham, Surrey TW20 0EX, UK, e-mail: D.Mcgregor@rhul.ac.uk (McGregor).

Modeling Path Dependence in Agricultural Adaptation to Climate Variability and Change

Netra B. Chhetri, William E. Easterling, Adam Terando, and Linda Mearns

Path dependence of farmers' technical choices for managing climate risk combined with farmers' difficulties in discerning climate change from natural variability might hamper adaptation to climate change. We examine the effects of climate variability and change on corn yields in the Southeast United States using a regional climate model nested within a global climate model (GCM) simulation of the equilibrium atmospheric CO_2 concentration of 540 ppm. In addition to a climate scenario with normal variance, we modify the GCM outputs to simulate a scenario with a highly variable climate. We find that climate variability poses a serious challenge to the abilities of farmers and their supporting institutions to adapt. Consistently lower corn yields, especially in the scenario with a highly variable climate, illustrate that farmers' abilities to make informed choices about their cropping decisions can be constrained by their inabilities to exit from their current technological regimes or path dependence. We also incorporate farmers' responses to climate change using three adaptation scenarios: no adaptation, "perfect knowledge," and a scenario that mimics diffusion of knowledge across the landscape. Regardless of adaptation scenario and variance structure, the most common result is a decline in corn production to the point where yield reductions of 1 percent to 20 percent occur across 60 percent to 80 percent of the region. The advantage of the perfect knowledge adaptation scenario declines through time compared to the diffusion-process adaptation scenario. We posit that the cost of path dependence to farmers, in the form of yield reductions, is likely unavoidable because the inherent variability of the climate system will result in adaptation choices that will be suboptimal for some years. *Key Words: adaptation, agriculture, climate change and variability, path dependence, Southeast United States.*

农民技术选择在管理气候风险上的路径依赖，和农民区别气候变化和作物自然变异的困难，妨碍了他们对气候变化的适应能力。我们用嵌套在全球气候模式（GCM）中的区域气候模式，来模拟大气中二氧化碳 540ppm 浓度平衡，研究气候变化的影响和美国东南部的玉米产量变化。除了正常变异的气候情况，我们修改了全球气候模式的输出来模拟一个极为多变的气候情景。我们发现，气候变化对农民的能力和他们赖以适应的辅助机构构成了严重挑战。特别是在高度多变的气候情况下持续地降低玉米产量，说明了农民作出明智种植决定的能力，受制于他们无法从目前的技术体制或路径依赖选择中退出的境况。我们还使用三种适应方案：没有适应，"完美的知识"，以及全景观的知识传播模拟方案，来体现农民对气候变化的反应。无论是哪种适应方案和方差结构，最常见的结果是玉米产量在该地区百分之六十至百分之八十的地方下降百分之一到百分之二十。相比扩散过程中的适应方案，完美知识适应情况下的优势随着时间消退。我们断定农民路径依赖的成本，将以减产的形式不可避免，这是因为气候系统的内在变化会导致一些年份里退而求其次的适应选择。*关键词：适应，农业，气候变化和变异，路径依赖，美国东南部。*

La dependencia de la trayectoria en las opciones técnicas de los agricultores para el manejo del riesgo, combinada con sus dificultades para discernir el cambio climático de la variabilidad natural, podrían malograr la adaptación al cambio climático. Estudiamos los efectos de la variabilidad y el cambio climático sobre la producción de maiz en el Sudeste de los Estados Unidos, utilizando un modelo climático regional como parte de la simulación del equilibrio de la concentración de CO2 atmosférico de 540 ppm en un modelo climático global (GCM, sigla en inglés). Además de un escenario climático de varianza normal, modificamos los resultados generados por el GCM para simular un escenario con un clima altamente variable. Encontramos que la variabilidad climática plantea un reto serio a las habilidades para adaptarse en los agricultores y las instituciones que los apoyan. Rendimientos consistentemente más reducidos, en especial en el escenario de clima muy variable, indican que las habilidades de los agricultores para hacer elecciones informadas acerca de sus decisiones de siembra pueden verse afectadas por sus inabilidades para dejar de lado sus actuales regímenes tecnológicos o dependencia de la trayectoria. Incorporamos también las respuestas de los agricultores al cambio climático utilizando tres escenarios de adaptación: ninguna adaptación, "conocimiento perfecto" y un escenario que imita la difusión de conocimiento a través del paisaje. Sin consideración al escenario de adaptación y estructura de la varianza, el resultado más común es un declive en la producción de maiz, al punto que se presentan reducciones del 1 al 20 por ciento en un 60 al 80 por ciento de la región. La ventaja del escenario del conocimiento perfecto para la adaptación declina con el tiempo en comparación con el escenario del proceso de difusión. Planteamos que para los agricultores el costo de la dependencia de la trayectoria, manifiesto en la caida de la producción, es poco menos que inevitable debido a que la inherente variabilidad del sistema climático resultará en la elección de adptaciones que estarán por debajo del óptimo durante varios años. *Palabras clave: adaptación, agricultura, cambio y variabilidad climática, dependencia de la trayectoria, Sudeste de los Estados Unidos.*

Farmers and their supporting institutions must respond to a never-ending barrage of challenges, some localized and short term and some global and long term. In both cases, history shows that farmers and those institutions are remarkably successful in introducing measures to respond and adapt to myriad challenges, some environmental and some socioeconomic. Global agricultural capacity has grown apace with demand throughout the latter twentieth and early twenty-first centuries. However, in spite of this success, farmers everywhere in the world remain vulnerable to the vagaries of climate. The Fourth Assessment Report of the Intergovernmental Panel on Climate Change (IPCC 2007) concludes that the Earth is committed to at least as much warming over the next ninety years as was experienced over the previous century, even if greenhouse gas concentrations are immediately curtailed to 2000 levels. This means that agricultural adaptation to climate change is both necessary and inevitable.

The ability of agricultural systems to adapt to climate change is unlikely to follow a smooth trajectory with time. One reason is due to the nature of climate change itself. Multiple climate model simulations show the potential for accelerating changes in climate with significant shifts in interannual variability (MacCracken et al. 2003; Stainforth et al. 2005). Although the average climatic conditions might be changing more or less monotonically, seasonal and interannual variability could be highly unstable, creating extreme climatic conditions. The impact of extreme climatic conditions on crop productivity will likely be far greater than effects associated with the average change in climatic conditions (Mearns, Rosenzweig, and Goldberg 1997). Another reason pertains to how accurately farmers and their supporting institutions will interpret signals of climate change and respond to them when they are embedded in a set of climate observations punctuated by a high degree of variability.

The ability of agricultural systems to adapt effectively might further worsen if those systems are path dependent on a suite of technologies that are rendered partially or totally ineffectual by the shift in climatic means and variability. The concept of path dependence or "lock-in"[1] is generally used in the analysis of adoption of competing innovations by end users, but in this article it is used to understand how path dependence might interact with climate variability to challenge the efficiency of agricultural adaptation in the future. In agriculture, path dependency occurs when a particular technological innovation becomes dominant and self-reinforcing, leading to a situation where it excludes competing and possibly superior alternatives. In the context of climate change, such rigidity implies an inability of farmers to respond appropriately to minimize losses or to take advantage of new climatic conditions. This need for appropriate and robust adaptation highlights the importance of a system that is flexible enough to make numerous informed midcourse adjustments to maintain optimal production as the climate changes. These corrections must be made in the absence of knowledge about how the climate is likely to evolve in the near future. Farmers must also establish the probability of being correct by introducing a set of adaptive choices available to them.

Does the interlocking nature of technological regimes interfere with smooth and efficient adaptive responses to climate change? In this study, we address this question through an assessment of the impacts of climate variability and change on corn (*Zea mays*) in the Southeastern United States. We use a coupled atmosphere–ocean global climate model, a regional climate model, and the Erosion Productivity Index Calculator (EPIC) agricultural production model (Williams, Jones, and Dyke 1984) to examine the impact that doubling of CO_2 emissions has on corn yields. We also investigate several possible scenarios of likely adaptive responses to changes in the climatic variability and mean in the context of path dependence as a regional process and a potential impediment to future agricultural production.

In the next section, we provide the theoretical premise of path dependency in the context of farmers' abilities to adapt smoothly to a changing climatic state. Also, we present the case that the existence of path dependency in agricultural systems is not only of theoretical importance; it has consequences for devising and implementing appropriate policies for adaptation to climate change. We discuss the study area and outline the methodology to test the concept of path dependency. We then present the results of the analysis. This is followed by a discussion of the prospect of regional agricultural adaptation to climate variability and change. In the final section, we summarize the overall findings in light of possible implications for agricultural systems to adapt to climate change and variability.

Theoretical Framework

Interest in path dependency emerged in the 1980s to counter neo-classical assumptions about the reversibility of economic decisions (Nelson and Winter 1982; Dosi 1984). It gained prominence after David's (1985) work on the economic history of technology and Arthur's (1989) work on nonlinear economic processes. The concept of path dependency was developed to describe how technologies and social systems could eventually become suboptimal solutions for new and emerging challenges due to norms associated with a particular technological regime and the sunk-in costs of investments in infrastructure for research and development (David 1985). This idea spawned interest in various fields, including geography, where it has been used to explain the fundamental features of the regional geography of economic activities. The most important characteristic of path dependency is its *nonergodicity*, a system's inability to detach itself from its past (Martin and Sunley 2006). In other words, a path-dependent system is one where the outcome evolves as a consequence of the system's own history (McGuire 2008).

Recent work on path dependency involves three overlapping ideas: path dependency as technological lock-in, as dynamic increasing returns, and as an institutional lag effect (Martin and Sunley 2006). Scholars trace the idea of path dependency as technological lock-in to David's work on adoption and the famous example of the continued dominance of the QWERTY keyboard since the 1880s, despite more efficient alternatives for keyboard arrangement. According to David (1985), lock-in exists due to "self-reinforcing" processes operating through three mechanisms: complementarities between components of technology and its uses, the economy of scale as associated with the incremental use of technology, and the inertia of sunk costs due to the difficulty of switching from prevailing to new technologies. Although reinforcing to each other, these mechanisms jointly create the conditions to privilege a set of options over new and more efficient alternatives, inhibiting the takeoff of superior technologies over the suboptimal ones, thereby generating the lock-in.

The second mechanism of path dependency, dynamic increasing returns to scale, was introduced by Arthur (1989) to illustrate that existing learning mechanisms reinforce the prevailing development paths due to shared comparative advantage. In addition, Arthur asserted that the development of technologies both influences and is influenced by the social, economic, and cultural setting in which it takes place. This leads to the idea that successful innovation and adoption of a new technology depend on the historical path of its development, including the characteristics of initial markets, the institutional and regulatory factors governing its introduction, and the expected outcomes by the users (Ruttan 1996; Kemp 2000; Berkhout 2002). Therefore, socioeconomic and institutional arrangements interact to guide technological innovations along the desired path guided by the trajectory of its development.

The third mechanism, institutional lag effects, was advanced by North (1990) and Setterfield (1993), with further elaboration by Ruttan (1996, 2005). They argued that institutional configuration is the most important factor to create lock-in and path dependency. That is, institutions and the economy coevolve in an interdependent way, in which seemingly discrete innovations are not only the products of the systems of technological innovation of which they are a part but also the products of a nested institutional arrangement that reinforces technological systems (Martin and Sunley 2006). According to Ruttan (1996, 2005), preexisting institutional and socioeconomic structures and the lag effects of activities undertaken in the past constitute the environment in which current activities occur. Scholars studying path dependency from the perspective of institutional lag effects stress that nascent institutional structures might not be the most efficient means to address new and emerging conditions such as climate change, and their evolving arrangement might be locked-in for a considerable period of time

(Liebowitz and Margolis 1995). Therefore, technological paths followed during one institutional structure can persist long after conditions have changed to a new state.

Although most literature on path dependence focuses on nonagricultural industries, lock-in also is observed in agricultural activities such as pest control strategies and breeding (Cowan and Gunby 1996; Wilson and Tisdell 2001; Vanloqueren and Baret 2008). In fact, we argue that the agricultural production system is a natural laboratory in which to examine potential path dependency. For instance, although recent years witnessed a higher level of awareness about the negative consequences of chemicals and pesticides and an increasing knowledge of alternative measures for crop protection, modern agriculture is still locked in to chemicals as the primary means of pest management (Cowan and Gunby 1996). Many farmers are not easily persuaded to switch to newer integrated pest management strategies due in part to uncertainty about their effectiveness, technological immaturity of the newer innovations, and the lack of coordination among various agencies that support them.

Plant breeding is another example where experience and knowledge from the past determine the trajectory of future research (McGuire 2008). For example, wheat varieties launched in the United States in the early 1990s rely on crop breeding research dating as far back as 1873 (Pardey and Beintema 2001). The accumulation of knowledge and the continuous interaction of all the drivers of innovation not only shape the current technological regime but also hinder the growth and development of new technological paradigms that are more applicable to the changing context in which the system operates (Håkansson, and Waluszewski 2002; McGuire 2008; Vanloqueren and Baret 2009). Likewise, if the existing system of agriculture is not flexible enough to make informed midcourse adjustments, it will have difficulty adapting to climate change.

We argue that path dependency is the cumulative outcome of technological trajectories adopted by farmers and promoted by extension services, agricultural policies, and agricultural research systems. Although technological choices in agriculture are rarely fixed, innovation and adoption of technologies often follow established pathways due to various factors. For example, in their discussion of pesticide lock-in, Vanloqueren and Baret (2009) have identified twelve factors that potentially can impede adoption of pesticide-resistant wheat cultivars in Wallonia (Belgium). These impeding factors exist at all levels, from farmers to input suppliers to national agricultural policies. The failure of pesticide-resistant wheat cultivars to become the mainstream cultivar of choice was not due to poor biological traits such as low yields or inferior bread quality but due to inhibiting lock-in processes that hindered the adoption by farmers of the more optimal configuration. Although new conditions such as tougher pesticide regulations, changing consumer preferences, and innovation of optimal technologies can break the lock-in, they also require a multidisciplinary effort at scales ranging from an individual farmer's field to the policy arena.

Method

This work is a continuation of an integrated assessment undertaken to explore the effects of global climate change and variability on agricultural production and economic welfare on a regional scale (see Mearns et al. [2003], for a complete discussion of the project). The study area is the Southeast United States (Figure 1). We employ the RegCM2 limited climate area model nested within the CSIRO Mark 2 GCM running an equilibrium experiment involving a near doubling of atmospheric CO_2 concentrations (330 ppm to 540 ppm). The RegCM2 domain consists of 414 grid cells over the Southeast United States at a resolution of 0.50 latitude by 0.50 longitude, whereas the GCM has a resolution of 50 latitude by 50 longitude. We note that given current emission trajectories, a doubling of CO_2 likely is the most optimistic scenario over the next 100 years. We chose to focus solely on corn production for the analysis because it is the most widely grown crop in the Southeast United States, is highly sensitive to meteorological stressors, and responds quickly to management changes.

Erosion Productivity Index Calculator

Due to the complexity of climate, soil, and crop systems, process-based computer simulation techniques are among the most practical approaches available to make assessments of climate change impacts on agriculture. We use the Erosion Productivity Index Calculator (EPIC; Williams, Jones, and Dyke 1984) model (currently known as Environmental Policy Integrated Climate) to estimate long-term climatic effects on corn yields. An attractive feature of EPIC is its ability to examine the effects of weather and agronomic practices on crop production and soil and water resources. The model integrates the major processes that occur in the

Figure 1. Southeast U.S. study area and RegCM2 grid.

soil, atmosphere, crop, and management systems. It runs on a daily time step at the scale of a single field. Input requirements include daily meteorological data, soil properties, and soil and crop management data. EPIC relies on radiation-use efficiency in calculating photosynthetic production of biomass and, as such, atmospheric CO_2 concentrations influence photosynthesis through this parameter (in the case of corn, we increase this rate by 10 percent for the doubled CO_2 scenario). The daily simulated potential biomass is adjusted to account for the level of stress from the input factors (e.g., water, temperature, nutrients [nitrogen and phosphorus], and soil aeration) in proportion to the magnitude of the most severe stress during that day. Stress days, which incorporate both stress duration and stress severity, are calculated during the growing season as the sum of (1 – daily stress factor). Thus, on a day when the stress factor is 0 (i.e., no growth), the model calculates one stress day. Similarly, if the stress factor is 1 (i.e., no stress) it calculates zero stress days. We assume that the number of stress days will go up with increased climate variability, causing negative consequences in the outcome of crop yields.

Adaptation Scenarios

We test the effects of technological lock-in in the context of highly variable climatic change using three adaptation scenarios: no adaptation, climatically optimized or clairvoyant adaptation, and adaptation based on the logistic model of technological substitution (we call this logistic adaptation). Typical adaptation actions that can be implemented in the EPIC crop model are changes in planting and harvest dates, changes in cultivar type, and, in extreme cases, irrigation and crop changes. Climatically optimized adaptation (the clairvoyant adaptation scenario) assumes that farmers make the correct adaptations in real time and in perfect timing with the climate change (Schneider, Easterling, and Mearns 2000). Although obviously not realistic, optimized adaptation is the most common approach used in modeling studies. The development of a model to represent logistic adaptation is described fully in Easterling, Chhetri, and Niu (2003). As shown in Figure 2, the model uses a logistic curve to represent varying rates of farmer adaptation to changing climate. In this scenario, a small fraction of farmers will change course and quickly adapt to a changing climate; the majority of farmers will take longer for various reasons, including technical lock-in or failure to realize that the climate has changed beyond the initial conditions; and a small fraction of farmers will wait many years to adjust production practices.

Because our model of logistic adaptation depicts changing behavior over many years, the analysis must include a time-evolving climate change scenario. The GCM output used to drive the RegCM output is from an equilibrium experiment and not a transient climate

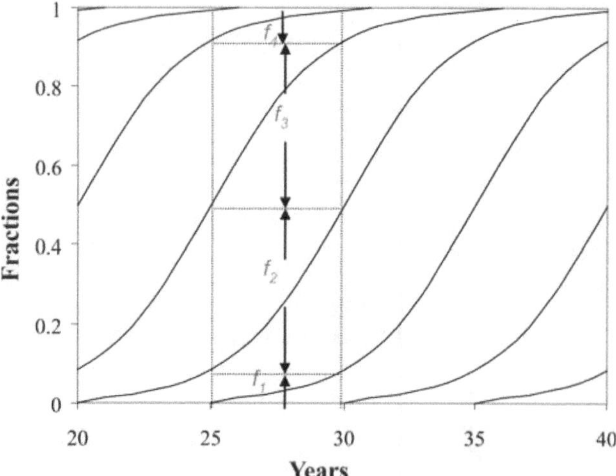

Figure 2. Logistic model depicting the time evolving fraction of farmers adapting to climate change: (f_1) cohort of early adopters who instantaneously recognize and adapt to a changing climate and climate variability, (f_2) cohort lagging on average five years behind f_1, (f_3) cohort lagging on average ten years behind early adopters, and (f_4) farmers lagging on average fifteen years behind the early adopters. Fractions f_2 and f_3 make up the largest fraction of farmers, and fractions f_1 and f_4 represent a minority of farmers.

change scenario. Therefore, we create a mock sixty-year scenario by dividing the difference between the RegCM doubled CO_2 climate and the baseline climate into twelve equal fractions representing five-year sampling intervals. We adjust the 1960 to 1995 baseline daily climate following the procedure of Mearns et al. (2003), using these twelve climate change segments for each EPIC climate parameter (i.e., temperature, precipitation, relative humidity, wind speed, and solar radiation). We aggregate these twelve climate segments into three twenty-year time series and assume that this represents the length of time required for adaptations to diffuse across the landscape under the logistic adaptation scenario (Easterling, Chhetri, and Niu 2003).

Altering Normal Climate

The first step is to create three possible annual variance outcomes given the monotonically changing climatic mean at each grid cell. Because the final outputs from the RegCM2 are monthly climatic means, it is necessary to develop estimates of daily meteorological data over many years to estimate a yearly variance structure. Using a stochastic weather generator we alter the variance associated with a parameter's monthly time series for each of the twelve climate segments used to simulate the corn yield outcomes. We follow the simple method suggested by Mearns (1995) in which the sample variance is inflated by a factor of 1.5 and 2.0 to simulate a more variable climate. Thus, the three possible variance outcomes are 100, 150, and 200 percent of the baseline climatic variance, adjusted for the twelve step changes in the climatic mean. Although there are more complicated methods for altering climate variability, Mearns, Rosenzweig, and Goldberg (1996, 1997) found the results were similar to their simpler method. The algorithm to modify the variance is as follows:

$$X't = \mu + \delta^{1/2}(Xt - \mu), \quad (1)$$

where $X't$ is the new value of the climate variable in question, μ is the mean of the time series, δ is the ratio of the new to the old variance, and Xt is the original value of the climate variable.

For the three adaptation scenarios (clairvoyant, logistic, and no adaptation), we produce model output from EPIC representing thirty-six years for each of the twelve climate change segments at every grid cell, using the variance-altered climate and the daily meteorological data from the stochastic weather generator. We ignore the initial year of the simulation to allow for adequate model spin-up, resulting in thirty-five years of EPIC simulations for each of the twelve climate segments.

Modeling Changing Variability

Although the stochastic weather generator alters the climatic variability, the resulting variance structures still represent constant variance scenarios through time (i.e., variance inflation factors of 1.0, 1.5, or 2.0). We do not believe this is realistic. The interannual variability likely will change through time, with some years experiencing highly variable conditions and others remaining close to the baseline variability. This speaks to the potential for path dependence to penalize farmers who fail to adapt appropriately to unknown future climatic variability.

To model this uncertain variance structure, we represent the annual climatic variance in the EPIC model runs as a stochastic process. First, we add a possible variance deflation factor of 0.5 to allow for the possibility of periods of lower climatic variance. With these four variance factors, we determine the variance for any particular year in a thirty-five-year EPIC model run through random sampling from a Gaussian probability distribution. Most years will experience climatic variability similar to that for the thirty-five-year baseline climate, allowing for adjustments in the climatic mean. The probability of any of the thirty-five years of the EPIC run experiencing 1.5 or 2.0 times the baseline variance is much smaller. For example, the probability that a grid cell will experience similar or less variable conditions than the baseline climate is approximately 0.82, and the probability that it will experience two times the original variance is less than 0.05. Using the randomly sampled variance multipliers, we produce thirty-five years of EPIC model runs for each climate segment under each farmer adaptation scenario (clairvoyant, logistic, and no adaptation) and aggregate the final crop yield results over the three twenty-year periods.

Results

Yield Change Projections

For each scenario, we examine the percentage of the study area with yield changes greater than or less than 10 percent, 20 percent, or 30 percent of the baseline yields. For all scenario combinations (i.e., combinations of climate change, adaptation type, and climate variance structure), corn yields less than 10 percent but greater than 20 percent of the baseline yields are the most common throughout the study area (Figure 4). This is especially so by the end of experiment, where the model projections show yield declines of this magnitude for over a third of the study area for each adaptation and variance scenario. In addition, yield losses are much more common than yield gains, although the most extreme losses are rare.

Two-sample differences of means tests do not indicate significant differences between the mean yields for clairvoyant and logistic adaptation scenarios, regardless of the variance structure (Table 1). There are significant differences ($p < 0.05$) between the no adaptation scenario and the clairvoyant adaptation scenario for all years and variance structures. The results are similar for the logistic adaptation scenario, with the exception of the first twenty-year period, where the test does not indicate statistically significant differences between the scenarios. The largest difference between the clairvoyant and logistic adaptation scenarios is in the first twenty years of the climate change experiment, and this difference declines through time. However, the yield differences among the no adaptation scenario and the other two scenarios increase from less than 1.5 percent to over 6 percent from year twenty to year sixty. By the end of the experiment, of the six possible adaptation and variance combinations, the no adaptation–random variance scenario produced the lowest corn yields for 81 percent of the grid cells (Figure 3).

We omit the no adaptation scenario from the analysis and aggregate the model output according to the two variance and two adaptation treatments applied to simulated yields (Figure 5). Substantial yield losses are more common than yield gains, but large yield swings (either gains or losses) are uncommon. The primary within-scenario difference is in the percentage area projected

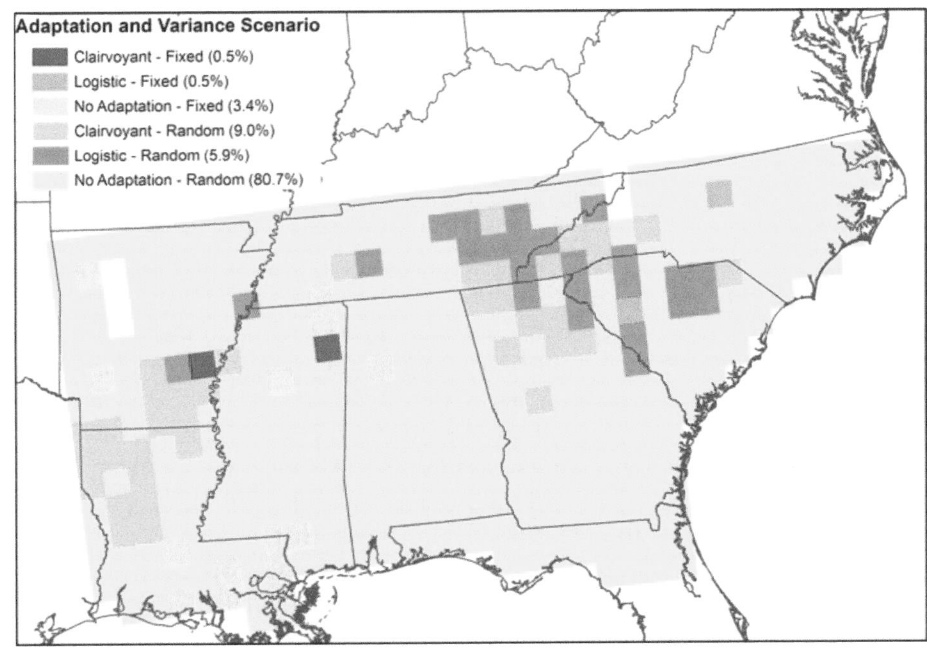

Figure 3. Spatial distribution of six adaption-variance scenarios according to the scenario with the lowest projected yields for each grid cell under the double CO_2 climate change experiment. The percentage of cells falling under each adaptation-variance scenario is displayed in parentheses in the legend.

Table 1. Comparison of grid-wise mean corn yield (mT Ha^{-1} and percent) between no adaptation, logistic adaptation, and clairvoyant adaptation under the fixed and random variance structures for the three twenty-year climate change periods

Climate/adaptation scenarios	Mean yield (mT Ha^{-1})	% Change in yield (from no adaptation scenario)	% Change in yield from logistic adaptation scenario
Year 20			
Fixed variance			
No adaptation	6.63		
Logistic adaptation	6.76	1.96	
Clairvoyant adaptation	6.87	3.60[a]	1.62
Random variance			
No adaptation	6.55		
Logistic adaptation	6.63	1.22	
Clairvoyant adaptation	6.72	2.60[a]	1.36
Year 40			
Fixed variance			
No adaptation	6.27		
Logistic adaptation	6.56	4.63[a]	
Clairvoyant adaptation	6.65	6.06[a]	1.37
Random variance			
No adaptation	6.16		
Logistic adaptation	6.42	4.22[a]	
Clairvoyant adaptation	6.51	5.68[a]	1.40
Year 60			
Fixed variance			
No adaptation	5.80		
Logistic adaptation	6.11	5.34[a]	
Clairvoyant adaptation	6.15	6.03[a]	0.65
Random variance			
No adaptation	5.71		
Logistic adaptation	6.01	5.25[a]	
Clairvoyant adaptation	6.04	5.78[a]	0.50

[a]Difference is statistically significant at $\alpha = .05$ level.

to experience modest increase in corn yields (Figure 5B). In this case, both the clairvoyant adaptation scenario and the fixed variance scenario projected higher corn yields than the alternative logistic or random variance scenarios.

Analysis of Variance (Two-Way)

We conduct a two-way analysis of variance (ANOVA) test on the corn yield projections to evaluate the relative impact of the two treatments (i.e., the adaptation scenario or the variance structure) on the mean corn yield for the three twenty-year climate change intervals (Table 2). There is evidence that the corn yield means for the different scenarios (adaptation or variance) are significantly different from each other ($p < 0.05$), with the possible exception of the variance scenario in the last twenty-year climate change period ($p = 0.10$). We fail to reject the null hypothesis of no interaction effect between adaptation choice and variance structure.

We reanalyze the model output using ANOVA but without the no adaptation scenario to remove outlier effects (Table 3). In this comparison, there still is no evidence for an interaction effect between the adaptation and variance scenarios. There are some critical differences among the main effects in this test compared to the previous test, however. For instance, only the first twenty-year period is close to showing significant differences among the means for the adaptation scenarios, whereas in all other periods, the p value does not warrant rejection of the null hypothesis of no difference among the means. The apparent convergence of crop yield means emerges as the simulation progresses through time. This suggests that, as long as some form of adaptation is occurring, farmers will perform better under the double CO_2 climate than if no adaptation occurs. At the same time, the perceived benefit of perfect

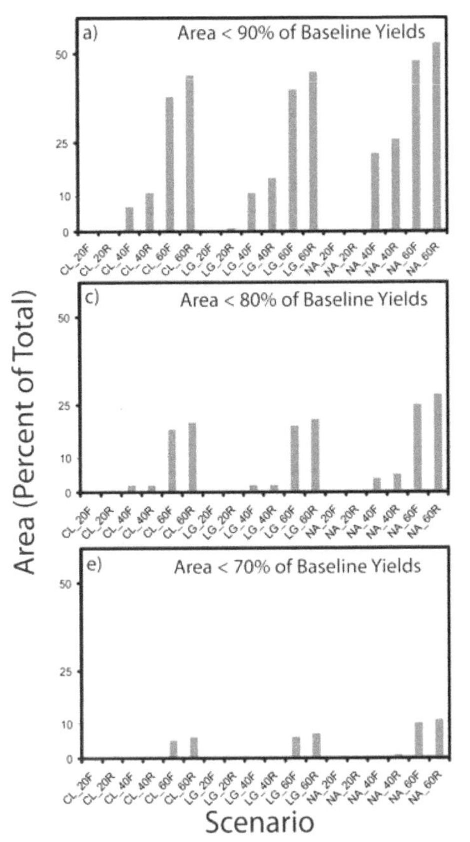

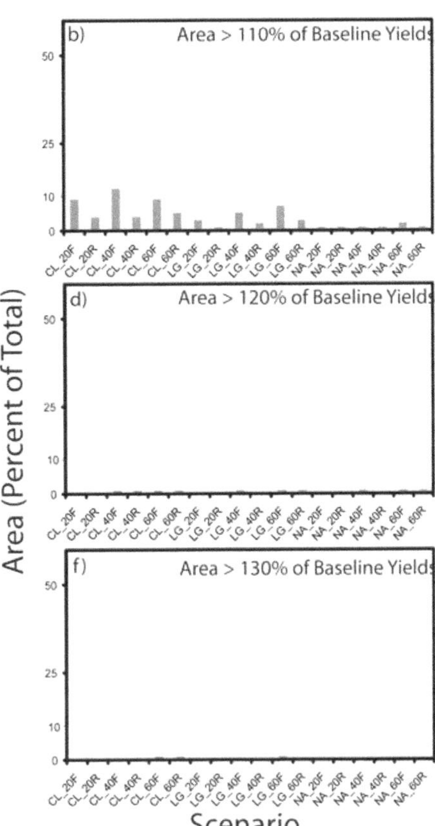

Figure 4. Grid-wise yield projection comparison for all adaptation scenarios. Bars represent percentage of total study area for a given scenario: (A) percentage area experiencing less than 90 percent of baseline yield amount for each grid cell, (B) percentage area experiencing 110 percent or more of baseline yields, (C) percentage area less than 80 percent of baseline, (D) percentage area greater than 120 percent of baseline, (E) percentage area less than 70 percent of baseline, and (F) percentage area greater than 130 percent of baseline. Scenarios are CL = clairvoyant adaptation; LG = logistic adaptation; NA = no adaptation; 20 = year twenty of climate change experiment; 40 = year forty; 60 = year sixty; F = fixed variance structure; R = random variance structure.

knowledge of the climate diminishes through time. In contrast, the ANOVA results for the variance scenarios are very similar to the results in Table 2. The tests show significant differences between the corn yield means for the random and fixed variance scenarios for the first forty years of the climate simulation. The difference between projected corn yields for the last twenty-year period is not significant at the 0.05 α level, but the pattern is similar to the previous two time periods.

Discussion

Investment in agricultural research and development has contributed substantially to meeting the global

Table 2. Analysis of variance results showing the main effects of the three adaptation scenarios and two variance structures on yield for the three twenty-year climate change periods

	Adaptation scenario clairvoyant, logistic, no adaptation ($N = 2{,}460$, $df = 2$)	Variance structure fixed, random ($N = 2{,}460$, $df = 1$)	Interaction effects ($N = 2{,}460$, $df = 2$)
Year 20			
μ	6.8, 6.7, 6.6	6.8, 6.6	—
F	5.9	5.72	0.17
p value	0.003	0.017	0.841
Year 40			
μ	6.6, 6.5, 6.2	6.5, 6.4	—
F	15.20	5.01	0.03
p value	0.000	0.025	0.969
Year 60			
μ	6.1, 6.1, 6.8	6.0, 5.9	—
F	12.07	2.73	0.01
p value	0.000	0.099	0.988

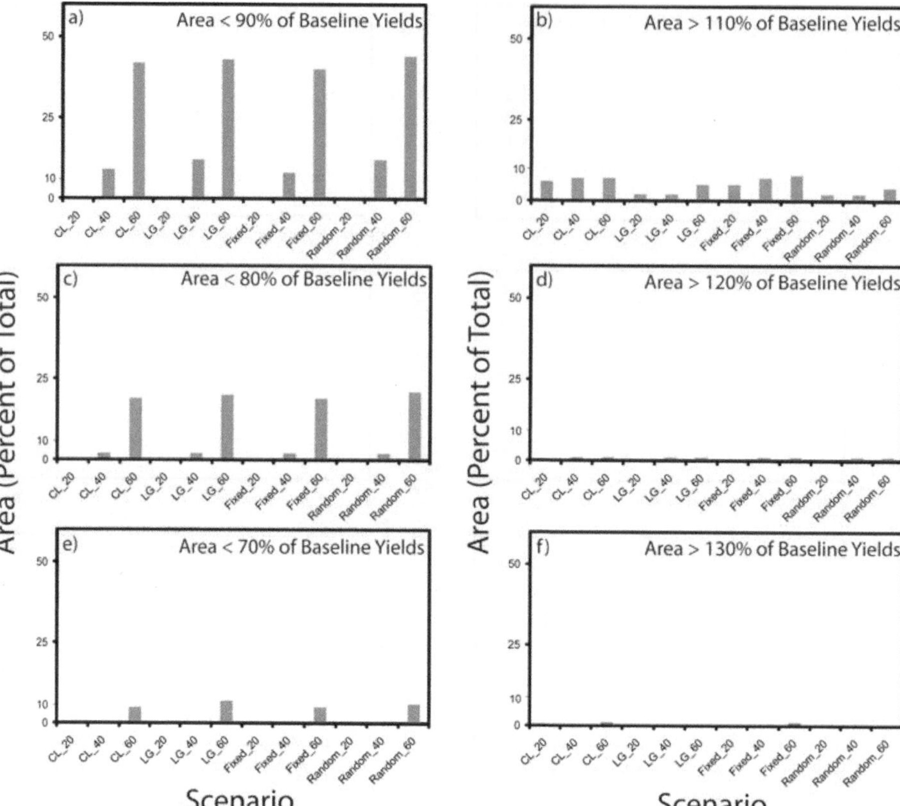

Figure 5. Same as Figure 4, except with aggregated variance (fixed variance or random variance) and adaptation (clairvoyant adaptation or logistic adaptation) climate change scenarios. Scenarios are CL = clairvoyant adaptation; LG = logistic adaptation; 20 = year twenty of climate change experiment; 40 = year forty; 60 = year sixty.

demand for food. This is achieved primarily through a strong focus on crop improvement, increased inputs of water and agrochemicals, and intense mechanization. Yet food security remains an unfulfilled dream for almost 800 million people (Food and Agriculture Organization of the United Nations 2005), and prospects for reducing this staggering figure appear grim as the world currently is consuming more than it is producing (von Braun 2007). In addition, the possible impact of climate variability and change in food production is raising the specter of a potentially perpetual food crisis in coming decades. The impact of increased climatic variability on crop yield could arise from several causes, including (1) increased sterility due to extreme temperature during anthesis (e.g., Wheeler et al. 2000); (2) moisture and heat stress during critical stages of crop growth and development (e.g., Easterling, Chhetri, and Niu 2003); and (3) reduction of input to agriculture

Table 3. Analysis of variance results showing the main effects of no adaptation scenarios and two variance structures on yield for the three twenty-year climate change periods

	Adaptation scenario clairvoyant, logistic ($N = 1{,}640$, $df = 1$)	Variance structure fixed, random ($N = 1{,}640$, $df = 1$)	Interaction effects ($N = 1{,}640$, $df = 1$)
Year 20			
μ	6.8, 6.7	6.8, 6.7	—
F	2.57	5.18	0.01
p value	0.109	0.023	0.925
Year 40			
μ	6.6, 6.5	6.6, 6.5	—
F	1.55	3.76	0.00
p value	0.213	0.053	0.958
Year 60			
μ	6.1, 6.1	6.1, 6.0	—
F	0.2	1.98	0.00
p value	0.652	0.159	0.993

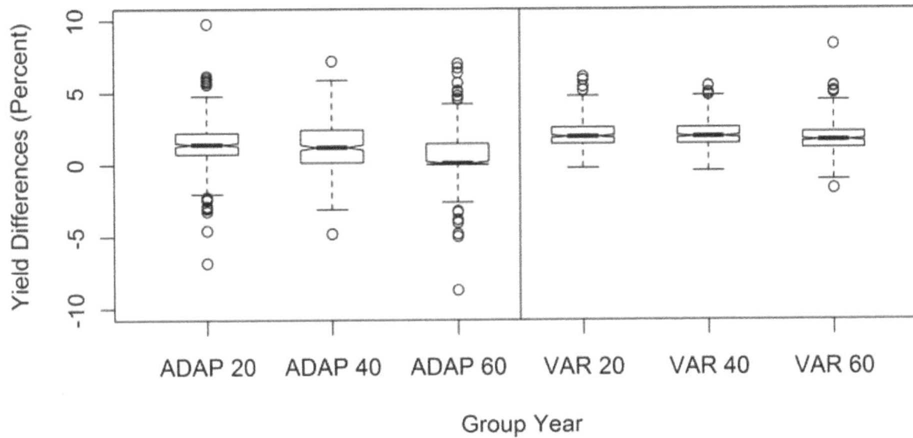

Figure 6. Box plot of grid-wise percentage differences in corn yields under each treatment for the three twenty-year time periods. The left panel shows the distribution of yield differences between the clairvoyant and logistic adaptation scenarios. The right panel shows the distribution of yield differences between the fixed and random variance scenarios. Positive values mean higher yields in the clairvoyant (left panel) and fixed variance (right panel) scenarios, respectively. The no adaptation scenario was removed from this portion of the analysis.

due to uncertainty associated with climate change (e.g., Antle et al. 2004). In the United States, the corn production losses due to extreme climatic conditions might double during the next thirty years, causing additional damages totaling an estimated $3 billion per year, which consequently will impact insurance and disaster relief programs (Rosenzweig et al. 2002). Addressing this challenge almost certainly will require a continuous flow of new technologies while adapting to emerging crop growing conditions brought about by climate variability and change.

The ability of society to make smooth transitions to new agricultural environments due to changing climatic conditions might worsen if farmers are locked in to a suite of rigid technological choices not suitable to the new climate. Although geographers have distinctly recognized the role that preexisting institutional and socioeconomic structures play in exerting influence on the evolution of technologies, no studies have yet explored the potential for technological lock-in to affect the efficiency of adaptation to climate change. In the case of agricultural adaptation, issues of scale take on added importance because lock-in will occur at regional scales. In our view, it is vitally important that geographers continue to research and explore the question of technological lock-in to avoid ambiguities brought about by the discussion in this topic. Following Martin and Sunley (2006), we believe that there is considerable value in applying the concept of path dependence in geographic inquiry. This emerging intellectual field can reveal its geographical foundation and implications, as exemplified by the centrality of regional lock-in to questions of agricultural adaptation to climate change.

Our results show that changes in temperature and precipitation affect both the mean and variance of corn yields. On average, corn yields increase as precipitation increases, but they decrease if temperature increases. Interestingly, the effect of increased precipitation variability is minimal compared to the effect of increased temperature variability. The results also indicate that, although the benefit from perfect knowledge of the climate decreases with time, the cost of a random variance structure in terms of yield reductions relative to the baseline remains nearly constant. The effects of the adaptation choice versus the variance structure differ in that the means of yields converge toward the same value for adaptation effects, but the difference among means remains nearly constant through time for the variance structure (Figure 6). Thus, the climate change scenario with a randomized variance structure consistently leads to lower yields compared to a fixed variance structure, regardless of the information available to the farmer. The consistently lower yields of the randomized variance scenario illustrate that farmers' abilities to make informed cropping decisions might be constrained by their inabilities to exit from their current system of strategies and technologies (i.e., they are locked in). The system's lock-in on "short shelf-life" sets of technologies will not maintain optimal production with a constantly changing variance structure.

Given the lower yield projections in the random variance structure, we posit that the cost of path dependence is a penalty for farmers who are subjected to a variance structure that is not consistent. Alternatively, this random variance structure shares similarity with regions currently experiencing large interannual variations in temperature and precipitation, such as the U.S. Great Plains. In these regions, boom and bust cycles of crop production are common and the risk of crop loss is higher compared to areas with a less variable climate. Thus, although the benefits of adaptation are clear, there is a cost of path dependence to farmers who lock in to a set of production choices that will likely be suboptimal for multiple years if the warming climate also is a more variable climate. Although society can make transitions to new technologies relatively quickly and efficiently, such as the transition from canals as a dominant mode of freight transportation to railways, the interlocking nature of technological regimes might discourage the deployment of efficient technologies needed by farmers. In the context of agricultural adaptation to climate change, such rigidity implies an inability of farmers to respond appropriately to minimize losses or to take advantage of new climatic conditions. This illustrates the importance of a system that is flexible enough to make numerous informed mid-course adjustments to maintain optimal production as the climate changes. These corrections must be made in the absence of knowledge about how the climate is likely to evolve in the near future and farmers must establish the probability of being correct by introducing a set of adaptive choices available to them.

The existing agricultural system is a result of decades of investment in research and development of agricultural technologies by the public and private sectors. The unprecedented growth of U.S. agricultural productivity over the past century is attributable to this investment, which includes such major developments as agricultural mechanization, chemical fertilizers and pesticides, genetic improvement in crops, and evolving agronomic practices. However, this system of investment in agricultural research has undergone significant changes in the last few decades due to developments in science, policy, and markets. Public investment, in real dollars, in U.S. agricultural research has declined steadily since the 1960s. As discussed earlier, the technological choices for farmers are not fixed, but innovation and adoption of technologies needed to address the challenges in the future are often guided by "self-reinforcing" mechanisms of complementarities, economies of scale, and the inertia of sunk costs. This is further complicated by the absence of perfect knowledge about the variability of climate over the timescales for which a set of adaptations must be effective.

Conclusion

The relative speed and efficiency with which farmers are able to change technologies and management practices in response to climate change will be an important determinant of adaptive success. It appears that farmers in the United States are susceptible to technological lock-in that forces them to stick with a particular suite of technologies, even when environmental or market conditions dictate technical change. Of particular interest here is the special case of lock-in of climate-related technology during a time of gradual climate change accompanied by different scenarios of change in interannual climate variability. We ask whether lock-in during a period of climate change accompanied by changes in interannual climate variability (i.e., change in frequency of extreme events) inhibits successful adaptation to avert crop yield loss.

To answer this question, we simulated future corn production in the Southeast United States in the context of a warming climate, using three adaptation scenarios (no adaptation, logistic adaptation, and clairvoyant or perfect knowledge adaptation) exposed to both a random and a fixed climatic variance structure. Logistic adaptation is a proxy for varying degrees of lock-in by farmers across a region. Clairvoyant adaptation is as if all farmers in a region are able to make technical changes in lock-step with climate change and variability. No adaptation is the control.

Across the study area, for all scenarios (i.e., combinations of climate change, adaptation type, and climate variance structure), the most common result is a moderate yield reduction (less than 10 percent decline from baseline yields), although the model output includes some areas with production declines of 30 percent or more. This decline is particularly pronounced by the end of the experiment. Corn yields decline in all adaptation scenarios and, by the end of the experiment, decrease 8 percent to 14 percent from the baseline yields with a fixed variance structure and decrease 10 percent to 15 percent with a random variance structure. We find consistently lower corn yields under a random variance structure compared to a fixed variance structure. Of the three adaptation scenarios tested, yield projections are consistently and significantly lower for the no adaptation scenario. No statistically significant differences are found between the clairvoyant and logistic adaptation scenarios, although the small differences that do occur (in the form of slightly higher yields for the clairvoyant scenario) decline through time. Although we conclude that farmers can ameliorate the effects of climate change by adjusting their crop growing decisions, there is a cost of path dependence to farmers who are locked in to a set of production choices that likely will be suboptimal for multiple years if the warming climate also is a more variable climate. The ability of agricultural systems to successfully transition to new crop-growing environments requires that farmers and their supporting institutions break their path dependency to rigid technological choices inappropriate for a nonstationary climate. This certainly requires a continuous flow of new technologies (or at a minimum, different technology) while adapting to emerging crop-growing conditions brought about by climate variability and change. This requires a multidisciplinary effort at scales ranging from an individual farmer's field to the policy arena to further increase resilience and reduce vulnerability to the changing climate.

Note

1. For the purposes of this study, we use the two terms *path dependency* and *lock-in* interchangeably.

Acknowledgments

This research was funded by the U.S. Environmental Protection Agency, NCERQA (Grant No. R824997-01-C), and the National Aeronautics and Space Administration, MTPE (Grant No. OA99073, W-19, 080). We thank the anonymous reviewers for the comments and encouragement. The views and conclusions contained in this article are those of the authors and should not be interpreted as representing the policies of the U.S. government.

References

Antle, J. M., S. M. Capalbo, E. T. Elliott, and K. H. Paustin. 2004. Adaptation, spatial heterogeneity, and the vulnerability of agricultural systems to climate change and CO_2 fertilization: An integrated assessment approach. *Climatic Change* 64 (3): 289–315.

Arthur, W. B. 1989. Competing technologies, increasing returns, and "lock-in" by historical events. *The Economic Journal* 99 (394): 116–31.

Berkhout, F. 2002. Technological regimes, path dependency and the environment. *Global Environmental Change* 12 (1): 1–4.

Cowan, R., and P. Gunby. 1996. Sprayed to death: Path dependence, lock-in and pest control strategies. *Economic Journal* 106 (436): 521–42.

David, P. 1985. Clio and the economics of QWERTY. *American Economic Review* 75 (3): 332–37.

Dosi, G. 1984. *Technical change and institutional transformation*. London: Macmillan.

Easterling, W. E., N. Chhetri, and X. Z. Niu. 2003. Improving the realism of modeling agronomic adaptation to climate change: Simulating technological substitution. *Climatic Change* 60 (1–2): 149–73.

Food and Agriculture Organization of the United Nations. 2005. *Food outlook no. 4-December 2005*. Rome, Italy: Food and Agriculture Organization of the United Nations.

Håkansson, H., and A. Waluszewski. 2002. Path dependence: Restricting or facilitating technical development? *Journal of Business Research* 55 (7): 561–70.

Intergovernmental Panel on Climate Change (IPCC). 2007. Summary for policymakers. In *Climate change 2007: The physical science basis*. Contribution of Working Group I to the Fourth Assessment Report of the Intergovernmental Panel on Climate Change, ed. S. Solomon, D. Qin, M. Manning, Z. Chen, M. Marquis, K. B. Averyt, M. Tignor, and H. L. Miller. New York: Cambridge University Press.

Kemp, R. 2000. Technology and environmental policy—Innovation effects of past policies and suggestions for improvement. Paper presented at OECD Workshop on Innovation and Environment, 19 June, Paris.

Liebowitz, S. J., and S. E. Margolis. 1995. Path dependence, lock in, and history. *Journal of Law, Economics and Organization* 11 (1): 205–26.

MacCracken, M. C., E. J. Barron, D. R. Easterling, B. S. Felzer, and T. R. Karl. 2003. Climate change scenarios for the U.S. National Assessment. *Bulletin of the American Meteorological Society* 84 (12): 1711–23.

Martin, R., and P. Sunley. 2006. Path dependence and regional economic evolution. *Journal of Economic Geography* 6 (4): 395–437.

McGuire, S. J. 2008. Path-dependency in plant breeding: Challenges facing participatory reforms in the Ethiopian Sorghum Improvement Program. *Agricultural Systems* 96 (1–3): 139–49.

Mearns, L. O. 1995. Research issues in determining the effects of changing climate variability on crop yields. In *Climate change and agriculture: Analysis of potential international impacts*, ed. C. Rosenzweig, J. T. Ritchie, and J. W. Jones, 123–46. Madison, WI: American Society of Agronomy.

Mearns, L. O., F. Giorgi, L. McDaniel, and C. Shields. 2003. Climate scenarios for the southeastern U.S. based on GCM and regional model simulations. *Climatic Change* 60 (1–2): 7–35.

Mearns, L. O., C. Rosenzweig, and R. Goldberg. 1996. The effect of changes in daily and interannual climatic variability on ceres-wheat: A sensitivity study. *Climatic Change* 32 (3): 257–92.

———. 1997. Mean and variance change in climate scenarios: Methods, agricultural applications, and measures of uncertainty. *Climatic Change* 35 (4): 367–96.

Nelson, R. R., and S. G. Winter. 1982. *An evolutionary theory of economic change*. Cambridge, MA: Harvard University Press.

North, D. C. 1990. *Institutions, institutional change and economic performance*. Cambridge, UK: Cambridge University Press.

Pardey, P. G., and N. M. Beintema. 2001. *Slow magic agricultural: R&D a century after Mendel*. International Food Policy Research Institute Report. Washington, DC: International Food Policy Research Institute.

Rosenzweig, C., F. N. Tubiello, R. A. Goldberg, E. Mills, and J. Bloomfield. 2002. Increased crop damage in the US from excess precipitation under climate change. *Global Environmental Change* 12 (3): 197–202.

Ruttan, V. 1996. Induced innovation and path dependence: A reassessment with respect to agricultural development and the environment. *Technological Forecasting and Social Change* 53:41–59.

———. 2005. Scientific and technical constraints on agricultural production: Prospects for the future. *Proceeding of the American Philosophical Society* 149 (4): 453–68.

Schneider, S. H., W. E. Easterling, and L. O. Mearns, 2000. Adaptation: Sensitivity to natural variability, agent assumptions and dynamic climate changes. *Climatic Change* 45 (1): 203–21.

Setterfield, M. 1993. A model of institutional hysteresis. *Journal of Economic Issues* 27 (3): 755–74.

Stainforth, D. A., T. Aina, C. Christensen, M. Collins, D. J. Frame, J. A. Kettleborough, S. Knight et al. 2005. Uncertainty in predictions of the climate response to rising levels of greenhouse gases. *Nature* 433 (7024): 403–06

Vanloqueren, G., and P. V. Baret. 2008. Why are ecological, low-input, multi-resistant wheat cultivars slow to develop commercially? A Belgian agricultural "lock-in" case study. *Ecological Economics* 66 (3): 436–46.

———. 2009. How agricultural research systems shape a technological regime that develops genetic engineering but locks out agroecological innovations. *Research Policy* 38 (6): 971–83.

von Braun, J. 2007. *The world food situation: New driving forces and required actions*. Washington, DC: International Food Policy Research Institute.

Wheeler, T. R., P. Q. Craufurd, R. H. Ellis, J. R. Porter, and P. V. Vara Prasad. 2000. Temperature variability and the yield of annual crops. *Agriculture, Ecosystems and Environment* 82:159–67.

Williams, J. R., C. A. Jones, and P. T. Dyke. 1984. The EPIC model and its application. In *Proceedings ICRISAT-IBSNAT-SYS Symposium on Minimum Data Sets for Agrotechnology Transfer*, ed. ICRISAT-IB-SNAT-SYS, 111–21. Hyderabad, India.

Wilson, C., and C. Tisdell. 2001. Why farmers continue to use pesticides despite environmental, health and sustainability costs. *Ecological Economics* 39 (3): 449–62.

Joint Effects of Marine Intrusion and Climate Change on the Mexican Avifauna

A. Townsend Peterson, Adolfo G. Navarro-Sigüenza, and Xingong Li

Changing climates are affecting biodiversity and natural systems, causing extinctions, range shifts, and phenological shifts. Efforts to forecast the spatial distribution and magnitude of these effects, however, have focused largely on direct effects of changing climates on species' distributional potential; recent work has considered secondary effects of warming climates via rising sea levels. Here, we present a first integration of the two dimensions of climate change effects on biodiversity, examining joint effects of marine intrusion and climate change on the distributional potential of seventy-six species of Mexican birds. The two phenomena are not related to one another—that is, a species seriously affected by one is not necessarily seriously affected by the other; however, the areas affected within species' distributions by the two phenomena tend to be complementary, compounding the negative effects. These results have implications for planning biodiversity conservation globally. *Key Words: biodiversity, bird, inundation, range, sea level rise.*

气候变化正在影响生物多样性和自然生态系统，造成物种灭绝，范围的变化，和物候的变化。人们试图预测这些影响的空间分布和严重程度，但是，大多数的研究主要集中气候变化对物种潜在分布的直接影响；最近的一些研究工作分析了气候变暖的副效应之一的海平面上升。在这里，我们首次提出了集成气候变化对生物多样性影响的两个维度的一种综合分析，研究海洋侵蚀和气候变化对墨西哥 76 种鸟类潜在分布的联合作用。这两种现象都是彼此不相关的，也就是说，某个物种受到一种因素的严重影响并不一定也受另一因素的严重影响，但是，在上述两个现象所影响到物种分布的地区内，所受影响往往是相辅相成的，混和而成特定的负面影响。这些研究结果对全球生物多样性保护规划会产生影响。*关键词：生物多样性，鸟类，被淹没，范围，海平面上升。*

Los climas en proceso de cambio están afectando la biodiversidad y los sistemas naturales, causando extinciones, transformación de escenarios naturales y desviaciones fenológicas. Sin embargo, los esfuerzos para predecir la distribución espacial y magnitud de estos efectos se han centrado en gran medida en los efectos directos de los climas cambiantes sobre el potencial de distribución de las especies; el trabajo reciente ha considerado los efectos secundarios de los climas en calentamiento a través de la elevación de los niveles del mar. Nosotros presentamos en este trabajo la primera integración de las dos dimensiones de los efectos del cambio climático sobre la biodiversidad, examinando los efectos conjuntos de la intrusión marina y el cambio climático sobre el potencial de distribución de setenta y cinco especies de aves mejicanas. Los dos fenómenos no están relacionados entre sí—esto es, una especie afectada seriamente por uno, no tiene por qué serlo seriamente por el otro; sin embargo, las áreas afectadas dentro de las distribuciones de especies por los dos fenómenos tienden a ser complementarias, acumulando efectos negativos. Los resultados tienen implicaciones para planificar la conservación de la biodiversidad globalmente. *Palabras clave: biodiversidad, aves, inundación, escenarios naturales, elevación del nivel del mar.*

Climate change is now recognized as a global phenomenon with far-reaching implications (Intergovernmental Panel on Climate Change [IPCC] 2007). Climate change is affecting natural and human biological systems by changing the environmental context to which those systems were adapted (Lovejoy and Hannah 2005). As such, species are already responding by phenological changes, poleward distributional shifts, upward elevational shifts, or extinction (Pounds and Crump 1994; Cresswell and McCleery 2003; Crozier 2003; Parmesan and Yohe 2003; van Lieshout et al. 2004). The great bulk of these effects, and indeed the main focus of efforts to model and forecast these effects, center on changes in the climate per se; other climate change–related phenomena such as rising sea levels and consequent marine intrusion into terrestrial regions remain little explored.

Sea level rise, and consequent marine intrusion, results primarily from thermal expansion and the exchange of water between oceans and other reservoirs (glaciers, ice caps, ice sheets, and other land water reservoirs), two processes that are mostly related to recent climate change (IPCC 2007). Because factors leading to sea level rise are complex, future projections vary. Best conservative estimates are of 0.5 m to 1.0 m (Carter et al. 2007), although others consider the complexity of the projections and range considerably higher (Oerlemans et al. 2005). Some estimates range much higher (Bindschadler 1998; R. Thomas et al. 2004; Rignot and Kanagaratnam 2006). Considerable reflection and analysis have focused on sea level rise from the human and economic perspective (Titus 1990; Mimura 1999; Hitz and Smith 2004; Bosello, Roson, and Tol 2007), yet few analyses have addressed biodiversity consequences of the phenomenon and only on regional scales for limited sectors of biodiversity (Daniels, White, and Chapman 1993; Galbraith et al. 2002; Gopal and Chauhan 2006; LaFever et al. 2007; McKee, Cahoon, and Feller 2007; Legra, Li, and Peterson 2008). Marine intrusion has the potential to erode or at least shift terrestrial species' distributions inland and to shift marine species as well; however, the dimensions of these potential effects are very poorly explored and little understood.

In this article, we explored, for the first time, the joint effects of climate change per se and sea level rise and consequent marine intrusion on elements of biodiversity. We reviewed the Mexican avifauna, seeking species of conservation concern with coastal distributions. We identified seventy-six species that were either endemic to Mexico or accorded some conservation status. We used a comprehensive database of known occurrences (Navarro-Sigüenza, Peterson, and Gordillo-Martínez 2003) and ecological niche modeling (Soberón and Peterson 2004) to produce distributional estimates for each species. We then used the ecological niche models to project species' ecological requirements onto future climate conditions (Araújo et al. 2005) and used a global marine intrusion estimate (Li et al. 2009). The conjunction of these three estimates offered us the opportunity to examine climate change and marine intrusion effects in tandem as regards this sample of the Mexican avifauna.

Materials and Methods

We chose species for analysis based on their status of conservation concern in the country. As such, we focused on species either endemic (sensu Navarro-Sigüenza and Peterson 2004) or accorded some endangerment status (BirdLife-International 2000; SEMARNAT 2002). Because endemic species are vulnerable to any landscape change occurring in Mexico, and endangered or threatened species are already under some level of threat, we consider this set of species to be the appropriate suite of species for this analysis. Evaluating only species with distributions at least close to one of Mexico's coasts, we ended with a total of seventy-six species for analysis (see Appendix). Among those species, fifty-nine are endemic and twenty are of conservation concern. Sample sizes of unique occurrence sites for these species ranged between 4 and 585, but samples for seventy species exceeded twenty unique occurrence points. Georeferencing of occurrence data was precise to 1 to 5 km.

We developed distributional estimates by means of ecological niche modeling (ENM) approaches, specifically using the genetic algorithm for rule-set prediction (Stockwell and Peters 1999) to profile each species' distribution in ecological space using seven "bioclimatic" variables (annual mean temperature, mean diurnal temperature range, maximum temperature of warmest month, minimum temperature of coldest month, annual precipitation, and precipitation of wettest and driest months; Hijmans, Cameron, and Parra 2005), at a spatial resolution of 0.01 degrees. This profile can then be projected back onto the geographic landscape to identify areas of potential distribution (Peterson 2003). Explicit assumptions regarding dispersal ability and the completeness of sampling of species' distributions are then necessary to convert estimates of potential distribution into estimates of actual distributions (Soberón and Peterson 2005). The approach and its assumptions and details are discussed at length in other publications (Soberón and Peterson 2004, 2005) and so are not detailed herein. As this analysis is mostly aimed at exploration of relationships with marine intrusion effects, we use only this one methodology and do not develop analyses using multiple niche modeling algorithms.

We then projected the set of climatic conditions identified in the ENMs as appropriate for each species onto modeled future climate conditions across Mexico. As our goal here was to assess climate–marine intrusion joint effects, we did not develop multiple future-climate scenarios but rather used a single future-climate model scenario (CCM3, a doubled-CO_2 view of future climates) that has been interpolated to fine spatial resolution (Hijmans, Cameron, and Parra 2005). Given that the species under analysis are generally of conservation concern, we assumed that they would have low dispersal ability, so we estimated future distributional potential as the intersection of present-day and future potential distributional areas; that is, that species would remain extant only in the future-suitable portions of their present-day ranges.

Geographical information systems (GIS) have been used in several previous studies to delineate potentially inundated areas resulting from projected sea level rises (Dasgupta et al. 2007; LaFever et al. 2007). In these analyses, inundation areas are identified if their elevation is below a projected sea level rise. The method has two important shortcomings: water connectivity to the ocean is not considered when inundation areas are delineated, and areas with elevations below the projected sea level rise might already be inland water bodies. Hence, simple GIS methods likely overpredict potential inundation areas. A new and more robust GIS analysis method developed by Li et al. (2009) overcomes these problems, as follows. Cells below a projected sea level rise are initially flagged. From the flagged cells, only those with connectivity to the ocean are selected. The selected cells are then checked to see whether or not they are part of existing inland water bodies. Only those cells that connect to the ocean and are not presently inland water are designated as inundation cells. The method was implemented as several steps in a GIS raster analysis framework to produce a global view of marine intrusion under sea level rise scenarios of 1 m and 6 m using the GLOBE digital elevation data from the National Geophysical Data Center with a spatial resolution of 1 km and vertical resolution of 1 m; details are provided in Li et al. (2009).

Finally, for each species, we estimated distributional areas for the present day as well as the present day reduced by (1) 1 m and (2) 6 m scenario marine intrusion estimates; (3) ENM assessments of range loss from climate change; and climate change in tandem with the (4) 1 m and (5) 6 m sea level rise marine intrusion estimates. These range estimates were organized into spreadsheets for analysis.

Results

Species' present-day geographic distributions areas ranged between 5,950 and 408,000 km^2. Estimates for two species are provided in Figure 1. These distributional areas were reduced by up to 71 percent by climate change and 62 percent by marine intrusion for particular species (see Appendix). Interestingly, the severity of effects from the two phenomena are not related to one another. A species seriously affected by climate change is not necessarily seriously affected by the marine intrusion and vice versa ($r^2 = 0.119$; Figure 2).

A next question is the spatial relationships between areas affected by climate change and those affected by marine intrusion for each species. That is, these areas could be complementary to one another (i.e., no overlap, affected area would be the sum of the two areas) or they could overlap completely (i.e., maximum overlap, affected area would be the maximum of the two areas). Using the joint effects distributional areas for each species, we evaluated its position along this spectrum (Figure 3). In general, the areas affected within species' distributions by the two phenomena tend to fall at the complementary end of the spectrum. That is, areas affected by climate change within the distribution of a species tend not to be those areas affected by marine intrusion.

Discussion

This analysis is best regarded as exploratory and heuristic. That is, we have used a simple future climate scenario (CCM3), a doubled-CO_2 concentration view of future climates (Govindasamy and Coquard 2003), rather than an ensemble of diverse scenarios of future conditions that would permit better assessment of sensitivity to model particulars (Peterson et al. 2002; Araújo et al. 2005). Similarly, we are using sea level rise scenarios with 1 km horizontal and 1 m vertical resolution. Although these are the best resolutions available at a global scale, they limit the detail with which the phenomena can be perceived and described; what is more, we neglect the potential for coastal ecosystems to respond to rising sea levels by migrating inland to colonize similar conditions along the "new" coastline. Finally, the ENM approach has its own potential limitations, in terms of both accuracy of the models and their transferability to future climate situations (Pearson and Dawson 2003; Thuiller, Brotons, et al. 2004a; Araújo and Guisan 2006; Araújo and Luoto 2007; Peterson, Pape, and Eaton 2007). Still, the purpose of this article is to develop a first survey of joint climate change–marine intrusion effects, which is unlikely to be compromised by these considerations.

The picture painted in this article is one of independent and unlinked effects of the two manifestations

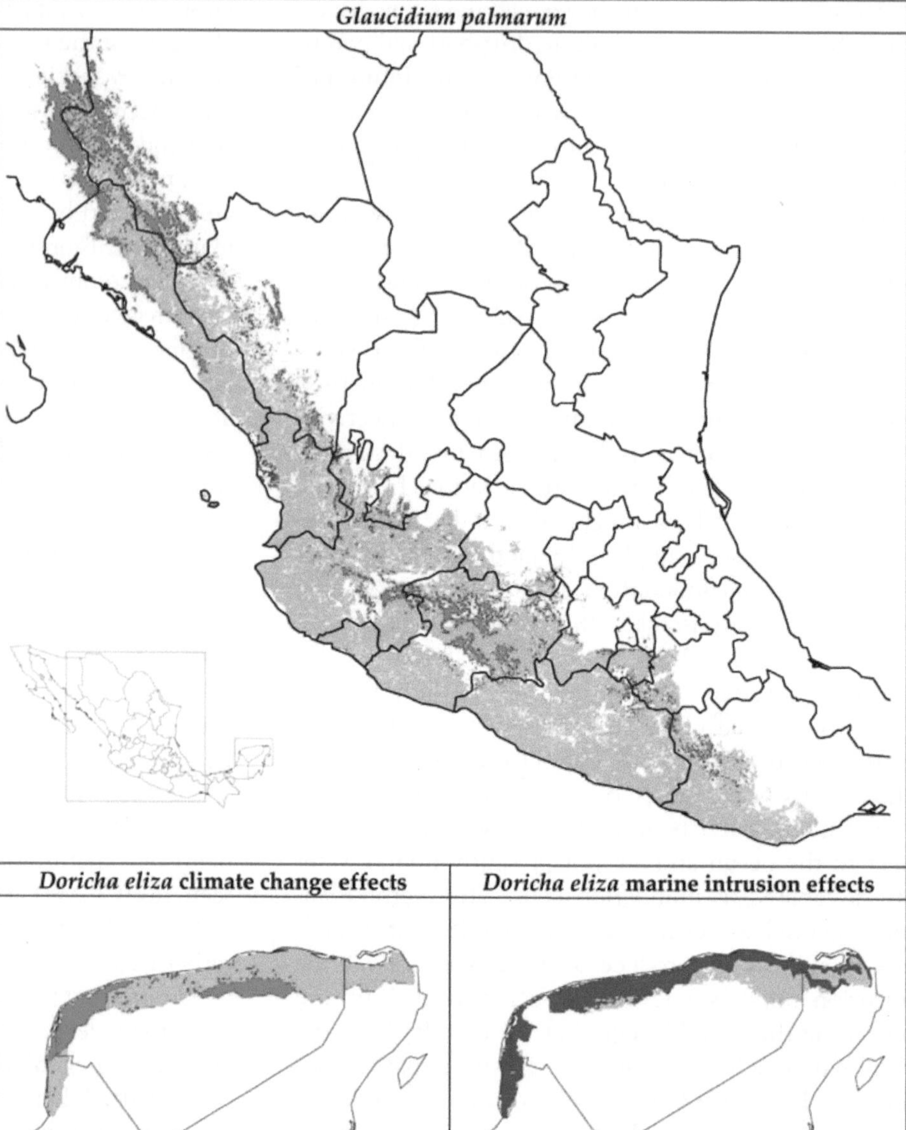

Figure 1. Two example species showing range loss owing to climate change (red) and to marine intrusion (blue); distributional areas not affected are shown in gray. (Top) Complementary effects in *Glaucidium palmarum* (note small blue area along coast). (Bottom) Overlapping effects in the Yucatan Peninsula portion of the distribution of *Doricha eliza*. Small inset map shows locations in Mexico.

of climate change on biodiversity. That is, species are affected by climate change directly or one of the consequences of climate change (sea level rise), but the degree to which they are affected is not related between the two. What is more, the areas within a species' distribution affected by the two tend to be complementary (i.e., nonoverlapping), rather than overlapping. As such, the picture that we reconstruct is that species are affected in a compounding manner by the two factors. These results, if strengthened by parallel signals from analyses of other sectors of biodiversity, would suggest that two rather independent phenomena threaten biodiversity: direct climate effects and indirect marine intrusion effects.

This first survey covers a subset of the Mexican avifauna that is of conservation concern (seventy-six species), which is about 10 percent of the country's resident avifauna. Earlier, climate-only surveys of climate change effects on biodiversity (C. D. Thomas et al. 2004) have been criticized on a number of grounds (Buckley and Roughgarden 2004; Thuiller, Araújo, et al. 2004). Certainly, one criticism should be that of not considering the full complexity of climate change implications. In this study, we show that marine intrusion also will likely affect biodiversity but in ways independent of and complementary to the effects of climate change per se. This phenomenon, in tandem with other complexities (e.g., shifting biotic interactions,

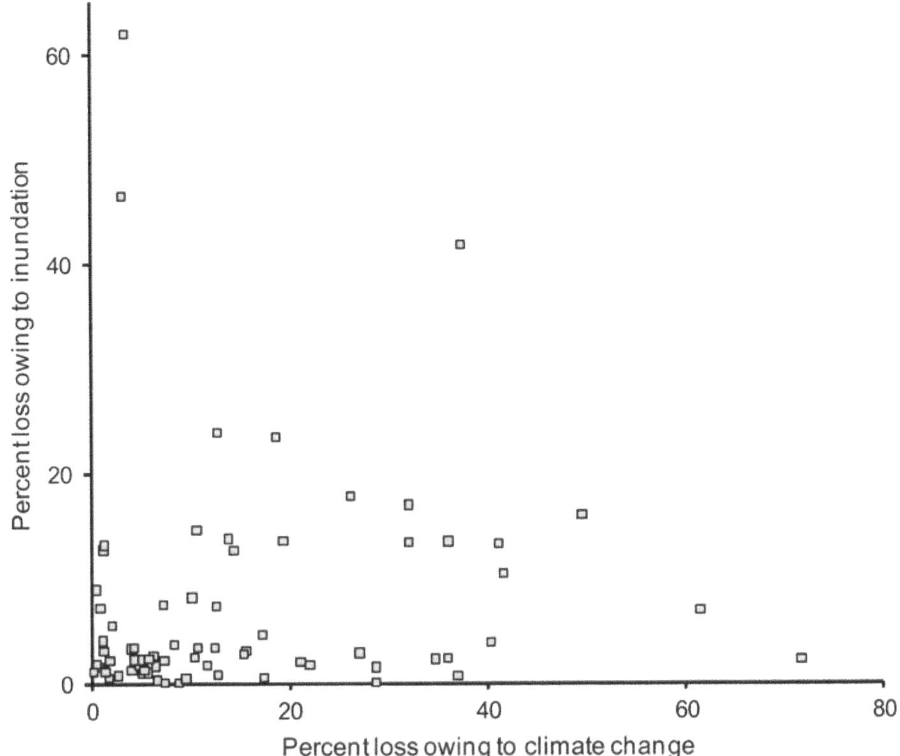

Figure 2. Summary of the relationship between proportion of percent distributional area lost to marine intrusion and proportion of percent distributional area lost to changing climates, across seventy-six species of Mexican birds (shown as squares).

increased ultraviolet light transmission by the atmosphere, effects of increased CO_2 on plant growth and productivity), can now be incorporated into such forecasting exercises to make their conclusions more robust and comprehensive.

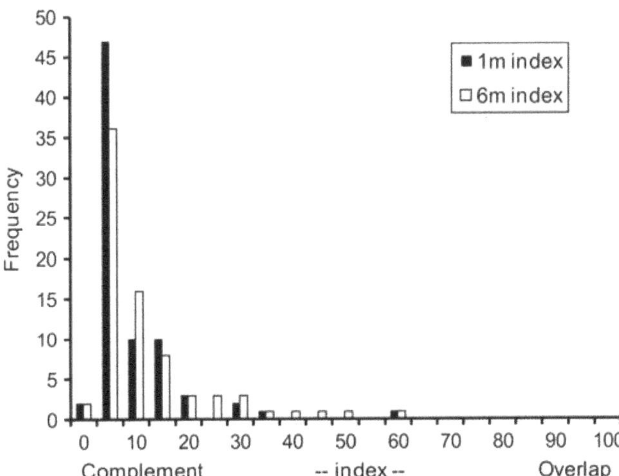

Figure 3. Distribution of complementary versus overlapping of areas affected by marine intrusion and changing climate among seventy-six species of Mexican birds. The index summarizes the distributional areas projected to be lost under projections of both inundation and climate change, relative to range areas lost under maximum overlap (left end of axis) versus maximum complementarity (right end of axis) of the two areas.

Acknowledgments

We would like to thank the reviewers for their comments and suggestions that have greatly improved the article. We also thank Alejandro Gordillo for coordinating the georeferencing of bird localities and CONABIO, CONACYT, and PAPIIT (IN-216408) for support for developing geographic analyses based on the Mexican Bird Atlas database. Thanks to the U.S. Department of Energy for support for development of the sea level rise and marine intrusion scenarios. Point data were obtained from the following institutions: American Museum of Natural History; Academy of Natural Sciences of Philadelphia; Bell Museum of Natural History; Museum Für Naturkunde, Berlin; Natural History Museum, Tring; Zoologische Forschungsinstitut Und Museum Alexander Koenig, Bonn; Übersee-Museum Bremen; Carnegie Museum of Natural History; California Academy of Sciences; Canadian Museum of Nature; Cornell University Museum of Vertebrates; Denver Museum of Natural History; Delaware Museum of Natural History; El Colegio de la Frontera Sur, Unidad Chetumal; Field Museum of Natural History; Colección Nacional de Aves, Instituto de Biología, UNAM; Instituto de Ecología Aplicada, Universidad Autonoma de Tamaulipas; Instituto de Historia Natural y Ecología de Chiapas; Iowa

State University; University of Kansas Natural History Museum; Los Angeles County Museum of Natural History; Natuurhistorische Museum Leiden; Louisiana State University Museum of Zoology; Museo de las Aves de México; Museo de La Biodiversidad Maya, Universidad Autónoma de Campeche; Museum of Comparative Zoology, Harvard University; Museum D'historie Naturelle de la Ville de Geneve; Moore Laboratory of Zoology, Occidental College; Michigan State University Museum; Museo Nacional de Ciencias Naturales Madrid; Museum Nationale D'histoire Naturelle Paris; Museum of Vertebrate Zoology, Berkeley; Museo de Zoología, Facultad de Ciencias, UNAM; University of Nebraska; Naturhistorische Museum Wien; Datos en Línea Red Mexicana de Información Sobre Biodiversidad (REMIB); Royal Ontario Museum; San Diego Natural History Museum; Sistema Nacional de Información Sobre Biodiversidad (SNIB, CONABIO); Texas Cooperative Wildlife Collections; Universidad Autónoma de Baja California; Colección Ornitológica, Facultad de Ciencias Biológicas, University of Arizona; University of British Columbia Museum of Zoology; University of California Los Angeles; Florida Museum of Natural History; University of Michigan, Museum of Zoology; Universidad Michoacana de San Nicolás de Hidalgo; University Museum of Zoology, University of Cambridge; University of Oklahoma; United States National Museum of Natural History; Burke Museum, University of Washington; Western Foundation of Vertebrate Zoology; and Peabody Museum of Natural History, Yale University.

References

Araújo, M. B., and A. Guisan. 2006. Five (or so) challenges for species distribution modelling. *Journal of Biogeography* 33 (10): 1677–88.

Araújo, M. B., and M. Luoto. 2007. The importance of biotic interactions for modelling species distributions under climate change. *Global Ecology and Biogeography* 16 (6): 743–53.

Araújo, M. B., R. J. Whittaker, R. J. Ladle, and M. Erhard. 2005. Reducing uncertainty in projections of extinction risk from climate change. *Global Ecology and Biogeography* 14 (6): 529–38.

Bindschadler, R. A. 1998. Future of the west Antarctic ice sheet. *Science* 282 (5388): 428–29.

BirdLife-International. 2000. *Threatened birds of the world*. Barcelona, Spain: Lynx Editions.

Bosello, F., R. Roson, and R. Tol. 2007. Economy-wide estimates of the implications of climate change: Sea level rise. *Environmental and Resource Economics* 37 (3): 549–71.

Buckley, L. B., and J. Roughgarden. 2004. Biodiversity conservation: Effects of changes in climate and land use. *Nature* 430:6995.

Carter, T. R., R. N. Jones, X. Lu, S. Bhadwal, C. Conde, L. O. Mearns, B. C. O'Neill, M. D. A. Rounsevell, and M. B. Zurek. 2007. New assessment methods and the characterization of future conditions. In *Climate change 2007: Impacts, adaptation and vulnerability*, ed. M. L. Parry, O. F. Canziani, J. P. Palutikof, P. J. van der Linden, and C. E. Hanson, 133–71. Cambridge, UK: Cambridge University Press.

Cresswell, W., and R. McCleery. 2003. How great tits maintain synchronization of their hatch date with food supply in response to long-term variability in temperature. *Journal of Animal Ecology* 72 (2): 356–66.

Crozier, L. 2003. Winter warming facilitates range expansion: Cold tolerance of the butterfly *Atalopedes campestris*. *Oecologia* 135 (4): 648–56.

Daniels, R., T. White, and K. Chapman. 1993. Sea-level rise: Destruction of threatened and endangered species habitat in South Carolina. *Environmental Management* 17 (3): 373–85.

Dasgupta, S., B. Laplante, C. Meisner, D. Wheeler, and J. Yan. 2007. The impact of sea level rise on developing countries: A comparative analysis. World Bank Policy Research Working Paper 4136 Washington, DC: World Bank.

Galbraith, H., R. Jones, R. Park, J. Clough, S. Herrod-Julius, B. Harrington, and G. Page. 2002. Global climate change and sea level rise: Potential losses of intertidal habitat for shorebirds. *Waterbirds* 25 (2): 173–83.

Gopal, B., and M. Chauhan. 2006. Biodiversity and its conservation in the Sundarban Mangrove ecosystem. *Aquatic Sciences* 68 (3): 338–54.

Govindasamy, P. B. D., and J. Coquard. 2003. High-resolution simulations of global climate, part 2: Effects of increased greenhouse cases. *Climate Dynamics* 21 (5–6): 391–404.

Hijmans, R. J., S. Cameron, and J. Parra. 2005. WorldClim, Version 1.3. Berkeley: University of California. http://biogeo.berkeley.edu/worldclim/worldclim.htm (last accessed 10 March 2010).

Hitz, S., and J. Smith. 2004. Estimating global impacts from climate change. *Global Environmental Change-Human and Policy Dimensions* 14 (3): 201–18.

Intergovernmental Panel on Climate Change (IPCC). 2007. *Climate change 2007: The physical science basis*. Cambridge, UK: Cambridge University Press.

LaFever, D. H., R. R. Lopez, R. A. Feagin, and N. J. Silvy. 2007. Predicting the impacts of future sea-level rise on an endangered Lagomorph. *Environmental Management* 40 (3): 430–37.

Legra, L., X. Li, and A. T. Peterson. 2008. Biodiversity consequences of sea level rise in New Guinea. *Pacific Conservation Biology* 14 (3): 191–99.

Li, X., R. J. Rowley, J. C. Kostelnick, D. Braaten, J. Meisel, and K. Hulbutta. 2009. GIS analysis of global inundation impacts from sea level rise. *Photogrammetric Engineering and Remote Sensing* 75 (7): 807–18.

Lovejoy, T. E., and L. Hannah, eds. 2005. *Climate change and biodiversity*. New Haven, CT: Yale University Press.

McKee, K. L., D. R. Cahoon, and I. C. Feller. 2007. Caribbean mangroves adjust to rising sea level through biotic

controls on change in soil elevation. *Global Ecology and Biogeography* 16 (5): 545–56.
Mimura, N. 1999. Vulnerability of island countries in the South Pacific to sea level rise and climate change. *Climate Research* 12 (2–3): 137–43.
Navarro-Sigüenza, A. G., and A. T. Peterson. 2004. An alternative species taxonomy of Mexican birds. *Biota Neotropica* 4 (2). http://specify5.specifysoftware.org/Informatics/bios/biostownpeterson/NP_BN_2004.pdf?q=Informatics/bios/biostownpeterson/NP_BN_2004.pdf (last accessed 10 March 2010).
Navarro-Sigüenza, A. G., A. T. Peterson, and A. Gordillo-Martínez. 2003. Museums working together: The atlas of the birds of Mexico. *Bulletin of the British Ornithologists' Club* 123A:207–25.
Oerlemans, J., R. P. Bassford, W. Chapman, J. A. Dowdeswell, A. F. Glazovsky, J. O. Hagen, K. Melvold, M. de Ruyter de Wildt, and R. S. W. van de Wal. 2005. Estimating the contribution of Arctic glaciers to sea-level change in the next 100 years. *Annals of Glaciology* 42 (1): 230–36.
Parmesan, C., and G. Yohe. 2003. A globally coherent fingerprint of climate change impacts across natural systems. *Nature* 421:37–42.
Pearson, R. G., and T. P. Dawson. 2003. Predicting the impacts of climate change on the distribution of species: Are bioclimate envelope models useful? *Global Ecology and Biogeography* 12 (5): 361–71.
Peterson, A. T. 2003. Predicting the geography of species' invasions via ecological niche modeling. *Quarterly Review of Biology* 78 (4): 419–33.
Peterson, A. T., M. A. Ortega-Huerta, J. Bartley, V. Sánchez-Cordero, J. Soberón, R. H. Buddemeier, and D. R. B. Stockwell. 2002. Future projections for Mexican faunas under global climate change scenarios. *Nature* 416:626–29.
Peterson, A. T., M. Pape, and M. Eaton. 2007. Transferability and model evaluation in ecological niche modeling: A comparison of GARP and Maxent. *Ecography* 30 (4): 550–60.
Pounds, J. A., and M. L. Crump. 1994. Amphibian declines and climate disturbance: The case of the golden toad and the harlequin frog. *Conservation Biology* 8 (1): 72–85.
Rignot, E., and P. Kanagaratnam. 2006. Changes in the velocity structure of the Greenland ice sheet. *Science* 311 (5763): 986–90.
SEMARNAT (Secretaría del Medio Ambiente y Recursos Naturales). 2002. Protección ambiental-Especies nativas de México de flora y fauna silvestres-Categorías de riesgo y especificaciones para su inclusión, exclusión o cambio-Lista de especies en riesgo; en dicha norma se determinan las especies de flora y fauna silvestres terrestres y acuáticas en peligro de extinción, amenazadas y las sujetas a protección especial [Risk categories and specifications for its inclusion, exclusion, or change for native wild flora and fauna species in Mexico]. Diario Oficial de la Federación, México. 6 March 2002. http://www.semarnat.gob.mx/leyesynormas/Normas%20Oficiales%20Mexicanas%20vigentes/NOM-ECOL-059-2001.pdf (last accessed 25 June 2010).
Soberón, J., and A. T. Peterson. 2004. Biodiversity informatics: Managing and applying primary biodiversity data. *Philosophical Transactions of the Royal Society of London B* 359 (1444): 689–98.
———. 2005. Interpretation of models of fundamental ecological niches and species' distributional areas. *Biodiversity Informatics* 2:1–10.
Stockwell, D. R. B., and D. P. Peters. 1999. The GARP modelling system: Problems and solutions to automated spatial prediction. *International Journal of Geographical Information Science* 13 (1): 143–58.
Thomas, C. D., A. Cameron, R. E. Green, M. Bakkenes, L. J. Beaumont, Y. C. Collingham, B. F. N. Erasmus et al. 2004. Extinction risk from climate change. *Nature* 427:145–48.
Thomas, R., E. Rignot, G. Casassa, P. Kanagaratnam, C. Acuna, T. Akins, H. Brecher et al. 2004. Accelerated sea-level rise from West Antarctica. *Science* 306 (5694): 255–58.
Thuiller, W., M. B. Araújo, R. G. Pearson, R. J. Whittaker, L. Brotons, and S. Lavorel. 2004. Biodiversity conservation: Uncertainty in predictions of extinction risk. *Nature* 430:6995.
Thuiller, W., L. Brotons, M. B. Araújo, and S. Lavorel. 2004. Effects of restricting environmental range of data to project current and future species distributions. *Ecography* 27 (2): 165–72.
Titus, J. G. 1990. Effect of climate change on sea-level rise and the implications for world agriculture. *Hortscience* 25:1567–72.
van Lieshout, M., R. S. Kovats, M. T. J. Livermore, and P. Martens. 2004. Climate change and malaria: Analysis of the SRES climate and socio-economic scenarios. *Global Environmental Change* 14 (1): 87–99.

Appendix. Summary of distributional area estimates for seventy-six species of Mexican birds, with estimates of present distributional area, and percentage of present distributional area projected to be retained under scenarios of climate change, 1 m of sea level rise, and 6 m of sea level rise, and estimated percentage of present distributional area projected to be retained under joint effects of climate change and 1 m and 6 m sea level rise

Species	Passerine (P), non-Passerine (N), mainly coastal lowland bird (C), mainly inland forest/ desert bird (F)	Endemic (E), quasiendemic (Q), or of conservation concern (T)	Present distributional area	Percentage area retained (climate change)	Percentage area retained (1 m rise)	Percentage area retained (6 m rise)	Percentage area retained (climate change and 1 m rise)	Percentage area retained (climate change and 6 m rise)
Coereba caboti	P, C	E	37,058	2.50	7.22	4.16	1.01	0.34
Vireo paluster	P, C	E, T	35,552	1.61	99.06	87.70	1.56	0.79
Passerculus beldingi	P, C	T	19,045	5.41	99.22	85.27	5.30	5.16
Vireo perquisitor	P, F	E, T	63,211	17.40	99.06	98.08	16.76	16.02
Campylopterus excellens	N, F	E, T	16,976	28.39	99.67	97.80	28.39	28.39
Toxostoma arenicola	P, F	E	26,445	38.55	99.90	93.16	38.54	38.24
Chaetura gaumeri	N. F	E	71,495	50.50	94.94	84.00	49.09	44.78
Uropsila leucogastra	P, C	Q	260,325	58.35	96.66	89.64	56.25	52.04
Doricha eliza	N, C,	E, T	20,518	62.65	87.16	58.27	57.02	37.36
Passerculus rostratus	P, C	E, T	39,814	58.81	99.57	86.80	58.63	51.42
Aimophila sumichrasti	P, F	E, T	10,989	59.62	97.81	96.21	59.62	59.62
Geothlypis beldingi	P, C	E, T	36,039	63.03	99.81	99.33	63.01	62.88
Amazona xantholora	N, F	Q, T	108,280	63.97	95.89	86.55	63.09	59.81
Deltarhynchus flammulatus	P, C	E, T	154,170	64.03	99.31	97.66	63.90	62.71
Toxostoma palmeri	P, F	E	181,585	65.33	99.86	97.77	65.24	63.54
Cyanocorax yucatanicus	P, F	E	131,667	67.99	94.79	83.07	67.06	64.22
Icterus auratus	P, C	E	90,597	67.97	96.65	86.62	67.24	64.47
Caprimulgus salvini	N, F	E	96,182	71.23	99.56	98.52	71.13	70.82
Phaethornis griseoventer	N, F	E	27,312	71.23	99.99	99.95	71.22	71.21
Arremonops rufivirgatus	P, F	Q	255,998	72.86	99.17	97.18	72.55	71.09
Arremonops sumichrasti	P, F	E	110,585	73.73	85.08	82.20	73.38	71.53
Caprimulgus badius	N, F	Q	46,198	81.30	92.71	76.62	75.97	63.89
Melanoptila glabrirostris	P, C	Q	111,657	80.56	95.78	86.46	77.28	69.01
Parula nigrilora	P, C	E	93,113	77.88	99.33	98.32	77.41	76.59
Icterus fuertesi	P, C	E, T	11,801	87.20	88.03	76.11	78.63	69.63
Forpus cyanopygius	N, C	E, T	119,515	78.83	99.83	98.01	78.83	78.77
Campylorhynchus rufinucha	P, C	E, T	12,192	86.11	92.90	86.25	79.79	74.61
Campylorhynchus yucatanicus	P, C	E, T	8,760	96.51	82.57	38.11	82.16	38.06
Glaucidium palmarum	N, F	E	273,521	82.50	99.93	99.52	82.44	82.05
Corvus sinaloae	P, C	E	79,719	82.66	99.69	95.37	82.61	80.95
Megascops lambi	N, F	E, T	6,082	87.39	94.79	92.73	83.72	82.13
Arremonops verticalis	P, C	Q	111,271	85.57	96.40	87.40	84.12	79.87
Polioptila californica	P, F	Q	113,309	84.30	99.67	96.95	84.18	82.02
Rhodinocichla schistacea	P, C	E	95,702	84.53	99.73	97.27	84.33	82.31
Melanerpes pygmaeus	N, F	Q	105,166	89.35	95.48	85.44	85.44	75.74
Campylorhynchus chiapensis	P, C	E, T	5,952	89.78	95.36	91.87	85.47	82.38
Streptoprocne semicollaris	N, F	Q	213,412	87.22	99.92	99.20	87.15	86.47
Polioptila albiventris	P, C	E	9,714	96.74	88.36	53.59	87.30	53.56
Toxostoma cinereum	P, F	E	96,628	87.48	99.65	96.62	87.33	85.41
Geothlypis flavovelata	P, C	E, T	39,943	89.22	98.60	96.66	88.01	86.13
Habia affinis	P, F	E	106,494	88.27	99.70	98.26	88.13	86.85
Uropsila pacifica	P, C	E	14,926	89.62	99.43	97.55	89.31	88.22
Passerina leclancherii	P, C	E	120,588	90.43	99.76	99.61	90.28	90.20
Amazilia yucatanensis	N, F	Q	408,720	92.75	97.73	92.51	90.48	85.25
Campylopterus curvipennis	N, F	E	61,701	91.15	99.97	99.95	91.13	91.11
Ortalis wagleri	N, C	E	89,515	91.63	99.73	96.36	91.47	88.70

(*Continued*)

Appendix. Summary of distributional area estimates for seventy-six species of Mexican birds, with estimates of present distributional area, and percentage of present distributional area projected to be retained under scenarios of climate change, 1 m of sea level rise, and 6 m of sea level rise, and estimated percentage of present distributional area projected to be retained under joint effects of climate change and 1 m and 6 m sea level rise (*Continued*)

Species	Passerine (P), non-Passerine (N), mainly coastal lowland bird (C), mainly inland forest/desert bird (F)	Endemic (E), quasiendemic (Q), or of conservation concern (T)	Present distributional area	Percentage area retained (climate change)	Percentage area retained (1 m rise)	Percentage area retained (6 m rise)	Percentage area retained (climate change and 1 m rise)	Percentage area retained (climate change and 6 m rise)
Toxostoma longirostre	P, F	E	212,572	92.62	99.56	97.89	92.28	90.70
Phaethornis mexicanus	N, F	E	17,557	92.51	99.99	99.99	92.50	92.50
Rhodothraupis celaeno	P, F	E	121,251	93.72	99.37	97.43	93.16	91.28
Granatellus venustus	P, C	E	232,174	93.51	99.54	98.39	93.17	92.08
Aimophila acuminata	P, F	E	166,173	93.32	99.94	99.74	93.29	93.20
Saltator vigorsii	P, C	E	128,613	94.18	99.60	97.73	93.86	92.10
Sporophila torqueola	P, F	E	305,010	94.27	99.85	98.97	94.17	93.49
Thryothorus felix	P, F	E	190,882	94.68	99.78	98.72	94.50	93.49
Polioptila nigriceps	P, C	E	166,281	94.80	99.85	97.76	94.69	92.72
Cacicus melanicterus	P, F	E	232,899	95.19	99.42	98.31	94.72	93.68
Attila pacificus	P, F	E	313,435	94.93	99.85	98.99	94.81	94.11
Sporophila sharpei	P, F	E	76,384	95.68	99.02	96.60	94.94	92.66
Trogon citreolus	N, F	E	141,140	95.71	99.29	97.76	95.16	93.71
Myiarchus yucatanensis	P, C	Q	113,488	98.70	96.03	86.83	95.27	86.16
Icterus cucullatus	P, C	Q	246,681	98.73	96.03	87.30	95.28	86.69
Melanerpes polygrammus	N, F	E	11,078	97.91	96.33	94.50	95.37	93.79
Campylorhynchus affinis	P, F	E	98,245	95.91	99.65	96.69	95.73	93.19
Amazona finschi	N, F	E, T	231,171	95.93	99.75	98.80	95.75	94.84
Amazona oratrix	N, F	Q, T	166,771	99.04	97.55	92.82	96.81	92.14
Campylopterus pampa	N, F	Q	114,731	99.43	97.15	91.10	96.85	90.86
Icterus pustulatus	P, C	E	134,136	97.22	99.59	99.28	96.90	96.63
Cyanocorax sanblasianus	P, C	E	54,116	98.13	99.73	97.88	97.96	96.25
Passerina rositae	P, F	E, T	10,235	98.17	99.83	99.56	98.03	97.78
Corvus imparatus	P, C	E	99,900	98.86	99.40	95.95	98.42	95.13
Thryothorus sinaloa	P, F	E	227,283	98.57	99.84	98.88	98.43	97.51
Stelgidopteryx ridgwayi	P, F	Q	136,954	98.72	99.71	96.90	98.47	95.69
Campylorhynchus humilus	P, C	E	88,326	99.17	99.08	98.50	98.50	98.00
Chlorostilbon auriceps	N, F	E	214,607	98.69	99.76	98.84	98.51	97.66
Melanerpes chrysogenys	N, F	E	163,862	99.42	99.66	98.13	99.15	97.69
Piaya Mexicana	N, F	E	224,222	99.70	99.81	98.90	99.54	98.67

Adapting Across Boundaries: Climate Change, Social Learning, and Resilience in the U.S.–Mexico Border Region

Margaret Wilder, Christopher A. Scott, Nicolás Pineda Pablos, Robert G. Varady, Gregg M. Garfin, and Jamie McEvoy

The spatial and human dimensions of climate change are brought into relief at international borders where climate change poses particular challenges. This article explores "double exposure" to climatic and globalization processes for the U.S.–Mexico border region, where rapid urbanization, industrialization, and agricultural intensification result in vulnerability to water scarcity as the primary climate change concern. For portions of the western border within the North American monsoon climate regime, the Intergovernmental Panel on Climate Change projects temperature increases of 2 to 4°C by midcentury and up to 3 to 5°C by 2100, with possible decreases of 5 to 8 percent in precipitation. Like the climate and water drivers themselves, proposed societal responses can also be regionalized across borders. Nevertheless, binational responses are confronted by a complex institutional landscape. The coproduction of science and policy must be situated in the context of competing institutional jurisdictions and legitimacy claims. Adaptation to climate change is conventionally understood as more difficult at international borders, yet regionalizing adaptive responses could also potentially increase resilience. We assess three cases of transboundary collaboration in the Arizona–Sonora region based on specific indicators that contribute importantly to building adaptive capacity. We conclude that three key factors can increase resilience over the long term: shared social learning, the formation of binational "communities of practice" among water managers or disaster-relief planners, and the coproduction of climate knowledge. *Key Words: adaptive capacity, climate change, U.S.–Mexico border, vulnerability, water.*

气候变化的空间和人文尺度被带入在气候变化带来尤其挑战的国际边界的救援。本文探讨美国和墨西哥的边境地区气候和全球化进程的双重影响。在此地区，快速的城市化，工业化和集约化农业导致主要气候变化的担忧，即水短缺的脆弱性。对于北美季风气候内的西部边境部分，政府间气候变化专门委员会预测气温到本世纪中叶将升高 2 至 4°C 和到 2100 年升高 3 至 5°C，同时伴随百分之 5 至 8 的降水下降。如同气候和水驱动因子本身，提出的社会反应也可是跨国界的区域化。尽管如此，两国的反应正面临复杂的体制景观。科学和政策的合拍必须位于司法管辖区的竞争体制和合法索赔的范围内。传统上认为在国际边界适应气候变化比较困难，但区域化适应性反应弹性还有可能增加。基于对建设适应能力作出重要贡献的特定的指标，我们评估了亚利桑那州一索诺拉区域的三个跨界协作案件。我们的结论是长远来讲三个关键因素可以增加弹性：共享的社会学习，水资源管理人员或救灾规划者之间形成的两国"实践社区"，以及气候知识的合拍。*关键词：自适应能力，气候变化，美国和墨西哥边境，脆弱性，水。*

Las dimensiones espaciales y humanas del cambio climático se hacen particularmente relevantes en las fronteras internacionales, lugares donde el cambio climático genera retos especiales. Este artículo explora la "doble exposición" de los procesos climáticos y globalizadores para la región fronteriza EE.UU.–México, donde la rápida urbanización, industrialización e intensificación agrícola resultan en vulnerabilidad por escasez de agua, como la preopcupación primaria por el cambio climático. En porciones de la frontera occidental ubicada dentro del régimen climático del monzón norteamericano, el Panel Intergubernamental de Cambio Climático proyecta incrementos de las temperaturas de 2° a 4°C para mediados de siglo y de hasta 3° a 5°C para el 2100, junto con una posible disminución de la precipitación de 5 al 8 por ciento. De la misma manera que ocurre con lo concerniente a clima y agua, las respuestas sociales que se proponen también pueden regionalizarse a través de las fronteras. Sin embargo, las respuestas binacionales se ven confrontadas con un paisaje institucional complejo. La coproducción de ciencia y políticas debe situarse en el contexto de jurisdicciones institucionales en competencia y reclamos de legitimidad. La adaptación al cambio climático se la entiende convencionalmente como más difícil en fronteras internacionales, pero regionalizar las respuesta adaptativas también puede aumentar potencialmente la resiliencia. Evaluamos tres casos de colaboración transfronteriza en la región Arizona-Sonora, con base en indicadores específicos que contribuyen de modo importante a construir capacidad adaptativa. Nuestra conclusión es que hay tres factores claves que pueden incrementar la resiliencia a largo plazo: aprendizaje social compartido, la formación de "comunidades de práctica" binacionales entre administradores del agua y planificadores de alivio por desastres y la coproducción de conocimiento sobre el clima. *Palabras clave: capacidad adaptativa, cambio climático, frontera EE.UU.–México, vulnerabilidad, agua.*

International borders bring into relief the complex and diverse spatial and human dimensions of climate change. The U.S.–Mexico border region is both emblematic—many countries share transboundary climatic regimes—and exceptional—infrequently does an international border juxtapose neighbors with such differing, highly uneven development, although some other border areas bear important similarities. For countries sharing land borders, the impacts of and adaptation to climate change in the transboundary context pose significant challenges (Pavlakovich-Kochi, Morehouse, and Wastl-Walter 2004). An increasing body of scholarship has emerged on the specter of global insecurity due to unstable and inequitable environmental governance practices. This research calls for a greater awareness of the security challenges—broadly interpreted—associated with managing scarce water and other resources in the context of climate change (Gerlak, Varady, and Haverland 2009; O'Brien, St. Clair, and Kristofferson 2010). Water security, particularly in a transboundary context, must increasingly consider climate change, hydrologic, economic, and institutional dimensions of access to and reliability of supply of water for expanding populations. International fora, referred to as "global water initiatives" by Varady et al. (2008, 1), have been established specifically to address this multidimensionality. Avoiding "hydroschizophrenia" and promoting "hydrosolidarity" among countries competing over contested and increasingly scarce water resources are principal goals of this body of work (Falkenmark 2001; Jarvis et al. 2005). Yet national aspects of these challenges typically remain the focus of policymaking and scholarship. We suggest that explicit attention to transboundary challenges of climate change could yield fresh and beneficial insights.

In the case of the United States and Mexico, developing national adaptive responses to climate change, without reference to political and social regimes across the 2,000-mile border, has often yielded less-than-optimal, even harmful outcomes. For example, when in 2008 the U.S. Department of Homeland Security extended its border wall at Nogales, without consulting Mexican officials, subsequent thunderstorm runoff flowing northward into Arizona became trapped and backed up, flooding numerous stores and homes in Mexico and causing significant property damage. Similar problems have occurred along the border, as when the United States unilaterally limited seepage losses in the All-American Canal, which conveys Colorado River water to San Diego, by lining the channel along the border west of Yuma, Arizona. In response, Mexico filed suit in international court to seek redress for the loss of groundwater recharge (from the canal seepage) that had for many decades served a major irrigation district and sustained critical wetlands habitat. In another example, Mexico's nonpayment of its water debt to the United States, per terms of the 1944 treaty governing sharing the waters of the Rio Grande, erupted in 2002 into a major geopolitical dispute.

In the past, such failures to address transboundary issues cooperatively have often characterized binational relations. Despite this history, recent initiatives in collaborative, transboundary environmental management—particularly for water and wastewater—have become more common, with the emergence of binational institutions such as the Border Environment Cooperation Commission.

This article argues for a transboundary approach to improve the adaptive capacity to climate change, especially for water resources management, in the Arizona–Sonora region. Adaptation to climate change is conventionally understood to be more difficult at international borders. Yet we maintain that regional adaptive responses across borders could increase resilience and decrease vulnerability to climatic changes. Such cross-border approaches can emerge through shared social learning and knowledge, by creating binational communities of practice, such as among water managers or disaster-relief planners, and by addressing inequities resulting from uneven development. We suggest that the strengthening of institutional networks and the coproduction of climate knowledge across borders enhance a binational region's long-term adaptive capacity and resilience.

Theoretical Approach

Although climate change introduces uncertainty and risk for water management, a process-oriented analysis focused on social learning—common understandings of challenges among individuals or institutions—allows a better understanding of this dynamism and uncertainty. Milly et al. (2008) questioned the validity of decision making based on static definitions of the bounds of climatic and hydrologic variability. Within transboundary contexts, adaptive responses to climate change are complicated. Risk and vulnerability, socially constructed concepts, are differentially conceived within cultures and across borders. Our Arizona–Sonora empirical analysis suggests that binational responses to water resources management must consider the context of competing institutional jurisdictions and legitimacy claims.

In seeking the key to the creation of sustainable policy informed by the best science, some scholars emphasize the process of knowledge transmission (Cash et al. 2003), whereas others focus on the creation of scientist–stakeholder networks (Lemos and Morehouse 2005; Pelling et al. 2008). Cash et al. (2003) argued that three criteria are key to mobilizing science and technology to achieve sustainability: the salience, or relevance of scientific information to decision makers; the credibility, or scientific adequacy of the information; and the legitimacy of the information or degree to which it reflects diverse stakeholder values and beliefs and is seen as unbiased. Cash et al. (2003) identified, in turn, three "functions" that lead to effective linkages between scientific knowledge and the production of sustainable policy, including the effectiveness of the communication of knowledge to policymakers, the translation (literal and figurative) of the knowledge in the scientist–decision maker interaction, and the mediation of conflicts to ensure transparency and rule enforcement. Lemos and Morehouse (2005) emphasized that the effective coproduction of scientific knowledge and the potential for developing meaningful policy relies on a synergistic relationship among stakeholders and researchers. In their interactions, these networks ideally move beyond discussion to adapt and transform processes (Lemos and Morehouse 2005, 61). Sustained and dynamic interactions among these networks can create "usable science" and effective policy.

Following Pelling et al. (2008), we ask how institutions shape capacity to build adaptive organizations within the Arizona–Sonora border region. We understand adaptive capacity to be a dynamic process based on social learning between and within institutions, rather than a static condition or set of attributes and outcomes (Pahl-Wostl 2007; Pelling et al. 2008). Shared social learning in a transboundary setting refers to the development of common conceptual understandings of climate challenges and regional vulnerability integrated over multiple institutional scales, from individuals and local agencies to state, federal, and binational actors and authorities. In their analysis of the processes associated with effective knowledge sharing between experts and decision makers, Cash et al. (2003) illuminated key aspects of the social learning process that are referenced in the following individual cases. Social learning can take place by individuals operating within a formal institution or collectively by institutions. Within professional communities (such as water managers, disaster-relief planners), informal communities of practice develop based on trust over sustained, iterative interactions and collaborative, peer-to-peer learning (Pelling et al. 2008). A sustained, dynamic, social learning process can stimulate adaptive capacity in regional water management institutions. Adaptive management itself is constructed—or very strongly conditioned—institutionally. By its very nature as an evolutionary, interactive ("learning"-based), and assimilative process, adaptation depends on how challenges are defined and desired outcomes set. Resilience refers to the capacity of socio-ecological systems to self-organize and to build capacity to learn and adapt, to undergo change while retaining the same functions and structures (Folke, Hahn, and Olsson 2005; Resilience Alliance, http://www.resalliance.org, last accessed 4 March 2010).

Communities are associations based on shared identity in which common values and practices are reinforced (Wenger 1999). *Networks* are informal constellations that cross boundaries of community identity and create new vehicles for information flow within or between organizations. Together, communities and networks can form communities of practice (Wenger 1999) that connect due to bridging ties of social capital (e.g., boundary people who bring different communities together into networks, and boundary objects, meetings or documents that join communities into a linked network). Communities of practice can develop adaptive pathways—new institutional priorities or ways of carrying out activities—within and among organizations.

One such pathway would be created if an organization changes its management structure or practices to accommodate different strategies needed to address climate change.

We first examine the challenges of vulnerability in the Arizona–Sonora region, targeting the area's diverse institutional composition and the problems posed for developing adaptive capacity. Second, we consider cases of innovation in regionalized practices that attempt to bridge the transboundary divide. Many of these are nascent in this region, responding to an evolving understanding of shared cross-border climate and water challenges. As a result, we address potential outcomes and emerging results, as all three cases presented are in early stages of development. Among the authors, one or more of us has been involved in these different collaborations or in researching them; thus, this assessment of adaptive potential also serves to illuminate the more (and less) adaptive aspects of these ongoing collaborative processes and may help to improve them as they develop. This assessment is not intended as a report card on outcomes of any of these processes, but rather, an exploration of the question: what components of binational collaboration are most critical to developing successful adaptation? Because all three cases possess the multiple climate–water–institutions dimensionality referred to earlier, assessment of their adaptive-capacity potential (hereafter, adaptive potential) permits us to apply social learning theory and transboundary analysis as central to science–policy coproduction within each of the three cases and to derive more generic understanding relevant for other cases. Evaluation of adaptive potential considers (1) augmentation strategies (Blackmore and Plant 2008) relying on water transferred from distant sources or using high-cost technologies and infrastructure, (2) information flows and data sharing, and (3) regionalized coproduction of knowledge and policy. Via our process-based understanding of adaptive potential, we employ three indicators, or measures: (1) dynamic, structured opportunities for social learning; (2) emergence of formal and informal networks; and (3) potential for development of adaptive pathways. These indicators permit us to assess the adaptive potential in three regionalized transboundary cases. In conclusion, we consider the implications of these regional strategies in three contexts: (1) coproduction of science and policy across national borders, (2) building of transboundary communities of practice, and (3) development of shared platforms for social learning within institutions.

Vulnerability in the U.S.–Mexico Border Region

Vulnerability is conditioned by socioeconomic, institutional, and political as well as environmental factors, including climate (Adger et al. 2006). Assessing vulnerability requires consideration not only of exposure to climate change but also of the risk associated with that exposure and the capacity of an individual, community, or nation to adapt to impacts of climate change (Adger et al. 2006). Vulnerability in the border region's water sector is thus a function of intensified socioeconomic processes—rapid growth, accelerated globalization—and environmental change. Socioeconomic vulnerability is also conditioned by age, ethnicity, gender, or class. For example, elderly people and African Americans in poor neighborhoods were most at risk to the devastation of Hurricane Katrina (Verchick 2008). In the border region, the high concentration of Hispanics, especially in poor U.S. counties and in unplanned Mexican colonias, increases vulnerability for these populations. People might be at higher risk to drought if water becomes scarce and therefore more expensive, if they lack sufficient resources to access or purchase nontraditional water sources. After storms, water trucks (*pipas*) that service marginal neighborhoods might not have access to homes via flooded streets. The region's capacity to respond to these and other high-vulnerabililty water-related challenges depends largely on its water management institutions.

The U.S.–Mexico border region—as a vulnerable area undergoing urbanization, industrialization, and agricultural intensification—is a textbook case of "double exposure" (Leichenko and O'Brien 2008) to climatic and globalization processes (Liverman and Merideth 2002; Ray et al. 2007). The U.S. Southwest and northwest Mexico, where global climate models project severe precipitation decreases and temperature increases, has been called "the front line of ongoing climate change" (Harrison 2009, 1; see Figure 1). Anticipated probable impacts include longer, more extreme droughts, higher water and energy demand, decreased inflows to rivers and streams, and increased urban–agricultural conflict over water (Intergovernmental Panel on Climate Change 2007; Seager et al. 2007).

Since the 1980s, the border region has grown faster than each country's national average. In the United States, an expanding leisure class of retirees, seasonal tourists, and other "amenity seekers" are influencing water management decisions about consumption and

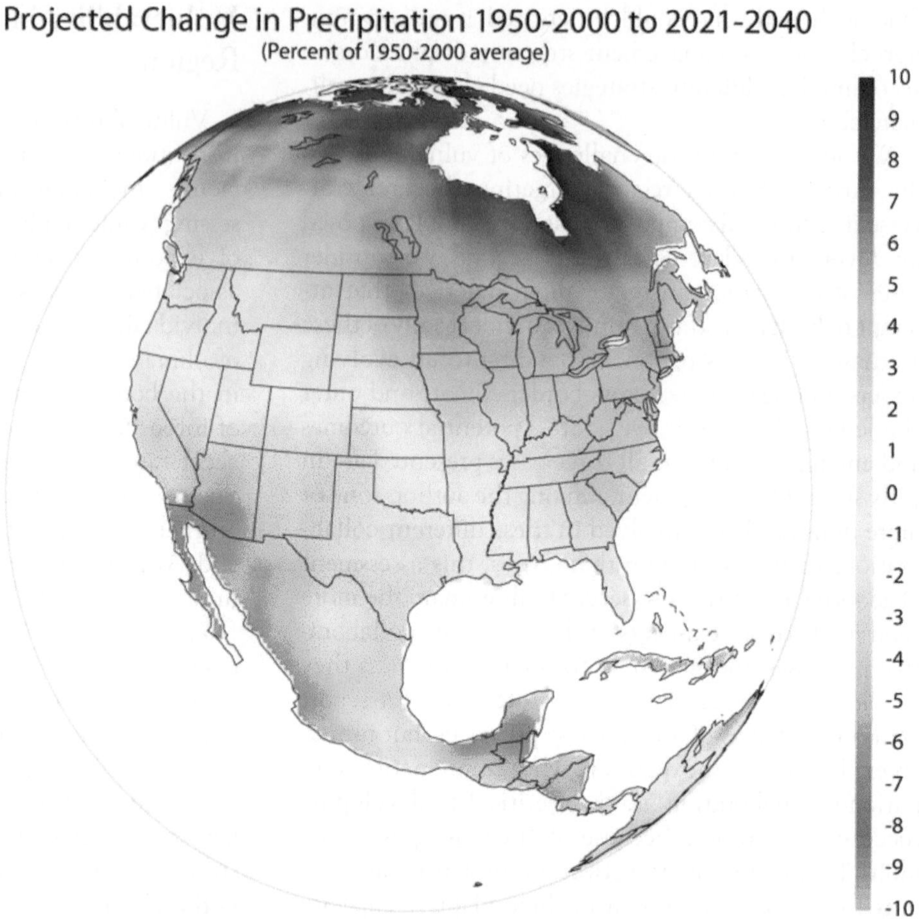

Projected Change in Precipitation 1950-2000 to 2021-2040
(Percent of 1950-2000 average)

Figure 1. Projected change in precipitation: 1950–2000 to 2021–2040: Projected change in precipitation for the 2021–2040 period minus the average over 1950–2000 as a percentage of the 1950–2000 precipitation. Results are averaged over simulations with nineteen different climate models. *Source:* Figure by Gabriel Vecchi, Geophysical Fluid Dynamics Laboratory, National Oceanic and Atmospheric Administration.

conservation. In Mexico, rapid urban growth, driven by availability of jobs created by hundreds of foreign-owned *maquiladoras*, has shifted water-use priorities away from the past farming and ranching economy. Although agriculture remains the largest user of water in Arizona (70 percent of total demand; Arizona Department of Water Resources 2009) and Sonora (86 percent of consumptive use; Comisión Nacional del Agua 2008), growth patterns are driving a shift of water to urban areas. In Mexico's northwest, one quarter of aquifers are severely overdrafted. In Sonora 95 percent of the population has potable water supply and 84 percent has sewerage service (Comisión Nacional del Agua 2008). Many households that have hookups experience daily interruptions to water service, however, and tap water is generally not of drinking quality.

Complex Binational Institutional Landscape

Water management in the border region is fragmented and complex with disparate characteristics in the two countries. The geopolitical relationship between the United States and Mexico complicates cooperation and agreement on water management. For example, U.S. immigration control or drug-trafficking policies often are made with little consultation with Mexico and exacerbate the geopolitical context within which binational water resources issues are considered.

Even otherwise uncomplicated tasks such as constructing a regional database are more difficult in this region, which lacks comparable data and a history of sharing such information (Comrie 2003). Over the past century, the two national governments have established several joint institutions for managing transboundary waters—such as the International Boundary and Water Commission (IBWC) and its Mexican counterpart, CILA; the La Paz Treaty of 1983; and the post-NAFTA Border Environment Cooperation Commission—but these institutions have only a narrow range of responsibility, much of it involving infrastructure construction (Varady and Ward 2009).

In Mexico, water management remains highly centralized in the National Water Commission,

Table 1. Indicators for assessment of potential adaptive capacity building in regionalized transboundary initiatives

Transboundary regionalized initiatives	Social learning	Formal networks	Informal networks	Potential to develop adaptive pathways within institutions	Overall assessment of potential adaptive capacity
Augmentation strategies: Desalination proposals	Low	Low	None	Low	Low
Data sharing and improved information flows: TAAP	High	High	Low	Medium	Medium
Coproduction of science and policy: *Binational Climate Summary* and urban/coastal vulnerability assessments and planning	High	High	High	Medium	High

Note: TAAP = U.S.–Mexico Transboundary Aquifer Assessment Program.

headquartered in the capital (Scott and Banister 2008; Varady and Ward 2009). Despite more than a decade of transferring water management formally to local municipalities and water user associations in irrigation districts, the impact of decentralization has been uneven and limited by lack of revenue-generating authority (Pineda 2002; Wilder and Romero-Lankao 2006; Wilder 2010). In spite of regional variations in water and climatic conditions, the federal government imposes a uniform administrative and management structure. Urban water managers have very limited access to the climate information necessary to plan more adaptively for climate change (Browning-Aiken et al. 2007). Although many rigorous environmental laws exist, their enforcement is uneven. Most seriously, in a weak economy, lack of funding constrains all levels of resource management. Short municipal terms, limited to three years, and lack of a civil service cause high personnel turnover, making sustained planning difficult to achieve over a multiyear horizon (Pineda 2002).

On the U.S. side, most water management institutions are decentralized, with multiple instances of overlapping missions and jurisdiction among various federal, state, and local agencies. Emblematically, dam and reservoir management and allocation is shared by the Army Corps of Engineers, Bureau of Reclamation, Federal Energy Regulatory Commission, federal power marketing agencies, state and local water management entities, public utilities, and irrigation districts (U.S. Climate Change Science Program 2009).

Given the challenges of socioeconomic and climatic vulnerability—increasing water demand, greater competition among users for a shrinking supply, and increasing economic intensification—water managers in the Arizona–Sonora border region are seeking ways both to augment and conserve water sources to ensure that supply can meet projected demand.

Next, we examine three cases of regionalized adaptation to transboundary water management (ranging in adaptive potential from low to high) based on the indicators we have identified—dynamic, structured opportunities for social learning; existence of formal and informal networks; and potential for developing adaptive pathways. The assessments are summarized in Table 1.

Case 1: Desalination as a Water Augmentation Strategy (Low Adaptive Potential)

Desalination of seawater has attracted both attention and financing by those who see it as a failproof source of water in the study region (Kohlhoff and Roberts 2007). As the cost of desalination has decreased, its appeal for augmentation has risen. Nevertheless, desalination does not rank high in our measures of adaptive potential. Although desalination, as a technological innovation, could meet increasing demand, it is unlikely to prompt sustainable change in water users' behaviors under climate change. In fact, desalination, if not coupled with conservation measures, enables a business-as-usual water culture—averse to social learning—and discourages sustainable water use. The region's major urban areas would become dependent on both desalination technology and good relations between U.S. and Mexican authorities—each of which could prove unreliable.

In the study region, desalination has important transboundary implications. For example, the municipality of Puerto Peñasco, a booming coastal resort town, plans to construct a desalination plant (Figure 2). There are twenty resorts in Puerto Peñasco, with approved permits for eighty-six more (Interview with desalination project coordinator, 10 November 2008). With limited access to surface water and exhausted aquifers, Puerto Peñasco must turn to desalination to sustain itself and enable growth.

The U.S. Trade and Development Agency financed a feasibility study for the desalination project, in part

Figure 2. Arizona–Sonora region of the U.S.–Mexico border. *Credit:* Rolando Diaz Caravantes.

due to the economic promise the $35 million plant has for U.S. consultants and contractors. States in the Southwestern United States are interested in more than feasibility studies, however; they favor transboundary arrangements offering access to the desalinated water. Arizona and Sonora have partnered to commission a study of a binational desalination plant, also in Puerto Peñasco, as a part of both states' future water augmentation strategies (HDR 2009). Nevada water augmentation plans speak of a similar strategy (Southern Nevada Water Authority 2009). The plans variously offer to pay for desalination plants in Mexico in exchange for shares of Mexico's Colorado River allocations or to convey water to points of use in Sonora and across the border in the United States (Glennon and Pearce 2007; Kohlhoff and Roberts 2007).

Many consequences of the proposed desalination—including the effects of brine "reject" discharge—are not known, and the results of an environmental impact study scheduled for completion in December 2008 have not been released. No existing federal law regulates how a desalination plant operates in Mexico (López-Pérez 2009). Although developing new sources of fresh water to augment existing groundwater sources would protect aquifers and potentially allow them to recover to nearer equilibrium levels, perceived limitless supplies of water likely would encourage urban growth. There could be additional impacts on the fragile estuaries and fisheries of the Gulf of California and potential disruption of significant ecosystems where the proposed aqueduct would traverse the desert. Moreover, because Arizona and Nevada would continue to use their full allotments plus desalinated supply—without reducing current use—no net gains to the aquifers or to Colorado River allocations likely would be realized.

Overall, then, we assess the augmentation strategies of desalination to be of low adaptive potential. Assessed against the identified indicators, the desalination proposals do not involve structured opportunities for social learning or changes in institutional culture or policy priorities. Data sharing would be in the context of formal contract-based exchanges, rather than more permeable, fluid, relational kinds of knowledge exchanges such as those identified by Cash et al. (2003). New communities of practice are not anticipated to emerge from desalination strategies and binational relationships will be straitjacketed within a bounded legal framework. The desalination strategies are not only unlikely to add to adaptive capacity, but they could lead to more of the entrenched, legalistic relations that have sometimes hampered cooperative, binational water management in the past. Absent a conservation strategy, these strategies enable a status quo water culture that views desalinated seawater as a limitless substitute for fresh water. Ironically, increased interdependence will ensue under the proposed desalination strategies, requiring improved cooperation between the United States and Mexico, yet these strategies do little to foster better communication and enhanced collaboration and therefore could actually increase vulnerability.

Case 2: Data Sharing and Improved Information Flows (Medium Adaptive Potential)

Within the border region, lack of data comparability and data sharing have long been challenges that hinder transboundary cooperation. Scientific knowledge about groundwater aquifers is particularly sparse. The U.S.–Mexico surface water treaty of 1944 and the commission structure to enforce it created institutions for water quantity allocation and water-quality monitoring. Transboundary groundwater, by contrast, has proved more difficult to govern (Feitelson 2006).

An emerging initiative, the U.S.–Mexico Transboundary Aquifer Assessment Program (TAAP), seeks to overcome these institutional and water-resource challenges through binational collaboration. Authorized by U.S. federal law and funded by annual budget appropriations, TAAP is implemented by the U.S. Geological Survey and the state water resources research institutes of Arizona, New Mexico, and Texas, with collaboration from Mexican federal, state, and local counterparts as well as IBWC and CILA. Three essential steps characterize TAAP: (1) building shared vision through joint setting of objectives and prioritized outcomes, a process based on learning among boundary people; (2) scientific assessment of groundwater resources; and (3) dual adaptive-management strategies that conform to each country's institutional environment while expanding binational information flows and data exchange.

Over TAAP's brief lifetime, mutually defined priorities for Arizona's and Sonora's common Santa Cruz and San Pedro aquifers have been identified as vehicles for water for growth, adaptation to climate change, local aquifer-recharge programs, and institutional assessment of groundwater management asymmetries. These priorities reflect fulfillment of two of the effective knowledge transmission criteria identified by Cash et al. (2003), as both salience (relevance of information shared) and credibility (scientific adequacy of the information) appear to be fully satisfied by TAAP processes for data sharing. It is explicitly recognized that binational aquifer assessment will support each nation's management of its share of transboundary aquifers. One implication is that water quality has received diminished attention, given that, upstream, Mexico considered it disadvantageous to identify sources of groundwater pollution. Additionally, TAAP takes a regional approach by emphasizing aquifer-level priority setting and assessment that account for differences between participating states on the U.S. side. However, the principal boundary object in this case (the physical aquifer spanning the border) is not subject to a shared learning approach to management as a result of contrasting laws and regulations for groundwater in the United States and Mexico.

Sharing of information—both as inputs to the scientific assessments and outputs from binational activities—is a critical social-learning feature of TAAP that confers it adaptive potential; however, much has yet to be realized. A negotiation process is underway within the IBWC/CILA umbrella, leading to a binational agreement to identify aquifers for assessment, permit exchange of information, initiate assessment activities, and disseminate results. In the United States, where groundwater is managed and regulated by state and local entities, a flexible mechanism was sought for direct cross-border collaboration with homologue entities. In Mexico, by contrast, federal authority regulates groundwater, and as a result of this asymmetry, agreement was sought within the IBWC/CILA framework. Operating within this institutional arrangement will present challenges for some TAAP stakeholders who are accustomed to pursuing water resources and institutional analyses unfettered by a commission structure and the need to review results prior to dissemination. Nevertheless, TAAP is already generating successful, binational examples of exchange of transboundary aquifer information; for Santa Cruz, a bilingual database of existing studies and reports has been created, and a similar one is in development for San Pedro. To date, users have been other stakeholders directly engaged in the TAAP process; a version for public, Web-based release is planned for the near future.

Such data sharing and improved information flow strategies rank in our assessment as of medium adaptive potential. For TAAP, they require substantial social learning and involve sustained interactions among primarily formal organizations. In this sense, the interactions among professional communities for the express purpose of data sharing take place within a more passive and bounded framework of interaction, with little emphasis on organic, informal network formation or a shared platform of ongoing social learning—hence our assessment that TAAP has medium, albeit very positive, adaptive potential. TAAP could lead to important improvements in sharing of climate and groundwater information across national boundaries and could potentially develop more systematic incorporation of new information sources into organizations' planning practices.

Case 3: Coproduction of Science and Policy with Binational Stakeholders (High Adaptive Potential)

The U.S.–Mexico border region is a fruitful place for collaboration among stakeholders and scientists who share a common interest in developing adaptive capacity to respond to climate change in the water sector. We focus this analysis on just two of several important stakeholder-based science initiatives within the binational Arizona–Sonora region that potentially can contribute in new ways to building adaptive capacity in the water sector.

The first is the development of a binational and bilingual climate outlook newsletter and Web site, called the *Border Climate Summary/Resumen del Clima de la Frontera* (henceforth *BCS*).[1] The *BCS* is based on the *Southwest Climate Outlook*, produced by the Climate Assessment of the Southwest program for over seven years (Jacobs, Garfin, and Lenart 2005). The *BCS* is being developed as part of collaborative research between climate and social scientists in Arizona and Sonora, in consultation with border-region stakeholders. Currently in its fifth quarterly edition, the *BCS* has three goals: (1) to give scientists a tangible foundation from which they can engage stakeholders in dialogues about hydroclimate data and information needs for decision making; (2) to integrate, in one place, value-added hydroclimate information from disparate sources in the United States and Mexico; and (3) to convey new science findings on topics germane to the interests of regional stakeholders.

The binational research team engages urban water managers and disaster-relief planners in a series of workshops that elicit stakeholders' feedback and suggestions to test and refine the *BCS* and to learn about other region-specific data, information, and research needs, such as a comprehensive hydroclimate information portal for the border region. Participants in a 2008 Sonoran workshop noted that they are keen to understand more about North American Monsoon dynamics, tropical cyclone prediction, groundwater resources, and other information that can be provided through *BCS* feature articles (Coles, Scott, and Garfin 2009). At the workshops, stakeholders iteratively provide feedback to the researchers, to inform and help create the science products and delineate the specific information they need to plan more adaptively for climate challenges. Workshop themes have focused on urban issues, national perspectives, and coastal vulnerabilities.

The *BCS* newsletter is a form of coproduction of knowledge that informs policy by increasing access to recent science results (in layperson-friendly language), building capacity for regional stakeholders to use climate information, involving them in the development of the content, and increasing coordination between information providers in both nations.

In related regional research, the binational team is working with regional stakeholders to develop vulnerability assessments for each participating urban area, taking into account the impacts of climate change on future water supply. Although planning toward mid- and long-term horizons (five, ten, or twenty years) is a well-developed practice among the Arizona water managers, very little planning is conducted in equally vulnerable Sonora urban areas due to lack of resources and inaccessibility of appropriate data. By working with stakeholders to jointly develop appropriate climate data (in part via the *BCS*) and understanding of urban vulnerabilities, to future water demand needs—via iterative workshops, surveys, and interviews—researchers hope to develop communities of practice, or informal transboundary networks of water and emergency management professionals. These networks ultimately might rely on common or shared sources of climate data, common understandings of urban vulnerability within the region, and best practices for planning adaptively to meet climate-related challenges.

All of these regionalized, transboundary processes to engage the scientific, stakeholder, and policymaking communities are ongoing, making it difficult to assess the ultimate level of achievement they will attain in yielding more adaptive regional solutions. We assess the *BCS* and the urban vulnerability–stakeholder engagement initiatives to be of high adaptive potential, due to the dynamic and structured social learning opportunities, the development of professional networks, and the coproduction of scientific knowledge (especially in the *BCS*) that could lead to new adaptive pathways within the participating institutions. Both the *BCS* and the urban vulnerability case studies engage iteratively with stakeholders via surveys and workshops that are designed explicitly to develop knowledge-sharing processes and products that meet the standards of salience, credibility, and legitimacy (Cash et al. 2003). The response and participation of stakeholders has been promising. For example, some 130 stakeholders representing twenty-one agencies and management institutions attended workshops held over the last eighteen months. About 1,500 persons receive the *BCS/RCF*. A formal initial evaluation of the *BCS* newsletter indicated that it fills a gap in needs for region-specific information, particularly by making bilingual information more accessible.

Conclusions

The regionalized approaches we have assessed have the potential to stimulate adaptive planning and management over a long term. On the other hand, the border region's convoluted and often divisive institutional arrangements, coupled with the legalistic framework that guides most decision making, can complicate and impede collaboration and cooperation. Although the literature on scientific knowledge transmission for effective policy offers helpful guideposts for evaluating the quality of social learning and knowledge-sharing processes (e.g., Cash et al. 2003; Pelling et al. 2008), it is challenging to identify objective measures of advances in social learning. Loose networks of stakeholders might come together for a time and coalesce around a shared platform of understanding and concepts about vulnerability and adaptation, only to be disrupted by elections that place new individuals in key stakeholder roles, or might be aided by a major occurrence (such as a hurricane) that underscores the vulnerability of communities within the region. Researchers rely on time-limited funding sources, making researcher interactions also part of the fragile fabric of scientist–stakeholder networks. In the transboundary U.S.–Mexico region, adaptations such as binational desalination plans could potentially reduce water supply vulnerability in the Southwestern United States while potentially increasing environmental vulnerability in Mexico. The transboundary nature of developing sustainable solutions is particularly difficult.

Nevertheless, we find that two of the three initiatives discussed hold promising adaptive potential. These strategies, if pursued, could increase social learning among urban water managers, emergency-preparedness planners, and coastal-resources planners. Both formal and informal networks are being advanced through sustained and iterative interactions among different resource managers within the Arizona–Sonora region, facilitated both by boundary people (e.g., the research team and local stakeholders in each site who plan and facilitate meetings) and by boundary objects (e.g., the workshops and the binational climate summary). Working together to produce and refine the binational climate summary with a regional focus on a shared climate regime (e.g., the monsoon) illustrates the coproduction of scientific knowledge that can influence policy within the region and encourage more sustainable planning. In the end, new communities of practice might emerge that institutionalize regional climate science and "climatic thinking" into their current and future water management practices, share institutional data within the community, and are committed to collaboration.

Moving beyond the entrenched patterns of divisive and bounded dealings on water management might increase regional resilience and offer communities more capacity to face looming changes. The obstacles associated with transboundary engagement are steep but the consequences of noncooperation are dire. Transboundary scientist–stakeholder collaboration might hold the key to confronting climate change in vulnerable borderlands.

Acknowledgments

The authors gratefully acknowledge the major funding sources that supported this research, including the National Oceanic and Atmospheric Administration (NOAA) Sectoral Applications Research Program (Grant NA08OAR4310704); the Climate Assessment for the Southwest Program (Grant NA16GP2578) at the University of Arizona, supported by the NOAA Climate Program Office; and the U.S.–Mexico Transboundary Aquifer Assessment Project, with funding from the U.S. Geological Survey. This research was also made possible in part under a grant from the Inter-American Institute for Global Change Research (IAI) project SGP-HD #005, which is supported by the U.S. National Science Foundation (Grant GEO-0642841). In addition, the Morris K. Udall and Stewart L. Udall Foundation has provided support since the mid-1990s for much of the environmental policy work underlying this article. At the Udall Center for Studies in Public Policy, we thank Robert Merideth for his editorial guidance.

Note

1. The *Border Climate Summary* is available in English and in Spanish (*Resumen del Clima de la Frontera*) at http://www.climas.arizona.edu/forecasts/border/summary.html. Links to the *BCS* are found on the Colegio de Sonora (www.colson.edu.mx) and Centro de Investigación Científica y de Educación Superior de Ensenada, Baja California (CICESE) Web sites (http://usuario.cicese.mx/~tcavazos/).

References

Adger, W. N., J. Paavola, S. Huq, and M. J. Mace, eds. 2006. *Fairness in adaptation to climate change*. Cambridge, MA: Massachusetts Institute of Technology Press.

Arizona Department of Water Resources. 2009. Statewide water demand in 2001–2005 and 2006. Phoenix: Arizona Department of Water Resources, Office of Resources Assessment Planning, Statewide Planning.

Blackmore, J. M., and R. A. J. Plant. 2008. Risk and resilience to enhance sustainability with application to urban water systems. *Journal of Water Resources Planning and Management* 134 (3): 224–33.

Browning-Aiken, A., B. J. Morehouse, A. Davis, M. Wilder, R. Varady, D. Goodrich, R. Carter, D. Moreno, and E. D. McGovern. 2007. Climate, water management, and policy in the San Pedro basin: Results of a survey of Mexican stakeholders near the U.S.-Mexico border. *Climatic Change* 85 (3–4): 323–41.

Cash, D. W., W. C. Clark, F. Alcock, N. M. Dickson, N. Eckley, D. H. Guston, J. Jager, and R. B. Mitchell. 2003. Knowledge systems for sustainable development. *Proceedings of the National Academy of Sciences* 100 (14): 8086–91.

Coles, A. R., C. A. Scott, and G. Garfin. 2009. Weather, climate, and water: An assessment of risk, vulnerability, and communication on the U.S.–Mexico border. Extended Abstract 1.4, Fourth Symposium on Policy and Socio-Economic Research, American Meteorological Society 89th annual meeting, Phoenix, AZ. http://ams.confex.com/ams/89annual/techprogram/paper_149938.htm (last accessed 3 March 2010).

Comisión Nacional del Agua [National Water Commission]. 2008. *Estadísticas del agua en México* [*Water statistics in Mexico*]. Mexico City, Mexico: CONAGUA.

Comrie, A. C. 2003. Climate doesn't stop at the border: U.S.–Mexican climatic regions and causes of variability. In *Climate and water: Transboundary challenges in the Americas*, ed. H. F. Díaz and B. J. Morehouse, 291–316. Dordrecht, The Netherlands: Kluwer Academic.

Falkenmark, M. 2001. The greatest water problem: The inability to link environmental security, water security, and food security. *Water Resources Development* 17 (4): 539–54.

Feitelson, E. 2006. Impediments to the management of shared aquifers: A political economy perspective. *Hydrogeology Journal* 14:319–29.

Folke, C., T. Hahn, and P. Olsson. 2005. Adaptive governance of socio-ecological ecosystems. *Annual Review of Environment and Resources* 30 (8): 18–33.

Gerlak, A. K., R. G. Varady, and A. C. Haverland. 2009. Hydrosolidarity and international water governance. *International Negotiation* 14 (21): 311–28.

Glennon, R., and M. J. Pearce. 2007. Transferring mainstem Colorado River water rights: The Arizona experience. *Arizona Law Review* 49 (2): 235–56.

Harrison, J. 2009. New report outlines current, future impacts of climate change. UA News Web site, interview with Jonathan Overpeck. Tucson: University of Arizona. http://uanews.org/node/26043 (last accessed 16 June 2009).

HDR. 2009. Investigation of binational desalination for the benefit of Arizona, United States and Sonora, Mexico: Final report to Central Arizona Project and Salt River Project, Phoenix, AZ, June 5.

Intergovernmental Panel on Climate Change. 2007. *Climate change 2007: The physical science basis. Contribution of Working Group I to the Fourth Assessment Report of the Intergovernmental Panel on Climate Change*, ed. S. Solomon, D. Qin, M. Manning, Z. Chen, M. Marquis, K. B. Averyt, M. Tignor, and H. L. Miller, 1–996. New York: Cambridge University Press.

Jacobs, K., G. Garfin, and M. Lenart. 2005. More than just talk: Connecting science and decisionmaking. *Environment* 47 (9): 6–21.

Jarvis, T., M. Giordano, S. Puri, K. Matsumoto, and A. Wolf. 2005. International borders, ground water flow, and hydroschizophrenia. *Ground Water* 43 (5): 764–70.

Kohlhoff, K., and D. Roberts. 2007. Beyond the Colorado River: Is an international water augmentation consortium in Arizona's future? *Arizona Law Review* 49:257–95.

Leichenko, R., and K. L. O'Brien. 2008. *Environmental change and globalization: Double exposures*. New York: Oxford University Press.

Lemos, M. C., and B. J. Morehouse. 2005. The co-production of science and policy in integrated climate assessments. *Global Environmental Change* 15:57–68.

Liverman, D. M., and R. Merideth. 2002. Climate and society in the U.S. Southwest: The context for a regional assessment. *Climate Research* 21:199–218.

López-Pérez, M. 2009. Desalination plants in Mexico: Operation, issues, and regulation. Paper presented at the

Arizona-Mexico Commission Water Committee Summer Plenary, Phoenix, AZ, June 5.
Milly, P. C. D., J. Betancourt, M. Falkenmark, R. M. Hirsch, Z. W. Kundzewicz, D. P. Lettenmaier, and R. J. Stouffer. 2008. Stationarity is dead: Whither water management? *Science* 319:573–74.
O'Brien, K. L., A. St. Clair, and B. Kristofferson, eds. 2010. *Climate change, ethics, and human security*. Cambridge, UK: Cambridge University Press.
Pahl-Wostl, C. 2007. Transitions towards adaptive management of water facing climate and global change. *Journal of Water Resources Management* 21 (1): 49–62.
Pavlakovich-Kochi, V., B. J. Morehouse, and D. Wastl-Walter, eds. 2004. *Challenged borderlands: Transcending political and cultural boundaries*. Burlington, VT: Ashgate.
Pelling, M., C. High, J. Dearing, and D. Smith. 2008. Shadow spaces for social learning: A relational understanding of adaptive capacity to climate change within organizations. *Environment and Planning* A 40: 867–84.
Pineda, P. N. 2002. La política urbana de agua potable en México: Del centralismo y los subsidios a la municipalización, la autosuficiencia y la privatización [Urban drinking water policy in Mexico: From centralization and subsidies to municipalization, self-sufficiency, and privatization.]. *Región y Sociedad* 14 (24): 41–69.
Ray, A. J., G. M. Garfin, M. Wilder, M. Vásquez-León, M. Lenart, and A. C. Comrie. 2007. Applications of monsoon research: Opportunities to inform decisionmaking and reduce regional vulnerability. *Journal of Climate* 20 (9): 1608–27.
Scott, C. A., and J. M. Banister. 2008. The dilemma of water management "regionalization" in Mexico under centralized resource allocation. *International Journal of Water Resources Development* 24 (1): 61–74.
Seager, R., M. Ting, I. Held, Y. Kushnir, J. Lu, G. Vecchi, H.-P. Huang, et al. 2007. Model projections of an imminent transition to a more arid climate in Southwestern North America. *Science* 316 (5828): 1181–84.
Southern Nevada Water Authority. 2009. *Water resource plan 09*. http://www.snwa.com/assets/pdf/wr_plan_chapter2.pdf (last accessed 3 March 2010).
United States Climate Change Science Program. 2009. *Decision-support experiments and evaluations using seasonal-to-interannual forecasts and observational data: A focus on water resources*. Washington, DC: U.S. Climate Change Science Program.
Varady, R. G., K. Meehan, J. Rodda, E. McGovern, and M. Iles-Shih. 2008. Strengthening global water initiatives. *Environment* 50 (2): 18–31.
Varady, R. G., and E. Ward. 2009. Transboundary conservation in context: What drives environmental change? In *Conservation of shared environments: Learning from the United States and Mexico*, ed. L. Lopez-Hoffman, E. McGovern, R. G. Varady, and K. W. Flessa, 9–22. Tucson: University of Arizona Press.
Verchick, R. M. 2008. Katrina, feminism, and environmental justice. *Cardozo Journal of Law and Gender* 14 (2): 791–800.
Wenger, E. 1999. *Communities of practice: Learning, meaning and identity*. Cambridge, UK: Cambridge University Press.
Wilder, M. 2010. Water governance in Mexico: Political and economic apertures and the shifting state-citizen relationship. *Ecology and Society* 15(2). http://www.ecologyandsociety.org/vol15/iss2/art22/ (last accessed 11 July 2010).
Wilder, M., and P. Romero-Lankao. 2006. Paradoxes of decentralization: Neoliberal reforms and water institutions in Mexico. *World Development* 34 (11): 1977–95.

Climate, Carbon, and Territory: Greenhouse Gas Mitigation in Seattle, Washington

Jennifer L. Rice

Hundreds of local governments in the United States have adopted greenhouse gas (GHG) reduction goals during the past several years, requiring critical examinations of the role of the state in climate governance and the effects of these programs on urban citizenship. Using a study of Seattle, Washington, a city at the forefront of implementing climate regulations through formal government institutions, this article examines how and why climate is incorporated into local environmental policy. By deliberately connecting the causes and consequences of global climate change to the local community, Seattle has been able to use climate as a conceptual resource for urban environmental policy via the climatization of the urban environment. Furthermore, a key mechanism for making climate governable in Seattle is the carbonization of urban governance, where a relationship between the production of GHG emissions and specific urban activities is established through the use of GHG inventories and emissions monitoring. These practices facilitate the act of territorialization, where material natures and state institutions are coconstituted through the production of carbon territories. A key effect of these practices is that Seattle has begun to enroll its residents as a new type of carbon-relevant citizen in the regulation of global climate, while also reaffirming its ability to regulate infrastructural design, commercial activities, and community development. These findings are discussed with respect to their implications for how we understand state practice in climate governance, as well as the relationship between "the state" and "nature" in environmental politics. *Key Words: climate governance, state practice, territorialization, urban environment.*

在过去几年中，有数以百计的美国地方政府采纳了温室气体（GHG）削减计划，要求对国家在气候治理上的角色和城市公民对这些方案的效果进行重要考核。本文以华盛顿州西雅图市为例，该城市正在通过正式的政府机构来实施气候管理条例，本文探讨了气候是如何以及为何被纳入到当地的环境政策中。通过特意连接全球气候变化对当地社区影响的原因和后果，将城市环境加以气候化，西雅图已经能够使用气候作为城市环境政策里的一个概念性资源。此外，为使西雅图的气候能够被治理的关键机制是城市管理的碳机制化，通过对温室气体排放进行清查和监测，建立起造成温室气体排放的生产和具体的城市活动之间的关系。这些做法促进了疆域法案，在那里物质自然和国家机构通过碳领域的生产而共同制宪。这些做法的主要作用是，西雅图已经开始将其居民吸收为全球气候章程下的碳相关公民，同时也籍此重申了其在规范基础设施，商业活动，和社区发展上的能力。本文也讨论了这些研究结果对我们如何理解下述问题上所产生的影响，即国家在气候治理上的做法，以及“国家”和“自然”在环保政治上的关系。*关键词：气候治理，国家惯例，疆域，城市环境。*

Por centenares, durante los pasados aaños los gobiernos locales de los Estados Unidos han adoptado metas de reducción de gases de invernadero (GHG, por la sigla en inglés), lo cual demanda exámenes críticos del papel del Estado en la administración climática y de los efectos de estos programas sobre la población urbana. Utilizando un estudio de Seattle, Washington, una ciudad de avanzada en la implememntación de regulaciones climáticas a través de instituciones gubernamentales formales, este artículo explora cómo y por qué el clima es incorporado en las políticas ambientales locales. Al conectar deliberadamente las causas y consecuencias del cambio climático global con la comunidad local, en Seattle se ha podido llegar a usar el clima como un recurso conceptual en política ambiental urbana por la via de la climatización del medio ambiente urbano. Además, un mecanismo clave para hacer gobernable el clima en Seattle es la carbonización adoptada por el gobierno urbano, al establecer una relación entre la producción de emisiones de GHG con actividades urbanas específicas, mediante el uso de inventarios de GHG y el monitoreo de emisiones. Tales prácticas facilitan la norma de territorialización , donde la naturaleza material y las instituciones del Estado se interconstituyen a través de la producción de territorios del carbono. Un efecto crucial de estas prácticas lo constituye el hecho de que Seattle ha empezado a comprometer a sus residentes en una nueva categoría de relevancia ciudadana por el carbono en la regulación del clima global, a la vez que también reafirma su habilidad para regular el diseño de la infraestructura, las actividades comerciales y el desarrollo comunitario. Estos descubrimientos son analizados en lo que tiene que ver con sus implicaciones sobre la manera como entendemos la práctica del estado en la administración del clima, lo mismo que la relación entre "el estado" y "la naturaleza" en términos de políticas ambientales. *Palabras clave: administración del clima, práctica estatal, territorialización, entorno urbano.*

During the past several years in the United States, something of a revolution has quietly taken place in climate governance. Many regions, states, and cities have forged ahead of the federal government by designing and implementing their own climate regulations, including greenhouse gas (GHG) reduction targets (Lutsey and Sperling 2008). Increasingly, some of these subnational mitigation programs can claim great successes. The city of Seattle, Washington, for example, has achieved an 11 percent reduction in per capita GHG emissions for 2005 relative to 1990 levels (City of Seattle 2007). This begs the question: Why are some city governments inventorying carbon emissions and setting GHG reduction goals without being mandated to do so? Furthermore, how is climate incorporated into urban environmental policy and with what implications for social life and the role of the state in climate governance?

Using an in-depth study of Seattle's climate mitigation efforts, I argue that through the *territorialization* of carbon—the active creation and quantification of bounded and ordered spaces of carbon-producing activities and simultaneous reproduction of local government jurisdictional capacities—Seattle is able to regulate, administer, and monitor policies on climate change. Using GHG inventories that attribute portions of the global carbon cycle to its local jurisdiction, Seattle exercises state power through territorial claims to material natures (i.e., GHGs) that do not necessarily reside within its boundaries, while reaffirming the city's governmental boundaries and functions. This has resulted in a new logic of environmental governance centered on the creation of carbon-relevant citizens (i.e., environmental subjects) who are enrolled in the process of governing global climate. I outline and describe this process as occurring through three stages, including the *climatization* of the urban environment, the *carbonization* of urban governance, and the *territorialization* of carbon in Seattle. In an era when many environmental issues are viewed as problems that reach far beyond the capacities of individual state institutions, government programs like the one in Seattle are demanding new examinations about the relationship between state territories and material natures emergent in climate governance.

Theoretical Framings: The Nature of the State in Climate Governance

The diversity of perspectives on how to understand political strategies related to climate change in current scholarship is drawing attention to the difficulty of theorizing modern state practice in relation to material natures. Because of the character of climate change, an environmental problem with consequences that are not confined to the same spaces as their causes, it is often argued that climate change challenges the capacities of states to protect their citizens due to increased interdependence of nation-states in mitigating emissions (Biermann and Dingwerth 2004). Political geographers, furthermore, have examined the ways that many state functions predicated on the importance of territory and sovereignty are (or are not) being challenged by the intensification of economic and social globalization and the rise of environmental concerns that cross territorial boundaries (e.g., Brenner 2004; Elden 2005).

In places like Seattle, however, local governments are asserting themselves as active institutions in the making of climate policy, despite neoliberal reforms that have favored market-led environmental regulations during the past several decades. Drawing on formulations of "the state" that emphasize the mundane practices and relations of the state in everyday life (Mitchell 1999; Marston 2004; Painter 2006) and the uneven and complex spatiality of the state (Coleman 2007; Secor 2007), it becomes evident that state power is expressed in diverse ways through many sites of governance, even in the execution of neoliberal policies. These theoretical orientations suggest that a more nuanced look at state practice, particularly as it occurs through the most routine and everyday activities of local governments, might shed light on the state–nature relationship being expressed through new subnational climate change programs.

In their discussion of the state–nature relationship, Whitehead, Jones, and Jones (2007) showed that the physical environment is brought into relation with modern state institutions through a twofold process: (1) the *centralization* of knowledge about nature using technologies of government that standardize, simplify, and abstract forms of nature; and (2) the *territorialization* of nature where legible and bounded spaces are utilized in the regulation and administration of state policies related to the environment. A central aspect of this state–nature relation is what Scott (1998, 9) called "state projects of legibility and simplification," where complex natures are made legible to state institutions using several techniques, including mapping of territories and standardization of measures. Stripple (2008) has further argued that states work to "border" climate change through a series of territorializing discourses about allocations of GHG reduction targets and accounting of carbon sinks among national territories, drawing a firm link among statecraft, territoriality, and climate change in environmental policy.

In this regard, Betsill and Bulkeley (2006, 152) have argued about the Cities for Climate Protection Program, that "political power and authority not only lie within nation-states, but can accrue to transnational networks operating through a different form of territoriality," suggesting that subnational climate programs are reframing (and perhaps, reclaiming) state practices that are often attributed to national governments as they engage with climate policy. Furthermore, in an effort to illustrate the relationship among municipal climate programs, state institutions, and climate change, some scholars argue that the rise of subnational actors in climate regulation exemplifies the creation of new forms of multilevel governance, where action on climate change occurs within and between international, national, regional, and local scales of authority (Bulkeley and Betsill 2003).

Importantly, this growing body of literature on local climate governance points to the diversity of actors and sites through which climate mitigation is occurring (Bulkeley and Moser 2007), the complex (and nonhierarchical) scalar associations of multiple levels of governments produced as "local" institutions engage the "global" political arena of climate mitigation (Bulkeley 2005), and how local governments create climate-related policies through complex negotiations with various interests groups, nonhuman actants, and their citizens (Rutland and Aylett 2008). Building on this body of research that critically opens up what constitutes global climate governance, and how climate policies and willing citizens are constructed through political discourse and practice, this article attempts to more explicitly illustrate the relationship between the state and material nature embedded in these accounts. This includes the specific and contingent relationship between the state and carbon that is embedded in territorial practices of local governments and the ways that these state practices provide the foundation for engaging urban citizens in mitigation efforts. Existing literature on local climate governance destabilizes widely held notions that national territories are (and should be) the primary mechanism by which climate is governed, but specific illustrations of the territorial practices that are occurring in these subnational climate programs are still needed. This article provides an explicit account of the relationship between climate and territory, as it has become articulated around carbon molecules and the creation of GHG inventories in Seattle.

Moreover, whereas increasing amounts of work show the importance of framing climate issues in terms of local cobenefits around economic advantages and public health (e.g., Koehn 2008), and the type of communities involved in climate mitigation (e.g., Zahran et al. 2008), scholars are only beginning to elucidate the wider role of the state in governing climate. As federal action on climate change has been stalled in the United States, local governments have been able to take advantage of this "policy room" (Rabe 2007) to use climate-related policies as a mechanism for garnering support for a wide variety of urban policies, indicating that climate change programs provide a unique opportunity for municipal governments to pursue and achieve diverse goals at the local level.

Agrawal (2005), however, drawing on the work of Michel Foucault, has also shown that expressions of state power related to the environment are often accompanied by the internalization of state objectives by citizen subjects. That is, a central part of state power is the making of *environmental subjects* that come to care for the environment in ways complementary to the goals of modern government. In an examination of the governmentality of Portland's climate policies, Rutland and Aylett (2008, 631) argued that "when individuals come to view themselves and their goals according to the same metrics as the state, and base their actions on these metrics, they become part of the network of self-regulating actors that is at the heart of the practice of governmentality." Given this connection between climate policy and individual action, it becomes important, particularly in the context of emerging climate institutions, to better understand how spaces of political authority are organized and controlled through everyday (often mundane and routine) state practices and subject formation related to collection of information and regulation of the physical environment. The remainder of this article explores these issues using the case study of Seattle's climate governance.

Governing Climate in Seattle, Washington

To explain how and why Seattle includes climate as part of its urban governance strategies, and the effect that this has on the city's residents, I argue that climate has come to be the target of urban governance in Seattle via a threefold process. First, the urban environment is *climatized*—that is, environmental policy in the city of Seattle has become centrally organized around issues related to the causes and consequences of climate change, making climate a key conceptual resource in environmental management. Second, urban governance must be *carbonized*, where all aspects of local government work are made carbon relevant through the use of GHG accounting tools. Carbon, as part of geographically inventoried emissions of GHGs that are created by activities that fall within the city's jurisdictional influence, becomes a governable aspect of the environment. As a result, carbon is *territorialized*—such that the authority of Seattle's government to address myriad social and environmental concerns is by virtue of its defined and delineated spaces of carbon production and responsibility, which serves to simultaneously reproduce Seattle's jurisdictional borders and authority. This tripartite set of conceptualizations has allowed Seattle's government to reaffirm its ability to regulate infrastructural design, commercial activities, and neighborhood development through formal and informal regulatory means that act directly on the city's citizens and enroll them in the process of mitigation. Although these transformations have been divided into three stages for the purpose of this analysis, they do occur concurrently and are intimately connected. These findings are based on a two-year study of the role of the state in Seattle's municipal climate action, where information was collected via archival research, sustained observation of city activities, and interviews with city workers during 2007 and 2008.

The Climatization of Seattle's Urban Environment

Through a gradual, yet deliberate, process of formal and informal political action during the past two decades, Seattle's urban environment has become centered on issues of climate change. Global climate change—its causes, consequences, and solutions—has become ever more local in Seattle, and local action on climate change has become more global in its political relevance. The process first began in 1992, with the passage of Resolution 28546, where the city "recognize[ed] the crisis of global warming . . . its effects upon our region, and urg[ed] the United States government to adopt measures designed to reduce emissions which contribute to global warming."[1] At a time when environmental issues like global climate change were firmly the domain of national and international communities, Seattle was working to link this global issue to the local community.

Five years later, Ordinance 118597 (adopted in May 1997) allowed for Seattle's participation in the Cities for Climate Protection (CCP) Campaign, where Seattle's electricity provider, Seattle City Light (SCL), would accept $43,000 from the International Council for Local Environmental Initiatives (ICLEI) to develop a local action plan to conduct a GHG audit and set a GHG reduction target for the utility. Seattle renewed its participation in the CCP program four more times (1998, 1999, and twice in 2000), for a total of $101,000 in additional funding from ICLEI. Each time, the ordinances make reference to the ways Seattle is attempting to "incorporate global climate change issues into City planning and educational efforts," again making formal links between local activities and global climate change and also asserting that the solutions for climate mitigation lie within local government and community actions, rather than solely the national or international sphere. The city's partnership with ICLEI proved to be important in the making of Seattle's climate policies, including how and why the city's climate programs were initially framed in terms of energy production.

In July 2001, Seattle committed to the "long-range goal of stabilizing atmospheric concentrations of greenhouse gases" (Resolution 30316), including reduction of the city's GHGs by 7 to 40 percent below 1990 levels by 2010. This resolution also claims that "many of the critical components of a local action plan for climate protection are already in place or under development in Seattle," directly linking many aspects of Seattle's existing urban landscape, such as the built environment and urban forests, to climate change. The resolution also requires that Seattle's Office of Sustainability and Environment (OSE) coordinate the city's GHG reduction initiatives, including defining and conducting a GHG inventory, developing and coordinating a plan to reduce GHGs, and assessing available projects and policies for meeting GHG reduction goals. A former elected official, who played a key role in implementing many of these early climate resolutions, provided the following statement about why she wanted to address climate change at the local level in Seattle[2]:

> I care a lot about our human impacts on the environment, and global climate change in particular, and wanting local governments to be a model. At that time it was very frustrating what was happening at the national level . . . so that's what motivated me. And I also felt as though we needed to create more livable cities too . . . and that is from solid waste, transportation, to climate change.

This statement shows that the motivation for local action on climate change in Seattle came from at least two places: (1) a desire to show leadership to the federal government and other cities on the issue of climate change and (2) the need to develop a narrative about environmental policy through which several other urban programs could be organized.

More recently, the climatization of Seattle took place at an extraordinary rate under the administration of Mayor Greg Nickels (2002–2009), including the creation of a Green Ribbon Commission to provide expert and community opinions on the creation and implementation of the city's Climate Action Plan, along with the formalization of climate change positions within the OSE. Upon putting climate change at the center of his environmental agenda, which occurred after recognition in late 2004 that climate change could have a dramatic effect on the city's water supplies (*Grist Magazine* 2005), Nickels targeted specific sectors for climate mitigation, including transportation emissions and energy efficiency. These areas are now central components of the city's Climate Action Plan, and climate is the central concern of the city's environmental programs. As a city worker stated:

> Before local action on global warming was . . . one of the many things that we were working on. Now, it's like our number one priority and we're allocating something like 70 percent of our overall resources to it. Climate change disruption and the local response thereto has become the organizing principle or the framework for sustainable development work in the city and in my office.

Climate has become a discursive, conceptual, and material resource for municipal action in a variety of areas in Seattle and an organizing principle in the execution of urban environmental governance. The city's green building program, for example, has become a central feature of Seattle's action plan to meet its GHG reductions goals, and GHG emissions are considered when evaluating transportation options in Seattle. The impetus for climatizing the urban environment in Seattle comes from the desire to show leadership on the issue to other governments, the ability of local climate programs to serve as an umbrella for several other urban issues (e.g., transportation planning and energy conservation), and fear of failure in providing key urban services to its residents because of the impacts of climate change (e.g., reliable water supplies). These efforts have also proven to be a unique opportunity for Seattle to reaffirm and rearticulate local government capacity in a variety of areas, as the solutions for global climate change are shown to exist within local environmental governance.

The Carbonization of Urban Governance

Using standards for city operations to meet GHG reductions internally, while also creating extensive GHG inventories for the wider community, nearly all aspects of urban governance in Seattle have become geographically defined and quantitatively tallied in terms of carbon (or more generally, GHG) emissions and reduction potential. Nearly all elements of traditional city work and life—providing transportation options, overseeing community development, and supplying services and utilities—have been made carbon relevant during the past decade in Seattle.

In 2000, Seattle adopted the "Earth Day" Resolution (30144), which required that SCL provide electricity with "no net greenhouse gas emissions." This meant that emissions from portions of Seattle's power supply (the Klamath Falls Project) and other operational activities (such as SCL automobiles and buildings) had to be mitigated through the use of conservation efforts. Where mitigation was not possible, the city would purchase carbon offsets. The city estimated that 272,727 tons of CO_2 would need to be offset annually at a cost of $1.4 million a year (Resolution 30256). Using offset project guidelines (Resolution 30359), several projects have allowed the SCL to reach its goal of being a "climate neutral," such as allowing cruise ships to plug into Seattle's shore power supply at port, rather than burning diesel fuel (Port of Seattle 2005). Because the City of Seattle's power supply is 90 percent hydropower—an energy supply with virtually no GHG emissions—achieving climate neutrality has been easier for Seattle than many other cities that rely more heavily on carbon-based electricity. The "no net emissions" policy did, however, formally recognize the role that government plays in environmental protection related to climate change through their everyday operations, service delivery, and regulatory authorities.

As part of its Climate Action Plan to reach the city's GHG reduction goals,[3] the city has formally quantified and geographically accounted for its carbon emissions through the use of GHG inventories. Deciding what is included in a carbon inventory requires that the city actively define its carbon jurisdiction because much GHG-related activity can cross the defined boundaries (e.g., an airport located inside city boundaries that serves several regional jurisdictions) or can be attributed to different spatial extents than those of the defined community (e.g., electricity production or waste disposal for a city that actually occurs outside city limits; City of Seattle 2007). As one city employee said about Seattle's community GHG inventory:

> The carbon foot-printing process for a geographic area as small as even a city is extremely complex. We spent a lot of time on that. We engaged the Green Ribbon Commission very deeply in making a lot of those policy judgment calls that go into defining your bubble and what's in and what's out.

The inventorying and monitoring process requires that Seattle collect and assess a significant amount of information about city activities in terms of their carbon emissions. Seattle has been keeping track of data related to energy use for decades (e.g., the city keeps records on fuel usage), but much of these data are now assessed in new ways to make it relevant for tracking GHG inventories (e.g., fuel use now used as a measure of GHG emissions). In the process of designing and implementing these carbon-related policies, Seattle has worked to construct climate as an object of urban governance via emissions accounting strategies and complex negotiations with various interest groups in Seattle, which are neither purely objective nor inevitable ways to inventory local emissions (see also Rutland and Aylett 2008).

Together, these ordinances and resolutions related to GHG reduction, offsetting, and inventorying have allowed climate to become a governable aspect of the urban environment in Seattle. In the process, nearly all aspects of Seattle's urban environment have become *carbon relevant*, which forms the basis of environmental policy within the city of Seattle and is a fundamental aspect of the Seattle's wider forms of urban governance. This began with efforts to make Seattle's electricity provider "carbon neutral," which required precise accounting of GHG emissions for energy production, and subsequently spread to other areas of urban governance, including land use and transportation planning. The importance of Seattle's carbonizing activities, particularly as they have emerged within and spread beyond the energy sector, is that GHG inventories allow the city to examine and monitor activities over which they have direct control (i.e., energy production) and those activities they do not directly control (i.e., people's transportation choices). This context sets the stage for Seattle's attempts to engage its local residents in climate policy, which is discussed in detail in the following section.

Territorializing Carbon in Seattle, Washington

Through the use of GHG inventories and centralization of local environmental policies around issues of climate change, Seattle has constructed political authority related to the climate through the territorialization of carbon. These state practices work to embed the abstract concept of climate change onto the urban environment and into city politics. GHG molecules, although part of the global atmosphere, are assigned to Seattle's jurisdiction by spatially referencing them to transportation, energy production and consumption, and other GHG-producing activities that occur within the city. Although it might seem that climate change represents an unterritorializable form of material nature because once emitted, emissions become part of a globally mixed atmosphere, the city of Seattle has created new strategies for incorporating GHGs into its jurisdictional territory. This analysis suggests that a rearticulation of territorialization with respect to climate change is occurring in Seattle, one that is not concerned with defining a territorially based "inside" and "outside" but with the attribution of material nature to specific places and activities that are within the boundaries of state institutions.

Because of the territorialization of carbon, Seattle has been able to reassert, and in many ways recast, its ability to regulate infrastructural aspects of the city (i.e., transportation structures) and social and economic practices (i.e., urban development). This process is fundamental to understanding how state power is exerted over diverse physical landscapes and social practices. Notably, this characterization of climate governance also illustrates the ways that state practices and material natures are coconstituted via territorializing actions of the local state. As climate is made governable by assigning GHGs to the local community, and the material nature of climate is rendered visible to state institutions via carbon governance, Seattle also reproduces its jurisdictional borders and governing capacity. As the problem of climate change becomes increasingly important in discourses of environmental security, carbon represents the key relation between the state and nature emergent in new government programs related to climate.

Importantly, however, this territorialization of carbon has facilitated, indeed required, that Seattle develop strategies to engage citizens in the effort to reduce GHGs. As a city worker stated:

> There's just the cold hard facts and the cold hard numbers which say that as a city government we have reduced our climate pollution 60 percent since 1990. That's huge, but we are not going to get there unless people start taking action. The reality is we as a city have stood up and said we're going to do this thing . . . we can't do it without the community . . . we can't literally push people onto the bus . . . we are going to have to rely on some of those behavioral changes.

To meet their GHG reduction goals, the city must motivate its residents. This is not an easy task and requires that the local government develop a suite of strategies to influence what have now been deemed carbon-related activities of the wider community. To do this, Seattle has created a series of programs to engage various areas of urban activity, targeted at individuals, neighborhoods, and businesses, and their corresponding spaces of carbon responsibility. Seattle Climate Action Now (Seattle CAN) is a community outreach program that helps residents understand and reduce their carbon footprints through home-based energy conservation programs, spreading information on climate change and encouragement for residents to utilize alternative forms of transportation. The Neighborhood Climate Protection Fund allows neighborhoods to apply for small grants to execute climate-related activities and projects in their own neighborhoods, and the Seattle Climate Partnership has been created to engage businesses by providing resources for local companies to assess and reduce their GHG emissions.

All of these programs, based on the ability of the city to regulate climate through the territorialization of global carbon, are creating a new form of carbon governmentality, where individuals are expected take on the responsibilities themselves of reducing their carbon footprint in accordance with the goals of the state. Get on a bus, turn down your thermostat, ride your bike to work: *Be a good carbon citizen.*

In Seattle, this effort is predicated on distinct and historically contingent territorial practices occurring via carbon and through the local state. Whereas some scholarship characterizes climate change as having the potential to dramatically undermine the capacity of governments to protect their citizens, this analysis suggests that states are actively working to manage the challenges of climate change through both territorial and bio-political means, which rely on individual action to address climate change, rather than larger structural changes in capitalist development (see also Grove forthcoming). The question emerges: As states increase their capacities to govern climate via territorial practices that have been executed by governments for centuries, do they run the risk of simply reproducing the socioeconomic conditions that enabled the problem of climate change to occur in the first place?

Carbon Territories: The Everyday State and the Logic of Urban Governance

This examination of how Seattle has made climate governable reveals that it is through carbon that climate can be made relevant to the practices of the state. The production of carbon territories and territorial ordering of climate via carbon is a central way that states can exercise political power related to climate. Whereas other scholarship on local climate governance has destabilized the primacy of hierarchical forms of environmental governance centered on national and international governments, this article further elucidates the ways that local government institutions work to incorporate GHGs into territorial state practices and are themselves reproduced in the process. Furthermore, as carbon territories enable governments to define and reduce their GHG emissions, urban citizens are necessarily enrolled in the process as carbon-relevant individuals who are expected to reduce their individual carbon footprints. Through these territorial practices frequently attributed to the nation-state, local governments are constructing new claims regarding who has the political authority to regulate GHG emissions and where state practices related to climate change can and should play out.

Furthermore, throughout the city of Seattle, climate and carbon are becoming the centralizing concept of urban environmental governance. Carbon matches the territorial logic of the state, precisely because it is measurable and quantifiable. The policies and practices related to the territorialization of carbon require increasing quantification, measurement, and verification of carbon-related activities, as Seattle moves from a "sustainability" paradigm of urban environmental governance to a climate-centered one. This ability to make climate legible, via the carbonization of urban governance, is perhaps why climate change is beginning to replace the concept of sustainability that has been so elusive in urban environmental governance for the past several decades. Although the actual practices of urban governance (i.e., transportation and land-use planning) have not changed significantly, the ways they are framed and evaluated are now territorially relevant and globally significant.

This analysis of the role of the state in climate governance also complements other research that claims that the state is still an important aspect of neoliberal environmental governance (e.g., Mansfield 2007), rather than a lame duck or shrinking institution. There is distinct room for state governments to take action related to climate, which is often in the most mundane of state activities—transportation, building codes, neighborhood development—rather than that of highly visible national and international agreements. With respect to climate governance, the state does not simply control and regulate material natures that happen to reside within state borders—the materiality of GHGs does not allow for it. Instead, the state actively produces a discourse about the materiality of climate as being referenced to carbon-producing activities in its jurisdiction that make it governable. As the state begins to reach the limits of its coercive power in climate mitigation (activities it can directly control), it works to achieve the consent of its residents to help accomplish its goals through programs that encourage residents to change their behavior. Most important, we see that making climate compatible with state practice through the creation of carbon territories also sets the stage for the making of willing citizens in the fight against global warming. In the process, however, it seems increasingly likely that state interventions in climate mitigation, which rely on the engagement of individuals to reach climate-related goals via territorially defined authority, might be reproducing the same political practices under which the problem of climate change was permitted to proliferate in the first place.

Acknowledgments

This research is part of a larger project on municipal climate governance funded by the National Science Foundation (Award #0802739) and the Morris K. Udall Foundation. I am very grateful to those individuals in Seattle who generously shared their time, knowledge, and experiences with me during interviews. A very special thanks to Paul Robbins, Sallie Marston, and J. P. Jones for their valuable guidance during this research and insightful comments on multiple drafts of the article. An earlier version of this article was presented at the Department of Geography at the University of Washington in April 2009, and I would like to thank those in attendance for their thoughtful questions and productive suggestions on this research. I would also like to thank the anonymous reviewers and the special issue editor for their helpful comments.

Notes

1. Full text of Seattle's Ordinances and Resolutions is available at http://clerk.ci.seattle.wa.us/~public/cbory.htm (last accessed 2 August 2010).
2. Names and affiliations are not provided with interview quotes to maintain confidentiality of interviewees. Quotes are from interviews with city workers whose jobs are directly related to the design, implementation, or both of Seattle's Climate Action Plan or other climate-related initiatives.
3. The city's GHG reduction goals are 7 percent below 1990 levels by 2012, 30 percent below 1990 levels by 2024, and 80 percent below 1990 levels by 2050.

References

Agrawal, A. 2005. *Environmentality: Technologies of government and the making of subjects*. Durham, NC: Duke University Press.

Betsill, M. M., and H. Bulkeley. 2006. Cities and the multilevel governance of global climate change. *Global Governance* 12:141–59.

Biermann, F., and K. Dingwerth. 2004. Global environmental change and the nation state. *Global Environmental Politics* 4:1–22.

Brenner, N. 2004. *New state spaces: Urban governance and the rescaling of statehood*. Oxford, UK: Oxford University Press.

Bulkeley, H. 2005. Reconfiguring environmental governance: Towards a politics of scales and networks. *Political Geography* 24:875–902.

Bulkeley, H., and M. Betsill. 2003. *Cities and climate change: Urban sustainability and global environmental governance*. London: Routledge.

Bulkeley, H., and S. C. Moser. 2007. Responding to climate change: Governance and social action beyond Kyoto. *Global Environmental Politics* 7 (2): 1–10.

City of Seattle. 2007. Seattle's community carbon footprint: An update. http://www.seattle.gov/climate/docs/Seattle%20Carbon%20Footprint%20Summary.pdf (last accessed 26 June 2009).

Coleman, M. 2007. Immigration geopolitics beyond the Mexico–US border. *Antipode* 39:54–76.

Elden, S. 2005. Missing the point: Globalization, deterritorialization, and the space of the world. *Transactions of the Institute of British Geographers* 30:8–19.

Grist Magazine. 2005. City city bang bang: An interview with Seattle Mayor Greg Nickels on his pro-Kyoto cities initiative. http://www.grist.org/article/little-nickels/ (last accessed 26 June 2009).

Grove, K. Forthcoming. Insuring "our common future?" Dangerous climate change and the biopolitics of environmental security. *Geopolitics* 15(3).

Koehn, P. H. 2008. Underneath Kyoto: Emerging subnational government initiatives and incipient issue-bundling opportunities in China and the United States. *Global Environmental Politics* 8 (1): 53–77.

Lutsey, N., and D. Sperling. 2008. America's bottom-up climate change mitigation policy. *Energy Policy* 36:673–85.

Mansfield, B. 2007. Articulation between neoliberal and state-oriented environmental regulation: Fisheries privatization and endangered species protection. *Environment and Planning A* 39:1926–42.

Marston, S. 2004. Space, culture, state: Uneven developments in political geography. *Political Geography* 23:1–16.

Mitchell, T. 1999. State, economy, and the state effect. In *State/culture: State formation after the cultural turn*, ed. G. Steinmetz, 76–97. Ithaca, NY: Cornell University Press.

Painter, J. 2006. Prosaic geographies of stateness. *Political Geography* 25:752–74.

Port of Seattle. 2005. Cruise ships plug into shore power at port of Seattle. http://www.portseattle.org/news/press/2005/07_23_2005_63.shtml (last accessed 26 June 2009).

Rabe, B. G. 2007. Beyond Kyoto: Climate change policy in multilevel governance systems. *Governance: An International Journal of Policy, Administration, and Institutions* 20 (3): 423–44.

Rutland, T., and A. Aylett. 2008. The work of policy: Actor networks, governmentality, and local action on climate change in Portland, Oregon. *Environment & Planning D: Society & Space* 26:627–46.

Scott, J. 1998. *Seeing like a state: How certain schemes to improve human condition have failed*. New Haven, CT: Yale University.

Secor, A. J. 2007. Between longing and despair: State, space, and subjectivity in Turkey. *Environment and Planning D: Society and Space* 25:33–52.

Stripple, J. 2008. Governing climate, (b)ordering the world. In *From Kyoto to the town hall: Making international and national climate policy work at the local level*, ed. J. L. Lundqvist and A. Biel, 137–54. London: Earthscan.

Whitehead, M., R. Jones, and M. Jones. 2007. *The nature of the state: Excavating the political ecologies of the modern state*. New York: Oxford University Press.

Zahran, S., S. D. Brody, A. Vedlitz, H. Grover, and C. Miller. 2008. Vulnerability and capacity: Explaining local commitment to climate-change policy. *Environment and Planning C: Government and Policy* 26 (3): 544–62.

Potential Impacts of Climate Change on Flood-Induced Travel Disruptions: A Case Study of Portland, Oregon, USA

Heejun Chang, Martin Lafrenz, Il-Won Jung, Miguel Figliozzi, Deena Platman, and Cindy Pederson

This study investigated potential impacts of climate change on travel disruption resulting from road closures in two urban watersheds in the Portland, Oregon, metropolitan area. We used ensemble climate change scenarios, a hydrologic model, a stream channel survey, a hydraulic model, and a travel forecast model to develop an integrated impact assessment method. High-resolution climate change scenarios are based on the combinations of two emission scenarios and eight general circulation models. The Precipitation-Runoff Modeling System was calibrated and validated for the historical period of 1988 and 2006 and simulated for determining the probability of floods for 2020 through 2049. We surveyed stream cross-sections at five road crossings for stream channel geometry and determined flood water surface elevations using the Hydrologic Engineering Centers River Analysis System (HEC-RAS) model. Four of the surveyed bridges and roadways were lower in elevation than the current 100-year flood water surface elevation, leading to relatively frequent nuisance flooding. These roadway flooding events will become more frequent under some climate change scenarios in the future, but climate change impacts will depend on local geomorphic conditions. Whereas vehicle miles traveled was not significantly affected by road closure, vehicle hours delay demonstrated a greater impact from road closures, increasing by 10 percent in the Fanno Creek area. Our research demonstrated the usefulness of the integration of top-down and bottom-up approaches in climate change impact assessment and the need for spatially explicit modeling and participatory planning in flood management and transportation planning under increasing climate uncertainty. *Key Words: climate change, integrated impact assessment, transportation, urban flooding.*

针对俄勒冈州波特兰市的两个市区流域，本研究探讨了气候变化对因道路封闭而造成交通中断的潜在影响。我们使用了总体气候变化情景，水文模型，河道的统计调查，水力模型，和交通预测模型，以此建立起一种综合性的影响评估方法。高分辨率的气候变化情景是基于两个排放情景和八个普通环流模式的组合。我们对降水径流模拟系统进行了标定，基于历史时期的1988年和2006年来模拟确定2020年到2049年的洪水概率。对于径流几何特征，我们调查了五个道路的径流截面，依据水文工程中心河流分析系统（HEC-RAS）模型，确定了洪水水面高程。接受调查的桥梁和道路中，有四个的海拔低于目前的100年洪水水面高程，会导致比较频繁洪水滋扰。在某些未来气候变化的情景下，这些巷道水浸事件将更为频繁，但气候变化的影响将取决于当地的地貌条件。研究表明，车辆的行驶里程没有受到封路的显著影响，车辆的延误时间受封路影响较大，在法诺河地区增长了百分之十。我们的研究表明了这种集成气候变化影响评估中自上而下和自下而上研究方法的实用性，同时也表明了在持续增长的气候不确定性的条件下，在洪水管理和交通规划实践中对明确的空间建模和参与规划的实际需求。*关键词：气候变化，综合影响评估，交通运输，城市洪涝灾害。*

En este estudio se investigaron los impactos potenciales del cambio climático en lo relacionado con interrupción de viajes por el cierre de caminos en dos cuencas urbanas del área metropolitana de Portland, Oregon. Utilizamos conjuntos de escenarios de cambio climático, un modelo hidrológico, un estudio de cauces, un modelo hidráulico y un modelo de predicción de viajes, para desarrollar un método integrado de evaluación de impactos. Los escenarios de cambio climático de alta resolución se basan en las combinaciones de dos escenarios de emisión y ocho modelos de circulación general. El Sistema de Modelado de Precipitación-Escorrentía fue calibrado y validado para el período histórico 1988–2006, simulado para determinar la probabilidad de inundaciones de 2020 hasta 2049. Se estudiaron tramos de corrientes sobre cinco cruces de carreteras para establecer la geometría de los cauces y se determinaron las elevaciones de la superficie de inundación mediante el uso del modelo del Sistema de Análisis de Ríos de los Centros de Ingeniería Hidrológica (HEC-RAS, sigla en inglés). Cuatro de los puentes y vías estudiados resultaron con elevaciones menores de lo que indica el cálculo actual del nivel del agua de inundación para 100 años, una molesta expectativa de inundaciones relativamente frecuentes. Es de esperar que estos eventos de inundación de las vías se harán más frecuentes bajo ciertos escenarios de cambio climático, pero los impactos de estos cambios climáticos dependerán de las condiciones geomórficas locales. Si bien las millas recorridas por los vehículos no aparecen afectadas significativamente por el cierre de las carreteras, las horas de demora demostraron un mayor impacto derivado de tales cierres, con un incremento del 10 por ciento en el área de Fanno Creek. Nuestra investigación demostró la utilidad de la integración de los enfoques de mayor a menor con los de lo inferior a superior, en la consideración del impacto de cambio climático, y también la necesidad de modelado espacialmente explícito y de planificación participativa en el manejo de problemas de inundación y de planificación del transporte, bajo creciente incertidumbre climática. *Palabras clave: cambio climático, evaluación integrada de impacto, transporte, inundaciones urbanas.*

According to the most recent Intergovernmental Panel on Climate Change report (IPCC 2007), anthropogenic climate change is projected to bring more frequent, heavier winter precipitation and earlier snowmelt in midlatitude areas. This is due to the fact that warmer air can hold more water vapor, thus accelerating the circulation of water between the atmosphere and land and oceans (Huntington 2006). As a result of rising intense precipitation and soil moisture content, water tables are likely to increase, leading to more frequent flooding in areas already affected by periodic flooding. This is evidenced by the recent flood events in 2006, 2007, and 2008 in the U.S. Pacific Northwest that severely limited surface transportation and thus disrupted the regional economy. Although there is ample need to investigate the vulnerability of regional transportation infrastructure and associated disruption of the transportation systems, very few studies have investigated the potential impacts or adaptation of climate change on the transportation sector (National Research Council [NRC] 2008; Koetse and Rietveld 2009) other than focusing on reducing greenhouse gas emissions (Black and Sato 2007; Chapman 2007).

Of all the possible climate impacts on transportation, the greatest in terms of cost is that of urban flooding (IPCC 2007). In the last ten years, there have been four cases when flooding of urban underground rail systems has caused damage worth more than 10 million euros and numerous cases of lesser damage (Compton, Ermolieva, and Linnerooth-Bayer 2002). The 1996 flood in the Boston metropolitan area caused $70 million in property damage, also disrupting business and personal travel. In the New York metropolitan area, torrential rainfall events in 2004 and 2007 sent water into the subway tunnels as the drainage system could not handle the excessive rainfall (Chan 2007).

To date, only a few studies have examined potential impacts or adaptation of climate change on transportation systems. These studies include regional economic impacts as a result of changing transportation modes in northern Canada (Lonergan, Difrancesco, and Woo 1993) and flood risk mapping for vulnerable roads and the cost of travel disruption in the Boston metropolitan area (Suarez et al. 2005), in the New York metropolitan area (Jacob, Gornitz, and Rosenzweig 2007), in Maryland (Sohn 2006), and more comprehensively along the East Coast of the United States (ICF International 2008). These studies, however, used either synthetic climate change scenarios (e.g., hypothetical sea level rise) or a couple of climate change scenarios that offered only limited climate impact assessment and quantification of associated uncertainties (Wurbs, Toneatti, and Sherwin 2001). Furthermore, they did not model hydrology and geomorphology explicitly at specific intersections of streams and roads, which can vary significantly at a local scale. Only recently have researchers started to investigate potential adaptation of railways in Sweden (Lindgreen, Jonsson, and Carlsson-Kanyama 2009) and in the United Kingdom (Dobney et al. 2009).

Studies on natural hazard impacts from future climate change have struggled to adequately assess impacts. This is largely due to a lack of adequate data, difficulty in interpreting the existing multidisciplinary data, the complexity of cascading effects resulting from flooding on the regional transportation systems (NRC 1999), and the focus on attempting to model only extreme events, which are inherently more difficult to predict and model (Pielke and Downton 2000; Changnon 2003). With this study, a multidisciplinary team of scientists and urban planners focus on the cumulative effects of a range of flood events, which are likely to increase in frequency as a result of climate change. In addition, we model the short-term transportation impact from temporary flooding in a few local roadways.

Study Area

Two creeks—Johnson Creek and Fanno Creek—in the Portland metropolitan area serve as our study sites (Figure 1). These two creeks were chosen because

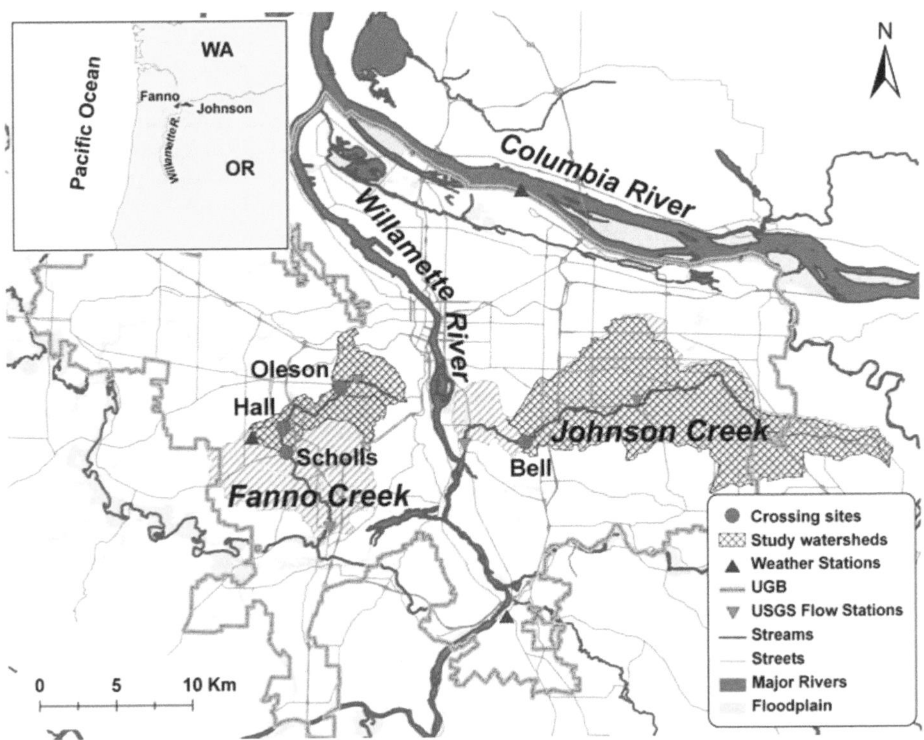

Figure 1. A map of the Portland metropolitan area and study watersheds. UGB = urban growth boundary.

they have historical flow data and exhibit high flooding potential; each also has high road density (Table 1). Daily vehicle miles of travel in Portland increased from 19.4 million to 29.2 million between 1990 and 2007 (Metro 2009). With different slopes and different degrees of urban development, the hydrological processes of the two watersheds are different (Fanno—highly urbanized and steep slopes; Johnson Creek—mixed land use with gentle slopes); each serves as a representative for other urban watersheds. Both streams are part of the Lower Willamette River watershed. The 42-km Johnson Creek stems from west of the Cascade Range and flows into the Willamette River just north of Elk Rock Island, whereas the 24-km Fanno Creek originates within the Portland city limits and drains into the Tualatin River at river mile 9.3.

Largely belonging to the marine West Coast climate, the study area exhibits wet, mild winters and dry, warm summers with annual precipitation amounts ranging from 1,000 mm at the mouth of both creeks to 1,500 mm in the headwaters. Precipitation variability is influenced by three- to five-year cycles of El Niño-Southern Oscillation events and the twenty- to thirty-year cycle of alternating phases of the Pacific Decadal Oscillation. Located at a relatively low altitude, snowmelt is not a big component of the hydrologic cycle in either watershed. Accordingly, streamflow is typically highest during winter months when flooding potentials are high (December to February). Both streams' discharge patterns exhibit the typical behavior of urban streams in the Pacific Northwest, with flashy and relatively high flow during winter rainfall periods and low flows dominated by

Table 1. Characteristics of each crossing basin

Subbasin	Area (km^2)	Mean elevation (m)	Mean slope (degrees)	Road density (km/km^2)	Land use (%) Urban	Water	Forest	Agricultural	Wetland
Oleson	12.8	158	7.75	12,399	78.30	0.00	21.61	0.08	0.00
Hall	26.7	125	5.75	11,927	83.58	0.00	15.60	0.16	0.66
Scholls	30.6	118	5.31	11,763	84.56	0.02	13.59	1.10	0.73
Bell	117.6	134	4.57	7,726	59.41	0.04	18.70	20.92	0.92

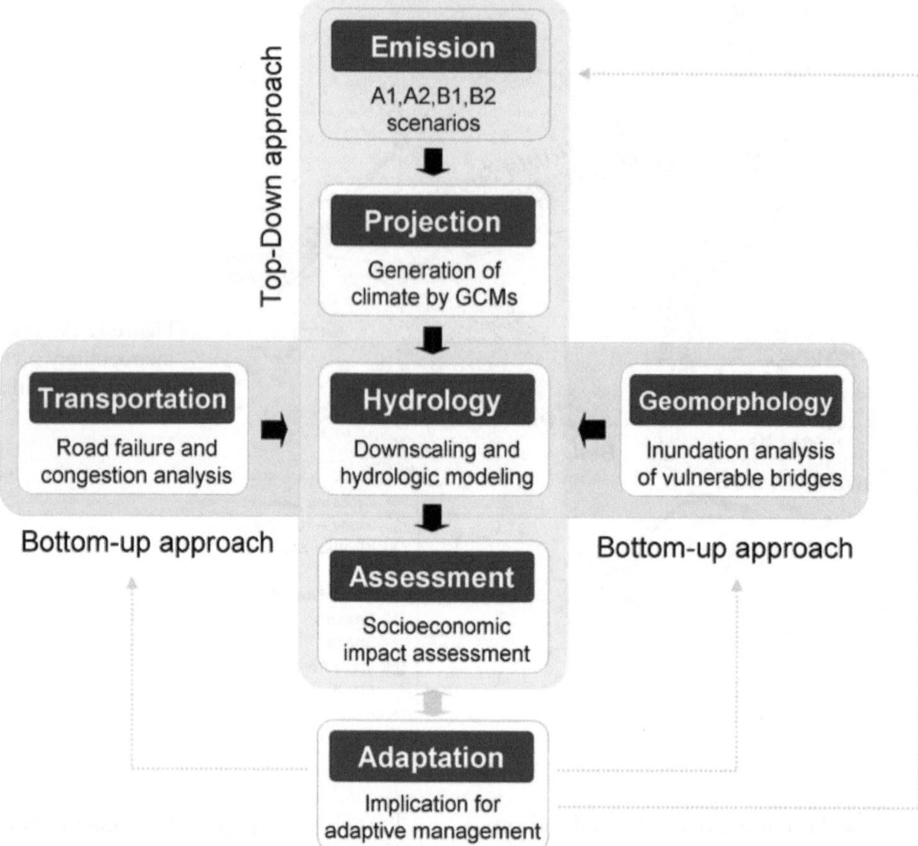

Figure 2. An integrated assessment of climate change impacts on transportation system The dotted arrows indicate feedbacks not modeled in this study. GCMs = global climate models.

groundwater discharge during the dry summer (Chang 2007).

Soil characteristics of the two watersheds are closely associated with bedrock lithology and elevation. The upper portion of Fanno Creek has moderately deep, somewhat poorly drained silt loam soils and significant areas of urban land dominated by impervious surfaces (Green 1983); the lower portion of the watershed includes deep, somewhat poorly drained and moderately well-drained silt loams (Green 1982). The upstream areas of Johnson Creek have moderately deep and somewhat poorly drained silt loams, and Lower Johnson Creek has more urban land and very deep, well-drained silt and gravelly loam soils (Gerig 1985). Soil permeability in both creeks reflects elevation gradients.

For all basins, urban land use is the dominant use, including more than 50 percent of each basin. In Fanno Creek, urban land use is increasing in downstream basins at the expense of forested areas. Because the upper portion of Johnson Creek is outside of the urban growth boundary, this basin contains a relatively high proportion of agricultural lands. However, a portion of Upper Johnson Creek was incorporated into the urban growth boundary in 2002.

Data and Methods

Integrated Impact Assessment

Figure 2 exhibits the methodological framework of the integrated assessment of climate change on urban flooding and transportation systems. This framework combines a traditional top-down impact assessment approach with a bottom-up vulnerability analysis. Developing climate change scenarios and downscaling for hydrologic impact assessment follow a traditional top-down approach. Regional stakeholders—the county transportation planner, the regional transportation group, local watershed councils, and community volunteers—were involved at the beginning of the project, informing researchers on the history of flooding and identifying vulnerable transportation nodes. They also provided feedback on our research

Table 2. Description of global climate models used in this study (Randall et al. 2007)

Model ID	Acronym	Country	Spatial resolution		Reference
			Atmosphere	Ocean	
CCSM3	CCSM3	United States	1.4° × 1.4°	1.0° × 1.0°	Collins et al. (2006)
CNRM-CM3	CNRM	France	1.9° × 1.9°	2.0° × 2.0°	Terray, Valcke, and Piacentini (1998)
ECHAM5/MPI-OM	ECHAM5	Germany	1.9° × 1.9°	1.5° × 1.5°	Jungclaus et al. (2006)
ECHO-G	ECHO-G	Germany/Korea	3.9° × 3.9°	2.8° × 2.8°	Min et al. (2005)
IPSL-CM4	IPSL	France	2.5° × 3.75°	2.0° × 2.0°	Marti et al. (2005)
MIROC3.2 (high res)	MIROC	Japan	1.1° × 1.1°	0.2° × 0.3°	K-1 Model Developers (2004)
PCM	PCM	United States	2.8° × 2.8°	0.7° × 1.1°	Washington et al. (2000)
UKMO-HadCM3	UKMO	United Kingdom	2.5° × 3.75°	1.25° × 1.25°	Gordon et al. (2000)

throughout the project period. Although assessments of climate change impacts, adaptation, and vulnerability become sophisticated (Carter et al. 2007), particularly at the regional scale (Knight and Jäger 2009), to our knowledge, this is the first attempt to apply an integrated assessment approach in the transportation sector.

Climate Scenarios

We used two different sets of climate data for climate modeling. First, observed daily precipitation and temperature from the Portland International Airport and Beaverton stations (Oregon Climate Service 2008) were used for hydrologic modeling. Second, statistically downscaled climate change data at a spatial resolution of 1/16 degree (Salathé, Mote, and Wiley 2007) from sixteen different climate simulations were used for climate impact assessment. We used the combinations of eight coupled atmosphere ocean global climate models and two greenhouse gas emission scenarios to explore uncertainty associated with global climate model (GCM) structure and greenhouse gas emission scenarios (Cameron 2006; see also Table 2). The years between 1970 and 1999 serve as a reference period, and the years between 2020 and 2049 serve as a future period representing the years around 2035.

The simulated precipitation and temperature data from the downscaled scenarios were compared with observed weather station data. When there were substantial biases in the downscaled data, these biases were corrected using quantile mapping (Wood et al. 2004). The bias-corrected data were then used as input to the hydrologic simulation model. Figure 3 shows changes in monthly precipitation and temperature for the study area. As shown in Figure 3, January and February precipitation amounts are generally projected to increase, whereas July precipitation is projected to decline regardless of emission scenarios. Temperature is projected to increase overall with higher increases under the A1B emissions scenario.

Hydrologic Modeling

The U.S. Geological Survey (USGS) Precipitation Runoff Modeling System (PRMS) model simulates runoff changes and resulting changes in flood frequency. This model has been used in climate impact assessment

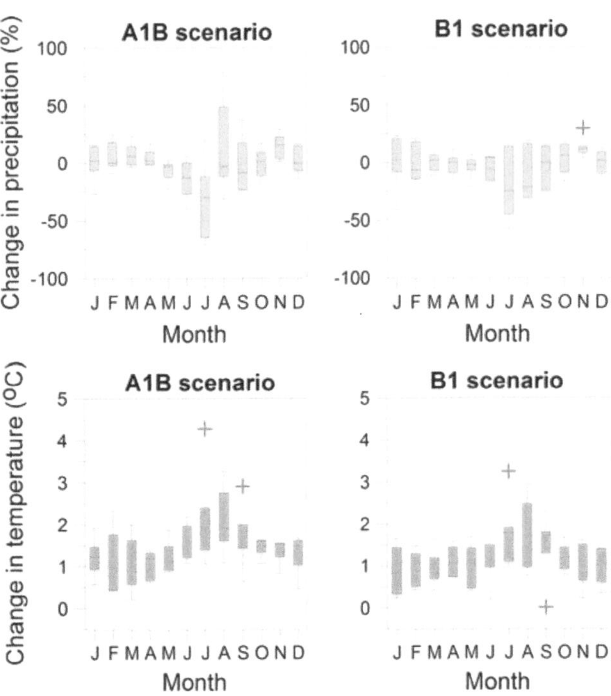

Figure 3. Changes in future (2020–2049) monthly precipitation and temperature from the reference period (1970–1999) under sixteen climate change scenarios (eight for A1B, eight for B1) for the Portland metropolitan areas. + indicates outliers that fall outside of whisker lengths.

Table 3. Precipitation Runoff Modeling System model parameters used for calibration

Parameter	Description	Range	Initial values	Calibrated values	Source
hru_elev	Mean elevation for each HRU, in feet	–300 to 30,000	—	117–602	D
hru_slope	HRU slope in decimal vertical feet/horizontal feet	0–10	—	0.01–0.09	D
cov_type	Cover type (0 = bare, 1 = grasses, 2 = shrubs, 3 = trees)	0–1	—	0–3	L
covden_sum	Summer vegetation cover density	0–1	—	0.01–0.9	L
covden_win	Winter vegetation cover density	0–1	—	0.01–0.8	L
wrain_intcp	Winter rain interception storage capacity, in inches	0–5	—	0.001–0.05	L
srain_intcp	Summer rain interception storage capacity, in inches	0–5	—	0.001–0.05	L
snow_intcp	Winter snow interception storage capacity, in inches	0–5	—	0.001–0.1	L
hru_percent_imperv	HRU impervious area, in decimal percent	0–1	—	0.1–0.6	L
soil_type	HRU soil type (1 = sand, 2 = loam, 3 = clay)	1–3	—	1–3	S
soil_moist_max	Maximum available water holding capacity in soil profile, in inches	0–20	—	5–9	S
soil_rechr_max	Maximum available water holding capacity for soil recharge zone, in inches	0–10	—	1–2	S
hamon_coef	Hamon evapotranspiration coefficient	0.004–0.008	0.0055	0.004–0.008	R
soil2gw_max	Maximum rate of soil water excess moving to ground water	0.0–5.0	0.15	0.12–0.15	R
smidx_coef	Coefficient in nonlinear surface runoff contributing area algorithm	0.0001–1.0000	0.01	0.001	R
smidx_exp	Exponent in nonlinear surface runoff contribution area algorithm	0.2–0.8	0.3	0.20–0.21	R
ssrcoef_sq	Coefficient to route subsurface storage to streamflow	0.0–1.0	0.1	0.05–0.44	R
ssrcoef_lin	Coefficient to route subsurface storage to streamflow	0.0–1.0	0.1	0.0001	R
ssr2gw_exp	Coefficient to route water from subsurface to groundwater	0.0–3.0	1.0	0.5–3.0	R
ssr2gw_rate	Coefficient to route water from subsurface to groundwater	0.0–1.0	0.1	0.006–0.02	R
gwflow_coef	Groundwater routing coefficient	0.000–1.000	0.015	0.003–0.07	R

Note: HRU = hydrological response unit; D = digital elevation map (10 m resolution); L = land use map; S = soil map; R = Rosenbrock method.

for a range of watersheds around the world (Burlando and Rosso 2002; Dagnachew, Christine, and Francoise 2003; Bae, Jung, and Chang 2008). PRMS uses observed daily mean precipitation and maximum and minimum temperature to simulate daily streamflow. PRMS is a semidistributed, physically based watershed model, based on hydrologic response units (HRUs; Leavesley et al. 2002), which is ideal for simulating changes in flow under different environmental scenarios, including climate change. HRUs, assumed to be homogeneous with respect to hydrologic response to climate condition, are partitioned by using a combination of slope, aspect, land use, soil type, and geology.

PRMS model parameters are calibrated focusing on sensitive parameters, which are based on the literature (Laenen and Risley 1997). Some parameters are directly estimated from the measurable basin characteristics in geographic information system (GIS) layers (see Table 3). The remaining parameters (shaded in Table 3) are estimated by Rosenbrock's (1960) automatic optimization method. PRMS is calibrated and verified for the four gauged sites for the period between 1988 and 2006 (see Table 4). The widely used Nash and Sutcliffe nondimensional model efficiency criterion was used to evaluate the performance of PRMS. The values in excess of 0.6 indicated a satisfactory fit between observed and simulated hydrographs (Wilby and Harris 2006). We applied the calibrated model for the ungauged cross-section survey sites using the regionalization method (Wagener and Wheater 2006).

Additionally, we used the Web-based GIS application StreamStats, which was developed by the USGS

Table 4. Precipitation Runoff Modeling System model performance for calibration and verification period

Catchment (area, km^2)	Calibration period (verification period)	r	RMSE	NSE
Upper Fanno (6.1)	2001–2003 (2003–2006)	0.84 (0.86)	1.31 (1.26)	0.69 (0.74)
Lower Fanno (80.5)	2001–2003 (2003–2006)	0.87 (0.80)	1.26 (1.67)	0.74 (0.62)
Upper Johnson (68.3)	1988–1998 (1999–2006)	0.82 (0.85)	1.73 (1.56)	0.66 (0.65)
Lower Johnson (132.1)	1988–1998 (1999–2006)	0.83 (0.81)	1.35 (1.17)	0.64 (0.62)

Note: RMSE = root mean square error = $\sqrt{\sum (O_i - S_i)^2/n}$, where n is a number of data; r = correlation coefficient (R) = $SS_{os}/\sqrt{SS_o \times SS_s}$, $SS_{os} = \sum (O_i - \overline{O})(S_i - \overline{S})$, $SS_o = \sum (O_i - \overline{O})^2$, $SS_s = \sum (S_i - \overline{S})^2$, where O is observed flow and S is simulated flow; NSE = Nash–Sutcliffe efficiency = $[\sum (O_i - \overline{O})^2 - \sum (O_i - S_i)^2] / \sum (O_i - \overline{O})^2$, where $\overline{O}$ is mean observed flow.

(http://water.usgs.gov/osw/streamstats/index.html), to determine the two-year, five-year, ten-year, twenty-five-year, fifty-year, and hundred-year discharge values for each of the five surveyed crossing locations. StreamStats uses a regional regression analysis to calculate the discharge for subbasins and an area-to-discharge relationship to determine the discharge to a user-selected location on a stream. The regression equations are specific to a particular region; our study areas fall completely within Oregon flood region 2b, which includes all basins between the crests of the Cascade and Coast Ranges with a mean basin elevation of less than 3,000 feet (Cooper 2005).

Channel Survey and Hydraulic Analysis

To evaluate future flooding impacts at actual road crossings identified by stakeholders, we conducted stream channel surveys at five different road crossings and conducted a hydraulic analysis with Hydrologic Engineering Centers River Analysis System (HEC-RAS), which has been used to find water surface elevations for historical floods (Benito, Diez-Herrero, and de Villalta 2003) and to model water flowing through bridges and culverts (Hotchkiss et al. 2008). With this approach, we determined the discharge value necessary to produce roadway flooding at each road crossing. We surveyed and modeled road flooding at five different bridge sites. There are three sites on Fanno Creek—Oleson Road, upstream Hall Boulevard, and the downstream Scholls Ferry Road crossing—and two sites on Johnson Creek—the Bell Avenue and Linwood Avenue crossings. Based on daily vehicle miles traveled, each road is classified as either an arterial or major arterial and, except for Oleson Road, also serves as a bus line. Four of the five locations have a history of road flooding during large storm events (Oregonian 2007). The Scholls Ferry Road bridge is a relatively large span with no history of flooding; this site was selected as a probable example of a correctly sized structure with respect to climate change–induced flooding. All sites are located between USGS gauging stations.

We collected the channel geometry data by following stream channel reference site protocol for wadeable streams (Harrelson, Rawlins, and Potyondy 1994). We surveyed four cross sections at each site with two upstream and two downstream of the bridge and captured the bridge geometry relative to the cross sections. At each cross section we noted the elevation at the top of the banks and the water surface elevation; all cross sections and bridge information were tied to a common datum to model water flowing through the reach. Using a flow meter, we measured stream velocity at the furthest upstream cross section and, for the purpose of calibrating the models, estimated the roughness (Manning's n) at each cross section (Chow 1959). In HEC-RAS we conducted a combined steady flow analysis using the bridge routine calibrated to our known water surface elevations for each cross section.

Transportation Impacts Methodology

To measure the potential disruption and costs of flooding on the transportation system, we used the Metropolitan Planning Organization's (MPO) four-step model and multimodal equilibrium traffic assignment procedures. The model produces current (2005) and future (2035) travel volumes based on growth, land use, and transportation assumptions in each time period. Household and employment estimates are assigned to a transportation analysis zone (TAZ), the "unit geography" for travel within the demand model. All the trips generated by the land use elements are aggregated and analyzed at the TAZ level. The travel model estimates the number of trips that will be made, the distribution patterns of the trips throughout the region, the likely mode used for the trip, and the actual roadways and transit lines used for auto, truck, and transit trips. Trip assignment applies the Frank–Wolfe algorithm (1956) to determine equilibrium flows in the transportation network. Trip paths are assigned between origins and destinations based on capacity, volume, and speed; trips will take the shortest and quickest path. Disruption on the path, such as flooding, will redistribute flows to other routes.

We acknowledge that the MPO's four-step travel forecast model provides a coarser level of output than is ideal for this type of research. Advance assignment methods such as dynamic traffic assignment (DTA) and microsimulation exist that more effectively capture traffic flow diurnal characteristics, the effects of queuing, and the duration of congestion that might result from a flooding event (Mahmassani 2001; Peeta and Ziliaskopoulos 2001). However, for the purpose of the study, we assumed that flooding events are forecasted and drivers have full information regarding the closure of the road crossings. Metro, the MPO, is currently developing DTA capabilities for use in exactly this type of application. These methods are more fully described in the NRC (2007) report.

The traffic analysis began with the identification of transportation network links that are expected to flood based on the findings from the climate and hydrological analysis. It is important to note that the analysis focused on route diversion only and did not alter origin–destination pattern. Initial one-hour midday and two-hour afternoon peak traffic assignments were run using Equilibre Multimodal/Multimodal Equilibrium software to establish baseline traffic volumes and link volume and capacity for 2005 and 2035 (Figure 4). Traffic assignments were then rerun with the flooded network links removed for both the Fanno Creek and Johnson Creek study areas. Using this output, a flood area of influence, comprised of TAZ clusters, was identified for each study area. Transportation evaluation measures were produced for each flood area of influence that included vehicle miles traveled, vehicle hours, and vehicle hours of delay for the one-hour midday and two-hour afternoon peak travel periods.

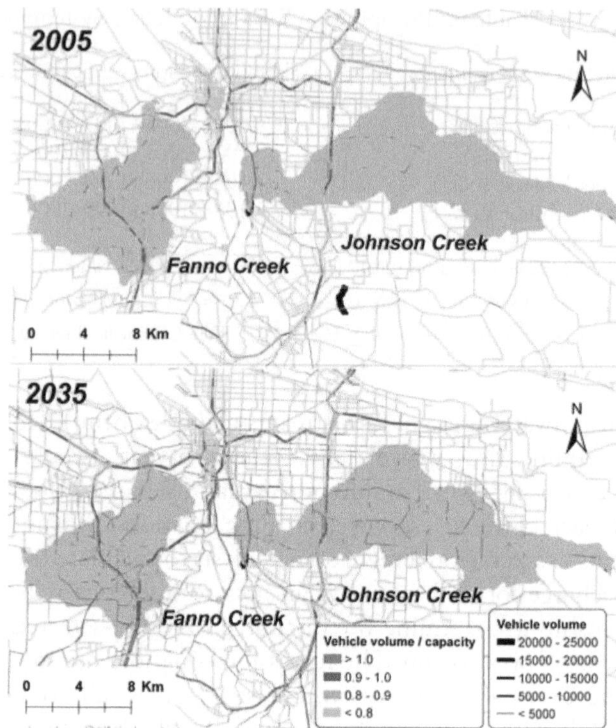

Figure 4. Changes in auto traffic volumes in the study area for 2005 and 2035.

Results and Discussion

Changes in Flood Frequency

We estimated the changes in flood frequency with different recurrence intervals by using the PeakFQ program developed by the USGS (Flynn, Kirby, and Hummel 2006). The PeakFQ provides estimates of instantaneous annual-maximum peak flows with recurrence intervals of two, five, ten, twenty-five, fifty, and one hundred years based on flood-frequency analyses recommended in Bulletin 17B (Interagency Advisory Committee on Water Data 1982). With this method, we constructed annual peak flows using simulated daily discharge values by the PRMS model for 1970 through 1999 and for 2020 through 2049 at the four study areas.

Recurrence flood flows for the reference period demonstrate that the simulated results by the PRMS using the downscaled climate simulations agree well with estimated flood flows from the USGS StreamStats (Figure 5). In particular, the below twenty-five-year recurrence flood flows more closely match with the USGS StreamStats results. Most inundated flood flows (dashed lines) occur between the ten-year and twenty-five-year recurrence flood flows except at the Scholls Ferry site. The over fifty-year recurrence flood flows are highly affected by the different GCM structures; however, the emission scenario impacts reveal relatively fewer differences than the GCM structure.

The recurrence flood flows for 2020 through 2049 represent the differences in the direction of change according to recurrence interval and the GCM simulations. As shown in Figure 6, ensemble averaged two-year, five-year, and ten-year recurrence flood flows

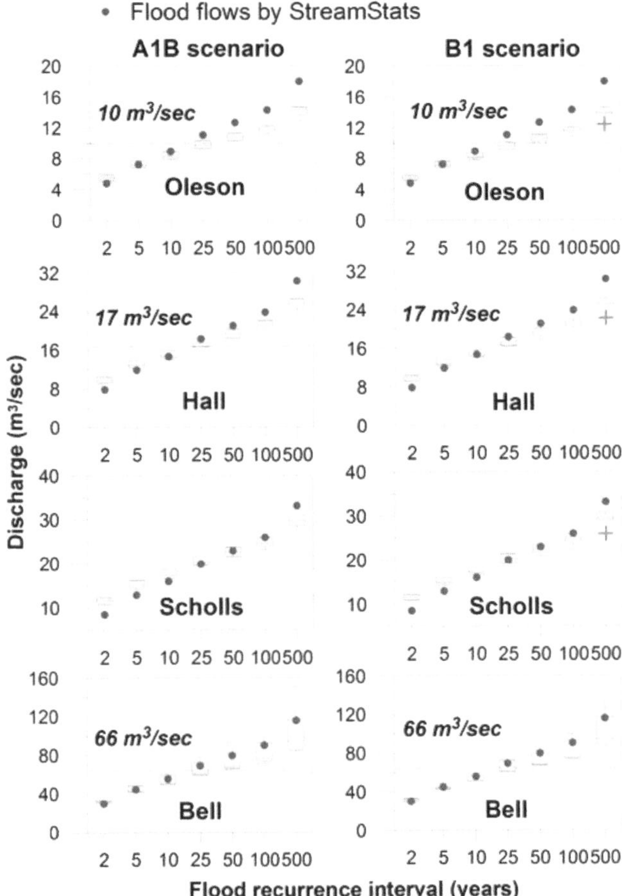

Figure 5. Comparison of estimated flood flows by USGS StreamStats and simulated flood flows by the Precipitation Runoff Modeling System model using sixteen climate change scenarios (eight for A1B, eight for B1) for each subbasin from 1970 to 1999. + indicates outliers that fall outside of whisker lengths.

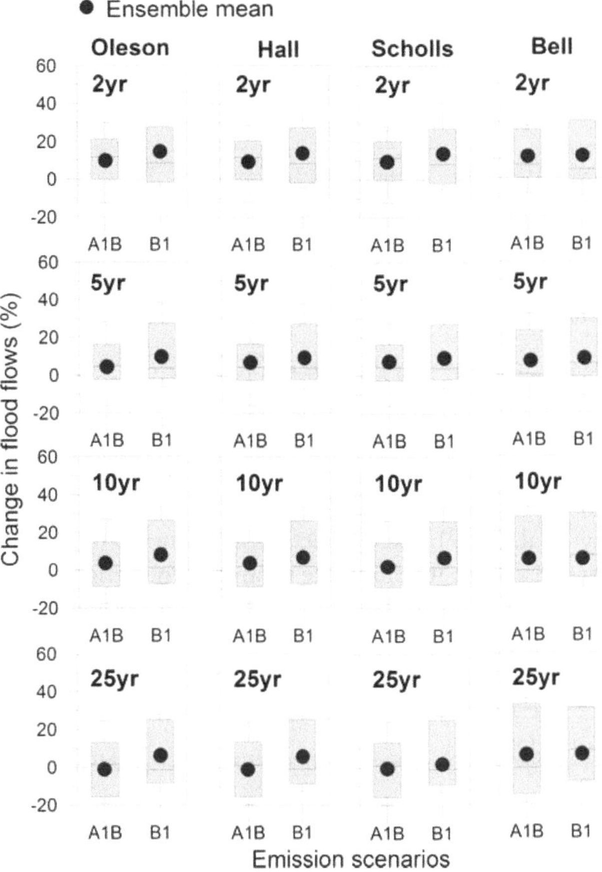

Figure 6. Changes in flood flows with recurrence flood intervals for two, five, ten, and twenty-five years under A1B and B1 scenarios.

at all sites increase, whereas the ensemble mean flood flow for twenty-five-year recurrence interval does not change under the A1B scenario. It does, however, increase under some climate change scenarios, suggesting that the magnitude and directions of change in the higher flood flows in the study areas are more affected by GCM structures than the emission scenarios.

Figure 7 shows the change in days that exceeded the two-year recurrence flood flows. These values show approximately thirty to sixty days for the reference period and thirty to eighty days for 2020 through 2049. Although the mean of the exceedance days increases at all sites, the GCMs and the emission scenarios remain as the main source of uncertainty, with higher uncertainty associated with the choice of GCM. Our finding is consistent with other previous studies. For example, the largest source of uncertainty was the GCM structure in U.K. catchments (Wilby and Harris 2006;

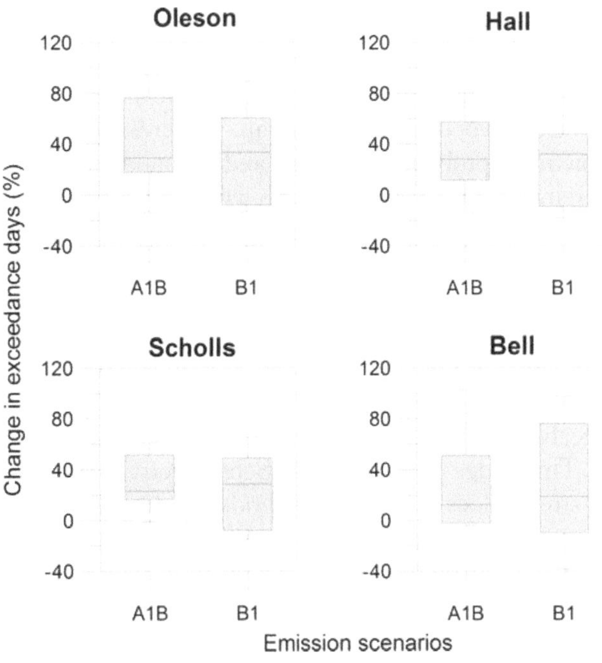

Figure 7. Changes in exceedance days with total days exceeding two-year recurrence flood interval for each subbasin.

Table 5. Exceedance probability of the threshold flood discharge under high, ensemble mean, and low climate change scenarios

Climate scenarios	Oleson		Hall		Bell	
	Ref	Future	Ref	Future	Ref	Future
A1B						
High	0.050	0.100	0.070	0.120	0.080	0.120
Mean	0.041	0.041	0.051	0.050	0.042	0.050
Low	0.025	0.002	0.035	0.005	0.020	0.002
B1						
High	0.050	0.130	0.060	0.180	0.070	0.100
Mean	0.040	0.038	0.050	0.060	0.042	0.051
Low	0.018	0.002	0.030	0.010	0.025	0.017

Kay et al. 2009). Exceedance probability of the inundated discharge for ensemble mean does not change remarkably (Table 5). However, the differences between high and low scenarios remain high, indicating that the quantification of uncertainty resulting from climate impact assessment is key in supporting new transportation system design.

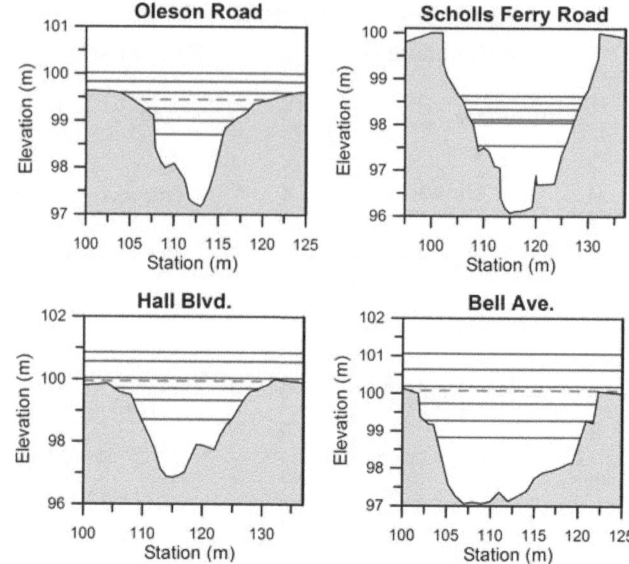

Figure 8. Stream channel cross sections with solid horizontal lines indicating the water surface elevations for, from bottom to top, the two-year, five-year, ten-year, twenty-five-year, fifty-year, and hundred-year events. The dashed line indicates the water surface elevation when water first begins flooding the road; this never occurs at Scholls Ferry Road. Linwood Avenue is nearly identical to Bell Avenue, so the stream channel cross section is not reported here.

Changes in the Probability of Road Flooding

The HEC-RAS model output shows that all cross sections with the exception of Scholls Ferry Road are inundated during the current twenty-five-year flood event (Figure 8). However, the crossings diverge somewhat for smaller events. The Oleson Road crossing floods much more frequently; this crossing area has an active floodplain upstream of the bridge but is controlled downstream by a wooden wall and riprap. This crossing is well known as a problem flood area, and our modeling simply reinforces the frequency with which this bridge can become impassable due to flooding. The Hall Boulevard crossing is flooded during less than a twenty-five-year magnitude event; this crossing has a much more extensive floodplain than the Oleson Road crossing and the channel is not constricted other than when passing through the bridge. Yet, the bridge opening itself is not large; hence, this road is subject to fairly frequent flooding. This bridge is crowned, as is Scholls Ferry Road, and the stream does not cover the bridge during any flood event; however, water does flow across the road in the floodplain during high water, which leads to closure of this crossing. With a large floodplain and a large bridge opening, the Scholls Ferry Road crossing does not flood at any discharge; however, the bike path adjacent to the stream that goes under the bridge is often inundated.

The bridge openings in Johnson Creek are much larger than most bridge openings in Fanno Creek, a legacy of channelization in Johnson Creek. Yet, each crossing site floods just prior to a twenty-five-year event. Like the Hall Boulevard site, these roadways begin flooding before water overtops the bridge itself. In the case of the Bell Avenue crossing, water will actually flow north of the stream channel, through a parking lot, and across the road approximately 3 m north of the bridge itself; hence, this road is closed more frequently than would be expected by our models.

Our results show that although floods depend on precipitation intensity, volume, and timing, they also rely on drainage basin and local geomorphic characteristics. However, given the predicted increases in flood return exceedance for smaller floods (Figure 6), nuisance flooding is likely to become more common at these cross sections. As illustrated in this study, restoration of floodplains will serve as a proactive adaptation strategy in reducing flood damage under a changing climate. In addition, future flooding potential could be further reduced as best management practices such as porous pavement or detention ponds are implemented. In particular, older neighborhoods (e.g., Lower Johnson Creek) would be good candidate areas for implementing these practices.

Impacts on Transportation Network

The Fanno Creek and Johnson Creek flood areas of influence generate an estimated 973,000 and 541,000 vehicle miles traveled (VMT), respectively, in the two-hour afternoon peak travel period. Together, these areas account for 24 percent of the total VMT generated in the 4:00 p.m. to 6:00 p.m. travel period. They are located in suburban locations where the arterial street network is fairly complete but the local street network is often discontinuous. Table 6 lists the street links prone to flooding and the average traffic volume they carry in the two-hour afternoon peak travel period for the base year 2005 and future year 2035.

An evaluation of the travel model output for the Fanno Creek and Johnson Creek flooding areas of influence forecasted negligible increases (less than 1 percent) in VMT in both travel periods for 2005 and 2035 when flooded links were removed from the street network (Figure 9). Vehicle hours delay (VHD) demonstrates a greater impact from flooding. Closing facilities in the Fanno Creek travel shed causes more than 200 additional hours of delay in the afternoon peak period. The base and future years show similar increases in hours, but the percentage increase is much higher for 2005. The Johnson Creek travel shed's increase ranges from 4 percent in the base year to 3.4 percent in the future. The impacts of closing these facilities due to flooding are minimal on overall miles traveled, but they do reveal that other roadways will be more congested, resulting in greater delays because the diverting vehicles and existing vehicles are all affected.

Table 6. Changes in average traffic volume as a result of road closures for 2005 and 2035

	Two-hour afternoon peak			
Facility	2005		2035	
SW Oleson Rd.	900 NB	1,200 SB	1,000 NB	1,400 SB
SW Hall Blvd.	3,800 NB	2,800 SB	4,400 NB	3,300 SB
SE Johnson Creek Blvd.	1,800 EB	1,600 WB	2,800 EB	2,300WB
SE Bell St.	400 NB	600 SB	500 NB	800 SB
SE Linwood St.	1,550 NB	1,600 SB	1,550 NB	1,750 SB

Note: NB = northbound travel; SB = southbound travel; EB = eastbound travel; WB = westbound travel.

The region-wide modeled transportation network provides several alternative routes to the flooded links. The macrolevel equilibrium traffic assignment assumes that travelers have perfect knowledge of the road conditions (without any incidents) and know which additional routes are available in time to make an informed decision about the path they will take. Therefore, the

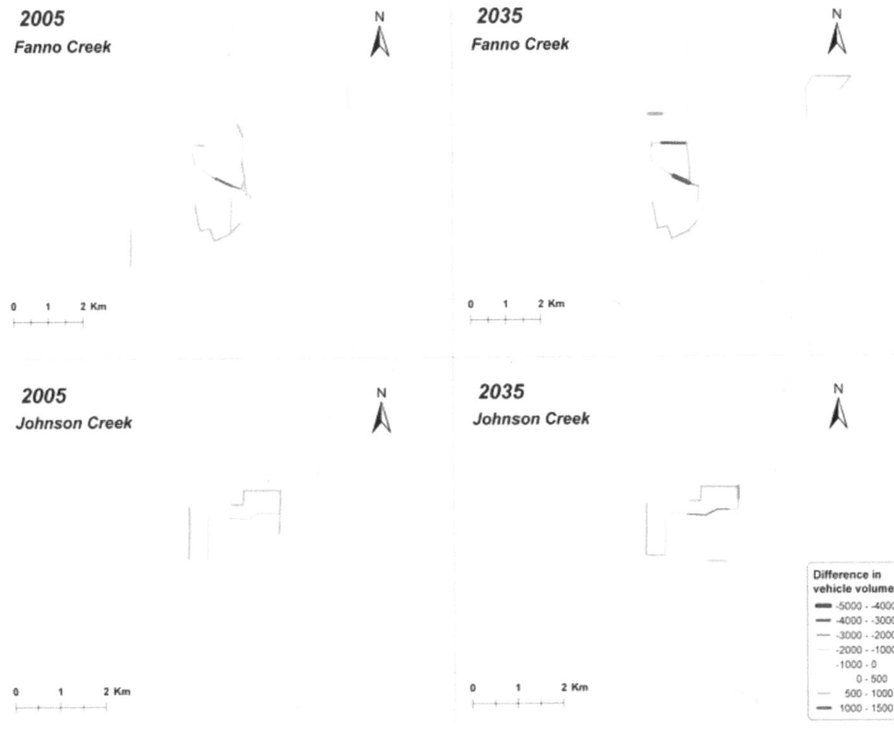

Figure 9. Changes in travel disruption as a result of road closure, 2005 and 2035 for Fanno Creek and Johnson Creek.

resulting statistics might underestimate the level of out-of-direction travel that would contribute to an increase in VMT during a short-term flooding event. Our findings are similar to other studies that show that the cost of delays and lost trips would be relatively small compared with damage to the infrastructure and to other property (Suarez et al. 2005; Kirshen, Knee, and Ruth 2008). This is due to the relatively large number of transportation networks and trips, a typical characteristic in mature metropolitan areas.

Conclusions

Global climate change will have significant impacts, particularly in urban areas where many socioeconomic activities are concentrated (Chang and Franczyk 2008). Many growing urban areas such as the U.S. Pacific Northwest will experience higher amounts and intensity of winter precipitation. In projecting future flood frequency, there are high uncertainties associated with GCM structure and emission scenarios. Despite this uncertainty, the five-year, ten-year, and other relatively small events are likely to increase in all study sites that have a history of chronic flooding. Stream channels will likely lag in adjusting to the new, slowly increasing discharge regimes and might not be able to adjust at bridge locations, which is likely to only further exacerbate roadway flooding. Although VMT in both periods show negligible increases, VHD demonstrated a greater impact from flooding. The estimates are, however, conservative, as the current approach assumes travelers' perfect knowledge regarding the closure of the road crossings.

Our results show that there is a nonlinear relation between precipitation change and urban flooding and that impacts on travel disruption are subject to local hydroclimate and geomorphic conditions. Although it is a specific case study, the integrated methodology used in this study can be applied to other urban areas facing similar transportation impact in a changing climate. If climate change, watershed hydrology, channel morphology, and transportation networks are readily available in a spatial database, it could even be possible to upscale a similar model to regional or national levels. However, the complex interactions of a changing precipitation regime, adjustments in channel morphology, and human response and adaptation to flooding still need to be further investigated. Both natural and human systems coevolve over time with complex feedbacks at multiple spatial and temporal scales in response to external changes. Thus, further research should include continuous monitoring of the system and the development of coupled system dynamics models.

Despite some of these limitations, this study is one of the few interdisciplinary attempts to assess potential impacts of climate change on the transportation sector. The integration of the top-down and the bottom-up approaches involving local stakeholders at the beginning of the project demonstrates a useful tool to assess climate change impacts at a local scale. A participatory regional integrated assessment tool found in other sectors such as water resources and agriculture (e.g., Holman et al. 2008) could be adapted for the transportation sector. Such integrated knowledge and spatially explicit modeling are essential for establishing proactive flood and transportation management planning and policies under increasing climate uncertainty.

Acknowledgments

This research was supported by the Oregon Transportation Research and Education Consortium (Grant number 2009–257). Additional support for Il-Won Jung was provided by the James F. and Marion L. Miller Foundation's postdoctoral fellowship in sustainability. We thank Eric Salathé of the University of Washington for providing downscaled climate change scenarios for the study area. We also appreciate Joe Marek of Clackamas County, who provided valuable insights throughout the project period. Many local volunteers helped with the stream channel survey for our study sites; in particular we wish to thank Brian Wegener, Tualatin Riverkeepers, and Greg Ciannella, Johnson Creek Watershed Council, for helping to organize the volunteer effort. The article was greatly improved by the constructive comments of two anonymous reviewers.

References

Bae, D. H., I. W. Jung, and H. Chang. 2008. Potential changes in Korean water resources estimated by high-resolution climate simulation. *Climate Research* 35:213–26.

Benito, G., A. Diez-Herrero, and M. F. de Villalta. 2003. Magnitude and frequency of flooding in the Tagus basin (Central Spain) over the last millennium. *Climatic Change* 58:171–92.

Black, W. R., and N. Sato. 2007. From global warming to sustainable transport 1989–2006. *International Journal of Sustainable Transportation* 1:73–89.

Burlando, P., and R. Rosso. 2002. Effects of transient climate change on basin hydrology: 1. Precipitation scenarios for the Arno River, central Italy. *Hydrological Processes* 16:1151–75.

Cameron, D. 2006. An application of the UKCIP02 climate change scenarios to flood estimation by continuous simulation for a gauged catchment in the northeast of Scotland, UK (with uncertainty). *Journal of Hydrology* 328:212–26.

Carter, T. R., R. N. Jones, X. Lu, S. Bhadwal, C. Conde, L. O. Mearns, B. C. O'Neill, M. D. A. Rounsevell, and M. B. Zurek. 2007. New assessment methods and the characterisation of future conditions. In *Climate change 2007: Impacts, adaptation and vulnerability. Contribution of Working Group II to the Fourth Assessment Report of the Intergovernmental Panel on Climate Change*, ed. M. L. Parry, O. F. Canziani, J. P. Palutikof, P. J. van der Linden, and C. E. Hanson, 133–71. Cambridge, UK: Cambridge University Press.

Chan, S. 2007. Flooding cripples subway system. *New York Times* 8 August. http://cityroom.blogs.nytimes.com/2007/08/08/flooding-cripples-subway-system/ (last accessed 1 July 2010).

Chang, H. 2007. Streamflow characteristics in urbanizing basins in the Portland metropolitan area, Oregon, USA. *Hydrological Processes* 21 (2): 211–22.

Chang, H., and J. Franczyk. 2008. Climate change, land cover change, and floods: Toward integrated assessments. *Geography Compass* 2 (5): 1549–79.

Changnon, S. D. 2003. Measures of economic impacts of weather extremes. *Bulletin of the American Meteorological Society* 54:1231–35.

Chapman, L. 2007. Transport and climate change: A review. *Journal of Transport Geography* 15:354–67.

Chow, V. T. 1959. *Open-channel hydraulics*. New York: McGraw-Hill.

Collins, W. D., C. M. Bitz, M. L. Blackmon, G. B. Bonan, C. S. Bretherton, J. A. Carton, P. Chang et al. 2006. The community climate system model, version 3 (CCSM3). *Journal of Climate* 19:2122–43.

Compton, K., T. Ermolieva, and J. C. Linnerooth-Bayer. 2002. Integrated flood risk management for urban infrastructure: Managing the flood risk to Vienna's heavy rail mass rapid transit system. In *Proceedings of the Second Annual International IASA-DPRI meeting: Integrated disaster risk management: Megacity vulnerability and resilience*. Laxenburg, Austria: International Institute for Applied Systems Analysis. http://www.iiasa.ac.at/Research/RMS/dpri2002/Papers/Compton.pdf (last accessed 18 June 2010).

Cooper, R. M. 2005. Estimation of peak discharges for rural, unregulated streams in Western Oregon. Scientific Investigations Report 2005–5116. Reston, VA: U.S. Geological Survey.

Dagnachew, L., V. C. Christine, and G. Francoise. 2003. Hydrological response of a catchment to climate and land use changes in tropical Africa: Case study south central Ethiopia. *Journal of Hydrology* 275:67–85.

Dobney, K., C. J. Baker, A. D. Quinn, and L. Chapman. 2009. Quantifying the effects of high summer temperatures due to climate change on buckling and rail related delays in South-east United Kingdom. *Meteorological Applications* 16:254–61.

Equilibre Multimodal/Multimodal Equilibrium, Version 3.0. Quebec, Canada: INRO.

Flynn, K. M., W. H. Kirby, and P. R. Hummel. 2006. User's manual for program PeakFQ, annual flood-frequency analysis using Bulletin 17B guidelines. Techniques and Methods 4-B4. Reston, VA: U.S. Geological Survey.

Frank, M., and P. Wolfe. 1956. An algorithm for quadratic programming. *Naval Research Logistics Quarterly* 3:95–110.

Gerig, A. J. 1985. *Soil survey of Clackamas County, Oregon*. Portland, OR: U.S. Department of Agriculture, Soil Conservation Service, and Forest Service.

Gordon, C., C. Cooper, C. A. Senior, H. Banks, J. M. Gregory, T. C. Johns, J. F. B. Mitchell, and R. A. Wood. 2000. The simulation of SST, sea ice extents and ocean heat transports in a version of the Hadley Centre coupled model without flux adjustments. *Climate Dynamics* 16:147–68.

Green, G. L. 1982. *Soil survey of Washington County, Oregon*. Portland, OR: U.S. Department of Agriculture, Soil Conservation Service, and Forest Service.

———. 1983. *Soil survey of Multnomah County, Oregon*. Portland, OR: U.S. Department of Agriculture, Soil Conservation Service, and Forest Service.

Harrelson, C. C., C. L. Rawlins, and J. P. Potyondy. 1994. Stream channel reference sites: An illustrated guide to field technique. General Technical Report RM-245. Fort Collins, CO: U.S. Department of Agriculture, Forest Service, Rocky Mountain Forest and Range Experiment Station.

Holman, I. P., M. D. A. Rounsevell, G. Cojacaru, S. Shackley, C. McLachlan, E. Audsley, P. M. Berry, et al. 2008. The concepts and development of a participatory regional integrated assessment tool. *Climatic Change* 90: 5–30.

Hotchkiss, R. H., E. A. Thiele, E. J. Nelson, and P. L. Thompson. 2008. Culvert hydraulics comparison of current computer models and recommended improvements. *Transportation Research Record* 2060:141–49.

Huntington, T. 2006. Evidence for intensification of the global water cycle: Review and synthesis. *Journal of Hydrology* 319:83–95.

ICF International. 2008. The potential impacts of global sea level rise on transportation infrastructure. Fairfax, VA: ICF International. http://www.bv.transports.gouv.qc.ca/mono/0965210.pdf (last accessed 12 July 2009).

Interagency Advisory Committee on Water Data. 1982. Guidelines for determining flood-flow frequency. Bulletin 17B of the Hydrology Subcommittee, Office of Water Data Coordination, U.S. Geological Survey, Reston, VA. http://water.usgs.gov/osw/bulletin17b/bulletin_17B.html (last accessed 3 June 2009).

Intergovernmental Panel on Climate Change (IPCC). 2007. *Climate change 2007: The physical science basis. Contribution of Working Group I to the Fourth Assessment Report of the Intergovernmental Panel on Climate Change*, ed. S. Solomon, D. Qin, M. Manning, Z. Chen, M. Marquis, K. B. Averyt, M. Tignor, and H. L. Miller. Cambridge, UK: Cambridge University Press.

Jacob, K., V. Gornitz, and C. Rosenzweig. 2007. Vulnerability of the New York City metropolitan area to coastal hazards, including sea level rise: Inferences for urban coastal risk management and adaptation policies. In *Managing coastal vulnerability*, ed. L. McFadden, R. Nicholls, and E. Penning-Roswell, pp. 141–58. Oxford, UK: Elsevier.

Jungclaus, J. H., M. Botzet, H. Haak, N. Keenlyside, J. J. Luo, M. Latif, J. Marotzke, U. Mikolajewicz, and E. Roeckner. 2006. Ocean circulation and tropical variability in the coupled model ECHAMS/MPI-OM. *Journal of Climate* 19 (16): 3952–72.

K-1 Model Developers. 2004. K-1 coupled GCM (MIROC) description. K-1 technical report, no. 1. Ed. H. Hasumi and S. Emori. Center for Climate System Research, University of Tokyo, National Institute for Environmental Studies, and the Frontier Research Center for Global Change. http://www.ccsr.u-tokyo.ac.jp/kyosei/hasumi/MIROC/tech-repo.pdf (last accessed 1 July 2010).

Kay, A. L., H. N. Davies, V. A. Bell, and R. G. Jones. 2009. Comparison of uncertainty sources for climate change impacts: Flood frequency in England. *Climatic Change* 92:41–63.

Kirshen, P., K. Knee, and M. Ruth. 2008. Climate change and coastal flooding in metro Boston: Impacts and adaptation strategies. *Climatic Change* 90:453–73.

Knight, C. G., and J. Jäger, eds. 2009. *Integrated regional assessment of global climate change*. New York: Cambridge University Press.

Koetse, M. J., and P. Rietveld. 2009. The impact of climate change and weather on transport: An overview of empirical findings. *Transportation Research Part D-Transport and Environment* 14:205–21.

Laenen, A., and J. C. Risley. 1997. Precipitation-runoff and streamflow routing models for the Willamette River Basin, Oregon. Water-Resources Investigation Report 95–4284. Reston, VA: U.S. Geological Survey.

Leavesley, G. H., S. L. Markstrom, P. J. Restrepo, and R. J. Viger. 2002. A modular approach to addressing model design, scale, and parameter estimation issues in distributed hydrological modelling. *Hydrological Processes* 16 (2): 173–87.

Lindgreen, J., D. K. Jonsson, and A. Carlsson-Kanyama. 2009. Climate adaptation of railways: Lessons from Sweden. *EJTIR* 9 (2): 164–81.

Lonergan, S., R. Difrancesco, and M.-K. Woo. 1993. Climate change and transportation in northern Canada: An integrated impact assessment. *Climatic Change* 24: 331–51.

Mahmassani, H. S. 2001. Dynamic network traffic assignment and simulation methodology for advanced system management applications. *Networks and Spatial Economics* 1 (3): 267–92.

Marti, O., P. Bracommot, J. Bellier, R. Benshila, S. Bony, P. Brockman, P. Cadulle et al. 2005. The new IPSL climate system model: IPSL-CM4. Institut Poerre Simon Laplace des Sciences de l'Environnement Global. http://www.ipsl.jussieu.fr/poles/Modelisation/NotesScience/note26.pdf (last accessed 1 July 2010).

Metro. 2009. Daily vehicle miles of travel (DVMT) for Portland. http://www.oregonmetro.gov/index.cfm/go/by.web/id=16340 (last accessed 7 July 2009).

Min, S. K., S. Legutke, A. Hense, and W. T. Kwon. 2005. Internal variability in a 1000-year control simulation with the coupled climate model ECHO-G. Part I. Near-surface temperature, precipitation and sea level pressure. *Tellus* 57A: 605–21.

National Research Council (NRC). 1999. The costs of natural disasters: A framework for assessment. Washington, DC: National Academy Press.

———. 2007. Metropolitan travel forecasting: Current practice and future direction. Transportation Research Board Special Report 288, National Academies Press, Washington, DC.

———. 2008. Transportation Research Board Special Report 290, National Academies Press, Washington, DC.

Oregon Climate Service. 2008. Station climate data, Zone 2 http://www.ocs.orst.edu/index.html (last accessed 2 October 2008).

Oregonian. 2007. Johnson Creek floods, closing nearby roads. http://blog.oregonlive.com/breakingnews/2007/12/johnson_creek_reaches_flood_st.html (last accessed 7 July 2009).

Peeta, S., and A. K. Ziliaskopoulos. 2001. Foundations of dynamic traffic assignment: The past, the present and the future. *Networks and Spatial Economics* 1 (3): 233–65.

Pielke, R. A., and M. W. Downton. 2000. Precipitation and damaging floods: Trends in the United States, 1932–97. *Journal of Climate* 13:3625–37.

Randall, D. A., R. A. Wood, S. Bony, R. Colman, T. Fichefet, J. Fyfe, V. Kattsov, et al. 2007. Climate models and their evaluation. In *Climate change 2007: The physical science basis. Contribution of Working Group I to the Fourth Assessment Report of the Intergovernmental Panel on Climate Change*, ed. S. Solomon, D. Qin, M. Manning, Z. Chen, M. Marquis, K. B. Averyt, M. Tignor et al., 589–662. New York: Cambridge University Press.

Rosenbrock, H. H. 1960. An automatic method of finding the greatest or least value of a function. *Computer Journal* 3:175–84.

Salathé, E. P., P. W. Mote, and M. W. Wiley. 2007. Review of scenario selection and downscaling methods for the assessment of climate change impacts on hydrology in the United States Pacific Northwest. *International Journal of Climatology* 27:1611–21.

Sohn, J. 2006. Evaluating the significance of highway network links under the flood damage: An accessibility approach. *Transportation Research Part A* 40: 491–506.

Suarez, P., W. Anderson, V. Mahal, and T. R. Lakshmanan. 2005. Impacts of flooding and climate change on urban transportation: A systemwide performance assessment of the Boston Metro Area. *Transportation Research Part D: Transport and Environment* 10 (3): 231–44.

Terray, L., S. Valcke, and A. Piacentini. 1998. OASIS 2.2 user's guide and reference manual, TR/CMGC/98-05. Available at CERFACS, 42 ave. G. Coriolis, 31057 Toulouse, France.

Wagener, T., and H. S. Wheater. 2006. Parameter estimation and regionalization for continuous rainfall-runoff models including uncertainty. *Journal of Hydrology* 320: 132–54.

Washington, W. M., J. W. Weatherly, G. A. Meehl, A. J. Semtner, T. W. Bettge, A. P. Craig, W. G. Strand et al. 2000. Parallel climate model (PCM) control and transient simulations. *Climate Dynamics* 16:755–74.

Wilby, R. L., and I. Harris. 2006. A framework for assessing uncertainties in climate change impacts: Low-flow scenarios for the River Thames, UK. *Water Resources Research* 42:W02419.

Wood, A. W., L. R. Leung, V. Sridhar, and D. P. Lettenmaier. 2004. Hydrologic implications of dynamic and statistical approaches to downscaling climate model outputs. *Climatic Change* 62 (1–3): 189–216.

Wurbs, R., S. Toneatti, and J. Sherwin. 2001. Modeling uncertainty in flood studies. *International Journal of Water Resources Development* 17 (3): 353–63.

Constructing Carbon Market Spacetime: Climate Change and the Onset of Neo-Modernity

Janelle Knox-Hayes

Climate change represents a new era in the development of capitalism, whereby humanity has become such a force of nature so as to destabilize its own environment and ultimately threaten its survival—neo-modernity. This article explores the creation of markets to control greenhouse gas emissions. Carbon markets are an important infrastructure to enable humanity to integrate nature into its sociopolitical and economic organization. The carbon markets are the embodiment of a process designed to reorganize human activities but also to organize and assimilate the natural environment. As with other eras, the key to success in neo-modernity is organizing complex and divergent human activities across space and time. Using an institutional approach, built on case studies and close dialogue with market participants and policymakers in the United States and Europe, this article analyzes the construction of carbon market infrastructure including how the markets organize environmental impacts in space and time. Particular attention is paid to the compressions of the spacetime of carbon commodities through the establishment of platforms, exchanges, and verifiers. The article concludes that markets are coordinating networks—the epitome of neo-modernity infrastructure and the beginning of a process through which the natural environment will become valued only in the context of further capitalist expansion. *Key Words: carbon markets, environmental finance, neo-modernity, spacetime.*

气候变化在资本主义发展历程中展现出一个新的时代，即人类已经变成一种自然的力量去破坏自己的环境并最终威胁自己的生存，我们称之为新现代化。本文探讨了创造控制温室气体排放量的新型市场。碳市场是一个重要的基础设施，使得人类得以将自然纳入其社会政治和经济的组织。碳市场是旨在不仅重组人类活动，而且同时组织和同化自然环境这一过程的具体体现。如同其它时代，新现代化的成功关键在于组织在空间和时间域里的复杂而且不同的人类活动。使用惯例办法，建立在案例研究以及与美国和欧洲的市场参与者和决策者的密切对话，本文分析了碳市场基础设施的构架过程，包括市场如何在空间和时间域里建立环境影响。本文特别注意了碳商品的时空压缩特性，该特性是通过建立平台，交易所，以及核查机构实现的。文章的结论是，市场正在协调网络，作为新现代化基础设施的一个缩影和这一过程的开始，籍此自然环境将变成只有在资本主义的进一步扩大范围的背景下才会被重视的因素。*关键词：碳市场，金融环境，新现代化，时空。*

El cambio climático representa una nueva era en el desarrollo del capitalismo, en la cual la humanidad ha llegado a convertirse en una fuerza natural tan efectiva que puede desestabilizar su propio entorno, amenazando en últimas su supervivencia—neo-modernidad. Este artículo explora la creación de mercados para controlar las emisiones de gases de invernadero. Los mercados de carbono son una importante infraestructura para capacitar a la humanidad en la integración de la naturaleza dentro de su organización sociopolítica y económica. Los mercados de carbono son la materialización de un proceso diseñado para reorganizar las actividades humanas, pero también para organizar y asimilar el entorno natural. Como ha ocurrido con otras eras, la clave del éxito para la neo-modernidad es organizar actividades humanas complejas y divergentes a través del espacio y el tiempo. Utilizando un enfoque institucional, construido a partir de estudios de caso y diálogo cercano con participantes del mercado y planificadores en Estados Unidos y Europa, este artículo analiza la construcción de infraestructura del mercado de carbono, incluyendo la manera como los mercados organizan los impactos ambientales en espacio y tiempo. Especial atención se le presta a las compresiones del espacio-tiempo de los bienes de carbono por medio del establecimiento de plataformas, intercambios y verificadores. El artículo concluye que los mercados son cadenas coordinadas—el epítome de la infraestructura de la neo-modernidad y el comienzo del proceso a través del cual el entorno natural se hará valioso solamente en el contexto de la futura expansión del capitalismo. *Palabras clave: mercados del carbono, finanzas ambientales, neo-modernidad, espacio-tiempo.*

Climate change marks a new challenge to the progression of capitalism. Capitalist development can be framed in two eras with distinctive (if overlapping) sociopolitical, economic, and cultural features—modernity and postmodernity (Harvey 1989). Each era can be defined by humanity's struggle to overcome a greater challenge to progress. Modernity's challenge was to emancipate the individual from the monarchy, religion, and tradition of the Middle Ages (Berman 1983). It established a new system of hierarchy and order, marked among other things by scientific pursuit of ultimate truth, industrialization, and individualism. Postmodernity constituted a readjustment, a backlash against the hierarchy, linear order, and supreme "truth" (including the extremes of fascism and totalitarianism) that arose in modernity. Integral to the framing of these eras is also the position or balance between humanity and nature. Prior to modernity, humanity was subject to the forces of nature. Under modernity, humanity strove to dominate and master nature. During postmodernity, humanity began to recognize the fragility of nature. The environmental movement was born out of a counternarrative to the damage of unrespited capitalism. Constant throughout each era, however, has been the drive to increase control of collective organization in space and time (Bell 1976; Jameson 1991). It is as though with mastery over space and time, humanity could overcome any threat or achieve any modernizing goal (Foucault 1977; Giddens 1990).

I argue for the possibility of framing a new era of capitalist development—a "neo-modernity." In some ways neo-modernity represents a continuation of the drive for progress, but like postmodernity it also represents a response or backlash to prior periods. Under neo-modernity, human civilization is coming to realize the dire consequence of anthropogenic climate change. As in previous eras, the key to addressing this great challenge seems to lie in exerting ever deeper mastery of human organization in space and time. In neo-modernity, this mastery takes the form of the coordinated decarbonization of the activities of billions of people. The distinction between postmodernity and neo-modernity lies in the identity and scope of capitalism. Postmodernity represents efforts to limit capitalist operation, restricting the use of conserved areas and limiting the extent of particular environmental damage.[1] The natural environment, if ever diminished in size, maintains a unique, noncapitalist identity. Neo-modernity, in contrast, represents efforts to fully integrate nature into the operation of capitalism. The existence of nature is now becoming neither a mere input to production nor a bounded area of restricted operation but rather a priced and controlled element of the system of capitalism.

This article posits the introduction of neo-modernity by exploring the creation of carbon markets. These markets are developing around the world as a governance mechanism to reduce greenhouse gas emissions (Hasselknippe 2003; Carr and Rosembuj 2007). A number of systems are in existence, including the European Union Emissions Trading System (EU ETS) and the Regional Greenhouse Gas Initiative in the Northeastern United States (MacKenzie 2007). I argue that carbon markets are coordinating networks—the epitome of neo-modernity infrastructure—that enhance our ability to organize our activities and to organize use of the environment in space and time. Carbon markets need artificial spacetime to reduce greenhouse gas emissions in part because carbon pollution is itself spatially and temporally unbounded. The markets are in this respect environmentally unique, but the concept of the markets is being expanded to encompass virtually all other intangible components of the natural environment from biodiversity to ecosystem services.

The arguments of the article are supported by the use of cases studies with market institutions, which used close dialogue (Clark 1998). More than 100 interviews with experts from banks, brokerages, intermediaries, legal firms, consultancies, power companies, and political institutions in Europe and the United States were conducted to understand the development, operation, and significance of carbon markets. The interviews provide insight into the social construction and operation of the markets, as well as the intent of market actors not readily available from open source data. Each interview has been qualitatively cross-checked with other interviews to verify the findings. The case study approach is well suited to describe and conceptualize the developing carbon markets and their relationships in space and time (Quattrone 2006). However, because perspectives and experiences of individuals are not always accurate representations of actions or facts, the data were triangulated with company documents and Web sites to confer rigor and credibility to the conceptualization (Strauss and Corbin 1998). A key prerequisite for this rigor and confidence in the findings is access to key market and decision makers (Goldstein 2003). To gain access these individuals were guaranteed anonymity; therefore, interlocutors are reported anonymously.

This article proceeds in four sections. The second section examines the meaning of spacetime and built infrastructure, highlighting the contributions of Manuel Castells and David Harvey. The third section explains the nature of spacetime in carbon markets. The fourth section examines the creation and operation of infrastructure to sustain carbon markets, including conventions, registries, and exchanges. The article concludes by suggesting that carbon markets are the beginning of a process through which the natural environment will become valued only in the context of further capitalist expansion.

Spacetime and the Infrastructure of Control

Seminal works by Harvey (1989) and Castells (1996) sought to theorize the essence of space and time in social interaction. Harvey and Castells are both concerned not so much with space and time as physical embodiments but as social phenomenon at the heart of human organization and as phenomena constructed through economic processes. The history of social and economic development has been one in which spatial relationships are compressed, and the rate of interaction and spatial transformation is accelerated. Harvey (1989) referred to this acceleration as time–space compression, a phenomenon that enables the shrinking of physical distance and enhances our ability to overcome time constraints. It is a process designed to speed economic scale and productivity.

Each era of capitalism has been concerned with mastering space and time or reducing spacetime barriers to production to establish a global economy. Castells (1996) referred to this as network society, which is a "mixing of tenses to create a forever universe, not self-expanding but self-maintaining, not cyclical but random, not recursive but incursive: timeless time, using technology to escape the contexts of its existence and appropriate selectively any value each context could offer to the ever-present" (433). The ability to appropriate value from any context (spacetime) and bring it into the present is a form of control over spacetime. As with any form of spacetime manipulation, the construction of timelessness, of instantaneous telecommunications and financial markets, requires the construction of infrastructure.

The physical manifestation of struggles to overcome space and time are preserved in the infrastructure of a society, whether the cables that link global time in the nineteenth century or the freeways that compress space in the twentieth century. Architecture and infrastructure are the material demonstration of the wealth and collective achievement of a civilization (Orlikowski 1992). Infrastructure has form, which records the composition, the struggles, and the development of the civilization that created it. It also has function, which organizes society and thereby embodies the civilization that creates it (Fligstein 2001).

Infrastructure development in the last 200 years has performed the function of compressing spacetime and building a network society. Modernity achieved the synchronization of clocks to create simultaneity through Einstein's theory of relativity (Galison 2003). Postmodernity used the construction of vast networks of roads, railways, automobiles, airplanes, and finally fiber-optic cables to link disparate localities around the globe and make them instantly accessible. Technological infrastructure has created a society of flows, a space of timeless time, capable of integrating both past and future time into the present. Virtual financial markets allow for the utilization of the future events with instruments such as derivatives (Tickell 2000). Social space and time has been virtually subsumed into network society.[2]

The last point of resistance is what Castells (1996, 467) referred to as "glacial time," or the natural environment. Regardless of social organization and spacetime compression, the environment exists according to its own time frame. It sits at the edge of Castells's forever time and holds back the spread of "eternal ephemerality." The environment has always been a stable point of reference in capitalist development against which humanity has positioned itself. Modernity represents the pursuit of mastery of the natural environment. Postmodernity to an extent represents the embrace of chaotic and organic characteristics of the natural environment, as well as the need to protect segments of it. However, neither era fully takes account of its effect on the environment. As a result, capitalist progress has drastically altered environmental cycles and faces destabilization under anthropogenic climate change. Capitalism's response—the integration of the environment into network society—is the beginning of neo-modernity. It is an era that will reorganize human and natural functions in space and time. Yet in the capitalist assimilation of the natural environment, neo-modernity is less a path to addressing environmental destruction and more a path to enabling the continued expansion of capitalism.

This article explores the infrastructure of the markets that allow for the integration of the environment to occur with ever greater control of space and time. Like

Einstein's clock network, environmental markets are a broader symbol of order in neo-modernity. Although this article explores only the development of carbon markets designed to reduce greenhouse gas emissions, the tenets of these markets enable them to be applied to other environmental areas such as waste, water, ecosystem services, and biodiversity.[3]

The Nature of Spacetime in Developing Carbon Markets

The postmodern condition or network society has been dedicated to producing technological innovation, economic growth, and operation of greater economies of scale, which require coordinating expansion of human productive activities. To achieve expansion it is necessary for financial markets to control and value future time because the value of future time creates coordinated and dedicated investment. In coming to terms with glacial time, it is necessary for the network society to coordinate the reduction of environmental impact as a result of human productive activities. To achieve reduction it is not only necessary for markets to value future time; it is necessary for them to value nontime. Reduction ultimately produces value for something that never happens. The carbon markets are the infrastructure of coordinating and globalizing emissions reductions.

Emissions reductions have neither real space nor real time because the emission never occurs. The reduction is rather a mere reflection of the counterfactual, of what might have otherwise occurred. Both its space and time must be constructed. The construction uses virtual space and blends time. The utilization and pricing of future events is not new, but in other eras it was restricted to events that at some point would happen. Secondary financial markets already manage future events with futures, swaps, and options. Carbon markets control things that never happen, by giving value to the prevention of a future occurrence (see Figure 1). The nonoccurrence is rewarded by giving it both virtual spacetime existence and artificial value.

Giving positive value to the absence of the emission creates automatic negative value for the existence of the emission. The utility of making the absence of emissions fungible is that it coordinates productive activities across otherwise disconnected temporalities and among otherwise disconnected actors. The absence of emissions in China is financed in Europe, with the recognition that eliminating an atmospheric

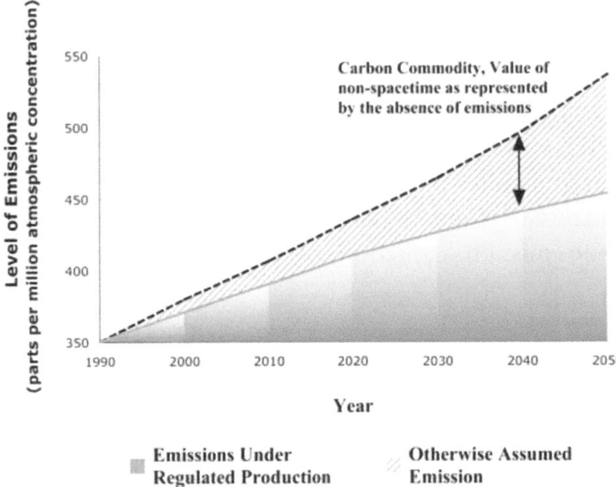

Figure 1. Illustration of the artificial spacetime of carbon commodities.

externality is universally beneficial (Bumpus and Liverman 2008). At face value, the exchange allows for the creation of the lowest cost emissions reduction. More important, in both places the environmental externality is valued and the development of emissions reduction activities is encouraged. Furthermore, in making a reduction a valued and tradable commodity, markets capture the interests of unassociated parties including banks, brokerages, and low-exposure firms seeking a corporate social responsibility or public relations benefit and create drive for additional reductions (Knox-Hayes 2009).

Carbon markets represent yet another level of management of spacetime. If carbon markets are successful in reducing emissions, additional markets for externalities like polluted water or ecosystem damage will be created. Controlling environmental impact by valuing externalities could be seen as a retreat of network expansion. Yet, breaking down the glacial time barrier by integrating environmental impact is a further incursion of network society, not only into human activities and organization but into environmental productivity as well (Lash and Urry 1994). The goal of a low carbon economy is, after all, still greater production and expansive economic growth, albeit with controlled environmental inputs and outputs.

The Infrastructure of Carbon Emissions Markets

Leaving aside ethical considerations or the potential for real success, there are practical concerns for building the infrastructure that can manage artificial spacetime. Valuing, connecting, and coordinating emissions reductions between places like China and Europe requires the development of advanced infrastructure—what might well become the hallmark of neo-modernity. In particular, it requires the creation of conventions, platforms, and registries and exchanges to create, verify, track, and trade emissions reduction commodities.

Conventions: Establishing and Verifying Emissions Reductions

The creation of a carbon commodity requires complex procedures and the construction of considerable infrastructure. The Clean Development Mechanism (CDM) is the primary source of reductions that can offset carbon emissions from regulated facilities. CDM carbon projects must demonstrate that the planned offsets are financially and environmental additional, that they would not occur without the project being developed or without the incentive provided by emission reductions credits. Establishing additionality is particularly controversial because it requires justifying a project against what would have otherwise happened. For example, a carbon aggregator can go to China and build a combined cycle gas-fired power plant that produces 5,000 tons of CO_2 emissions per year. The aggregator then argues that if the gas-fired power plant were not built, a less expensive coal-fired power plant would be built. The coal-fired power plant would produce 15,000 tons of CO_2 emissions per year. The aggregator thereby claims a reduction value of 10,000 tons CO_2 equivalent per year, but this value is measured against something that does not actually happen.

Producing additional reductions relies more on social infrastructure—the creation of conventions—than on the construction of physical infrastructure. One of the biggest contentions with the CDM is that additionality is very difficult to prove (Greiner and Michaelowa 2003). Some studies show that the use of additionality in the CDM likewise does not lead to projects that directly enhance sustainable development (Olsen 2007). Determining what would have otherwise occurred, in either an ecological context or a developmental context, is always subjective. One of the interlocutors suggests that subjectivity creates widespread distrust for the process: "The CDM does not specify up front what is or is not a qualified project. It is an unreliable process. Subjectivity comes in with additionality because it is so amorphous. That is why relationships with verifiers are important, but it creates distrust" (Director at Exchange, Chicago, 22 May 2008).

The determination of additionality is left to the discretion of designated operational entities who validate and verify the projects. Yet the verification is built on protocol and conventions, which are negotiated by party members. As a result, the CDM requires the negotiation of considerable procedure and conventions, which complicate emissions reductions systems. A Europe-only system is easier to establish, monitor, and control. There is a level of technical certainty as to what constitutes an emissions allowance. Offsets, in contrast, require structuring non-spacetime, which is contentious. The economic argument for allowing offsets is that they allow emissions reductions at a lower price, but this is not the complete picture. Packaging altered spacetime as a commodity links and coordinates activities between otherwise disconnected areas. The United Nations Framework Convention on Climate Change (UNFCCC) created the CDM to engage and integrate the developing world into an emissions governance regime. The linking and coordinating effect goes even further than this, however. Certified emissions reductions (CERs) that do not qualify for certification under the CDM often become verified emissions reductions (VERs), which are sold in voluntary markets to a range of interested parties.

The direct value of an emission reduction is that it meets a regulatory requirement and avoids a penalty or fine. However, it also creates a system of indirect value. The voluntary markets thrive off of the purchase of CERs and VERs by companies for corporate social responsibility or public relations and by individuals seeking to reduce their carbon footprints (Hamilton, Sjardin, Marcello, and Xu 2008). The reductions link and coordinate activities among disparate temporalities. For example, offsetting a flight between London and New York can help finance the development of a wind farm in Latin America. Ultimately the value of these reductions is a measure of the social value placed on addressing climate change. That value is transported and shared through the creation of spacetime altered commodities. The value of these reductions must be maintained, however, through confidence in the rigor of verification, accounting, and tracking.

Platforms and Registries: Tracking Reductions

Commodities like gold are stored in reserves around the world. Carbon reduction commodities are held in electronic registries because they have no physical

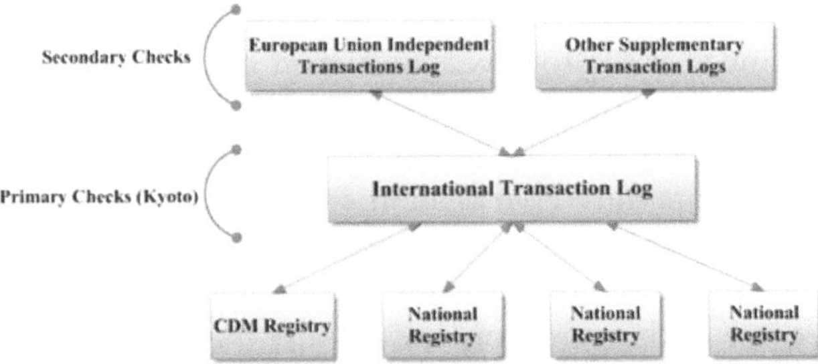

Figure 2. Links between UNFCCC and national registries. CDM = Clean Development Mechanism.

manifestation. These registries require both the construction of convention and infrastructure. The registry must track and record the entire life cycle of a credit, including its creation, ownership, transfer, and expiry. The registry performs two critical functions: (1) it substantiates the credit's existence and (2) it ensures the integrity of each credit, so that it is not double counted for compliance. To do this the registry assigns each carbon unit a unique serial number and records its issue date (vintage), point of production, ownership, and the registries in which it has been held.

Smooth transfer of credits between different regions requires an integrated system of registries. Under the Kyoto Protocol, each Annex 1 (compliance) country has its own registry and the CDM has a registry. The registries are linked through an International Transaction Log (ITL), which verifies transactions first through national registries. The ITL is also linked to the Community Independent Transaction Log (CITL), which serves as the community registry for the EU ETS (UNFCCC Secretariat 2009). The ITL performs a secondary check on all transactions through the CITL (Figure 2). The linkage of these registries requires considerable convention and coordination, but it also allows companies as well as individuals to set up user accounts in national registries and access carbon account information from their desktops.

The registry system for the Kytoto Protocol has taken six years to develop and has become a massive infrastructure, requiring the coordination of its thirty-six Annex 1 countries and the CDM Executive Board. The registries can lean on already established Internet infrastructure but still require new servers, hubs, technicians, and software. In addition, the conventions that create, record, transfer, and expire reductions have taken years to develop. As suggested in the previous section, the value of a reduction credit is that it allows for the connection of different spacetimes and redirection of future development. All of the coordinated effort of the Kyoto member states is therefore dedicated toward building infrastructure that enables the connection and transfer of directed spacetimes. One of the technological service providers interviewed commented on the nature of registries and the importance of common frameworks of operation:

> Every step of the life cycle works through the registry. It keeps records public even when credits retire . . . you need registry-interoperability because each standard has its own registry. . . . A global registry would be ideal, assuming, and it would be critical, to guarantee principal information. (Market Director at a market technology service provider, New York, 12 May 2008)

There are a number of standards that can produce reduction credits. The only way to track these electronic commodities is to link the registries that record their existence. The ITL is as such a considerable representation of neo-modernity. It enables users to access, engage, and move specific environmental temporalities through user accounts from virtually anywhere in the world. It is an infrastructure that accomplishes greater awareness of and control of human activity in space and time. In addition to the registries for regulated carbon markets, registries are being created to encompass voluntary carbon markets as well as markets for other developing environmental commodities including forestry and biodiversity.

Exchanges: Trading Reductions

Between 70 percent and 80 percent of carbon reductions are sold over-the-counter (OTC), which means buyers and sellers identify themselves, often through brokers and working with intermediates or banks (Capoor and Ambrosi 2008). OTC trading is inefficient in that it takes time for the brokers to match buyers and sellers. As with other types of commodity and financial exchanges, carbon exchanges provide a virtual space in which buyers and sellers can meet and trade directly. A number of exchanges have developed to trade carbon including the Chicago Climate Exchange, the European Climate Exchange, Eurex, the Asian Carbon Trade Exchange, and Bluenext. Established exchanges such as the New York Mercantile Exchange (NYMEX) and Nordpool are also beginning to trade carbon products. Exchanges play three main roles: They increase access and build liquidity, they hedge counter party risk, and they develop sophisticated instruments to provide price forecasting.

The primary role of the exchange is to move emissions reductions in space and time. Like registries, the exchanges operate from virtual platforms, so that carbon can be bought and sold from desktops located virtually anywhere. These exchanges enable EU ETS compliance parties to trade European Union allowances and to move CER credits from the developing world to the developed world. The exchange provides a more liquid marketplace, because buyers and sellers can find each other anonymously. It enhances the accessibility of trading because many more institutions and people have access to the marketplace through an exchange account, which opens carbon trading beyond big financial institutions.

The second role of the exchange is to hedge counterparty risk. Often the exchange serves as a clearinghouse and as such bears the risks of failure to deliver. When serving as a clearinghouse, the exchange does not trade seller to buyer, but rather they collect the entire sell and buy bids, group the risks, and clear them as a total unit. The exchange takes the counterparty risk, as each buyer or seller is dealing exclusively with the exchange as opposed to other buyers and sellers. If a single buyer or one seller fails to deliver, the impact is minimized through collectivized risk. For large exchanges like the NYMEX, this is the most effective way to hedge risks. The exchange bears the risk but receives a considerable service fee to compensate.

The third role of the exchange is to enable the development of more sophisticated financial instruments such as futures options and swaps. These instruments allow companies to hedge their risks by locking in a future price on the reduction with an option to either to sell or buy at that price. Futures products incur a more expensive transaction price, but they are a way for companies to hedge the risk of carbon liabilities in their cost accounting because futures lock in a specific price for the commodity. An interlocutor at a large financial exchange highlighted the importance of price signaling:

> The two main roles of the exchange are price reference and to allow easy access to market participants. The third role is to set up a stable infrastructure. . . . You need to have stable infrastructure with long-term objectives to have a secondary price. The time frame of building a nuclear plant is five to eight years for investment to integrate the price of CO_2. (Director of Business at an exchange, Paris, 17 June 2008)

As the director suggests, futures instruments have indirect benefits as well. Having a future price on carbon sends a signal to the rest of the market that there is continued value to invest in carbon alternatives. Futures instruments allow individual participants to gauge the confidence of the group as a whole. The futures instruments thus serve as another form of communication between different localities or a collectivization of different local interests. This is particularly important because of the regulatory uncertainty that comes with carbon markets. Most systems are undetermined past 2012 or 2020. The liquidity that a future price provides stabilizes the market. However, only a limited number of financial players such as large exchanges and banks are capable of providing the future price. Their coordination through the development of exchanges expands the infrastructure of neo-modernity, not only in space but in time as well.

Conclusion

Capitalist development has proceeded with ever greater control of human organization in space and time. During modernity, human organization was restricted to a linear movement of time across space. Economic expansion required the development of a system to synchronize time, to make the operation of ever greater economies of scale possible. During postmodernity, human organization became a two-dimensional movement across time and across space, bringing the past into present and stabilizing the present with the future. Neo-modernity is an attempt to become multidimensional—organizing human (as well as environmental activity) in both time and non-time as well as in both space and non-space, making capitalist development instantaneous and virtual, with enough extension to direct and shape future trajectories (Table 1). Neo-modernity also represents a change in the relationship between capitalism and the environment. Whereas a postmodern natural environment is a bounded space from which capitalist production is restricted or prevented, a neo-modern natural environment contains a calculated existence that ultimately is exchangeable.

Carbon markets are the beginning of an infrastructure designed to accomplish this organization. Carbon market infrastructure produces control over carbon production activities by giving their absence value and making this absence traceable and tradable. The infrastructure is a means as well as a product of considerable coordination among capitalist societies. If climate change is to be mitigated, the development of this infrastructure might well play a critical role. However,

Table 1. The defining characteristics of the eras of capitalism

	Human–natural environment relationship	Organization in space and time
Modernity	Drive to conquer nature; search of ultimate truth, individualism, industrialization	Space–time compression; bringing past into present
Postmodernity	Attempt to conserve components of nature and limit the scope of damage; incorporation of organic, chaotic multiplicity in social and scientific pursuits	Network society; utilizing past, present, and future
Neo-modernity	Conscription of nature into socioeconomic productivity	Virtual networks; controlling future events by valuing nonspacetime

ultimately carbon markets are designed to continue capitalist development and expansion. The implication of these markets and what they signal for the balance between humanity and nature should be of greater consideration. Many of the market participants interviewed suggested that pricing and generating new technology is the key to solving climate change.

> As a more widespread approach, the market approach drives down cost to society, and allows the private sector to seek out the least cost methods of reducing emissions. It allows and focuses the economy on finding the least cost emissions reductions and creates investments into new technologies. We are reducing emissions by putting a price on carbon capture as an externality. The market decides the price. If we had a $100 per ton price, we wouldn't have a climate change problem. (Director of carbon asset development company, New York, 2 November 2007)

There is widespread belief that climate change is a result of failure to adequately price environmental externalities. If we could price the natural environment adequately we could prevent its destruction. Fundamentally, this ignores the underlining imbalance between capitalism and the natural environment. Capitalism is driven to greater rates of spacetime operation, which overruns environmental cycles of generation and regeneration. Furthermore, as many interlocutors highlighted, carbon is just the beginning. The goal is to conscript and manage all aspects of the natural environment under pricing systems.

> This is the first time that environmental objectives are being achieved and driven by a market. Finance people are making money out of saving the environment. That is the only way to get large investment flows. . . . There is a problem with waste, water resources, perhaps water trading rights. Forestry is also big and connected to the CO_2. The question is how to get finance flow from banks in London to the rainforest? You turn that environmental asset into financial value. I can see market leaders, academics, high-level policymakers at the EU/UN level pushing this agenda forward and thinking how else can we create environmental markets? What sort of instruments can put in place environmental objectives? (Partner, legal firm, London, 3 July 2008)

Climate change represents a warning about the impact of capitalist productivity on the natural environment. The danger of the neo-modern approach is that climate change becomes not a problem of too much capitalism but too little. The economic system cannot account for impact on an independent natural environment. Rather than address this problem, the solution is to convert the environment into something that capitalism can value and trade—conservation of forest credits or survival of biodiversity units. Transferring the environment into capitalism is seen as the only viable way to save it. Neo-modern capitalism thereby claims ownership of the existence of the natural environment and demands that it produce profit.

Carbon markets suggest that capitalism can address climate change while continuing the logic of never-ceasing economic growth and expansion. These markets restructure the human–nature relationship, such that all environmental impacts and attributes can ultimately be controlled by capitalism. Credits to protect biodiversity are already being sold and the next phase of carbon market development under negotiation in the UNFCCC process is a program called Reducing Emissions from Degradation and Deforestation (REDD; Miles and Kapos 2008). The program ties forest conservation to the finance of carbon markets by valuing the carbon sink of forests. At least in the short term, REDD might be the only way to conserve some of the world's largest and oldest remaining forests. Yet it sets a precedent that the value of a tropical rainforest, or indeed the biodiversity it contains, only exists to the extent that it is calculated and controlled as part of the system of capitalism. Pricing the natural environment converts all intrinsic value to exchange value. It distorts the meaning of the natural environment. Can a $5 per ton price (or a $100 per ton price for that matter) ever truly capture the value of 10,000 years of forest growth, and at what point will it no longer be sufficient to protect the forests?

This in its essence is neo-modernity. Climate change represents a challenge to further capitalist expansion (as fueled by carbon fuel sources), but it also signifies a tipping point in the human–nature balance. The solution we have created to address climate change is to assimilate nature into socioeconomic processes. This is accomplished through tighter social organization and greater spacetime control. Capitalist ingenuity and coordination might overcome the challenge of climate change by building vast markets or networks to manage environmental impact. The price signal that creates social value for the absence of carbon emissions might eventually be significant enough to produce a global transition to alternative energy sources. However, the goal of energy transition is still to break down barriers (in this case climatic) to the expansion of capitalism. Attaining this goal comes at a cost. If the markets are successful in creating a low-carbon economy, capitalism will prolong expansion with the illusion that we have overcome climate change, one of the greatest environmental problems humanity has faced. In the meantime the natural environment will continue to lose meaning and value beyond the extent to which it can be integrated and controlled.

Acknowledgments

The author would like to thank the many institutions and individuals who participated in the study and made this research possible. The author would also like to thank Gordon Clark, John Walsh, Amy Glasmeier, Dariusz Wójcik, Pratima Bansal, and Jarrod Hayes for helpful comments on the article. Support for this research was provided by the National Science Foundation (0802799) and the Jack Kent Cooke Foundation. None of these should be held responsible for errors, omissions, or any opinions expressed herein.

Notes

1. I recognize that the term *postmodernity* carries contested connotations with significance for art, literature, and culture. The framing proposed here focuses on relationships between society and nature in time and space and is not intended to engage with cultural and literary debates. Rather, the framing is intended to highlight the changing nature of capitalism's relationship to the natural environment, as epitomized by carbon emissions markets, and to comment on the significance of these changes. With respect to the natural environment, the term postmodernity could also be substituted by neo-romanticism. It encompasses recognition of the need to protect the environment from complete capitalist damage.
2. It should be noted that the argument here is not intended to imply that network society is universal or equally shared. Castells (1996) recognized that there are still communities without access to the network, and control of the network is anything but egalitarian. Production (or as directly explored here, market construction) is furthermore still very much a material project that requires proximity and social connection (Knox-Hayes 2009).
3. TZ1, an Australian registry, has already begun to develop and sell habitat conservation credits (Fogatry 2007).

References

Bell, D. 1976. *The cultural contradictions of capitalism*. London: Heinemann.

Berman, M. 1983. *All that is solid melts into air: The experience of modernity*. London: Verso.

Bumpus, A. G., and D. M. Liverman. 2008. Accumulation by decarbonization and the governance of carbon offsets. *Economic Geography* 84 (2): 127–55.

Capoor, K., and P. Ambrosi. 2008. *State and trends of the carbon market, 2007*. Washington, DC: World Bank Carbon Finance Unit.

Carr, C., and F. Rosembuj. 2007. World Bank experiences in contracting for emissions reductions. *Environmental Liability* 2:114–19.

Castells, M. 1996. *The rise of the network society*. Malden, MA: Blackwell.

Clark, G. L. 1998. Stylized facts and close dialogue: Methodology in economic geography. *Annals of the Association of American Geographers* 88 (1): 73–87.

Fligstein, N. 2001. *The architecture of markets: An economic sociology of twenty-first-century capitalist societies*. Princeton, NJ: Princeton University Press.

Fogatry, D. 2007. Green credits potential boon for emissions markets. *Reuters News Service* 17 October. http://www.guardian.co.uk/business/feedarticle/7886437 (last accessed 20 October 2008).

Foucault, M. 1977. *Discipline and punish: The birth of the prison*. London: Allen Lane.

Galison, P. 2003. *Einstein's clocks, Poincarâe's maps: Empires of time*. London: Sceptre.

Giddens, A. 1990. *The consequences of modernity*. Cambridge, UK: Polity.

Goldstein, K. 2003. Getting in the door: Sampling and completing elite interviews. *PS: Political Science and Politics* 35 (4): 669–72.

Greiner, S., and A. Michaelowa. 2003. Defining investment additionality for CDM projects—Practical approaches. *Energy Policy* 31 (10): 1007–15.

Hamilton, K., M. Sjardin, T. Marcello, and G. Xu. 2008. *Forging a frontier: State of the voluntary carbon markets 2008*. Washington, DC: Ecosystem Marketplace and New Carbon Finance.

Harvey, D. 1989. *The condition of postmodernity: An enquiry into the origins of cultural change*. Oxford, UK: Blackwell.

Hasselknippe, H. 2003. Systems for carbon trading: An overview. *Climate Policy* 3 (1002): 43–57.

Jameson, F. 1991. *Postmodernism, or, the cultural logic of late capitalism*. London: Verso.

Knox-Hayes, J. 2009. The developing carbon financial service industry: Expertise, adaptation and complementarity in London and New York. *Journal of Economic Geography* 9 (6): 749–77.

Lash, S., and J. Urry. 1994. *Economies of signs and space*. London: Sage.

MacKenzie, D. 2007. Finding the ratchet: The political economy of carbon trading. *London Review of Books* 29 (7): 29–31.

Miles, L., and V. Kapos. 2008. Reducing greenhouse gas emissions from deforestation and forest degradation: Global land-use implications. *Science* 320 (5882): 1454.

Olsen, K. 2007. The clean development mechanism's contribution to sustainable development: A review of the literature. *Climatic Change* 84 (1): 59–73.

Orlikowski, W. J. 1992. The duality of technology: Rethinking the concept of technology in organizations. *Organization Science* 3 (3): 398–427.

Quattrone, P. 2006. The possibility of the testimony: A case for case study research. *Organization* 13 (1): 143.

Strauss, A. L., and J. M. Corbin. 1998. *Basics of qualitative research: Techniques and procedures for developing grounded theory*. Thousand Oaks, CA: Sage.

Tickell, A. 2000. Dangerous derivatives: Controlling and creating risks in international money. *Geoforum* 31 (1): 87–99.

UNFCCC Secretariat. 2009. Registry systems. In Registry Systems under the Kyoto Protocol. http://unfccc.int/kyoto_protocol/registry_systems/items/2723.php (last accessed 15 November 2009).

Climate Change and the Global Financial Crisis: A Case of Double Exposure

Robin M. Leichenko, Karen L. O'Brien, and William D. Solecki

Despite widespread and growing public recognition of the linkages between environmental change and economic activities, geographic research efforts to date have paid only limited attention to the connections and interactions between climate change and globalization. As a consequence, critical linkages, feedbacks, and synergies between these two processes often go unnoticed. In this article, we draw on the framework of double exposure to motivate new research on the connections between climate change and globalization. The double exposure framework provides a generalized approach for analysis of the interactions between global environmental and economic changes, paying particular attention to the ways that the two interacting processes spread risk and vulnerability over both space and time. Focusing on the linkages between climate change and the ongoing financial crisis and using the case example of California's Central Valley, this article shows how application of the double exposure framework can provide new insights into social vulnerabilities and can inform efforts to respond and adapt to both processes. The article also illustrates how the double exposure framework can be used to inform and promote place-specific, geographic analyses of the complex interactions between large-scale environmental and economic shifts. *Key Words: California, climate change, economic crisis, geographic research, globalization.*

尽管公众正在日益广泛地意识到环境变化和经济活动之间的联系，迄今为止，地理研究工作中对气候变化和全球化的联系和相互作用尚未足够重视。从而导致这两个过程之间的关键联系，反馈，协同作用往往被忽视。在这篇文章中，我们勾画出一种双重曝光框架，以推动针对气候变化和全球化之间联系的新研究。这种双重曝光框架提供了一种通用的方法，以分析全球环境和经济变化之间的相互作用，特别注意到这两个相互作用的过程在空间和时间上的风险和脆弱性散布。以加州中央山谷地区的情况为个例，本文将重点放在气候变化和正在进行的金融危机之间的联系上，展示了如何应用双重曝光框架来提供对社会脆弱性的新见解，以及如何指导应对和适应这两个进程的各项工作。对于大尺度的环境和经济变化之间的复杂相互作用，文章还举例说明了双重曝光框架是如何被用于宣传和推广地域相关分析和地理研究的。*关键词：美国加州，气候变化，经济危机，地理研究，全球化。*

A pesar del amplio y creciente reconocimiento público de la conexión entre cambio ambiental y las actividades económicas, hasta el momento los esfuerzos de la investigación geográfica solo han puesto limitada atención a las relaciones e interacciones entre cambio climático y globalización. El resultado es que los lazos críticos, retroalimentación y sinergias entre estos dos procesos a menudo pasan inadvertidos. En este artículo nos basamos en el marco de la doble exposición para motivar nueva investigación sobre las conexiones entre cambio climático y globalización. El marco de la doble exposición provee un enfoque generalizado para el análisis de las interacciones entre cambios ambientales y económicos globales, dándole atención particular a la manera como los dos procesos interactivos despliegan por igual riesgo y vulnerabilidad a través del espacio y tiempo. Al concentrarse sobre los lazos entre cambio climático y la actual crisis financiera y utilizando el caso del Valle Central de California, este artículo muestra cómo la aplicación del marco de la doble exposición puede generar nuevas perspicacias alrededor de las vulnerabilidades sociales e ilustrar esfuerzos para responder y adaptarse a ambos procesos. El artículo ilustra también cómo el marco de la doble exposición puede usarse para informar y promover análisis geográficos, para lugares específicos, sobre las complejas interacciones entre las transformaciones ambientales y económicas a gran escala. *Palabras clave: California, cambio climático, crisis económica, investigación geográfica, globalización.*

Climate change and globalization are weaving together the fates of households, communities, and people across all regions of the globe. Both processes are enhancing connections across space and time, such that actions taken in one location have increasingly visible effects on other locations, often in ways that are hard to predict. Both processes are seen as creating not only new opportunities but also growing risks and increasing uncertainty. Whereas geographic research has paid much attention to climate change and globalization as separate and distinct processes, only limited attention has been directed to the interactions between the processes. A thread of studies has explored exposure to multiple stressors, winners and losers from both processes, dynamic and teleconnected vulnerabilities, and relationships between neoliberal policies and the capacity to adapt to climate change (e.g., O'Brien and Leichenko 2003; O'Brien et al. 2004; Eakin 2006; Liverman and Vilas 2006; Keskitalo 2008; Adger, Eakin, and Winkels 2009; Silva, Eriksen, and Ombe 2009). Yet many questions remain regarding potential feedbacks between processes of globalization and climate change, adaptation to climate change under conditions of rapid socioeconomic change, and resilience to the risks and uncertainties associated with both processes.

The linkages between climate change and globalization are readily evident when we consider the connections between the current global financial crisis and new initiatives in the areas of environmental and climate policy. There is, in fact, widespread public recognition, articulated by U.S. President Barack Obama, United Nations Environment Programme Executive Director Achim Steiner, journalist Thomas Friedman, and others, that the solution to the financial crisis and climate crisis go hand in hand: Efforts to transform energy systems and dramatically reduce greenhouse gas emissions can create new jobs and a better environment (Friedman 2008; Administration of Barack H. Obama 2009; Barbier 2009). Despite considerable public and media discussion of these issues, there have been few formal analyses of the interactions between climate change and the financial crisis. In particular, there has been little attention to how the mechanisms associated with global financial markets mediate climate change outcomes across spatial and temporal scales or how joint responses to both processes might influence adaptation efforts.

In this article, we apply the "double exposure" framework developed by Leichenko and O'Brien (2008) to motivate new research on the connections between climate change and the current financial crisis. The double exposure framework provides a generalized approach for analysis of the interactions between global environmental and economic changes, paying particular attention to the ways that the two interacting processes spread risk and vulnerability over both space and time. As such, the framework is reflective of broader efforts by geographers and other social scientists to enhance understandings of vulnerability and adaptation, including factors influencing the resilience of communities, regions, and socio-ecological systems to shocks and stresses associated with processes of global change (e.g., Turner et al. 2003; Eakin and Luers 2006; Berkes 2007; Nelson, Adger, and Brown 2007; Polsky, Neff, and Yarnal 2007; Acosta-Michlik et al. 2008; Eakin and Wehbe 2009; Eriksen and Silva 2009).

In the next part of the article, we discuss how and why the financial crisis and climate crisis are seen as increasingly linked and describe and critique a proposed "Green New Deal" for addressing them together. We then present the basic elements of the double exposure framework and consider how the framework can be applied to assess interactions between the global financial crisis and climate change. Focusing on the case of the drought in California and the collapse of the housing market in California's Central Valley, we illustrate potential tensions among the impacts and responses to climate change and the financial crisis. We show how these tensions are mediated by government structures and policies and how deregulation and globalization have increased vulnerability not only of the housing sector but also of people and places to climate-related risks. We conclude by arguing that the intersections and interactions of multiple risks and stressors call for new analyses of proposed responses to both processes.

Two Crises, One Solution?

The current global financial crisis, similar to most large-scale disaster events, reveals underlying conditions that were largely hidden prior to the event, along with conditions that were quite apparent—although to some only in hindsight (see Reinhart and Rogoff 2008; Yohe and Leichenko 2010). The crisis began with high rates of loan default among subprime mortgages in selected U.S. housing markets, but it quickly spread to other locations and sectors. As with most disasters, the financial crisis can be attributed to multiple factors. There was undoubtedly human error and bad judgment involved, and perhaps even criminal behavior.

Nevertheless, the globalization of financial markets, the deregulation of the financial industry, and the global marketing of mortgage-backed securities were key factors that enabled these risk-taking behaviors (Newman 2008, 2009). This transfer of financial risk across both space and time created a mechanism for the crisis to spread rapidly and deeply and extend beyond the confines of any local, regional, or national origins.

It is possible to identify numerous connections between the global spread of financial risk and the current climate crisis. For example, the globalization of finance contributed to increased availability of low-interest-rate loans (i.e., "cheap" money), which in turn helped spawn energy-intensive, automobile-oriented suburban developments throughout the world (Leichenko and Solecki 2005, 2008). These new developments, which are often concentrated in environmentally sensitive, "amenity" landscapes, such as coastal zones, forested hill slopes, and exurban desert regions, not only contribute to increased greenhouse gas emissions but also have the potential to increase vulnerabilities to many types of extreme climatic events, including hurricanes, wildfires, and droughts. Although climate variability, in itself, is considered a "normal" risk, increasing global temperatures are likely to lead to increased climate variability and more extreme events, including less predictability of weather, more droughts, more storms, and more floods in some areas.

Another connection can be illustrated through the example of the globalization of corporate debt. In December 2008, just a few months into the financial crisis, the *Wall Street Journal* profiled the case of Parkes, Australia (population 10,000), which invested close to $10 million in a credit derivative known as a "synthetic collateralized debt obligation," a form of insurance against defaults on corporate debt (Whitehouse and Ng 2008). As the financial crisis spread from subprime mortgages to other sectors, corporations began reneging on their debts. As a result, Parkes must pay close to $12 million to honor its insurance commitments. The town has thus lost a significant portion of its financial reserves, including money that it had designated to rebuild its water supply amid the country's enduring drought. The case of Parkes illustrates how the financial crisis is undermining state and local government capacity to respond and adapt to climate change. The case also shows how financial globalization, although illustrating a high degree of spatial dynamism, becomes rooted in place, such that local and state governments often must serve as a buffer to mediate the impacts.

The preceding examples illustrate the negative side of double exposure, but the potential for positive connections between responses to the financial crisis and the climate crisis have not gone unnoticed. With reference to the social and economic programs enacted by Franklin D. Roosevelt during the economic depression in the 1930s, the idea of a Green New Deal has been sounded as a way to simultaneously address the financial crisis, the climate crisis, and the energy crisis (Barbier 2009). One specific solution offered by the U.K.-based Green New Deal Group (2009) calls for reregulating finance and taxation and at the same time drastically reducing the use of fossil fuels. Such efforts would need to be backed by a strong legislative framework and price signals from the market that accelerate the development of low-carbon technologies, such as retrofitting of buildings to improve energy efficiency, and development of alternative energy sources.

The idea of simultaneously addressing both the financial crisis and climate crisis through a combination of government regulation and free-market incentives represents an innovative and transformative strategy. However, what is left out of this proposed Green New Deal is attention to the myriad underlying factors that make people vulnerable to interacting shocks in the first place. The concept of *adaptation*, which involves strategies or behaviors to reduce the impacts of a particular shock or stressor, is generally absent from Green New Deal or "green revolution" thinking. Moreover, the social programs that went hand in hand with economic programs during the Great Depression and are necessary to ensure resilience to financial and climatic changes are not addressed through these proposed solutions. We return to these issues later in the article.

The Double Exposure Framework

Despite widespread recognition of the linkages between climatic change and economic activities, and growing efforts to develop integrated frameworks for analysis of multiple stressors (as cited earlier), geographic research on connections and interactions between climate change and globalization remains limited. The majority of research on climate change emerges from the physical or ecological sides of the discipline, with emphasis on dynamics of the atmosphere, hydrosphere, and biosphere. In contrast, studies of globalization tend to emphasize political, economic, and cultural phenomena such as liberalization of trade, formation of transnational commodity chains, and emergence of a global mass media. Although much globalization research addresses environmental issues, this literature does not typically consider how globalization influences or interacts with larger processes of climate change. As a consequence, critical linkages, feedbacks, and synergies between globalization and climate change often go unnoticed and in turn are undertheorized.

The double exposure framework, as developed by O'Brien and Leichenko (2000) and Leichenko and O'Brien (2008), provides a general approach for analysis of many types of interactions between environmental change and globalization. One important difference between the double exposure framework and other vulnerability approaches, as cited earlier, is that other frameworks generally do not take into account the full extent of the potential interactions between climate change and other global change processes, particularly concerning how the outcomes of multiple processes interact across space and over time. There is also relatively little recognition within other frameworks of how the two processes together transform the context in which people and places experience and respond to changes of many types. Many vulnerability frameworks stress the importance of context for explaining both differential outcomes and vulnerabilities, yet the frameworks seldom recognize the extent to which the context itself is dynamic, dramatically changing as the result of both global environmental change and globalization. Within the double exposure framework, changing contextual conditions can affect exposure and responses to future global change processes, resulting in new patterns of vulnerability and new challenges for social and ecological resilience (Leichenko and O'Brien 2008).

The double exposure framework's point of departure is that multiple global change processes are occurring both simultaneously and sequentially, creating either negative or positive outcomes for individuals, households, communities, and social groups. Within the framework, global environmental change and globalization manifest as either gradual or sudden changes (i.e., stressors or shocks) that have differential effects across a particular exposure frame. Depending on the focus of the research, an exposure frame might be a spatial, political, or ecological region; an economic sector; or a network of institutions. Exposure results in measurable outcomes, which might, in turn, affect the processes as well as the context in which future changes are experienced.

In each case, exposure to global change processes is influenced by the characteristics of the change (e.g., direction, rate, magnitude, intensity, and spatial extent) and by factors in the contextual environment (e.g., institutional, economic, social, political, biophysical, cultural, and technological conditions). Responses, which might include actions taken either in anticipation of or following from exposure, are conditioned by factors in the contextual environment, as well as by the individual attributes of each affected actor (e.g., education, values, beliefs, cognition). Outcomes depend on both the degree of exposure to each global process and on the actions taken by the affected individuals or other actors.

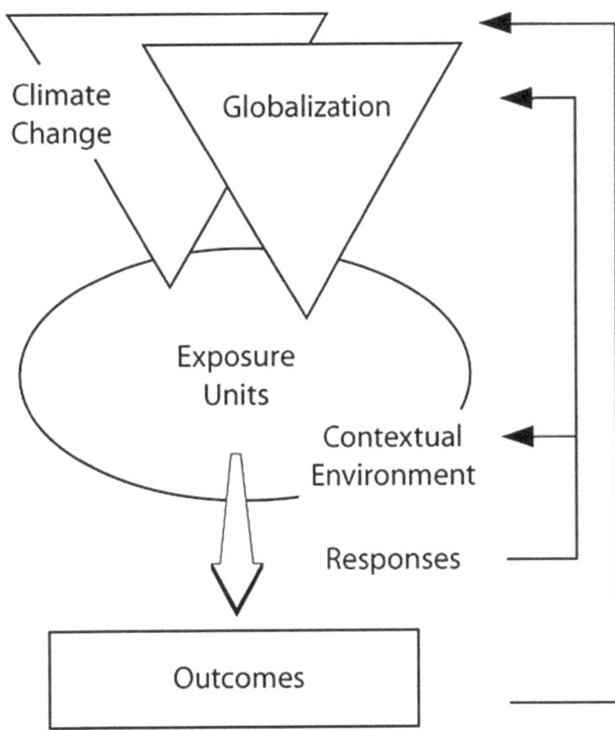

Figure 1. The double exposure framework (adapted from Leichenko and O'Brien 2008).

Figure 1 provides a simple illustration of the main components of the double exposure framework. Processes of global environmental change and globalization are represented as partially overlapping triangles. These processes manifest in a specific contextual environment, portrayed as an oval. The extent or magnitude of exposure to the processes is depicted as the intersection between the triangles and the oval. An arrow leading from the contextual environment to a square representing outcomes symbolizes responses to the processes. Outcomes are depicted as separate from the contextual environment to emphasize that any outcome reflects measurable conditions at a specific point in time.

The framework incorporates dynamic linkages between the components. Processes can alter the contextual environment, responses can affect the processes, outcomes can affect responses, and so forth. Dynamics are also incorporated in the framework through recognition that processes and outcomes are often reflexive. Within the figure, the arrow leading from responses and outcomes back to the process triangles depicts these types of circular linkages, which are termed "feedbacks." Although Figure 1 focuses on a single exposure frame, it is important to note that outcomes and responses that occur within one exposure frame might have widespread influence on other exposure frames both across space and over time.

The double exposure framework articulates a number of potential pathways of interaction between the two processes (Leichenko and O'Brien 2008). The pathway of *outcome* double exposure highlights overlapping impacts of both globalization and climate change on a particular exposure unit, whether it be a region, sector, or social group, showing how the combined effects of both processes often exacerbate existing patterns of spatial and social inequality and vulnerability. This pathway identifies what can be referred to as "double winners" and "double losers." The pathway of *context* double exposure shows how one process can influence the capacity to respond to shocks and stresses associated with the other process, often leading to negative outcomes. By incorporating the temporal dynamics of global change processes, context double exposure provides insights on how long-term resilience can be undermined by current changes to the contextual environment. The pathway of *feedback* double exposure demonstrates how the contextual changes, responses, and outcomes associated with either or both processes can contribute to drivers of the processes, thereby perpetuating cycles of double exposure and posing challenges to long-term sustainability. By emphasizing the dynamic interactions among processes, responses, and outcomes, the framework aims to elicit new insights and research questions, beyond those associated with separate framings of each global change process (Leichenko and O'Brien 2008). As illustrated in the following case example, the double exposure framework can also provide insights on systemwide economic shocks and how interconnections between multiple processes spread and magnify risks and vulnerabilities.

The Global Financial Crisis and Climatic Risk: California's Central Valley

The double exposure framework serves as an analytical tool for investigating the connections between environmental change and globalization. The framework, which focuses on specific pathways of interaction, can be used to assess the consequences of both the financial crisis and increased risks associated with climate change. In what follows, we illustrate how the framework can contribute to a better understanding of the complex interrelationships among drought, the collapse of the housing market, and the fiscal crisis in the California. Drawing on the case of California's Central Valley, our application of the double exposure framework shows how the interactions between the two crises have exacerbated social vulnerabilities and undermined efforts to respond and adapt to both processes.

California is the largest food producer in the United States, supplying nearly half of the country's fruits, nuts, and vegetables. The agricultural sector is also important to California's economy, contributing to an estimated 6.5 percent of the state's value added in 2006 (including the direct, indirect, and induced linkages; Roland-Holst and Kahrl 2008). Yet this engine of economic growth—itself a hallmark of the globalization of the agricultural sector—has relied heavily on cheap and flexible immigrant labor to produce specialty crops, contributing to persistent poverty in agricultural areas (Mitchell 1996; Martin, Fix, and Taylor 2006; Guthman 2008). Furthermore, most high-value agriculture in California depends on irrigation, which is at risk due to salinization of soil and groundwater resources (Schoups et al. 2005).

California's agriculture is also considered highly vulnerable to drought (Roland-Holst and Kahrl 2008). Indeed, much of the state is already subject to frequent drought. Currently, the state is entering a fourth year of drought (as of March 2010) that has devastated many agricultural communities in the Central Valley. Severe heat waves during the summer of 2009 exacerbated the impacts. At the same time, environmental policies aimed at preservation of threatened species have reduced the export of water from the Delta region of California and have limited any options for additional water allocations in response to the drought (Howitt, MacEwan, and Medellin-Azuara 2009). The recent drought has contributed to lower profits, increased unemployment, and rising food prices, with many of these adverse effects clustered in the Central Valley.

Under climate change, it is likely that the agricultural sector in the Central Valley will have to adapt to new and more complex conditions (Roland-Holst and Kahrl 2008). These conditions include reduced water availability in California and the adjoining U.S. Southwest region (e.g., from increased evapotranspiration rates, reduced rainfall, or changes in runoff from snowmelt) and other looming climate-related threats to agriculture, including outbreaks of pests and disease. Winter warming alone could dramatically reduce the area's fruit and nut production. Although drought, heat waves, and winter warming are not necessarily due to climate change, they nonetheless represent the types of conditions that are likely to occur more frequently as the climate changes (California Climate Change Center 2006). Although California agriculture is often seen as having high capacity to adapt to climate change through engineering and technology, this capacity comes at a significant economic cost, particularly costs associated with changes in California's groundwater storage and extensive water transfers among regions and users (Tanaka et al. 2006). As discussed by Moser (2009), there are also many barriers to adaptation in California beyond technological and financial constraints, including, for example, lack of institutional capacity and political cooperation among jurisdictions and social and cultural values that emphasize private property and individual rights.

Central Valley farmers have responded to the lack of rain and to limited availability of other state or federal supplied water by fallowing fields and planting low-maintenance and low-labor crops (McKinley 2009). As a stopgap measure to avoid a summer agricultural catastrophe, Governor Arnold Schwarzenegger ordered the release of 100,000 acre-feet of water to Central Valley farmers in July 2009 from the State Water Project with the proviso that it be repaid by the end of November of the same year. This countered a "zero allocation" policy for farmers who purchase water from the Central Valley Project, which is a federally managed irrigation project. The possibility of reduced or no water supplies has raised the likelihood of increased food costs, not only in the Central Valley but also in state, national, and international markets (Sung 2009).

Yet drought and climate change are not the only crises facing farmers in this region. Cutter and Finch (2008) highlighted the Central Valley as one of the U.S. regions with the highest social vulnerability to all types of natural hazards. This region has also been experiencing some of the most rapid residential housing growth in the country. Real estate speculation has long been a key driver of California's economic growth (Guthman 2008), and investments in the Central Valley grew dramatically during the past decade such that the region is currently bearing the full brunt of the ongoing housing crisis. The median sales price of homes in the Central Valley has plummeted, and there are thousands of foreclosures on the market. Several of the major cities in the Valley have experienced some of the greatest housing price declines of any cities in the United States, with losses in some cases exceeding 40 percent. The tax-assessed value of homes in the Valley declined by 11 percent during the first half of 2009 (Borenstein 2009), and local housing markets are not expected to recover until foreclosures have been absorbed (Streitfeld 2008). The collapse of the housing sector in the Central Valley, as well as other parts of the state, has directly impacted credit markets and local economies. This has made the capital resources needed to respond more difficult to acquire. Collapse of the local housing market has also devastated the tax revenues for municipalities, further restricting response capacity.

The already limited capacities of local governments have been further hampered by a state budget resolution in the summer of 2009, which was built around the state borrowing money from localities with a process of paying back these funds as California's economy improves. Moreover, the State of California has a limited capacity to intervene in the Valley, both because it cannot raise property taxes due to institutional and legal constraints and because there are few funds available for social services that might benefit vulnerable populations. The state has also instituted "new public management" approaches, commonly associated with neoliberalism and globalization, that entail devolution of resource management responsibilities to local levels (Christensen and Lægreid 2001; Moser 2009). In the case of drought management, devolution has shifted some of the responsibility to the local level, even though the financial crisis has meant that local institutions have reduced capacity to manage regional water systems.

Pathways of Double Exposure

Some households and communities in the Central Valley can be considered double exposed to both climate extremes and financial shocks, with the outcomes of the two shocks reflecting more than just additive effects. Local and state commentators regularly describe the Central Valley as a region in crisis and at the leading edge of economic disaster. In fact, U.S. Representative Dennis Cardoza from Merced is pushing for federal legislation to declare the Central Valley an economic disaster area ("Cardoza Pushes to Have Valley Declared an Economic Disaster Area" 2009). The three pathways of double exposure help to untangle this complex situation and offer insights on response strategies that might alleviate long-term vulnerability to multiple stressors.

The pathway of outcome double exposure, which focuses on overlapping exposure to multiple processes of global change, is perhaps most visible. This pathway highlights that households impacted by drought-related income and livelihood changes might also be subjected to instability and to a loss of shelter or equity linked to the housing and financial crises. Populations at the economic and employment margins are particularly vulnerable to the overlapping consequences of multiple crises. Within the Central Valley, this includes those working in industries directly linked to agriculture, such as fieldworkers, processing handlers, food packers, and truckers, as well as the local, small businesses that serve these various groups. Outcome double exposure is illustrated by the case of Mendota, a Central Valley city dependent on agricultural employment. Current rates of unemployment in Mendota are nearly 40 percent—more than triple the rate in California overall—and there is little prospect for near term improvement, both in terms of an end to drought (and related water restrictions) and to a broader turnaround in economic conditions (Fagan 2009; McKinley 2009).

The pathway of context double exposure explores transformations in the contextual environment as the result of either or both processes of global change. This interaction between the two processes might sometimes create conditions that undermine adaptive capacity, thereby limiting future responses to one or both crises. In California, the financial crisis has affected credit markets such that farmers are unable to borrow money from banks for investments in high-value, climate-risk-tolerant crops, or for more efficient water supply and irrigation systems. Importantly, changes in water availability, combined with changes in soil characteristics and other climatic variables, can affect land prices and the average value of farmland as well (Schlenker, Hanemann, and Fisher 2007). For the first time since the passage of Proposition 13 in 1978—the landmark state of California property tax-cutting initiative—assessed land values (not market values) in the region have declined, directly impacting local tax revenues. As a consequence, local tax revenues have collapsed at precisely the time when the demand for services has increased due to both the drought and the economic downturn. The financial crisis has also undermined California's efforts to implement measures to promote adaptation to climate change, many of which require financial, technical, and institutional resources that go beyond "normal" management and maintenance expenses (Moser 2009). This is especially true in the case of water infrastructure, where adaptation might entail increasing the resilience of water supply systems to extreme climate events via costly new water collection and conveyance infrastructure or through changes in allocation via new regulatory approaches, review of systems of legacy entitlement, and public–private cost sharing (Roland-Holst and Kahrl 2008).

The pathway of feedback double exposure emphasizes the relationships between the responses and outcomes of one crisis and the drivers of another. For example, changes in state and federal water allocations in response to the drought will severely reduce Central Valley income, employment revenues, and cropped areas, thereby increasing pressures on other sectors of the economy (Howitt, MacEwan, and Medellin-Azuara 2009). In the face of housing value reductions of more than 40 percent during the period from mid-2008 to mid-2009 and an unemployment rate of over 20 percent, largely due to contraction of the dairy sector, Merced County officials have turned to local small businesses, requesting that they try to employ more local residents (Choi 2009). Feedback double exposure also draws attention to the ways that development of the housing sector and urbanization processes in the Central Valley interact with patterns of energy consumption. Because suburban expansion has been an important driver of growth in greenhouse gas emissions, the ongoing contraction of the housing sector amidst the financial crises might present an opportunity to alter this trend. Such a response, along the lines suggested by proponents of a Green New Deal, might include an emphasis on construction of green buildings, expansion of alternative energy production, and water conservation. To be effective and sustainable at the local level, however, such efforts must incorporate adaptation planning and attention to the social and institutional supports needed to ensure resilience to both climatic and economic shocks and stresses.

Conclusion

Dramatic economic and climate changes are underway in locales throughout the world, and individuals, households, communities, sectors, and regions will be confronted with impacts, whether it is through shifts in investments and sectoral upheavals or through more extreme weather events. The double exposure framework provides geographers and others utilizing the approach a way to move beyond descriptive statements about the scope of change and to more fully analyze and explain the processes underway in places like California generally and the Central Valley specifically.

Although California has long been a beacon of hope for a nation of people seeking a better life with steady work and a roof over their heads, the example of the Central Valley highlights the dramatic changes that are currently experienced in this era of double exposure. The Central Valley with its rich agricultural land and wide-open space has long exemplified a typically American blend of reality and myth. Indeed, the first New Deal brought many families of "Joads" to the Valley seeking a better life. The intensification of agricultural production and the rapid rise of residential population growth—the backbone of the regional economy throughout much of the post–World War II era—are now severely shaken. More important, the underlying baselines on which they were built have started to shift. The environmental conditions from which the agricultural economy emerged are changing, and the economic base of continued investment capital from state, national, and global sources has become much more uncertain.

What is also striking about this moment is that although the agricultural and residential real estate development sectors historically have been seen as largely separate in the Central Valley, double exposure to climate risk and financial instability have threaded them together throughout the region, causing hardship for many households. The double exposure framework helps to reveal how these pressures came together in the Central Valley and set off a cascade of local financial pressures and, in turn, social transformations that are still emerging and deepening. The inability of local governments to effectively respond to the current crisis via jurisdictional mandate, local initiative, or state-assisted action is perhaps most ominous and portends a broader renegotiation of the social contract between government and local residents (O'Brien, Hayward, and Berkes 2009). Similar evidence of such a renegotiation was also present in the case of Hurricane Katrina and the government's inability to effectively respond to that climate-related disaster event (Leichenko and O'Brien 2008).

The case of the Central Valley reveals that climate change adaptation efforts—as well as efforts to mitigate climate change—need take into account interactions with other types of socially created risks. The double exposure framework shows that addressing the processes that have contributed to increased risks, whether through state policies or a Green New Deal, is necessary but insufficient. Controlling global financial markets through regulations and incentives and limiting climate change by reducing greenhouse gas emissions addresses only part of the problem. There is a simultaneous need to address shifts in the underlying context that make people vulnerable to shocks and to respond to uneven outcomes that contribute to increased inequality and future vulnerability. Without a careful analysis of interactions among processes, responses, and the outcomes of multiple stressors, today's "solutions" might prove to be inadequate responses to the social, economic, and environmental challenges of global change.

References

Acosta-Michlik, L., K. Kumar, R. J. T. Klein, and S. Campe. 2008. Application of fuzzy models to assess susceptibility to droughts from a socio-economic perspective. *Regional Environmental Change* 8:150–61.

Adger, W. N., H. Eakin, and A. Winkels. 2009. Nested and tele-connected vulnerabilities to environmental change. *Frontiers in Ecology and the Environment* 7 (3): 150–57.

Administration of Barack H. Obama. 2009. *Remarks at Southern California Edison's Electric Vehicle Technical Center in Pomona, California*. 19 March. Washington, DC: Office of the Federal Register, National Archives and Records Administration, U.S. Government Printing Office. http://www.gpo.gov/fdsys/pkg/DCPD-200900170/pdf/DCPD-200900170.pdf (last accessed 3 November 2009).

Barbier, E. B. 2009. *Rethinking the economic recovery: A global green New Deal*. Geneva, Switzerland: United Nations Environment Programme, Economics and Trade Branch. http://www.unep.org/greeneconomy/portals/30/docs/GGND-Report-April2009.pdf (last accessed 3 November 2009).

Berkes, F. 2007. Understanding uncertainty and reducing vulnerability: Lessons from resilience thinking. *Natural Hazards* 41:283–95.

Borenstein, D. 2009. Property tax revenue loss hits cities. *Bay Area News Group* 26 July. http://www.allbusiness.com/government/government-bodies-offices-regional-local/12573839-1.html (last accessed 3 November 2009).

California Climate Change Center. 2006. Our changing climate: Assessing the risks to California. CEC-500-2006-077. Sacramento, CA: California Climate Change Center, Cardoza pushes to have valley declared an economic disaster area. *The Record*. 19 June. http://www.recordnet.com/apps/pbcs.dll/article?AID=/20090619/A_NEWS/90619014/-1/rss02 (last accessed 3 November 2009).

Choi, A. S. 2009. Can entrepreneurs save this town? *Business Week* 5 June. http://www.businessweek.com/print/magazine/content/09_66/s0906058698290.htm (last accessed 3 November 2009).

Christensen, T., and P. Lægreid, eds. 2001. *New public management: The transformation of ideas and practice*. Hampshire, UK: Ashgate.

Cutter, S., and C. Finch. 2008. Temporal and spatial changes in social vulnerability to natural hazards. *PNAS* 105 (7): 2301–06.

Eakin, H. 2006. *Weathering risk in rural Mexico: Climatic, institutional, and economic change*. Tucson: The University of Arizona Press.

Eakin, H., and A. Luers. 2006. Assessing the vulnerability of social-environmental systems. *Annual Review of Environment and Resources* 31:365–94.

Eakin, H., and M. Wehbe. 2009. Linking local vulnerability to system sustainability in a resilience framework: Two cases from Latin America. *Climatic Change* 93: 355–77.

Eriksen, S., and J. Silva. 2009. The vulnerability context of a savanna area in Mozambique: Household drought coping strategies and responses to economic change. *Environmental Science and Policy* 12:33–52.

Fagan, K. 2009. Mendota: A town scraping bottom. *San Francisco Chronicle* 26 July. http://www.sfchronicle.us/cgi-in/article.cgi?f=/c/a/2009/07/26/MNQ718IAAI.DTL (last accessed 3 November 2009).

Friedman, T. L. 2008. *Hot, flat and crowded: Why we need a green revolution—And how it can renew America*. New York: Farrar, Straus & Giroux.

Green New Deal Group. 2009. A green New Deal: Joined-up policies to solve the triple crunch of the credit crisis, climate change and high oil prices. London, UK: New Economics Foundation. http://www.neweconomics.org/sites/neweconomics.org/files/A_Green_New_Deal_1.pdf (last accessed 3 November 2009).

Guthman, J. 2008. Neoliberalism and the making of food politics in California. *Geoforum* 39:1171–83.

Howitt, R., D. MacEwan, and J. Medellin-Azuara. 2009. Economic impacts of reductions in delta exports on Central Valley agriculture. *Agricultural and Resource Economics Update* 12 (3): 1–4.

Keskitalo, C. H. 2008. *Climate change and globalization in the Arctic: An integrated approach to vulnerability assessment*. London: Earthscan.

Leichenko, R., and K. O'Brien. 2008. *Environmental change and globalization: Double exposures*. New York: Oxford University Press.

Leichenko, R., and W. Solecki. 2005. Exporting the American dream: The globalization of suburban consumption landscapes. *Regional Studies* 39 (2): 241–53.

———. 2008. Consumption, inequity, and environmental justice: The making of new metropolitan landscapes in developing countries. *Society and Natural Resources* 21 (7): 611–24.

Liverman, D. M., and S. Vilas. 2006. Neoliberalism and the environment in Latin America. *Annual Review of Environment and Resources* 31:2.1–2.37.

Martin, P., M. Fix, and J. E. Taylor. 2006. *The new rural poverty: Agriculture and immigration in California*. Baltimore: Urban Institute Press.

McKinley, J. 2009. Drought adds hardships to California. *New York Times* 21 February: A1.

Mitchell, D. 1996. *The lie of the land: Migrant workers and the California landscape*. Minneapolis: University of Minnesota Press.

Moser, S. 2009. Adaptation planning in California: Process, progress, challenges and opportunities. Paper presented at the IHDP Global Environmental Change and Human Security (GECHS) Synthesis Conference, Oslo, Norway.

Nelson, D., W. N. Adger, and K. Brown. 2007. Adaptation to environmental change: Contributions of a resilience framework. *Annual Review of Environment and Resources* 32:395–419.

Newman, K. 2008. The perfect storm: Contextualizing the foreclosure crisis. *Urban Geography* 29: 745–84.

———. 2009. Post-industrial widgets: Capital flows and the production of the urban. *International Journal of Urban and Regional Research* 33 (2): 750–54.

O'Brien, K., B. Hayward, and F. Berkes. 2009. Rethinking social contracts: Building resilience in a changing climate. *Ecology & Society* 14 (2): 12.

O'Brien, K., and R. Leichenko. 2000. Double exposure: Assessing the impacts of climate change within the context of economic globalization. *Global Environmental Change* 10 (3): 221–32.

———. 2003. Winners and losers in the context of global change. *Annals of the Association of American Geographers* 93 (1): 89–103.

O'Brien, K., R. Leichenko, U. Kelkar, H. Venema, G. Aandahl, H. Tompkins, A. Javed, et al. 2004. Mapping vulnerability to multiple stressors: Climate change and globalization in India. *Global Environmental Change* 14 (4): 303–13.

Polsky, C., R. Neff, and B. Yarnal. 2007. Building comparable global change vulnerability assessments: The vulnerability scoping diagram. *Global Environmental Change* 17:472–85.

Reinhart, C. M., and K. S. Rogoff. 2008. Is the 2007 U.S. sub-prime financial crisis so different? An international historical comparison. *American Economic Review: Papers & Proceedings* 298 (2): 339–44.

Roland-Holst, D., and F. Kahrl. 2008. *California climate risk and response: Executive summary*. Berkeley: University of California at Berkeley. http://www.next10.0rg/next10/pdf/report_CCRR/75_01_5%20ClimateRiskandResponseWEB06.pdf (last accessed 3 November 2009).

Schlenker, W., W. M. Hanemann, and A. C. Fisher. 2007. Water availability, degree days, and the potential impact of climate change on irrigated agriculture in California. *Climatic Change* 81:19–38.

Schoups, G., J. W. Hopmans, C. A. Young, J. A. Vrugt, W. W. Wallender, K. K. Tanji, and S. Panday. 2005. Sustainability of irrigated agriculture in the San Joaquin Valley, California. *Proceedings of the National Academy of Sciences* 102 (43): 15352–56.

Silva, J. A., S. Eriksen, and Z. A. Ombe. 2009. Double exposure in Mozambique's Limpopo River Basin. *Geographical Journal* 176 (1): 6–24.

Streitfeld, D. 2008. In the Central Valley, the ruins of the housing bust. *New York Times* 24 August: BU1.

Sung, A. 2009. California drought impact worsens. *Supermarket News* 2 March: 32.

Tanaka, S. K., T. Zhu., J. R. Lund, R. E. Howitt, M. W. Jenkins, M. A. Pulido, M. Tauber, R. S. Ritzema, and I. C. Ferreira. 2006. Climate warming and water management adaptation for California. *Climatic Change* 76: 361–87.

Turner, B. L. II, R. E. Kasperson, P. A. Matson, J. McCarthy, R. W. Corell, L. Christensen, N. Eckley et al. 2003. A framework for vulnerability analysis in sustainability science. *Proceedings of the National Academy of Sciences* 100 (14): 8074–79.

Whitehouse, M., and S. Ng. 2008. Insurance deals spread pain of U.S. defaults world-wide. *Wall Street Journal* 23 December: A1.

Yohe, G., and R. Leichenko. 2010. Adapting a risk based approach. *Annals of the New York Academy of Sciences* 1196:29–40.

Integrity of the Emerging Global Markets in Greenhouse Gases

Barry D. Solomon and Michael K. Heiman

This article considers the integrity of the emerging emissions allowance markets for greenhouse gases (GHG) under the international emissions trading system created by the Kyoto Protocol and the parallel European Union Emissions Trading Scheme. In particular, we suggest that accepted definitions of emissions baselines, initial allocation of emission credits, verification of "additionality" for GHG reduction beyond what would have occurred without trading, assuring permanence for offset-generating projects, preventing leakage of emissions-generating activities from protected project areas, monitoring, and reporting requirements are problematic. As with previous reforms that also rely on market allocation and expansion of private property rights, GHG trading deflects change in the social relations of production required for more sustainable production while commodifying access to the atmosphere and nature's ability to recycle carbon. Alternatives to this modern-day GHG indulgence system are considered, especially an upstream carbon tax, designed to force change in energy procurement and use. Although we found that most of the conditions required for the efficient operation of the international GHG markets are not being met thus far, further changes in the trading systems are being made. *Key Words: carbon trading, clean development mechanism, EU Emissions Trading Scheme, global climate change.*

根据京都议定书制定的国际排放交易系统和与此平行的欧洲联盟排放交易计划，本文探讨了新兴温室气体（GHG）排放津贴市场的完整性。本文特别指出，我们认为：接受排放基线的定义，初始分配排放额度，核查“额外”的温室气体削减量，上述情况已经超出了没有碳交易所能发生的一切：确保了抵消类项目的持久性，防止了保护项目领域内活动所造成的排放泄漏，监测并汇报了有问题的要求。正如以前的改革同样也依赖于市场分配和私人财产权利的扩充，温室气体交易偏转了社会关系生产的变化，需要更可持续的生产，同时需要对大气碳排途径和自然碳循环的能力进行商品化。本文认为，那些现代温室气体控制系统的替代物，尤其是上游的碳税，目的是迫使能源采购和使用产生必要的变化。虽然我们发现，为国际温室气体市场的有效运作所需的大部分条件迄今尚未得到满足，但贸易体系正在进行进一步的改进。*关键词：碳交易，清洁发展机制，欧盟排放交易计划，全球气候变化。*

Este artículo se refiere a la integridad de los cupos del mercado de emisiones de gases de invernadero (GHG, sigla en inglés) que ahora surgen dentro del sistema internacional para el negocio de emisiones creado en el Protocolo de Kyoto y el paralelo Esquema de Negocio de Emisiones de la Unión Europea. En particular, nos referimos a lo problemáticas que son las definiciones adoptadas de las líneas de base de emisiones, los créditos iniciales de la asignación de emisiones, lo mismo que la verificación de la "adicionalidad" en la reducción de GHG más allá de lo que habría ocurrido sin este tipo de negociación, el asegurar la permanencia para proyectos de generación, la prevención de fugas de emisiones por actividades que las generen en áreas protegidas de proyectos, el monitoreo y los requisitos para los informes. Como ha ocurrido con reformas anteriores también dependientes de la asignación de mercado y de la ampliación de derechos de propíedad privados, la venta de GHG desvía el cambio en las relaciones sociales de la producción que se requieren para una producción más sostenible, mientras comodifica el acceso a la habilidad de la atmósfera y la naturaleza para reciclar el carbono. Se toman en cuenta las alternativas disponibles frente a este moderno sistema de indulgencia de GHG, en especial la de un impuesto al carbono en la propia fuente del problema, diseñado para forzar el cambio en lo que se refiere a generación y uso de energía. Aunque encontramos que hasta el momento la mayoría de las condiciones requeridas para la eficiente operación del comercio internacional del carbono no se están cumpliendo, si se están presentando cambios en el sistema de funcionamiento de este comercio. *Palabras clave: comercio del carbono, mecanismos de desarrollo limpio, Esquema del Comercio de Emisiones de la UE, cambio climático global.*

As global greenhouse gas (GHG) emissions continue to rise, many prominent scientists have called for emission cuts of 50 to 80 percent by 2050 to stave off the worst effects of climate change (Intergovernmental Panel on Climate Change 2007). Economists have argued that if scientists and policymakers determined it important to reduce GHG emissions that a market-incentive (neoliberal) system is more effective, efficient, and equitable than traditional command and control regulation. The two best market-based candidates are carbon taxes (Nordhaus 1992) and emissions allowance trading (Solomon 1995, 1999), the latter more closely aligned with neoliberal practice.

With the 2005 ratification of the Kyoto Protocol to the United Nations Framework Convention on Climate Change (UNFCCC), the political debate on policy instruments has all but ended. This is because Kyoto encourages emissions trading of permits (allowances) to be earned from several "flexibility mechanisms," and the European Union (EU) and U.S. government are focusing on the cap-and-trade approach to reduce emissions. Academic debates notwithstanding, existing GHG market mechanisms include the *Clean Development Mechanism* (CDM) allowing emissions reduction projects in developing countries not yet subject to the Kyoto Protocol, thereby generating certified emission reductions (CERs) to offset carbon dioxide (CO_2) and other GHG emissions from industrialized nations; *Joint Implementation* (JI) projects generating *Emission Reduction Unit* (ERU) allowances from the former Soviet-bloc nations and in developed countries; and *Land Use, Land-Use Change, and Forestry* projects generating more restricted *Removal Units* (RUs).

These trading mechanisms are predicated on a neoliberal view that the marginal cost for emission reduction is generally less in the developing world than it is in the industrial nations. Furthermore, as emissions and reductions of GHGs—particularly carbon—are spatially neutral with regard to impact, progress toward GHG reduction would be more efficient and effective if the developing world was allowed to sell carbon allowances through offset projects and forest conservation to the industrialized world.

Besides the emerging international GHG market, commonly referred to as *International Emissions Trading* (IET), there has been a *European Union Emissions Trading Scheme* (EU ETS) in place since 2005 (a second phase begun in 2008 runs through 2012) and mandatory state systems in effect in Australia since 2003 and several U.S. states since 2008. In addition, voluntary GHG markets have existed in the United States, Canada, the United Kingdom, Denmark, Australia, and Japan, some restricted to the electric power sector (Bailey 2007a; Bumpus and Liverman 2008).

Although the ratifiers of the Kyoto Protocol represent about three fourths of total GHG emissions, IET extends to many nations without reliable emissions baselines. Moreover, it covers multiple sectors (electric power plants, mobile sources, buildings, chemicals, other industry, forestry, etc.); six GHGs—CO_2, methane (CH_4), nitrous oxide (N_2O), sulfur hexafluoride (SF_6), hydrofluorocarbons (HFCs), and perfluorocarbons (PFCs)—and does not require continuous emissions monitoring.

As noted later, several challenges should be considered regarding the operation of any international market for GHGs. GHG trading is modeled after the successful sulfur dioxide (SO_2) cap-and-trade system of the Acid Rain Program under the U.S. Clean Air Act Amendments of 1990 (Solomon 1995, 1998). The success of the U.S. market for SO_2 allowances has been a function of its relative simplicity and transparency. It covers just one nation, one pollutant (SO_2), and largely one sector (coal-fired power plants) responsible for two thirds of total emissions; it utilizes reliable emissions baselines; it has little likelihood of emissions leakage to other sources or nations; and sources are required to use continuous emissions monitors. However, the achievements of the U.S. system are contested by some analysts because problems of "additionality" emerge when old power plants that would have shut down anyway are allowed to sell their "right to pollute," and uncertainty is introduced in the investment climate through fluctuating allowance prices (Lohmann 2006). Geographers have increased their analysis of the emerging GHG and carbon markets by focusing on issues of scale; place; trading patterns; and the efficiency, equity, and effectiveness of new policy instruments for the governance of nature (Solomon 1995, 1998; Bailey 2007a, 2007b; Bumpus and Liverman 2008; Bailey and Maresh 2009). This article is in the same vein, taking a critical view of market rollout with the hope of improving the imperfect institutions, while also cognizant of the limitations posed by the neoliberal agenda to extend commodification of ecosystem services and of nature in general (Castree 2008).

We examine here the emerging neoliberal global markets in GHGs, with a focus on tradable emission allowances, and then consider the alternative instrument of a direct upstream tax on carbon and other GHG sources, favored by many economists and environmental interests. The next section briefly reviews the Kyoto requirements, the EU ETS, and its early market experience. This is followed by discussion of requirements for well-functioning markets and how they apply to the global GHG markets. Subsequent sections explore GHG accounting, implementation challenges in several GHG markets, and alternatives to cap-and-trade with accompanying global offset trading that are designed to force change in energy procurement and use, an upstream carbon tax in particular. We close the article with some conclusions and final discussion of the emerging market.

Kyoto Protocol Requirements and the European Union Response

The Kyoto Protocol obligates the EU and thirty-seven industrialized nations that ratified it to reduce their GHG emissions by an average of 5 percent over 1990 levels during the First Commitment Period of 2008 through 2012. This initial focus on "common but differentiated responsibilities" was in recognition of historic responsibility by industrialized nations for anthropogenic GHG accumulation in the atmosphere (UNFCCC 1998). Actual assigned emission levels ranged from 110 percent for Iceland to 92 percent (with an 8 percent reduction) for most EU nations. The emissions allocation of 100 percent for Russian and Ukraine was especially contentious as it allowed those nations to accumulate millions of excess allowances (one allowance equals one metric ton of CO_2 equivalent emissions), even as their inefficient GHG-intensive industries were already contracting in the post-Soviet era absent any mandate.

The most forceful and comprehensive response to Kyoto Protocol commitments came from the EU through adoption of a cap-and-trade program among the twenty-five (now twenty-seven) member nations. The program design of the EU ETS is in sharp contrast to the baseline-and-credit and the project-based approaches of the CDM and JI markets (Table 1). Covering more than 12,000 installations, including thermal combustion generators over 20-megawatt capacity, and large emitters from major manufacturing sectors, the EU ETS initially addressed less than half of the EU's CO_2 emissions and but a third of its GHG emissions, although member states can add other GHGs and industrial sectors starting with Phase II (2008–2012). Thus participating nations can meet their Kyoto commitments through actual GHG reductions, as well as by purchasing emission reduction allowances from governments or private entities through the EU ETS (internal sources) or from outside the EU via the CDM, JI, and possibly in the future through land use and forestry projects. As the latter must meet additionality criteria and be supplemental to a country's domestic efforts, their use has been restricted (Skjaerseth and Wettestad 2009).

Lasting from 2005 through 2007, Phase I of the EU ETS has been criticized for its generous free allocation rather than auction of emission allowances to key industries and the fact that each nation could set its own cap when meeting its Kyoto commitment (Bailey 2007b; Ellerman and Buchner 2008; Clo 2009). This allowed emitters to reap windfall profits as emissions declined due to economic conditions and already anticipated production modifications—including plant closures—while providing little incentive to move toward sustainable production (Lohmann 2006). Free allocation also deprived member states of a revenue source for investment in renewable energy and energy efficiency. Moreover, the extreme volatility in the price for emission allowances (ranging more than an order of magnitude over the trading period before collapsing to near zero toward its end) contributed to initial market uncertainty affecting the whole range of energy investments, including fossil fuel, nuclear, and renewable energy sources (F. Harvey 2009; Schiermeier 2009).

Because the EU ETS serves as a model for pending U.S. legislation to control GHGs, and it will continue

Table 1. Mechanisms of the major international greenhouse gas markets

Market name	Emissions cap	Allowance allocation	Crediting program	Allowance banking and borrowing
IET	No	Nations	CDM, JI, land use, land-use change and forestry, EU ETS allowances, etc.	Yes (with limits)
EU ETS	Yes	EU	EU ETS allowances, CDM, JI	Yes (with limits)
CDM	No	None	Project-based baseline and CER credits	No
JI	No	None	Project-based baseline and ERU credits	No
Land use, land-use change, and forestry	No	None	Project-based baseline and RU credits	No

Note: IET = international emissions trading; CDM = clean development mechanism; JI = joint implementation; EU ETS = European Union Emissions Trading Scheme; CER = certified emission reduction; ERU = emission reduction units; RU = removal units.

through at least 2020 (European Commission 2009a), consideration of its limits and promise is warranted. Defenders of the EU ETS and of cap-and-and trade in general note that Phase I was intended as a trial to prepare for actual Kyoto implementation. Thus, volatility should be expected, particularly as the EU underwent electricity market liberalization during that period (Ellerman and Joskow 2008; Oberndorfer 2009). In addition, continuing volatility during the first two years of Phase II was due in part to the global economic downturn. Furthermore, the allowance market collapse toward the end of Phase I was attributed to an inability of firms to carry over allowances, a situation since rectified with banking and borrowing allowed in Phase II (Pew Center on Global Climate Change 2009). Addressing the allocation debate proved more difficult, as only a maximum of 10 percent of the allowances are permitted to be auctioned by nations during Phase II, although the goal is for 100 percent permissible auctioning by 2013 in Phase III. Even so, several achievements can be noted: establishment of an EU-wide carbon price, bringing together national emission-reduction strategies into a unified governance scheme, the progress made since its introduction, and the stimulation of global carbon markets (Capoor and Ambrosi 2009).

Conditions for Well-Functioning Markets

A successful market system should maximize economic efficiency and innovation while avoiding social disparity. The new GHG markets are no exception. In this section, we review six basic requirements for the formation and functioning of efficient markets and briefly comment on how the GHG markets have measured up thus far:

- *The tradable commodity must be homogenous and fungible*. This requires all GHG emissions and emission cuts to be treated equally after accounting for their differential global warming potentials (see later). EU ETS allowances were designed to be fungible with those in IET, but Phase I included only CO_2 and did not count sinks for forestry projects because of vexing challenges of additionality, accurate measurement and monitoring, carbon leakage, and permanence (Table 2). All six major GHGs can be traded under IET, through which individual nations can meet their Kyoto targets. However, interpollutant trading has not been directly allowed under either market.
- *There must be a large enough number of market participants, and with cost differences, to generate an active market*. More than 12,000 energy and industrial facilities are covered by the EU ETS, although they account for only 46 percent of the EU's CO_2 emissions (Oberndorfer 2009). IET, in turn, theoretically can be applied to three fourths of global GHG emissions, although many potential allowances are only available through the CDM on an incremental, case-by-case basis through allowance generation in developing countries after addressing the problem of establishing reliable emission baselines. This condition is being met in the two main GHG markets, IET and EU ETS.

Table 2. Types of allowance-generating projects allowed in the greenhouse gas market

Project category	GHG reduced	Which market	Examples
Energy	CO_2, CH_4	IET; EU ETS (CH_4 in Phase II)	Hydroelectricity, wind power, biomass energy, power plant fuel switching, industrial energy efficiency, sealing of leaky natural gas pipelines, landfill gas recovery
Cement	CO_2	IET; EU ETS	Increased efficiency in cement manufacturing
Agriculture	CH_4, CO_2, N_2O	IET; EU ETS (CH_4 and N_2O in Phase II)	Gas recovery from manure, alternative fertilizers
Forestry	CO_2	IET	Afforestation, reforestation, prevention of deforestation and forest degradation, improved forest management
Others	N_2O, HFC, PFC, SF_6	IET; EU ETS (in Phase II)	Process changes in the manufacture of fertilizers, nylon, aluminum smelting, electricity transmission facilities, destruction of refrigerants

Note: GHG = greenhouse gas; IET = international emissions trading; EU ETS = European Union Emissions Trading Scheme; HFC = hydrofluorocarbon; PFC = perfluorocarbon. *Source:* Capoor and Ambrosi (2009); UNEP Risoe Centre on Energy, Climate and Sustainable Development (2009); URC (2009).

- *The market must have ease of entry and exit.* Global GHG trading is voluntary and new firms and other entrants can join the markets, as long as they meet the stipulated criteria; this condition is also being met.
- *Absence of significant market power.* This is not yet a clear problem either, as no European nation accounts for even 3 percent of global CO_2 emissions (other than the Eurasian nation of Russia) and the only nations that account for over 10 percent, the United States and China, are not yet full participants in IET. However, as noted later, China has been the principal generator of CDM allowances.
- *Widespread availability of market information/minimal transaction costs.* Although transaction costs to buy or sell allowances for emissions trading are generally low because of logs established by the UN Climate Change Secretariat and the ready availability of brokers (UNFCCC 2008), expenses are high to generate allowances under JI and especially under the CDM when determining baselines, additionality, and emissions verification (van Kooten, Shaikh, and Suchanek 2002; Muller 2007; Schneider 2009).
- *No social or environmental externalities.* This is a significant issue because CO_2 and other GHGs are underpriced in most countries (if valued at all). Thus, the cost of generating emissions is not fully internalized. Moreover, although the immediate impact of CO_2 emissions is not spatially specific, GHG emission allowance purchase can lead to pollution hot spots and social disparity when additional pollutants that concentrate locally accompany continued production.

Greenhouse Gas Accounting

The Kyoto Protocol requires Annex I (developed) nations to meet strict GHG emissions obligations. They are allowed, but not required, to use the IET; CDM; JI; and land use, land-use change, and forestry projects to meet these obligations, thereby necessitating accurate and sophisticated emissions accounting. Under the EU ETS, there are additional binding guidelines for monitoring, reporting, and verification of GHG emissions (European Commission 2007).

Annex I nations are also required to prepare and submit annual GHG emissions inventories that are reviewable by other treaty parties. Although relatively straightforward for CO_2 and the three industrial gases, inventory development is more difficult for N_2O and CH_4 because the sources and sinks are less understood. These national registries are also required to ensure accurate accounting of any CERs, ERUs, and RUs, noted earlier, to facilitate GHG trading in both the IET and EU ETS systems. The allowances can be generated beyond the baseline "assigned amount units" allocated to each Annex I party to the treaty. For each project, changes in GHG emissions must be estimated, monitored, evaluated, reported, verified, and certified (Lawrence Berkeley National Laboratory 2000). Because the global warming potential of non-CO_2 GHGs is much higher, the potential allowance generation (and financial gain) is commensurately higher, underscoring the importance and challenge of scientifically accurate accounting systems (MacKenzie 2009).

To have integrity, GHG markets thus have to address several challenging issues in allowance generation. These include establishing reliable emissions baselines, assigning allowances, monitoring continuing emissions, assuring additionality for offset projects beyond what would have happened in the absence of trading, and assuring permanence for GHG removal or carbon sequestration while also preventing leakage of GHG-emitting activities to other geographic areas due to restrictions imposed by a project. Although these issues affect all such markets, they are most vexing for the CDM; JI; and land use, land-use change, and forestry programs. The EU ETS has already addressed baselines and allocated allowances to sources, and national programs that seek to participate in the IET must address these on a case-by-case basis.

Initial Implementation Experience and Problems Uncovered

In 2008, project-based CO_2 trades, including the CDM, JI, and voluntary markets, reached US$7.2 billion, with the CDM responsible for more than 90 percent of that amount (Capoor and Ambrosi 2009). Allowance generation was primarily based on energy projects (mainly hydroelectricity), wind power, biomass energy, energy efficiency, and fuel switching (UNEP Risoe Centre on Energy, Climate and Sustainable Development 2009). In addition, the EU ETS logged $92 billion in allowance trades, for a total for all transactions double that for 2007. This growth in the carbon market was fully expected as 2008 was the first year of the Kyoto Commitment Period. Thus far, the major seller of allowances has been China, and the largest buyer has been the United Kingdom (Bailey 2007a). Prices have been volatile, having fallen from over 28 to 15 euros as of late 2009 in step with the global economic recession as European firms had sufficient allowances to get by during the downturn and also due to generous allowance allocations in Phase I (Capoor and Ambrosi 2009; Clo 2009; Schiermeier 2009; Skjaerseth and Wettestad 2009). Overall the EU is on track to meet its Kyoto treaty target but not without the use of IET and external offsets, especially the CDM (European Commission 2009b).

The U.S. Environmental Protection Agency (EPA) formally declared CO_2 and the five other GHGs public health pollutants on 7 December 2009, thus setting the stage for direct regulation under the U.S. Clean Air Act (EPA 2009). A U.S. GHG reporting rule was previously issued on 22 September 2009, although regulatory control of GHG emissions in the United States has yet to be acted on. Preference for market-based solutions, which can successfully work alongside direct regulation of GHGs as has occurred with the U.S. Acid Rain Program, recognizes the dominant view among economists that costs are lower when firms are free to choose their method of compliance (Solomon 1998). Program effectiveness is improved and equity among emitters optimized when emitters are given an incentive to reduce emissions below any standard set, whether through sale of allocated or auctioned emission allowances or through a carbon tax (Parry and Pizer 2007). Nevertheless, market approaches have their critics as various implementation problems arise.

For allowances to be assigned, any GHG reduction has to be beyond what is mandated or the emitter would have done under current or foreseeable market conditions. Unfortunately, verifying such additionality is a difficult process subject to manipulation, particularly with CDM projects (Lohmann 2006; Schneider 2009). As of October 2009 approximately 1.7 billion CERs, each representing one ton of equivalent CO_2 reduction, among 1,868 projects, have been registered for implementation through 2012. Of those already implemented, the majority of allowances went to China (59 percent) and India (11 percent), primarily for a few large projects eliminating large quantities of industrial HFC gas release in China that was expeditiously accomplished at modest expense, yet generated numerous allowances given the huge CO_2 equivalent of these GHGs (MacKenzie 2009; UNFCCC 2009). Flaring CH_4 gas from solid waste landfills and oil wells are other high-value CER projects with additionality that has been challenged. By contrast, very few projects were implemented in Africa, despite the expectation that the CDM would help transition this region to a low-carbon economy, with the intent to use CDM financing to promote sustainable development (UNFCCC 2009).

The verification and certification of GHG emissions reduction under the CDM, in particular, has come under serious challenge. Indeed, from November 2008 thru February 2009, the United Nations temporarily suspended the accreditation of the main company that validates carbon-offset projects in the developing world due to serious flaws with project management, internal auditing, and inappropriate sectoral expertise (Schiermeier 2008). Critics claim that companies involved in verifying these projects too often have a business interest (and thus a conflict of interest) when also involved with project management (Lohmann 2006).

Assuring permanence—another prerequisite for a functioning carbon market—rests on the assumption that a technological fix for CO_2 emissions will emerge if we assign the proper price and allow the market to make the appropriate adjustment. However, whereas it is relatively easy to anticipate the life cycle and impact of a solar, wind, or energy efficiency project, it is much more difficult to assign allowances with carbon sequestration, accepted by many as an unavoidable necessity if we are to slow climate change. Outside of a few unique natural gas recovery demonstration projects in Norway, The Netherlands, Algeria, and Australia, CO_2 injection from coal demonstration projects in the United States and northern Europe, and a few coal-based CO_2 capture projects that have been or will be used for enhanced oil recovery in Canada and China, geological sequestration of compressed or liquid CO_2 at a commercial scale is unlikely in the near future for the hundreds of millions of tons required (e.g., there were 31.5 billion tons of CO_2 emissions from fossil fuel use worldwide in 2007; Pearce 2008b; Chu 2009).

With biological carbon sequestration (e.g., reforestation and afforestation), neutralization of the offset emissions occurs over many years as the planted forest matures; yet not counting future carbon savings as a current value erodes offset prices. In addition, with biological sequestration and assignment of offset allowances through avoided deforestation, reforestation, or biofuels plantations, we run into a social impact problem where nearby residents dependent on pastures or forests for their livelihood are often displaced, leading in turn to carbon leakage as they move into new areas (Lohmann 2006). Although the Kyoto Protocol discourages certification of CO_2 allowances from biofuels linked to deforestation and the EU ETS does not yet accept any land use and forestry projects, IET allows afforestation, reforestation, and avoided deforestation through the CDM. Only a handful of projects generating less than 1 million tons total in allowances have been certified (UNFCCC 2009). However, with 20 percent of global GHG emissions linked to agriculture, land use, and deforestation, attention to such was high on the agenda of the December 2009 UN Climate Change Conference in Copenhagen as ways are sought to minimize social dislocation and carbon leakage, allowing land use and forestry projects to play a more forceful role in future carbon markets. At a minimum, significant challenges in monitoring and verifying carbon sequestration are raised (Macdicken 1997; Brown et al. 2000).

Up from $120 billion in 2008, the global CO_2 allowance market might well exceed $400 billion on U.S. entry and reach over $2 trillion by 2020, which might make carbon the largest commodity market in the world, surpassing even petroleum (*New Carbon Finance* 2009). In anticipation, an entirely new realm for capital accumulation has emerged with bondsmen, lawyers, allowance verifiers, bioengineers, and especially bankers, hedge fund managers, and other financers active. The irony is that with the possible exception of the World Bank as a broker and the United Nations as a certifier, we have here the same players or their contemporaries who were responsible for the collapse of the housing market leading to the 2008–2009 global financial crisis, principally Goldman Sachs, CitiGroup, Morgan Stanley, JP Morgan, Credit Suisse, Deutsche Bank, Barclays Bank, and most of the other major investment banks, even Lehman Brothers and Bear Sterns before their collapse (Pearce 2008a). Here primary and future carbon offsets—whether from a wind farm in Germany, a palm oil plantation in Indonesia, or a forest in Uganda from which the squatters have been evicted—are bundled and sold in a secondary market with investors selling the allowances to current emitters, banking them for future sale or selling them again as derivatives in a tertiary market. According to the industry trade journal *Hedgeweek* (2008), a record 16.7 million tons of carbon futures was traded in October 2008 at the height of the financial crisis, seventeen times higher than the preceding October and 90 percent higher than the previous August 2008 record, with expansion reflecting investment opportunities as we have moved into the Kyoto Commitment Period.

The Carbon Tax Alternative

No doubt these traders play a vital role, adding necessary flexibility to the market. Yet the difficulty of attending to the aforementioned issues with additionality, permanence, leakage, and social impact, in addition to the complexity of the market mechanism with a potential for junk carbon bonds replacing junk mortgage bonds as capital seeks new areas for investment, have led many economists and policy analysts to suggest that a carbon tax rather than a cap-and-trade mechanism be the principal means whereby nations address GHG emissions. For example, with only 2,100 large suppliers or handlers, including coal mines, natural gas processors, and petroleum refineries in the United States, an upstream tax on the well-profiled carbon content of fossil fuels as they come out of the ground or enter the distribution grid might be easier to assign, more transparent to the consumer, less subject to manipulation, and better able to implement personal income and geographic impact neutrality through corresponding reductions in income and select corporate taxes, than the cap-and-trade system envisioned by pending legislation in the U.S. Congress in late 2009 (Metcalf and Weisbach 2009; Carbon Tax Center 2009; Holt and Whitney 2009). That bill would require a detailed accounting from all major GHG sources from resource extraction, through product manufacturing, to GHG-linked consumption. Furthermore, the offset system accompanying a cap-and-trade program creates private wealth at public expense. In addition, border tax adjustments on imports from nations not assessing carbon costs should be easier to justify under the General Agreement on Tariffs and Trade when a carbon tax is in place than would be the case with a cap-and-trade program (Metcalf and Weisbach 2009).

These attributes notwithstanding, a main argument against a carbon tax is that the U.S. public would not accept another tax, especially in the current economic climate—a charge that tax advocates suggest can be addressed both through initial impact neutrality and perhaps a symbolic name change to "atmosphere user fee"—and the cap-and-trade system itself imposes a hidden tax on consumers. Acknowledging the regressive nature of consumption taxes, revenues can be returned though direct grant to all residents regardless of their production tax status—much as Alaska currently distributes its oil revenue.

More significant is the realization that a tax, although providing certainty in price, is less likely to provide certainty with the level of emission reduction achieved, the latter being a major advantage of cap-and-trade systems. Here, too, tax advocates counter that taxes can be easily adjusted for over the multiyear implementation period as the tax rises in step with the expected benefits, whereas price volatility accompanying allowance trading generates much greater investment risk (U.S. Congressional Budget Office 2008). Finally, and perhaps decisively, the cap-and-trade system is already firmly entrenched through the EU ETS and accepted under the Kyoto Protocol. Thus, it is unlikely that the United States and other nations negotiating a follow-up to the Protocol would strike out on their own with a program that could not easily be synchronized with prevailing international practice. Even Scandinavian nations, where a carbon tax has been introduced, have also opted to join the European GHG trading system to address emissions beyond those already taxed or where sectoral loopholes were granted, especially in Norway.

Discussion and Conclusions

We have shown that most of the six conditions for well-functioning markets are not being met in the international GHG market (both EU ETS and IET). Only two of these, the requirement for a large number of market participants and ease of market entry and exit, appear to be satisfied. Absence of significant market power, a third condition, is questionable, with China as the main allowance generator in the CDM program. It is unclear if this situation will continue, given the dramatic drop-off with HFC projects and increasing questions about additionality of hydroelectric power projects in China (Pearce 2008a; Capoor and Ambrosi 2009; Schneider 2009). The other three conditions for well-functioning markets are not being met, given the great diversity of allowance-generating projects and restrictions on some of these, especially in Europe; the high transaction costs of allowance generation; and the widespread undervaluation of externalities associated with GHG emissions.

On a broader scale, whether through allocation or auction, turning the atmosphere, or more precisely the right to pollute the atmosphere, into a tradable commodity is merely the latest phase of capitalist accumulation (D. Harvey 2005; N. Smith 2007). Here resources previously held in common were forcibly privatized and turned into commodities, from land in sixteenth-century England or eighteenth-century India to modern-day biodiversity through patents on tropical pharmaceuticals, debt-for-nature swaps, water privatization in Bolivia, wetlands mitigation trading, marine fisheries, and now both pollution of the atmosphere as well as securing the tropical forests set aside for carbon offsets (Mansfield 2004; Robertson 2006; K. Smith 2007; Castree 2008).

Addressing anthropogenic climate change will require major structural adjustment in our way of life, change that most political systems have so far been unable to acknowledge, let alone prepare their constituents for. Most likely recognition of the urgency to act will be too slow to hold off serious global dislocation. Reflecting on the global economic crisis of 2008–2009, Wallerstein (2009, 18) advised that the primary agenda now for the Left is to organize at every level and in a thousand ways to "encourage the decommodification of as much as we can decommodify." The same might be said for climate change where we have trading in ephemeral commodities with real consequences as common resources are privatized.

While capitalist crisis is structurally inevitable, each collapse also offers new opportunities. Although social transformation, either by design or in reaction, is already underway in many areas most affected by climate change, for the developed world what we actually have before us is a neoliberal attempt, however imperfect, to address climate change within the confines of contemporary social and economic relations. With the United States, China, and other leading GHG emitters likely to accede to some form of cap-and-trade in subsequent rounds of the UN climate change negotiations, ratcheting down the cap while acknowledging the historical emissions and addressing the social inequity engendered will remain the most acceptable public policy option for the time being. Ultimately much more will be required.

References

Bailey, I. 2007a. Market environmentalism, new environmental policy instruments and climate policy in the United Kingdom and Germany. *Annals of the Association of American Geographers* 97 (3): 530–50.

———. 2007b. Neoliberalism, climate governance and the scalar politics of EU emissions trading. *Area* 39 (94): 431–42.

Bailey, I., and S. Maresh. 2009. Scales and networks of neoliberal climate governance: The regulatory and territorial logic of European Union emissions trading. *Transactions of the Institute of British Geographers* 34 (4): 445–61.

Brown, S., M. Burnham, M. Delaney, M. Powell, R. Vaca, and A. Moreno. 2000. Issues and challenges for forest-based carbon-offset projects: A case study of the Noel Kempff climate action project in Bolivia. *Mitigation and Adaptation Strategies for Global Change* 5 (1): 99–121.

Bumpus, A. G., and D. M. Liverman. 2008. Accumulation by decarbonization and the governance of carbon offsets. *Economic Geography* 84 (2): 127–55.

Capoor, K., and P. Ambrosi. 2009. *State and trends of the carbon market 2009*. Washington, DC: The World Bank.

Carbon Tax Center. 2009. Introduction: What's a carbon tax? New York: Carbon Tax Center. http://www.carbontax.org/introduction/–what (last accessed 20 February 2010).

Castree, N. 2008. Neoliberalising nature: The logics of deregulation and reregulation. *Environment and Planning* A 40 (1): 131–52.

Chu, S. 2009. Carbon capture and sequestration. *Science* 325 (5948): 1599.

Clo, S. 2009. The effectiveness of the EU emissions trading scheme. *Climate Policy* 9 (3): 227–41.

Ellerman, A. D., and B. K. Buchner. 2008. Over-allocation or abatement? A preliminary analysis of the EU ETS based on the 2005–06 emissions data. *Environmental & Resource Economics* 41 (2): 267–87.

Ellerman, A. D., and P. L. Joskow. 2008. *The European Union's emissions trading system in perspective*. Philadelphia: Pew Center on Global Climate Change. http://www.pewclimate.org/docUploads/EU-ETS-In-Perspective-Report.pdf (last accessed 9 July 2009).

European Commission. 2007. 2007/589/EC: Commission Decision of 18 July 2007 establishing guidelines for the monitoring and reporting of greenhouse gas emissions pursuant to Directive 2003/87/EC of the European Parliament and of the Council. Brussels, Belgium: European Commission. http://vlex.com/vid/guidelines-reporting-greenhouse-emissions-36467981(last accessed 9 July 2009).

———. 2009a. Directive 2009/29/EC of the European Parliament and of the Council of 23 April 2009 amending Directive 2003/87/EC so as to improve and extend the greenhouse gas emission allowance trading scheme of the Community. Brussels, Belgium: European Commission. http://eur-lex.europa.eu/LexUriServ/LexUriServ.do?uri=OJ:L:2009:140:0063:0087:en:PDF (last accessed 27 October 2009).

———. 2009b. Emissions trading: EU ETS emissions fall 3 % in 2008. Press release, 15 May 2009. http://europa.eu/rapid/pressReleasesAction.do?reference=IP/09/794 (last accessed 9 July 2009).

Harvey, D. 2005. *A brief history of neoliberalism*. Oxford, UK: Oxford University Press.

Harvey, F. 2009. Fall in CO_2 price threat to "green" investment. *Financial Times* 2 February:15.

Hedgeweek. 2008. New monthly record for European carbon derivatives market. 7 November. http://hedgeweek.com/2008/11/07/new-monthly-record-european-carbon-derivatives-market (last accessed 20 February 2010).

Holt, M., and G. Whitney. 2009. Greenhouse gas legislation: Summary and analysis of H.R. 2454 as reported by the House Committee on Energy and Commerce. Washington, DC: Congressional Research Service. http://fpc.state.gov/documents/organization/125498.pdf (last accessed 9 July 2009).

Intergovernmental Panel on Climate Change (IPCC). 2007. *Climate change 2007: Mitigation of climate change*. New York: Cambridge University Press.

Lawrence Berkeley National Laboratory. 2000. *Best practices guide: Monitoring, evaluation, reporting, verification, and certification of climate change mitigation projects*. Washington, DC: United States Agency for International Development.

Lohmann, L., ed. 2006. *Carbon trading: A critical conversation on climate change, privatization, and power*. Uppsala, Sweden: Dag Hammarskjold Foundation.

Macdicken, K. G. 1997. Project specific monitoring and verification: State of the art and challenges. *Mitigation and Adaptation Strategies for Global Change* 2 (2–3): 191–202.

MacKenzie, D. 2009. Making things the same: Gases, emission rights and the politics of carbon markets. *Accounting, Organizations and Society* 34 (3–4): 440–55.

Mansfield, B. 2004. Rules of privatization: Contradictions in neoliberal regulation of North Pacific fisheries. *Annals of the Association of American Geographers* 94 (3): 565–84.

Metcalf, G., and D. Weisbach. 2009. The design of a carbon tax. *Harvard Environmental Law Review* 33 (2): 499–556.

Muller, A. 2007. How to make the clean development mechanism sustainable: The potential of rent extraction. *Energy Policy* 35 (6): 3203–12.

New Carbon Finance. 2009. Carbon market roundup Q1 2009. London: New Energy Finance.

Nordhaus, W. D. 1992. An optimal transition path for controlling greenhouse gases. *Science* 258 (5086): 1315–19.

Oberndorfer, U. 2009. EU emission allowances and the stock market: Evidence from the electricity industry. *Ecological Economics* 68 (4): 1116–26.

Parry, I., and W. Pizer. 2007. Emissions trading versus CO_2 taxes versus standards. In *Assessing U.S. climate policy options*, ed. R. Kopp and W. Pizer, 80–86. Washington, DC: Resources for the Future.

Pearce, F. 2008a. Dirty, sexy money: Carbon is the new commodity, but can trading it like a currency really save the planet? *New Scientist* 19 April:38–41.

———. 2008b. Let's bury coal's carbon problem. *New Scientist* 29 March:36–39.

Pew Center on Global Climate Change. 2009. Emissions trading in the European Union: Its brief history. http://www.pewclimate.org/docUploads/emissions-trading-in-the-EU.pdf (last accessed 9 July 2009).

Robertson, M. 2006. The nature that capital can see: Science, state, and market in the commodification of ecosystem services. *Environment and Planning D: Society and Space* 24 (3): 367–87.

Schiermeier, Q. 2008. UN suspends leading carbon-offset firm. *Nature* 456 (7223): 686–87.

———. 2009. Prices plummet on carbon market. *Nature* 457 (7228): 365.

Schneider, L. 2009. Assessing the additionality of CDM projects: Practical experiences and lessons learned. *Climate Policy* 9 (3): 242–54.

Skjaerseth, J. B., and J. Wettestad. 2009. The origin, evolution and consequences of the EU emissions trading system. *Global Environmental Politics* 9 (2): 101–22.

Smith, K. 2007. *The carbon neutral myth: Offset indulgences for your climate sins*. Amsterdam, The Netherlands: Carbon Trade Watch. http://www.carbontradewatch.org/pubs/carbon_neutral_myth.pdf (last accessed 30 October 2009).

Smith, N. 2007. Nature as accumulation strategy. In *Socialist register 2007*, ed. L. Panitch and C. Leys, 16–36. London: Merlin Press.

Solomon, B. D. 1995. Global CO_2 emissions trading: Early lessons from the U.S. Acid Rain Program. *Climatic Change* 30 (1): 75–96.

———. 1998. Five years of interstate SO_2 allowance trading: Geographic patterns and potential cost savings. *The Electricity Journal* 11 (4): 58–70.

———. 1999. New directions in emissions trading: The potential contribution of new institutional economics. *Ecological Economics* 30 (3): 371–87.

UNEP Risoe Centre on Energy, Climate and Sustainable Development. 2009. CDM/JI pipeline analysis and database. http://cdmpipeline.org/ (last accessed 9 July 2009).

United Nations Framework Convention on Climate Change (UNFCCC). 1998. Kyoto Protocol to the United Nations Framework Convention on Climate Change. http://unfccc.int/essential_background/kyoto_protocol/items/1678.php (last accessed 21 February 2010).

———. 2008. Kyoto Protocol emissions trading system goes global. Press release 14 October 2008. http://unfccc.int/files/press/news_room/press_releases_and_advisories/application/pdf/20081014_press_release_itl_citl.pdf (last accessed 21 February 2010).

———. 2009. CDM statistics. Bonn, Germany: UNFCCC. http://cdm.unfccc.int/Statistics/index.html (last accessed 27 October 2009).

U.S. Congressional Budget Office. 2008. *Policy options for reducing CO_2emissions*. http://www.cbo.gov/ftpdocs/89xx/doc8934/toc.htm (last accessed 29 January 2010).

U.S. Environmental Protection Agency (EPA). 2009. Endangerment and cause or contribute findings for greenhouse gases under the Clean Air Act. http://www.epa.gov/climatechange/endangerment.html (last accessed 15 February 2010).

van Kooten, G. C., S. L. Shaikh, and P. Suchanek. 2002. Mitigating climate change by planting trees: The transaction cost trap. *Land Economics* 78 (4): 559–72.

Wallerstein, I. 2009. Reimagining socialism. Reply to B. Ehrenreich and B. Fletcher, Jr. *The Nation* 23 March:17–18.

Climate Change, Capitalism, and the Challenge of Transdisciplinarity

Joel Wainwright

A new sense of urgency has grown among many scientists for policies to address climate change, resulting in unprecedented investments by scientists in public education and, in some cases, political activism. As climate scientists have investigated future climate scenarios—and potential social responses to environmental changes—they have become, *ipso facto*, social scientists. This article examines these changes to reflect on how they could reshape geography, a discipline that appears well positioned to advance transdisciplinary research. In light of the intellectual and political urgency of transdisciplinary climate research, why have we seen so little substantive collaboration across the science/social science divide? The answer, I argue, stems from differences between research in natural science, on one hand, and the social sciences and humanities, on the other. These problems need not cause paralysis, but to address them they must be understood. The article attends to this challenge by reflecting on Albert Einstein's arguments concerning science and capitalism. *Key Words: Albert Einstein, capitalism, climate change, disciplinary knowledge, geographical thought.*

制定应对气候变化的政策，是许多科学家已经感到的一种新的紧迫感，这一情况造成了科学家在公共教育，在某些情况下，甚至是政治活动上空前的投资。随着气候学家们不断地研究未来的气候状况，以及对环境变化的潜在社会反应，他们已成为在事实上的社会科学家。本文探讨了这些变化，以期反映他们是如何重塑地理科学，做好准备以便更好地推进跨学科研究。有鉴于跨学科的气候研究所面临的学术和政治紧迫性，为什么我们很少看到跨越科学和社会科学鸿沟的全面实质性合作？至于答案，我认为，源于两者研究性质的差异，一方面是自然科学研究，另一方面是社会科学和人文科学研究。这些问题并非一定会造成学术僵硬，但是，为了解决这些问题，我们必须对此加以了解。本文迎接这一挑战，论述了爱因斯坦关于科学和资本主义的争论。*关键词：爱因斯坦，资本主义，气候变化，学科的知识，地理思维。*

En muchos científicos ha aparecido un nuevo sentido de urgencia por políticas que aboquen el cambio climático, resultando en inversiones sin precedentes causadas por aquéllos en educación pública y, en algunos casos, en activismo político. A medida que los estudiosos del clima investigan escenarios climáticos futuros—y respuestas sociales potenciales a los cambios ambientales—estos especialistas se convierten, *ipso facto*, en científicos sociales. Este artículo examina estos cambios para establecer cómo se podría con ellos reconfigurar la geografía, disciplina que aparece bien posicionada para adelantar investigación transdisciplinaria. A la luz de la urgencia intelectual y política por investigación transdisciplinaria del clima, ¿por qué ha sido tan escasa la colaboración sustantiva al otro lado de la frontera ciencia/ciencia social? La respuesta, según mi opinión, brota de de las diferencias existentes entre la investigación en ciencia natural, por un lado, y las ciencias sociales y las humanidades, por el otro. Estos problemas no son causa de parálisis, pero para abocarlos primero debemos entenderlos bien. El artículo enfrenta este reto reflexionando en los argumentos de Albert Einstein en relación con la ciencia y el capitalismo. *Palabras clave*: *Albert Einstein, capitalismo, cambio climático, conocimiento disciplinario, pensamiento geográfico.*

> There is *polemos* when a field is determined as a field of battle because there is no metalanguage, no locus of truth outside the field, no absolute and ahistorical overhang; and this absence of overhang—in other words, the radical historicity of the field—makes the field necessarily subject to multiplicity and heterogeneity. As a result, those who are inscribed in this field are necessarily inscribed in a *polemos*, even if they have no special taste for war.
>
> —Derrida (2002, 12)

Today we are experiencing a shift in conceptions of climate change, with potentially great implications for geography. On one hand, a newly strengthened scientific consensus concerning climate change has congealed. This consensus—manifest in the Intergovernmental Panel on Climate Change (2007) assessment report and its Nobel Peace Prize—is supported by a growing body of observations, enhanced global climate models, and a clearer understanding of the underlying mechanisms of change. Concurrently, a heightened sense of urgency has emerged among many scientists for state policies to address climate change, resulting in unprecedented investments by scientists in public education and, in some cases, political activism. As many climate scientists have studied and spoken out about future climate scenarios—and *social* responses to environmental changes—they have become, *ipso facto*, social scientists. Consider, for instance, the 30 July 2009 testimony by the esteemed scientist John Holdren (present Director of the U.S. Office of Science and Technology Policy) before the U.S. Senate. Dr. Holdren claims that "we know what we can and must do to avoid the worst of the possible outcomes of climate change" (2009, 2). The things we "must do" involve substantial social changes. Although this growing attention to social processes and political advocacy has generated much discussion among climate scientists, it has received relatively little scholarly attention. Such consideration is overdue.

This article aims to contribute to the emerging literature concerning climate change, disciplinarity, and geographical knowledge. The central argument is that climate change could change the discipline in terms of our object of study, our approach, and even geography's position in the modern university. This broad claim (concerned with the future, no less) cannot be demonstrated empirically. In the conclusion I propose a particular response to this situation that centers on capitalism. The form of my argument is therefore polemical. Polemic comes from the Greek word for war or conflict, transliterated *polemos*. As Derrida explained (see epigram), there is—and must be—*polemos* wherever there is thinking about the boundaries and character of a field of knowledge, or discipline. Even if we would wish it otherwise—those of us "with no taste for war," in Derrida's words—there is a certain necessity in debating the conditions that sustain a discipline (or "field"), so long as that field lacks a "metalanguage" or "locus of truth outside the field." One of my aims in this article is to reaffirm the inherent lack of such a metalanguage or locus of truth for geography. All is not lost, however. A positive result of this condition is our multiplicity and heterogeneity—qualities necessary for the thinking that is needed in an era of climate change.

Let me posit the following. Despite the extraordinary urgency of *transdisciplinary* climate research—by which I mean thinking that transcends the conventional boundaries around disciplines—the modern university has largely failed to rise to the challenge. I do not deny the production of brilliant research about climate change. Rather, I postulate that research to date has resulted mainly in technical advances and that our understanding of the physical processes that are driving climate has run far ahead of our explanations of the social processes driving the physical processes. But of course it is the social processes that must change (as attested by Holdren).

Why then has there been so little collaboration across the natural science–social science divide in climate change research? A partial answer stems from two differences between research in the natural sciences, on one hand, and the social sciences and humanities on the other.[1] The first concerns fundamental concepts. Although climate scientists engage in (often contentious) debates about the meaning of their results, they rarely reopen the "black boxes" (Latour 1987, 2) that scientists take for granted in their research. Consider, for instance, carbon. Two physical scientists might engage in heady debate about the precise role of CO_2 or CH_4 in forcing a certain atmospheric process, but it is hard to imagine that carbon's basic qualities—its atomic number or weight, chemical properties, and so on—would be called into question.[2] By contrast, two social scientists discussing, say, the hegemony of carbon-emissions markets in climate policy discourse (as indeed we do: Miller 2004; Liverman 2009) would need to agree on the meaning of *hegemony*, *markets*, *climate policy*, *discourse*, and so on. This turns out to be no mean feat, because distinct interpretations of these and related concepts reflect different conceptions of the world (Gramsci 1971). Our black boxes keep popping open. This is not to deny that social thinking can be more or less precise and rigorous. Yet one social thinker's rigor is often another's mere ideology. Why must it be so? Because we are always involved in social life and, thus, in the constant reuse and remaking of social concepts through language. And there is no metalanguage that lies outside of social life with which to objectively calibrate these concepts. Consequently, debates over the meaning of the building-block concepts for social thought are, by necessity, complex and interminable.

My use of the words *necessity* and *interminable* in the previous sentence point to a second relative difference between the natural sciences and social knowledge. There is no way to use a concept apart from one's place in a constantly changing world. Social knowledge is therefore invariably bound up with the historical processes that produce it as such (Gramsci 1971). The boundaries around accepted methods of ascertaining social knowledge—which is one way to conceptualize a discipline—are also in constant flux. (This is what Derrida meant in referring to "the radical historicity of the field"; see epigram.) Science, too, is historical (Marx [1867] 1976; Latour 1987; on climate science: Miller 2004; Liverman 2009), but its historicality complicates the task of forming new knowledge much less directly. Because we unconsciously inherit our social concepts, as well as our means of calibrating their use, social thinking at its best proceeds by accounting for its conditions of possibility via a kind of recursive process of reflecting on our thought. Gramsci (1971) called this approach *absolute historicism*. Most human geographers today call it *social theory*. Whatever we call it, it invariably enriches and complicates the task of social analysis.

Sixty years ago, Albert Einstein addressed these challenges in a concise and little-known essay (Einstein [1949] 1976), written to inaugurate the first issue of the socialist magazine *Monthly Review*. His essay confronts the question of whether and how his status as a natural scientist facilitates his venture into social thinking. Because this question concerning the relations between science and social knowledge lies at the heart of the present debate about geography and climate change, his essay merits careful reading today.

"It might appear," Einstein ([1949] 1976) began, "that there are no essential methodological differences between astronomy and economics: scientists in both fields attempt to discover laws of general acceptability" (123). But, he explains, there are two key differences. The first is that the active involvement of conscious human activity in social relations introduces profound complexities for social analysis. Taking economics as his illustration of a social science, Einstein ([1949] 1976) wrote, "The discovery of general laws in the field of economics is made difficult by the circumstance that observed economic phenomena are often affected by many factors which are very hard to evaluate separately" (123). These complications make the task of predicting human affairs—such as climate scientists might do by modeling social and economic responses to climate change—extremely complex, if not impossible.

Einstein used a curious illustration to make this point, one that elegantly foreshadowed the core argument of his essay and carries profound implications for the climate debate today. He examined the way economics is complicated by the fact that neither its object of study (the economy) nor our core economic concepts can be separated from the history of conquest and empire that facilitated the emergence of global capitalism. Einstein wrote:

> The discovery of general laws in the field of economics is made difficult by the circumstance that ... the experience which has accumulated since the beginning of the so-called civilized period of human history has ... been largely influenced and limited by causes which are by no means exclusively economic in nature. For example, most of the major states of history owed their existence to conquest. The conquering peoples established themselves ... as the privileged class of the conquered country. They seized for themselves a monopoly of the land ownership and appointed a priesthood from among their own ranks. The priests, in control of education, made the class division of society into a permanent institution and created a system of values by which the people were thenceforth, to a large extent unconsciously, guided in their social behavior. (123)

In this remarkable passage, Einstein underscored the two differences between natural science and the study of humanity discussed earlier. First, the complex interdigitation of social knowledge with unequal social relations makes it difficult to discern causal relations (what Einstein called "general laws"). Second, Einstein emphasized the historical embeddedness of economics in the processes that shape our thought, what he called the creation of a "system of values" that "unconsciously ... guid[e] social behavior."

Einstein could have stopped there, leaving the problems of economics to economists, but after making these acute observations, he drew the conclusion that scientists have a responsibility to engage with worldly affairs, one they should embrace with humility and an awareness of science's limitations:

> Science ... cannot create ends and, even less, instill them in human beings; science, at most, can supply the means by which to attain certain ends. But the ends themselves are conceived by personalities with lofty ethical ideals and—if these ends are not stillborn, but vital and vigorous—are adopted and carried forward by those many human beings who, half unconsciously, determine the slow evolution of society. For these reasons, we should be on our guard *not to overestimate science* and scientific methods when it is a question of human problems; and we should not assume that experts are the only ones who have a right to express themselves on questions affecting the organization of society. (Einstein [1949] 1976, 124, italics added)

Einstein thus qualified his own capacity to speak to social issues by locating his critique within the histories of contested knowledge, yet proceeded to express his views "on questions affecting the organization of society." Where does this strategy leave geographers who wish to "express [our]selves" on the social dimensions of climate change?

Geography and Climate Change

The debate over the appropriate response by geographers to climate change (and global environmental change more generally) has produced a considerable literature (Smith [1984] 2008; Kates 1987; Balling 2000; Turner 2002; Thrift 2002; Hobson 2008; Hulme 2008; Demeritt 2009; Harden 2010). These and other papers build on earlier debates about the character of geography qua discipline and the challenge of overcoming long-standing divisions between scientists and nonscientists. Since the formation of the modern university in the nineteenth century, few qualities have proven as permanent as the science–nonscience distinction and the organization of knowledge around it—even as the underlying mission of the university has all but crumbled.[3] As a discipline with scientists and nonscientists working side by side on a vast range of questions and problems, geography stands alone. With the possible exceptions of anthropology and psychology (and only then in certain instances), there are no disciplines where scientists and "humanists" have sustained shared curriculum and research for long periods. But these other cases, notably, at least share a common object ("man"; the psyche). Geographers have also disagreed on our object: spatial relations, human–environment relations, the region, and so on.

Climate change is fomenting a new round of debate about our object, purpose, and place in the university—that is, about geography's very disciplinarity. To this point, Turner (2002) argued that "geography's different identities have never been unified within a logic that matches the full scholarly practice of its membership and is consistent with the rationale on which knowledge is organized" (53), a condition that Turner finds problematic. He concluded that modern geography "searches for intellectual coherence and acceptance but votes for diversity of practice" (62) and needs reconciliation.

Unfortunately, Turner's paper leaves unexplored—and this is true throughout the literature—the potential conditions of possibility for a bridging within a discipline that is, perhaps, essentially transdisciplinary (or "science-to-humanities," in Turner's words). It is precisely this unexplored terrain that has been illuminated by the problem of thinking about climate change. One unexpected element illuminated by climate change is that, strangely enough, the *object* of study might not have been the problem after all: Witness the geographers from one end (called "science") of the discipline to the other ("humanities") who are writing about climate change today. Perhaps the *polemos* over our object might end?

The prospect of a new era—a golden one for geography?—has gripped many geographers with enthusiasm. Consider Balling's (2000) argument for the "geographer's niche in the greenhouse millennium." Balling's central claim is that "the entire global warming/greenhouse issue is perfectly suited to our discipline" (115), for two reasons. First, "geographers have a role in developing and refining the numerical models of climate that are used in the simulation studies," particularly in grappling with the scalar and spatial complexities of these models (115). Second, geographers are well attuned to the challenge of "conduct[ing] effective tests of the theoretical models with empirical data." Balling provided a persuasive argument and one that I think many geographers would like to believe. Who could deny that it is pleasant to discover—in the midst of the challenges wrought by climate change—a disciplinary silver lining? By a stroke of luck, it turns out that the planet's problems are our professional gain.

Balling's (2000) argument that climate change is a boon for geography is rooted in the assumption that geographers are good at spatial analysis, which therefore allows us to grapple with "the fundamental questions that underlie . . . global warming" (115). This is an attractive argument but one that raises a number of questions. Do we in fact know which are the "fundamental questions" underlying global warming? If so, how? Is this a question that can be ascertained scientifically? How, moreover, do we know that these fundamental questions call for the sort of spatial analysis that some geographers do?

With these questions in mind I turn to Demeritt (2009), who cautioned against "over-emphasizing the unique pedigree and potential of geography as a synthetic environmental science" (127). Demeritt made two arguments that throw cold water on the sort of enthusiastic celebration of geography in an era of climate change we find in Balling (as well as Hulme 2008[4]). The first is that geography no longer has a clearly defined object (such as the region or spatiality) to serve as the basis for collaboration across the science–social divide. A propos Balling, it is indubitable that climate change is spatially complex. Although geographers might grapple with the spatial complexity of natural and social processes, this "spatial complexity" is itself multiple;

and because we think in different conceptual languages, the heterogeneity of approaches to spatial complexity leaves a considerable task of translation. Climate change is indeed spatially complex, but this does not ensure that geographers can or will study it collaboratively.

This leads to a second argument from Demeritt's paper. The considerable differences between physical and human geographers and the specialized training received by graduate students in each subfield create challenges for collaboration:

> [H]uman and physical geography are increasingly alienated from one another. Within those two very broad churches, further subdisciplinary specialization . . . sometimes make it seem as if the only thing that the geographers in many academic departments share in common is that their mail all gets delivered to the same address. . . . It is true that geography departments contain many of the separate specialists that science policy makers are trying to bring together through interdisciplinary programmes. . . . But it is not clear that the discipline will necessarily enjoy any greater success bringing those parts together than collaborations based outside geography departments. (128)

Demeritt's argument is fair, yet he might overstate the novelty of the situation. Lamenting the "increasing" division of human and physical geography is an old theme (see Kates 1987).

Outside of the discipline, much of the recent writing on interdisciplinary research around climate change carries a celebratory tone. Consider, for instance, the recent National Academy of Sciences (NAS 2009) report calling for the restructuring of national climate research. Their assessment is that "a transformational change is needed in how climate change research is organized and in how the results are incorporated into public policy" (1). They outlined six priorities for change; four directly concern our knowledge about the social *causes*, and especially the *ramifications*, of climate change. They call for "reorganiz[ing] the program around integrated scientific-societal issues" (2); more research into "adaptation, migration, and vulnerability" (3); and the creation of a comprehensive "national assessment of the risks and costs of climate change impacts" (3). All this sounds like sweet music to geographers.

Yet there are elements of the NAS (2009) report that I suspect would rub many human geographers, at least, the wrong way. Take the NAS's last suggestion:

> *Coordinate Federal Efforts to Provide Climate Information, Tools, and Forecasts Routinely to Decision Makers.* Demand is growing for credible, understandable, and useful information for responding to climate change. To satisfy this demand, a national climate service is needed to facilitate two-way communication between scientists and decision makers, monitor climate trends, and issue information and predictions to support decision makers. (4)

On the face of it, this seems like a terrific idea. Who could argue against closer coordination between scientists, who know better than anyone how climate is changing, with decision makers, who are in a position to effect change?

For his part, Einstein cautioned scientists not to uncritically offer their expertise to state elites nor to overestimate the value of their scientific knowledge. Bearing his argument in mind, the drive to facilitate stronger "two-way communication between scientists and decision makers" is questionable. This conception of a two-way communication leaves no place in the conversation for nonscientists, workers, farmers, the unemployed, and so forth; in short, everyone other than scientists and decision makers. And who, indeed, are decision makers? This is a euphemism for state officials and elites (i.e., the wealthy). Here comes the "privileged class" that Einstein warned us about.

Let me be clear. I see the NAS (2009) call for a "transformational change" in climate research as a progressive step. Yet, we should ask this: Exactly what kind of transformation in our knowledge about climate change is needed? Kates (1987, 532) argued that the underlying strength of geography in the modern university lies in our emphasis in training students to ponder "the great questions" surrounding human–environmental change. I agree. Yet, what are the great questions concerning climate change? Can we be confident that we are training our students to answer them?

To address these questions and to respond justly to the challenge of climate change, I contend, will require us to be open to collaborative, transdisciplinary, radical thinking. (By *radical* I mean both getting to the roots of the matter and politically contentious.) Because this argument is—of necessity—polemical, it is only responsible to conclude by sketching the boundaries of such a possible project. Thus, to close this essay, I aim to clarify some of the great questions concerning climate change and capitalism. I do so for three reasons. First, I am following Einstein's cue. Second, if Readings (1997) is correct, the crisis of the university is partly a result of the university's uncertain place in global capitalist society. Third, if we geographers wish to pride ourselves in pondering the great questions of our time, we must be willing to ask whether the very form of social and economic life—capitalism—is not an underlying cause of climate change.

Capitalism and Climate Change

The historical coincidence of the emergence of global capitalism with the transformation of our planet's atmosphere is no accident. Capitalism is at the heart of the challenge of confronting climate change, and any serious attempt to address global climate change must contend with global capitalism. Why? Allow me to summarize a complex body of ideas under two points (see Marx [1867] 1976; Walker and Large 1975; Harvey 1982; Smith [1984] 2008; Bello 2008; Karatani 2008).

The Accumulation Drive

Capitalism is defined by the production and sale of commodities to extract surplus value and facilitate accumulation of money. Marx's ([1867] 1976) general formula for capital is M – C – M′. This means, first, that money (M) is invested by a capitalist to purchase labor power and means of production to produce commodities (C). This process (M – C) is *production*. As Einstein ([1949] 1976) wrote:

> We see before us a huge community of producers the members of which are unceasingly striving to deprive each other of the fruits of their collective labor—not by force, but on the whole in faithful compliance with legally established rules. . . . [T]he means of production . . . may legally be, and for the most part are, the private property of individuals. (167)[5]

Yet production comprises only one half of the process we call capitalism. The commodities must be sold, or purchased, returning money to the capitalist who put up the money to begin the process (to get more money at the end of the process). The second moment of the general formula for capital (C – M′) is *consumption*. This is when the value congealed in the commodity via production is realized as money (M′). The money that is returned to the capitalist is typically reinvested into the process again, to facilitate further accumulation. Capitalism is best conceptualized not as a thing but as a growth-oriented process of commodity production and consumption driven by the constant need to realize more value (as money). Capital's drive for circulation and accumulation (M – C – M′) is the underlying source of the incessant demand for growth and expansion that we associate with capitalist economies. A society organized on capitalist lines cannot deny this drive for capital accumulation. Accumulation begets accumulation, without end or purpose: This is the source of capitalism's undeniable dynamism.

This has several implications for climate change. Perhaps the most obvious is that expansion and accumulation require the constant conversion of the planet into means of production. To put it crudely, capital's drive converts the world into commodities. Although individual capitalists are often swayed by environmental ideals, as a social class, capitalists cannot but treat nature as an enormous collection of resources. Yet the increasing concentration of CO_2 in the Earth's atmosphere (from ~280 ppm before the emergence of capitalism to ~385 ppm today) suggests that there are planetary limits to capitalism's growth drive.[6] This is not to deny the possibility of social and technical responses to climate change that mitigate its effects. But unless these responses address the principle cause of climate change (i.e., energy use to fuel the global capitalist economy), they are severely limited. Heidegger ([1955] 1969) wrote, "The world now appears as an object open to the attacks of calculative thought. . . . Nature becomes a gigantic gasoline station, an energy source for modern technology and industry" (50). This conception of the world is incompatible with an effective global response to climate change.

Inequalities

It is no accident that the world has become dramatically more unequal with the emergence of global capitalism. Capitalism's inherent drive for accumulation produces inequalities of wealth and power.[7] Einstein ([1949] 1976) saw this, too:

> Private capital tends to become concentrated in few hands, partly because of competition among the capitalists, and partly because technological development and the increasing division of labor encourage the formation of larger units of production at the expense of smaller ones. The result of these developments is an oligarchy of private capital the enormous power of which cannot be effectively checked even by a democratically organized political society. This is true since the members of legislative bodies are selected by political parties, largely financed or otherwise influenced by private capitalists who, for all practical purposes, separate the electorate from the legislature. The consequence is that the representatives of the people do not in fact sufficiently protect the interests of the underprivileged sections of the population. (168)

Capitalism's tendency to deepen inequalities of wealth and power is tightly linked to the challenge of confronting climate change, because a meaningful response to climate change will require sacrifices, transnational alliances, and trans-class cooperation. Inequality is fatal to these at two levels. Within a capitalist economy, inequalities in wealth and power make it difficult to build coalitions around shared sacrifice. Inequality also entrenches the capacity of the wealthy—who benefit disproportionately from economic growth—to prevent the conversion of our carbon-intensive economy into a more sustainable alternative. In Einstein's ([1949] 1976) words, the state fails to "protect the interests of the underprivileged sections of the population," which is most people, because "political parties [are] largely financed or otherwise influenced by private capitalists" (168). Consider the effectiveness of U.S. energy companies in funding "climate skepticism" and lobbying politicians against a carbon tax (Compston and Bailey 2008, 265). Their effectiveness is rooted in private wealth.

Second, between capitalist economies, the massively unequal dispensations of wealth and power in the world prevent the kind of global compromise that will be necessary to address climate change. In their trenchant analysis of international carbon production and climate change politics, Roberts and Parks (2007) showed that the failure to achieve any global agreement to reduce carbon emissions is "rooted in the problem of global inequality: inequality in who is suffering the problem, who caused it . . . , who is expected to address [it], and who currently benefits disproportionately from the goods produced by the global economy" (135).[8] As long as the world is capitalist, these inequalities will persist: so too, then, global arriers to addressing climate change.

In light of these two fundamental criticisms, it might seem that nothing short of a radical transformation of the social and economic life is necessary for survival. To conclude with this argument might seem unduly polemical, but it is, I think, true. I take courage here from Einstein's conclusion to his essay: "I am convinced there is only one way to eliminate these grave evils, namely through the establishment of a socialist economy, accompanied by an educational system which would be oriented toward social goals" ([1949] 1976, 131).

I recognize that this conclusion raises numerous questions that go beyond the scope of this article. Let me mention four. What might constitute an education that is "oriented toward social goals" in Einstein's sense? What conception of the world might create the conditions for a just world economy defined by low-carbon livelihoods? If sustainability requires that we transcend capitalism, how can we avoid the violence and tragic failures of the communisms of the twentieth century? How can we contribute to a radical reimagination of our world, one more just and sustainable?

These might seem like questions for a political party, not an academic discipline. Yet, I contend, they are the sort of great questions that Kates identified at the heart of the geographical tradition. Taken together, they define the sort of transdisciplinary project that I have fended for in this article. They define a project that belongs to no discipline but one that requires critical geographical thinking. The character of this thinking might be mainly nonscientific, but that is not to say that scientists have no role. On the contrary. As Einstein demonstrates, the most thoughtful criticisms might come from natural scientists.

Acknowledgments

The article benefitted from the thoughts and criticism of Kiran Asher, Jason Box, Leila Harris, Will Jones, David Lansing, Geoff Mann, Kristin Mercer, Ellen Mosley Thompson, Morgan Robertson, and four anonymous reviewers. It is dedicated to my colleagues in climate science, whose labors calibrate our hopes and fears.

Notes

1. Scholars of the social sciences and humanities study much the same things, albeit differently, and usually without much collaboration. In Bloom's (1985) words, the two fields "represent the two responses to the crisis caused by the definitive ejection of man . . . from nature, and hence from the purview of natural science or natural philosophy, toward the end of the eighteenth century" (357). For Bloom, the difference between them "comes down to the fact that social science really wants to be predictive, meaning that man [*sic*] is predictable, while the humanities say that he is not" (357). Compare Latour (1987, 2004) and Kagan (2009).
2. This is emphatically not to deny the necessity of examining scientific practices (see Latour 1987, 2004; for a critique, see Wainwright 2005) nor the social and political qualities of climate science (see Jasanoff 2004; Miller 2004; Hulme 2009). Rather, it is to emphasize these substantive *relative differences* between science and social science and the humanities (Bloom 1985; Kagan 2009). I agree with Marx ([1867] 1976) that "there is but a single science, . . . not separate sciences of nature and society," to cite Smith's gloss ([1984] 2008, 276; compare Karatani 2008, 571).
3. Many people who study climate change do not work in academia. Here I discuss the university ideal because

(1) most geographers who study climate change do so within formal universities (i.e., institutions structurally based on this ideal), and (2) the effects of this ideal impinge on the conditions of our work and thinking. For contrasting views on the crisis of the university, compare Bloom (1985), Kates (1987), and Readings (1997). Bloom's book became a best seller because it savaged liberal hegemony in U.S. higher education, but there is more to his study than this. The political implications of Readings's study—the best of these—are also ambiguous. Only Kates addressed geography and human–environment issues.

4. Drawing on Latour, Hulme (2008, 6) contended that geographers "have a unique role to play" in studying climate change because of the need for studying climate as a "hybrid phenomenon"; that is, one that that is neither natural nor cultural. Hulme's argument presumes that geographers often cross these boundaries to study social and natural objects, but I am not so sure.
5. Although his essay contains no citations, Einstein's critique of capitalism draws from Marx's *Capital* ([1867] 1976). The essay's middle section could be read as a lucid summary of *Capital*. Thus, Einstein provided a brilliant exception to an argument Marx advanced in a footnote from *Capital*: "The weaknesses of the abstract materialism of natural science . . . are immediately evident from the abstract and ideological conceptions expressed by its spokesmen whenever they venture beyond the bounds of their own specialty" (Marx [1867] 1976, 494).
6. There are also economic and social limits. Capital's drive cannot but produce periodic crises (Marx [1867] 1976; Harvey 1982) like the present one (rooted in overproduction and financial liberalization). Economic crises typically compel states to intervene to stimulate consumption (C – M′), a tendency that runs contrary to the response needed for climate change.
7. For theory, see Marx ([1867] 1976) and Harvey (1982); for data, see Wade (2004). It is indubitable that total socialized value ("wealth") and global median income have increased. But even with rising incomes, growing inequality can undermine the capacity for collective action by reducing willingness to share sacrifices.
8. Consider the July 2009 agreement by the world's leading states to limit global temperature increases to 2°C, while failing to place any substantive limits on the emissions that cause warming. This agreement perfectly reflects Mark Twain's (1897) quip that "Everybody talks about the weather, but nobody does anything about it."

References

Balling, R. 2000. The geographer's niche in the greenhouse millennium. *Annals of the Association of American Geographers* 90 (1): 114–22.

Bello, W. 2008. Will capitalism survive climate change? *Bangkok Post* 29 March. http://www.waldenbello.org/content/view/88/30/ (last accessed 9 July 2009).

Bloom, A. 1985. *The closing of the American mind*. New York: Touchstone.

Compston, H., and I. Bailey, eds. 2008. *Turning down the heat: The politics of climate policy in affluent democracies*. London: Palgrave Macmillan.

Demeritt, D. 2009. Geography and the promise of integrative environmental research. *Geoforum* 40:127–29.

Derrida, J. 2002. *A taste for the secret*. New York: Polity.

Einstein, A. [1949] 1976. Why socialism? In *Out of my later years*, 123–31. New York: Citadel.

Gramsci, A. 1971. *Selections from the prison notebooks*, ed. Q. Hoare and G. N. Smith. New York: International.

Harden, C. 2010. A tipping point for geography. *AAG Newsletter* 45 (3): 3.

Harvey, D. 1982. *The limits to capital*. Chicago: University of Chicago.

Heidegger, M. [1955] 1969. *Discourse on thinking*. New York: Harper & Row.

Hobson, K. 2008. Reasons to be cheerful: Thinking sustainably in a (climate) changing world. *Geography Compass* 2 (1): 199–214.

Holdren, J. 2009. Statement of Dr. John P. Holdren, Director, Office of Science and Technology Policy, Executive Office of the President, before the Committee on Commerce, Science and Transportation, United States Senate, 30 July. Washington, DC: The White House. http://www.whitehouse.gov/files/documents/ostp/press_release_files/HoldrenTestimony.pdf (last accessed 1 October 2009).

Hulme, M. 2008. Geographical work at the boundaries of climate change. *Transactions of the Institution of British Geographers* 33:5–11.

———. 2009. *Why we disagree about climate change*. Cambridge, UK: Cambridge University Press.

Intergovernmental Panel on Climate Change. 2007. Climate change 2007: Synthesis report: Summary for policymakers. http://www.ipcc.ch/pdf/assessment-report/ar4/syr/ar4_syr_spm.pdf (last accessed 6 July 2009).

Jasanoff, S., ed. 2004. *States of nature: The co-production of science and social order*. London and New York: Routledge.

Kagan, J. 2009. *The three cultures: Natural sciences, social sciences, and the humanities in the 21st century*. Cambridge, UK: Cambridge University Press.

Karatani, K. 2008. Beyond capital-nation-state. *Rethinking Marxism* 20 (4): 569–95.

Kates, R. 1987. The human environment: The road not taken, the road still beckoning. *Annals of the Association of American Geographers* 77 (4): 525–34.

Latour, B. 1987. *Science in action: How to follow scientists and engineers through society*. Cambridge, MA: Harvard University Press.

———. 2004. *Politics of nature: How to bring the sciences into democracy*. Cambridge, MA: Harvard University Press.

Liverman, D. 2009. Conventions of climate change: Constructions of danger and the dispossession of the atmosphere. *Journal of Historical Geography* 35:279–96.

Marx, K. [1867] 1976. *Capital*. New York: Penguin.

Miller, C. 2004. Climate science and the making of a global political order. In *States of nature*, ed. S. Jasanoff, 46–66. London and New York: Routledge.

National Academy of Sciences. 2009. Restructuring federal climate research to meet the challenges of climate change. http://www.nap.edu/catalog.php?record_id=12595#toc (last accessed 6 July 2009).

Readings, B. 1997. *The university in ruins*. Cambridge, MA: Harvard University Press.

Roberts, J. T., and B. C. Parks. 2007. *A climate of injustice: Global inequality, North–South politics, and climate policy*. Boston: MIT Press.

Smith, N. [1984] 2008. *Uneven development: Nature, capital, and the production of space*. Athens: University of Georgia.

Thrift, N. 2002. The future of geography. *Geoforum* 33, 291-298.

Twain, M. 1897. Source uncertain. *Hartford Courant* 24 August 1897. http://quotationsbook.com/quote/41294/ (last accessed 6 July 2009).

Turner, B. L. 2002. Contested identities: Human-environment geography and disciplinary implications in a restructuring academy. *Annals of the Association of American Geographers* 92:52–74.

Wade, R. 2004. Is globalization reducing poverty and inequality? *World Development* 32 (4): 567–89.

Wainwright, J. 2005. Three recent works by Bruno Latour. *Capitalism Nature Socialism* 16 (1): 115–27.

Walker, R., and D. Large. 1975. The economics of energy extravagance. *Ecology Law Quarterly* 4:963–85.

Contested Sovereignty in a Changing Arctic

Hannes Gerhardt, Philip E. Steinberg, Jeremy Tasch, Sandra J. Fabiano, and Rob Shields

Climate change is challenging the notions of permanency and stability on which the ideal of the sovereign, territorial state historically has rested. Nowhere is this challenge more pressing than in the Arctic. As states expand their sovereignty claims northward in pursuit of potential opportunities (in many cases made possible by climate change), these same states are being confronted with the region's increasing territorial indeterminacy (which also is exacerbated by climate change). To investigate how climate change is challenging the territorial imaginaries around which notions of sovereignty historically have been based, we turn to three debates in the contemporary Arctic: the question of sovereignty in the Northwest Passage, conflicts over territorial control in the Arctic Ocean, and the potential for enhanced multilateral governance. Through our study of these debates we engage the Arctic both as a region that is undergoing climate change's most extreme impacts and as a laboratory for understanding how these and similar impacts might modify the spatial organization of political authority across the world. *Key Words: Arctic, climate change, ice, North Pole, sovereignty.*

气候变化对主权的永久性和稳定性提出了挑战，这两个特性是领土国家历史性概念的基础。北极所面临的挑战最为迫切。一些国家扩大了其北上主权的要求，籍此谋求潜在的机会（这一可能性在许多情况下是由气候变化造成的），同样是这些国家，正在面临该地区领土不确定性的增加（同样由于气候变化而加剧）。为了探讨气候变化是如何挑战这些历史上主权概念所依赖的领土理念，我们对当代北极所涉及的三次辩论进行了研究：西北通道的主权争议，北冰洋领土控制的冲突，以及增强多边治理的可能性。通过我们对这些辩论的研究，我们认为，北极既是一个正在经历气候变化最极端影响的地区，同时也是一个实验室，籍此我们可以了解这些影响和其它类似的影响将如何改变世界各地政权的空间组织结构。*关键词：北极的，气候变化，冰，北极，主权。*

El cambio climático está poniendo en tela de juicio las nociones de permanencia y estabilidad sobre las cuales históricamente ha descansado el ideal del estado territorial soberano. En ninguna otra parte este reto es tan apremiante como en el Ártico. A la vez que los estados expanden sus reclamos de soberanía hacia el norte en pos de oportunidades potenciales (que en muchos casos han sido posibles por el cambio climático), estos mismos estados se ven confrontados con la creciente indeterminación territorial de la región (la cual es también exacerbada por el cambio climático). Para investigar el modo como el cambio climático está retando los imaginarios territoriales alrededor de los cuales históricamente se han basado las nociones de soberanía, volvemos sobre tres de los debates relacionados con el Ártico contemporáneo: el asunto de la soberanía en el Paso del Noroeste, los conflictos de control territorial en el Océano Ártico y el potencial de un gobierno multilateral fortalecido. A través de nuestro estudio de estos debates, abocamos el Ártico tanto como una región que está experimentando los impactos más agudos del cambio climático, como una suerte de laboratorio que nos permita comprender cómo estos y otros impactos similares podrían modificar la organización espacial de la autoridad política a través del mundo. *Palabras clave: Ártico, cambio climático, hielo, Polo Norte, soberanía.*

Global climate change is nowhere more pronounced in the Northern Hemisphere than in the Arctic region, where sea ice and glaciers are diminishing at an unprecedented rate. These changes will have profound effects on livelihoods around the world, as melting glaciers lead to sea-level rise and reduced albedo increases the rate of global warming trends (see Figure 1).

Much of the world views these changes with consternation, but northern states have come to realize that climate change in the Arctic presents not only socioecological challenges but economic and geopolitical opportunities. Arctic states—particularly Canada, Denmark (Greenland), Norway, Russia, and the United States, the five states with significant coastal claims in the Arctic Ocean—recognize that a warming climate

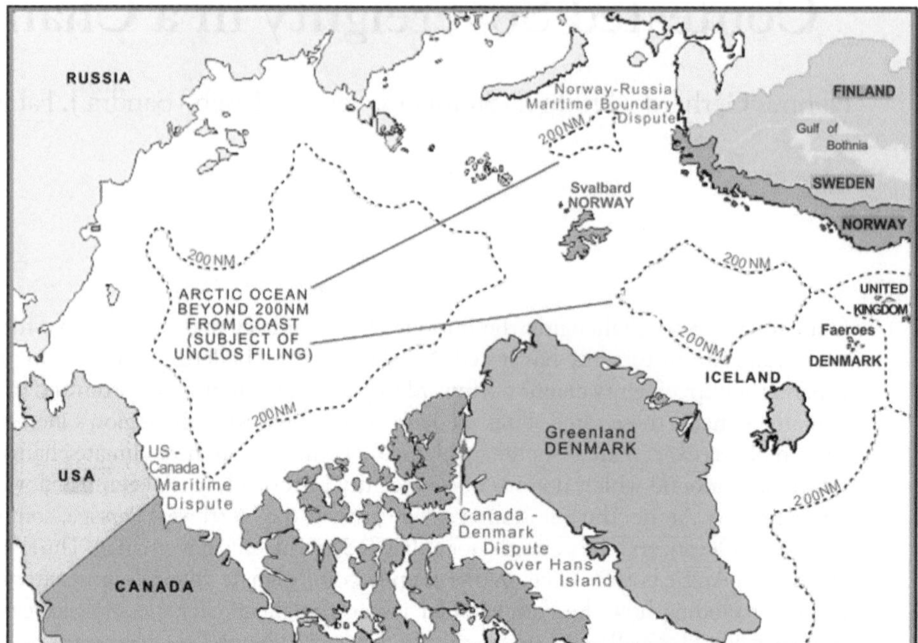

Figure 1. The Arctic (map by J. Tasch, based on C. Jayasuriya).

can bring a range of transformations to the region, including the opening of new shipping routes; easier access to reserves of natural gas, oil, and precious minerals; and new options for projecting military presence.

In fact, the economic opportunities presented by a thawing Arctic might be overstated, both with reference to the region's navigational potential (Arctic Council 2009) and its mineral wealth (Powell 2008). Likewise, many of the environmental conditions associated with the Arctic and discussed in this article existed prior to recent changes in the regional climate. Nonetheless, the translation of scientific knowledge into everyday understandings and practices is mediated through institutions and iconography that can proceed in parallel, becoming partially detached from the changes and continuities that are occurring on the ground, particularly when these occur at a difficult-to-grasp scale (e.g., global climate change) or a difficult-to-access region (e.g., the Arctic; Slocum 2004). And, as scholars of critical geopolitics have demonstrated, changes in media representations and everyday understandings of seemingly distant problems are often sufficient to alter the political landscape and influence policy choices (Ó Tuathail 1996).

As Arctic states grapple with changing conceptualizations of their northern frontiers and consider expanding economic activities there, unresolved questions regarding the limits of state sovereignty in the Arctic Ocean constrain their efforts. It is our contention that ongoing contestation of sovereignty in the Arctic is rooted in the region's indeterminate and unstable geophysical characteristics. Historically, ideals of sovereignty have assumed a basic elemental distinction between land, recognized as amenable to sedentarization and hence territorialization, and water, designated as resistant to these assertions of control (Steinberg 2001, 2009). Even those who argue that a fundamental shift in the relationship between land power and sea power occurred in the late nineteenth century (e.g., Mackinder 1904; Schmitt 2006) expand the ontological distinction between Earth's two essential surface features—boundable land and unboundable water—into an epistemological principle.[1] This pairing of two apparently fixed and totalizing binaries—land versus water and boundable territory versus unboundable nonterritory—resonates with a long-standing tendency to assert difference and belonging by fusing "metageographic" reifications of territorial landmasses with appeals to "commonsense" materiality (Lewis and Wigen 1997). However, commonsense materiality, which is always contestable, is particularly problematic in the Arctic (Lopez 2001; Davidson 2005), especially amidst climate change.

At the most basic level, the binary division of Earth into land and water is confounded in the Arctic by the presence of ice, a liminal substance that combines and confuses properties of the two. Beneath the cover of ice, it is not always clear where land ends and water begins (Pyne 1998; Laidler 2006), a situation that has long frustrated state constructions of power and knowledge in polar regions, whether gazing from the bow of a research vessel or analyzing an image taken by a satellite (Wråkberg 2004; Yusoff 2005). Furthermore, today's ice could be tomorrow's water. Like the ever-shifting tectonic plates of Massey's (2005, 2006) landscape (cf. J. T. Wilson 1963), Arctic materiality, stable in neither time nor space, complicates the ideal of permanence that underpins modern concepts of sovereignty, just as it complicates the practice of polar science (Frazar 2008).

Given this disparity between the idealized materiality of the system of territorially delimited sovereign states and the actual materiality of the Arctic, it is perhaps not surprising that practices of sovereignty and statehood in the region have diverged from the modern norm. For many northern coastal peoples, for example, materiality is not limited to land and water but incorporates sea ice, seasonal land-fast ice, and the shifting spaces in between, together with localized connections among varying types of wind and water currents, animal behaviors, tidal stages, snow, temperature, and the processes that operate underneath the ice cover (Kerttula 2000; Aporta 2002; Jolly et al. 2002). Such varying conceptions of the Arctic's materiality can lead to political institutions that stray even further from the modern ideal.

Because the Arctic has been governed from non-Arctic capitals, the region has seen a number of variations on the practice of sovereignty. For instance, British sovereignty in the Canadian Arctic historically was mediated by the relatively autonomous territorial control exercised by the Hudson's Bay Company. The Spitzbergen Treaty of 1920 gives Norway sovereignty over Svalbard but with certain provisions that limit Norway's control over economic resources. In several countries in the region, indigenous peoples have some level of autonomy. Further complicating these gradations of sovereignty, northern peoples often assert circumpolar bonds that transcend state structures and cross state borders, as in the case of the Barents Euro-Arctic Region, which brings together nonstate, state, and suprastate (the European Union) actors for regional governance. Even the United Nations Convention on the Law of the Sea (UNCLOS; United Nations 1982), a document that fundamentally reinforces the idealized political–juridical separation of water from land, establishes a special regime for areas with "particularly severe climatic conditions and the presence of ice covering such areas for most of the year" (Article 234), where coastal states are granted additional jurisdiction with respect to the environment in their Exclusive Economic Zones (EEZs).

In short, the history of sovereignty in the Arctic reflects ongoing attempts to square the multifaceted and unstable nature of Arctic materiality with a constructed ontology (or, to use Schmitt's term, a *nomos*) that assumes a fixed division between land and water.[2] Yet, as states pursue this goal, the Arctic's climate is becoming more volatile. Past ambiguities and inaccessibility can no longer be relied on by coastal states, and this is leading to a scramble for new material foundations on which to base sovereign claims.

In Canada, one of several Arctic states in which current and past governments have directed resources to the northern frontier, officials have been particularly eager to classify the region's dynamic materiality. In support of these efforts, some have suggested that Canada is less a continental landmass than an archipelago: a nation of land and water (and the liminal ice that mediates this division; Vannini et al. 2009). Indeed, if ice creates ambiguity, it allows water to be construed as solid territory. As Canadian jurist Donat Pharand has written, when arguing for drawing straight baselines around Canada's coasts, "The quasi-permanency of the ice over the enclosed waters bolsters the physical unity of land and sea" (Pharand 2007, 19). A senior official with Canada's Department of Foreign Affairs and International Trade (DFAIT) elaborated on this point:

> We're dealing with virtually the world's only archipelago, certainly the world's only large archipelago, which has ice-covered areas throughout its surface. The question is what is the status of that ice vis-à-vis the land around it. . . . At some point we may end up before an international court [and] we will bring evidence that shows the people of the Canadian North, or Canadian citizens, in the winter time have treated the ice exactly the same as the land, and we'll make a very strong argument for that. (Interview, June 2008; see also Byers 2010; Fenge 2007–2008)

The focus on ice has been so intense that Canadian nationalists have not considered the other aspects of Article 234, "severe climatic conditions," which some models forecast will become even more extreme with a hitherto unexperienced Arctic hurricane season (Kolstad and Bracegirdle 2008). This suggests that geopolitical arguments are driven by popular idealizations of a place—territorial imaginaries[3]—as well as legal precedents and statutes. Historically, territorial imaginaries have served as foundations for constructions of sovereignty by referring to territory as solid and stable—boundable units of land that can be charted, demarcated, and mapped to signify the "body" of the nation. In the Arctic, states often seek to perpetuate these traditional and assumed ways of defining and delimiting sovereign jurisdiction, notwithstanding the region's increasing material instability brought about by climate change.

To expound on this proposition we turn to three debates in the contemporary Arctic: the question of sovereignty in the Northwest Passage, conflicts over control in the Arctic Ocean, and the potential for enhanced multilateral governance in the region. Through our study of these debates, we engage the Arctic both as a region that is undergoing climate change's most extreme impacts and as a space for understanding how these impacts can modify the territorial organization of political authority in the region and beyond.

Navigating Sovereignty in the Northwest Passage

In September 2007, sea ice in the Canadian Arctic archipelago shrank to its lowest levels since satellite measurements began (European Space Agency 2007), and many have predicted that within the next decades commercial-size vessels will be able to make summer transit unassisted (Institute of the North, U.S. Arctic Research Commission, and International Arctic Science Committee 2004). These developments have led to anxious sovereignty assertions by the Canadian government, which maintains that the Passage is a component of the internal waters of Canada (Office of the Prime Minister 2007). Others with an interest in Arctic shipping (most notably the United States) have repeated long-held positions that the Passage is an international strait that, although bisecting Canadian territory, connects two bodies of water (Bureau of International Information Programs 2007). The resolution of the dispute has important implications for transit in the region. If the Passage is classified as Canadian internal waters, Canada has unlimited rights to restrict other nations' vessels. Contrarily, if these same waters are classified as an international strait, Canada's rights to regulate transit passage are significantly reduced.

In one sense, this is a narrow dispute of international law, as governments duel over the juridical status of the Northwest Passage. Yet the very designation of this tangle of waterways in the Arctic archipelago as a singular Northwest Passage belies a more complex attempt to spatialize a liminal environment (Shields 1991). The designation of the region as the Northwest Passage both increases the space's governability (it is now a distinct and named space) and can decrease its visibility and viability (it is no longer a place or a set of places but simply a mere "passage" to somewhere else that matters). In addition, this attempt at "fixing" the Northwest Passage as a distinct space is accompanied by attempts to fix it with a distinct nature: what it is (ice or water), what it was, and what it will be. Thus, the Canadian DFAIT official quoted earlier, who was referring specifically to the Northwest Passage and why it should be governed as Canadian internal waters, was basing his argument on the assertion that historically the Passage was geophysically and socially "just like the land." Conversely, although it has been navigated only rarely, and mostly not under power but by drifting in pack ice over a winter, U.S. Navy Commander James Kraska has asserted that the Passage is a "strait used for international navigation" and therefore it is fundamentally a part of the ocean and should be governed according to the principles of UNCLOS (Kraska 2007, 257).

Yet, despite such efforts to ascribe permanence to Arctic space, states have also at times acknowledged that the Arctic's materiality, and hence its social function, is ever shifting. In 1970, Canada passed the Arctic Waters Pollution Prevention Act (Government of Canada 1985), which, in acknowledgment of the region's exceptional environment, restricts uses of adjacent northern seas, and which formed the basis for Article 234 of UNCLOS.

As the Arctic Waters Prevention Pollution Act and Article 234 suggest, the binary division of space into land and water and the conceptual mapping of land to territory and water to nonterritory fail to resonate with the region's actual materiality. It is this materiality, furthermore, that resident and nonresident travelers in the region have long been confronted by and to which they have adapted. In the past, attempts to use the Passage as a shipping corridor were obstructed when it was frozen, even as, conversely, the same winter conditions are incorporated by the Inuit in traditional seasonal movements and routines, whether on dogsled or snowmobile. Notwithstanding the DFAIT official's assertion that "the people of the Canadian North . . . have treated the ice exactly the same as the land," northern peoples are keenly aware of the distinctions between ice and land, and among various forms of ice (Oozeva et al. 2004). In this sense, the Canadian tendency to assign the icy passage to the land side of the land–water binary is no more accurate (in terms of the passage's material form, its historic uses, or the ways in which it is perceived through everyday spatial practices) than is the American inclination to assign the Passage to the water side. The reification of this space by statespersons for its geophysical properties—with all of the associated assumptions of permanence and transparency—is thus quite arbitrary, and this is especially so in the dynamic and indeterminate environment of the Arctic.

Claiming the North Pole

The dilemma facing statespersons in geophysically "fixing" the Northwest Passage, and the consequences this has for staking claims to territorial sovereignty, finds its parallel in the Arctic Ocean region as a whole. As long as the Arctic Ocean was predominantly frozen but drifting ice pack, international interests in staking territorial claims over the vast Arctic were pushed to the side. This indefinite arrangement, however, was challenged in August 2007 when a team of Russian scientists and legislators, together with an Australian adventure tour operator and a Swedish pharmaceuticals magnate, planted Russia's flag on the seabed at the North Pole.

Although the Russian flag planting was subsequently seen by much of the world as a media stunt (Canada's Foreign Minister Peter MacKay rebutted, "This isn't the 15th century. You can't go around the world and just plant flags and say, "We're claiming this territory"'; CanWest Media 2007), following their return, Russian government promoters of the expedition heralded it as a symbolic move to affirm and extend Russia's historic presence in the Arctic (Komsomolskaya Pravda 2007), a history that dates back some 500 years (Brigham 1991). Following Article 76 of UNCLOS, which permits coastal states under certain conditions to claim seabed rights past the 200 nautical mile limit that marks the furthest extent of a state's EEZ, the flag planting was mobilized by Russia as one component of a broader campaign by which Russia is using seismic and bathymetric data to assert that the land beneath the Arctic Ocean, up to the North Pole, is an extension of its landmass.[4] Thus, even though Russia's renewed interest in the region was due to changes in the geophysical environment and its concern over demographic and economic challenges in its Far North, Russia chose not to place its actions within the context of climate change or the dynamic materiality of the Arctic. Instead, Russia appealed to the properties of solid land (the seabed) to identify a natural and timeless contiguity between Russian territory and its continental shelf.[5] In other words, Russia appropriated the symbolism implied by the planting of its flag to issue a public proclamation that upheld the static nature of state sovereignty in a changing world: Even as the Arctic's increasing climate change–induced liquidity creates new incentives for incorporating it within the territories of sovereign states, state territorialization continues to rest on a "solid" foundation below the changes occurring above.

In constructing the Arctic seabed as a fixed piece of territory that naturally extends Russian soil, and by supporting the replication of a classic means of claiming territory (flag planting), Russia appeared to suggest that the Arctic was, to paraphrase the previously quoted Canadian DFAIT official, "exactly the same as the land." However, Russia's Foreign Minister Sergei Lavrov subsequently explained that the goal of the mission was not imperial: "The aim of the expedition is not to stake Russia's claim but to show that our shelf reaches to the North Pole" (CNN 2007). Lavrov thus constructed the Arctic less as a contested space than as a tangible space of opportunity, a construction that implicitly referenced both the legal specifics of the UNCLOS EEZ regime (see note 1) and a long-standing attitude that expansion into contiguous frontier regions is a component of the "natural" extension of the Russian nation (Griffiths 1991). Still other Russian officials strayed further from the conceptualization of the Arctic Ocean as essential Russian territory. Shortly after the mission, Sergei Balyasnikov of Russia's Arctic and Antarctic Institute announced, "For me this is like planting a flag on the moon" (CNN 2007).

If the Russians seemed divided regarding whether the Arctic Ocean was claimable land, unclaimable water, or a special category of space into which states could extend their authority, the Canadian response was equally ambiguous. Even while dismissing the Russian flag planting as inappropriate, Foreign Minister MacKay announced, "The question of sovereignty of the Arctic is not a question. It's clear. It's our country. It's our property. It's our water. . . . The Arctic is Canadian" (CanWest Media 2007; see also Steinberg 2010).

In fact, both states have a history of claiming sovereignty over Arctic waters. Beginning in 1907, in the absence of an international regime defining the status of the Arctic Ocean, Canada and then Russia contended that their borders should extend to the North Pole, via their respective northern coastal waters (Horensma 1991; Dufresne 2007). Subsequently, both countries have abandoned this "sectoral" position, restricting their claims to land in their northern

Figure 2. The Territories, from *The Atlas of Canada* (Natural Resources Canada [2006]; Reproduced with the permission of Natural Resources Canada, 2010.)

waters and placing the Arctic Ocean itself within the juridical classification system codified in the Law of the Sea. Nonetheless, the statements by the two foreign ministers suggest that the idea of the Arctic as a special, claimable space, notwithstanding that its central feature is the Arctic Ocean, continues to influence national policies (Dodds 2008, 2010; Morozov 2009). Indeed, in the Canadian government's *Atlas of Canada*, the sectoral principle remains unquestioned on the map of the Territorial North (Figure 2).

Perhaps the most consistent proponent of the opposite view has been the United States, which has long asserted that the juridical status of the Arctic is akin to any other region: Land can be claimed by individual states, but water—including frozen water—is beyond state territory (Horensma 1991). The United States maintained this position in its reaction to the flag-planting episode. As an official with the U.S. Arctic Research Commission said, "[The Arctic] is an ocean surrounded by continents, not the other way around. So I think from a government perspective, we don't get exercised by the flags being planted [on the seabed]" (Interview, June 2008).

Yet 100 years earlier, Admiral Robert Peary, Matthew Henson, and a team of rarely acknowledged Inuit claim to have planted the first American flag atop the sea ice at the North Pole. If so, the flag would have been placed not on solid, spatially fixed land but on a mathematically determined spot marked on a maze of mobile and shifting ice floes. These floes, in turn, would have swiftly carried away any physical symbol of individual hubris or national territorial claim. The futility of this flag placement from the point of view of sovereignty, and the audaciousness associated with that performed by Russia 100 years later, points to the problems inherent in asserting sovereignty in a space whose physical properties are dynamic in both space and time.

Beyond Sovereignty?

Since the age of exploration, the Arctic imaginaries embraced by metropolitan capitals of northern states have remained relatively stagnant, shaped by cultural memory and an established political ontology of space. Even today, these imaginaries are primarily defined by historical practices rather than the region's changing environment—although it is the latter that marks the emergence of a new era of economic opportunity. Yet, attempts to construct sovereign space in the Arctic have been complicated by the disparity between existing imaginaries and the emergent geophysical and social realities of the region. An alternative to these approaches could be a system of global governance at the margins of sovereign power, which would be based on a fundamental shift in Arctic imaginaries. As Mikhail Gorbachev proclaimed, "[The] Arctic is not only the Arctic Ocean. . . . It is the place where the Eurasian, North American, and Asian Pacific regions meet, where the frontiers come close to one another and the interests of states . . . cross" (cited in Keskitalo 2004, 43). It follows that such a space could be governed not as a bounded, fixed entity (the territorial state model) or its conceptual antithesis (the ostensible ungoverned space of the world ocean), but as a fluid space of crossings. Indeed, Gorbachev's speech inspired the 1991 Arctic Environmental Protection Strategy, which subsequently led to the formation of the Arctic Council.

This Arctic imaginary drives the vision of the Inuit Circumpolar Council (ICC), which, in April 2009, issued a Declaration on Sovereignty (ICC 2009). Citing the complex materiality of this "vast circumpolar region of land, sea, and ice" (Article 1.1) and the many changes that the region faces due to climate change (Article 3.6), the Declaration is pointedly not a declaration of independence. The ICC's constituent members declare their continuing loyalty to predominantly non-Arctic nation-states, but they also assert that their sovereign rights must be respected and the region's multiple, dynamic spaces be collectively governed (Shadian 2006). This sentiment is echoed in the increasingly assertive voice of the European Union, which, notwithstanding its differences with the ICC's vision of the Arctic Council's powers and composition, similarly questions the strictly state-centered focus of the Arctic (e.g., Ilulissat Declaration 2008), drawing instead on the vision of the Arctic Ocean region as a commons that is best governed transnationally (Young 2009).

Although Arctic states have at times resisted this alternative construction of sovereignty, it has been institutionalized in the Arctic Council, an intergovernmental forum established in 1996, in which six indigenous peoples' organizations (some of which themselves represent circumpolar constituencies) have near-equal standing with eight Arctic member states. The Arctic Council, however, has no legal authority to bind its members and its mission is limited to environmental protection and sustainable development (Bloom 1999). Indeed, many critics contend that the Arctic Council's "weak institutional structure, soft law status, and ad hoc funding system" limit its efficacy and that alternative proposals should be explored for an Arctic treaty (Koivurova 2008, 14; see also Stokke and Hønneland 2007; Rothwell 2008). Yet, notwithstanding these critics, the Arctic Council has been uniquely successful in two areas, both of which connect directly with the region's status as a space of unstable materiality and indeterminate sovereignty: fostering cooperation in science, particularly in climate change and environmental monitoring (Hoel 2007), and formalizing indigenous representation and access to high-level policymakers at the circumpolar level (E. Wilson and Øverland 2007). In this way, the Arctic Council provides the broadest forum for residents, organizations, authorities, and observers to discuss possibilities for collective governance.

In fact, the Arctic Council's efficacy is dependent on recognizing the Arctic as geophysically and sociopolitically distinct and on recognizing that these two aspects of Arctic exceptionalism are linked: "From the perspective of legitimizing the environmental protection mandate of the Council . . . it can and must present itself as safeguarding this special relationship with the still relatively undisturbed environment of the Arctic indigenous peoples" (Koivurova 2008, 25). A similar point is tacitly acknowledged by the ICC in its decision to begin the Declaration on Sovereignty with a reference to the Arctic's materiality. An Inuit elected official from Canada's northern territory of Nunavut also draws connections between rethinking the nature of Arctic space (and what it means to live in such a space) with changing constructions and assertions of sovereignty in the region: "Because we [residents of the North] use the land, that has to be one of the foremost factors for retaining sovereignty. Prime Minister Harper [says], 'Use it or lose it.' Well, that [is] very offensive for us, that flippant little one line. I mean, what are *we* if we're not using it?" (Interview, June 2008).

Although not specifically addressing his comment, one could imagine this elected official responding to the DFAIT official who suggested that Canada assert its sovereignty over the North by pointing to a history of

Northern peoples who "have treated the ice exactly the same as the land." As the elected official suggests, the challenge facing Canada (and other Arctic states) is not to cast the region's complex and dynamic environment as "exactly the same as the land" (i.e., a "used" space)—whether the pretext be floating ice or seabed—and then apply that designation to justify its incorporation into the sovereign, territorial state. Nor should the region be externalized as "unused" ocean. Rather, the Arctic region, with its unique and liminal environment, is a dynamic region where new systems of governance can be employed that push the limits of the state form and enable new possibilities for cooperation and inclusion within and across state borders. Thus, by positioning themselves as "an integral part of the very lands and waters" they use and occupy (Kleivan 1992, quoted in Tennberg 2000, 32), the Inuit, instead of directly challenging the role of distant states in the Arctic, question the scalar assumptions of sovereignty itself. With such a stance, the Inuit can be viewed as adopting a flat ontology (Marston, Jones, and Woodward 2005), whereby degrees and spaces of jurisdiction are downplayed in favor of the actual interactions between the various parties who maintain specific interests in the region, a phenomenon that also can be seen in other regional cooperative arrangements, such as the proposed Nordic Saami Convention that would allow Saami to travel freely between Norway, Sweden, and Finland and intermittent proposals to allow free travel of indigenous peoples across the Bering Strait.[6] Whereas the Inuit see themselves as primary stakeholders in such deliberations, their linkages to relevant state, corporate, and nongovernmental organization (NGO) actors are at times held as nonscalar connections to disparate entities with varying levels of power and resources. One such example of a scale-defying politics is demonstrated in the formation of the "Many Strong Voices" program, consisting of a partnership of government and NGO groups that link Arctic interests to those of Small Island Developing States (SIDS) in the context of rapid climate change.

At its most profound level, this assertion of a flat ontology in the Arctic challenges not just the ideal of the sovereign state with boundaries that are fixed in time and space (see Shadian forthcoming) but also the modern notion of a world that is divided into distinct societies. In this vein, Craciun (2009) suggested that circumpolarity can ground the growing calls within cultural studies for a "planetary" perspective that rethinks global society as a series of connections, a perspective echoed by Lynge (2008) in his identification of the Arctic (and Greenland in particular) as a space of peaceful global crossings. The notion and the experience of circumpolarity can direct the planetary orientation away from intermediate (or heterotopic) spaces that simultaneously connect and divide (e.g., oceans, borders, ships, etc.) and toward an everyday space that is undergoing continual reterritorialization by those who reside there, those who are attempting to extend their reach, and those who are just passing through. It follows that if one wishes to develop new perspectives for understanding and intervening in a world of connections, where, for instance, increased carbon emissions in one country can trigger a series of climactic changes that could lead to population migrations, new trade and production patterns, and further environmental change at all scales, one could do well to look at ongoing struggles over governance in the circumpolar Arctic.

Conclusion

The Arctic has never fit well within the spatial template of the state system, which is based on a foundational, permanent distinction between enclosable land and free-flowing water. Today, climate change is bringing this divergence, which long had been at the margins of political consciousness, to the core, in Arctic states and beyond. On the one hand, climate change is opening opportunities in the Arctic, giving states new incentive to clearly define the region within the spatial ontology of the state system, whether as developable space that can be enclosed within territories or as transit space that is exempt from state power. On the other hand, these same geophysical changes that are spurring increased interest in the region are making it all the more difficult for Arctic stakeholders to designate specific points in Arctic space as either definitively "inside" or "outside" state territory.

Although this situation appears ripe for conflict, the examples of the Arctic Council and the ICC suggest that it also might present new opportunities for cooperation and for rethinking the sovereign system of mutually exclusive territorial polities. By bringing the sovereignty debate to this materially liminal zone, climate change is forcing the state system to confront its accepted suppositions about the relationship among land, state, territory, and nation. Out of this confrontation could come new experiments with implications that would span far beyond the northernmost latitudes, with meaningful consequences for the evolution of global governance.

Acknowledgments

This article uses data and perspectives associated with National Science Foundation Geography & Spatial Sciences and Arctic Social Sciences Grant 0921436/0921424/0921704, ("Collaborative Research: Territorial Imaginaries and Arctic Sovereignty Claims," Gerhardt, Steinberg, and Tasch, Principal Invesitgators) as well as a grant from the International Council for Canadian Studies ("Contested Materialities in Northwest Passage Sovereignty Claims," Steinberg, Principal Investigator).

Notes

1. Although the United Nations Convention on the Law of the Sea recognizes territorial waters in which states are sovereign, these waters, like internal waters, gain this status from their proximity to land (United Nations 1982, Articles 2 and 3), and principles of state territoriality apply there in spite of their material liquidity. Likewise, although the Exclusive Economic Zone extends beyond territorial waters, this is an area in which a coastal state has sovereign rights but not *sovereignty* (United Nations 1982, Article 56).
2. Antarctica presents many of the same challenges and is another example of a space where sovereignty is modified to accommodate the unusual geophysics of a region. Nonetheless, there are a number of factors that make the Arctic different from Antarctica, including the extent to which already existing sovereign authority reaches into the Arctic, the Arctic's nonterrestrial nature, and the fact that the Arctic is inhabited (Young 1992). Thus, we restrict our study to the Arctic.
3. We use the term *territorial imaginary* instead of the more commonly used term *geographical imaginary* to emphasize the importance of idealizations of a place's geophysical materiality as well as its idealized cultural attributes.
4. Article 76 of UNCLOS describes the manner by which claims to the continental shelf beyond 200 nautical miles involve submissions by the coastal state (including a definition of the shelf as well as the data on which this is based) to the twenty-one-member Commission on the Limits of the Continental Shelf. A subcommittee of the Commission then considers the submission to determine whether it has been defined in accordance with Article 76 and forwards its recommendations to the full Commission, the Secretary-General, and a nonvoting representative of the applicant state. If a state disagrees with the subcommittee's recommendations it can assemble a revised or new proposal for resubmission to the Commission.
5. More than forty-five years ago, geophysicist J. T. Wilson (1963) posited that the Lomonosov Ridge, a narrow rise of continental crust stretching between Greenland and Siberia through the North Pole and on which Russia is basing its Arctic Ocean claims, was likely a displaced sliver of crust that "originally" had been a part of Eurasia, north of Scandinavia and Russia. Whether a piece of crust, which by happenstance is located adjacent to a country's shelf, is also attached to that shelf further problematizes the complex and underlying uncertainty of nations' Arctic Ocean territorial claims.
6. It should be noted that such a flat ontology cannot be attributed to the Inuit as a whole. There are currently numerous and partly contradictory Inuit political projects unfolding. In Greenland, the recent vote for self-rule is seen as an initial step toward full sovereignty. The government of Greenland is thus much more in line with a Westphalian scalar ontology based on state sovereignty (see Nuttall 2008).

References

Aporta, C. 2002. Life on the ice: Understanding the codes of a changing environment. *Polar Record* 38:341–54.

Arctic Council. 2009. Arctic marine shipping assessment. http://arcticportal.org/uploads/4v/cb/4vcbFSnnKFT8A-B51XZ9_TQ/AMSA2009Report.pdf (last accessed 30 May 2009).

Bloom, E. 1999. Establishment of the Arctic Council. *American Journal of International Law* 93:712–22.

Brigham, L. 1991. Introduction. In *The Soviet maritime Arctic*, ed. L. Brigham, 1–15. London: Belhaven.

Bureau of International Information Programs, U.S. Department of State. 2007. Ambassador Wilkins discusses U.S. Canadian relations and Northern issues in webchat, 16 February. http://www.canadanorth.usvpp.gov/webchat_wilkins_021607.asp (last accessed 15 September 2007).

Byers, M. 2010. *Who owns the Arctic: Understanding sovereignty disputes in the North*. Vancouver: Douglas & McIntyre.

CanWest Media. 2007. Russians plant flag on North Pole seabed, 3 August. http://www.canada.com/reginaleaderpost/story.html?id=55f5b74e-3728-4245-97ed-f6713c3f1bb8&k=40585 (last accessed 15 May 2009).

CNN. 2007. Russia plants flag on Arctic floor, 4 August. http://www.cnn.com/2007/WORLD/europe/08/02/arctic.sub.reut/index.html (last accessed 30 December 2007).

Craciun, A. 2009. The scramble for the Arctic. *Interventions* 11:103–14.

Davidson, P. 2005. *The idea of north*. London: Reaktion.

Dodds, K. 2008. Icy geopolitics. *Environment and Planning D: Society & Space* 26:1–6.

———. 2010. Flag planting and finger pointing: The law of the sea, the Arctic, and the political geographies of the outer continental shelf. *Political Geography* 29 (2): 63–73.

Dufresne, R. 2007. Canada's legal claims over Arctic territories and waters, PRB 07–39E. Ottawa, Canada: Parliamentary Information and Research Service, Library of Parliament. http://www2.parl.gc.ca/Content/LOP/ResearchPublications/prb0739-e.htm (last accessed 4 February 2010).

European Space Agency. 2007. Satellites witness lowest Arctic ice coverage in history, 14 September. http://www.esa.int/esaCP/SENYTC136F_index_0.html (last accessed 15 September 2007).

Fenge, T. 2007–2008. Inuit and the Nunavut Land Claims Agreement: Supporting Canada's Arctic sovereignty. *Policy Options/Options Politiques* (December–January): 84–88.

Frazar, H. 2008. Core matters: Greenland, Denver, and the GISP2 ice core. In *High places: Cultural geographies of mountains, ice, and science*, ed. D. Cosgrove and V. della Dora, 64–83. New York: Palgrave Macmillan.

Government of Canada. 1985. Arctic waters pollution prevention act of 1970, revised statutes 1985, c. A-12. http://laws.justice.gc.ca/en/ShowFullDoc/cs/A-12///en (last accessed 30 December 2007).

Griffiths, F. 1991. The Arctic in the Russian identity. In *The Soviet maritime Arctic*, ed. L. Brigham, 83–107. London: Belhaven.

Hoel, A. 2007. Climate change. In *International cooperation and Arctic governance*, ed. O. Stokke and G. Hønneland, 112–37. London and New York: Routledge.

Horensma, P. 1991. *The Soviet Arctic*. London and New York: Routledge.

Ilulissat Declaration. 2008. http://arctic-council.org/filearchive/Ilulissat-declaration.pdf (last accessed 14 December 2008).

Institute of the North, U.S. Arctic Research Commission, and International Arctic Science Committee. 2004. Report of Arctic marine transport workshop, 28–30 September. http://www.arctic.gov/publications/arctic_marine_transport.pdf (last accessed 15 November 2009).

Inuit Circumpolar Council (ICC). 2009. A circumpolar Inuit declaration on sovereignty in the Arctic. http://www.inuitcircumpolar.com (last accessed 19 May 2009).

Jolly, D., F. Berkes, J. Castleden, and T. Nichols. 2002. We can't predict the weather like we used to: Inuvialuit observations of climate change, Sachs Harbour, Western Canadian Arctic. In *The Earth is faster now: Indigenous observations of Arctic environmental change*, ed. I Krupnik and D. Jolly, 54–91. Fairbanks, AK: Arctic Research Consortium of the United States.

Kerttula, A. 2000. *Antler on the sea: The Yupik and Chukchi of the Russian Far East*. Ithaca, NY: Cornell University Press.

Keskitalo, E. C. R. 2004. *Negotiating the Arctic: The construction of an international region*. London and New York: Routledge.

Kleivan, I. 1992. The Arctic Peoples' Conference in Copenhagen, November 22–23, 1973. *Etudes/Inuit/Studies* 16:227–36.

Koivurova, T. 2008. Alternatives for an Arctic treaty: Evaluation and a new proposal. *Review of European Community & International Environmental Law* 17:14–26.

Kolstad, E., and T. Bracegirdle. 2008. Marine cold-air outbreaks in the future: An assessment of IPCC AR4 model results for the Northern Hemisphere. *Climate Dynamics* 30:871–85.

Komsomolskaya Pravda. 2007. The Arctic: A 21st century hot spot, 23 April. http://www.kp.ru/daily/23892/66464/ (last accessed 15 April 2009).

Kraska, J. 2007. The Law of the Sea Convention and the Northwest Passage. *International Journal of Marine and Coastal Law* 22:257–81.

Laidler, G. 2006. Inuit and scientific perspectives on the relationship between sea ice and climate change: The ideal complement? *Climatic Change* 78:407–44.

Lewis, M., and K. Wigen. 1997. *The myth of continents: A critique of metageography*. Berkeley: University of California Press.

Lopez, B. 2001. *Arctic dreams*. New York: Vintage.

Lynge, F. 2008. Full circle. *Ambio* Special Report 14 (November): 514–16.

Mackinder, H. 1904. The geographical pivot of history. *The Geographical Journal* 23:421–37.

Marston, S. A., J. P. Jones, III, and K. Woodward. 2005. Human geography without scale. *Transactions of the Institute of British Geographers, New Series* 30:416–32.

Massey, D. 2005. *For space*. Thousand Oaks, CA: Sage.

———. 2006. Landscape as a provocation: Reflections on moving mountains. *Journal of Material Culture* 11: 33–48.

Morozov, Y. 2009. *The Arctic: The next "hot spot" of international relations or a region of cooperation?* New York: Carnegie Council for Ethics in International Affairs. http://www.cceia.org/resources/articles_papers_reports/0039.html (last accessed 4 February 2010).

Natural Resources Canada. 2006. The Territories. In *The atlas of Canada*. http://atlas.nrcan.gc.ca/site/english/maps/reference/provincesterritories/northern_territories (last accessed 30 May 2009).

Nuttall, M. 2008. Climate change and the warming politics of autonomy in Greenland. *Indigenous Affairs* 1–2: 44–51.

Ó Tuathail, G. 1996. *Critical geopolitics*. London and New York: Routledge.

Office of the Prime Minister, Government of Canada. 2007. Prime Minister Stephen Harper announces new Arctic offshore patrol ships, 9 July. http://www.pm.gc.ca/eng/media.asp?category=1&id=1742 (last accessed 15 September 2007).

Oozeva, C., C. Noongwook, G. Noongwook, C. Alowa, and I. Krupnik, eds. 2004. *Sikugmengllu eslamengllu esghapalleghput: Watching ice and weather our way*. Washington, DC: Arctic Studies Center.

Pharand, D. 2007. The Arctic waters and the Northwest Passage: A final revisit. *Ocean Development and International Law* 38:3–69.

Powell, R. 2008. Configuring an "Arctic commons"? *Political Geography* 27:827–32.

Pyne, S. 1998. *The ice: A journey to Antarctica*. Seattle: University of Washington Press.

Rothwell, D. 2008. The Arctic in international law: Time for a new regime? *Brown Journal of World Affairs* 15: 241–53.

Schmitt, C. 2006. *The* nomos *of the Earth in the international law of* Jus Publicum Europaeum. New York: Telos.

Shadian, J. 2006. Remaking Arctic governance: The construction of an Arctic Inuit polity. *Polar Record* 42:249–59.

———. Forthcoming. From states to polities: Reconceptualising sovereignty through Inuit governance. *European Journal of International Relations*.

Shields, R. 1991. *Places on the margin: Alternative geographies of modernity*. London and New York: Routledge.

Slocum, R. 2004. Polar bears and energy-efficient lightbulbs: Strategies to bring climate change home. *Environment and Planning D: Society & Space* 22:413–38.

Steinberg, P. 2001. *The social construction of the ocean*. Cambridge, UK: Cambridge University Press.

———. 2009. Sovereignty, territory, and the mapping of mobility: A view from the outside. *Annals of the Association of American Geographers* 99:467–95.

———. 2010. You are (not) here: On the ambiguity of flag planting and finger pointing in the Arctic. *Political Geography* 29 (2): 81–84.

Stokke, O., and G. Hønneland, eds. 2007. *International cooperation and Arctic governance*. London and New York: Routledge.

Tennberg, M. 2000. *Arctic environmental cooperation: A study in governmentality*. Aldershot, UK: Ashgate.

United Nations. 1982. Convention on the law of the sea, A/Conf.62/122. http://www.un.org/Depts/los/convention_agreements/texts/unclos/unclos_e.pdf (last accessed 30 December 2007).

Vannini, P., G. Baldacchino, L. Guay, S. Royle, and P. Steinberg. 2009. Reterritorializing Canada: Arctic ice's liquid modernity and the imagining of a Canadian archipelago. *Island Studies Journal* 4:121–38.

Wilson, E., and I. Øverland. 2007. Indigenous issues. In *International cooperation and Arctic governance*, ed. O. Stokke and G. Hønneland, 27–49. London and New York: Routledge.

Wilson, J. T. 1963. Evidence from islands on the spreading of ocean floors. *Nature* 197: 536–38.

Wråkberg, U. 2004. Delineating a continent of ice and snow: Cartographic claims of knowledge and territory in Antarctica in the 19th and early 20th century. In *Antarctic challenges: Historical and current perspectives on Otto Nordenskjöld's Antarctic expedition*, ed. A. Elzinga, D. Turner, T. Nordin, and U. Wråkberg, 123–43. Göteborg, Sweden: Royal Society of Arts and Sciences.

Young, O. 1992. *Arctic politics: Conflict and cooperation in the circumpolar north*. Hanover, NH: University Press of New England.

———. 2009. The Arctic in play: Governance in a time of rapid change. *International Journal of Marine and Coastal Law* 24:423–42.

Yusoff, K. 2005. Visualizing Antarctica as a place in time: From the geological sublime to "real time." *Space & Culture* 8:381–98.

Kiavallakkikput Agviq (Into the Whaling Cycle): Cetaceousness and Climate Change Among the Iñupiat of Arctic Alaska

Chie Sakakibara

The Iñupiat of Arctic Alaska identify themselves as the "People of the Whales." The flesh of the bowhead whale (*Balaena mysticetus*) is high in vitamins and other components that traditionally sustained human physiology in a climate that is unsuitable for agriculture. Not surprisingly, the People of the Whales depend on the bowhead for sustenance and cultural meaning. The bowhead remains central to Iñupiat life and culture through the hunting process, the communal distribution of meat and other body parts, and associated ceremonials and other events to sustain cultural well-being, which I call the Iñupiat whaling cycle. For this study, I coined the term *cetaceousness* as a hybrid of *cetaceous* and *consciousness*, which links human awareness with cetaceans or whales. I use this term to refer to human–whale interactions at all levels. Particularly in Alaska, cetaceousness is a social and emotional process for the Iñupiat to communicate with the whales. Based on my ethnographic fieldwork in Barrow and Point Hope, Alaska, from 2004 through 2007, this study reveals how collective uncertainty about the environment is expressed and managed in Iñupiat practices and, by extension, how deeply global warming penetrates the cultural core of their society. To do so, I illustrate different aspects of Iñupiat–bowhead whale relationships or the ways people make whales a central feature of their lives. By influencing the bowhead harvest and the Iñupiat homeland, climate change increases environmental uncertainties that both threaten and intensify human emotions tied to identity. This emotional intensity is revealed in the prevalence of traditional and newly invented whale-related events and performances, the number of people involved, the frequency of their involvement, and the verve or feelings with which they participate. Thus, this study investigates how collective uncertainty about the future of the environment would be expressed and managed in Iñupiat practices and, by extension, how deeply climate change penetrates the cultural core of their society. My findings demonstrate how the Iñupiat retain and strengthen their cultural identity to survive unexpected difficulties with an unpredictable environment by reinforcing their relationship with the whales. *Key Words: adaptation, Arctic Alaska, bowhead whale (*Balaena mysticetus*), climate change, cultural identity, global warming, humanistic geography, Iñupiat.*

阿拉斯加北极的因纽皮特人将自己称为"鲸鱼人。"北极露脊鲸（Balaena mysticetus）的肉里富含维生素和其它成分，在这种传统上不适合农业的地区，它是人类得以维持生存的重要因素。毫不奇怪，在生存条件和文化意义上，"鲸鱼人"都依赖于北极露脊鲸。籍因狩猎过程，鲸肉和其它鲸鱼身体部位的商品交易，与此相关的庆祝活动和其他活动，在因纽皮特人的生活和文化中，北极露脊鲸依然处于中心地位，我将之称为因纽皮特人的鲸鱼周期。在这项研究中，我创造了一个混合词"鲸鱼意识"，将"鲸鱼"和"意识"混合，将人的意识与鲸豚或鲸鱼相连接。我使用这个词来说明在各个层次上人与鲸鱼的相互作用。特别是在阿拉斯加，"鲸鱼意识"是因纽皮特人与鲸鱼的社会和情感沟通过程。根据我在阿拉斯加州巴罗和普因提后普所进行的人种学实地调查（从 2004 年到 2007 年），这项研究揭示了环境如何在因纽皮特人的日常生活实践中得以表达和管理，推而广之，全球变暖是如何深入地渗透到他们社会文化的核心。为此，我描述了因纽皮特人和北极露脊鲸之间的关系，或者人们将鲸鱼置于他们生活中心的其它方式，以上种种的不同方面。通过影响北极露脊鲸的收成和因纽皮特人的家园，气候变化增加了环境的不确定性，威胁并且加强了人类对自我身份认识的情感连接。这种情感的强度在下列事件中得以展现：传统和新发明的与鲸鱼相关的活动和表演的流行程度，参与的人数，人们的参与频率，和人们参与的热情和情感投入。因此，本研究探讨了未来环境的总体不确定性将如何在因纽皮特人的实践中得以表达和管理，推而广之，全球变暖是如何深入地渗透到他们社会文化的核心。我的研究结果展示了因纽皮特人如何保持和加强其文化身份，通过加强它们与鲸鱼的关系，从而得以克服因不可预知的环境而造成的意想不到的困难。*关键词：适应，北极阿拉斯加，北极露脊鲸（*Balaena mysticetus*），气候变化，文化认同，全球变暖，人文地理，因纽皮特人。*

Los Iñupiat de la Alaska ártica se identifican a sí mismos como "La Gente de las Ballenas." La carne de la ballena boreal (*Balaena mysticetus*) es rica en vitaminas y otros componentes que tradicionalmente han sostenido la fisiología humana en un clima inadecuado para la agricultura. No es de sorprenderse, entonces, que la Gente de las Ballenas dependa de las boreales para sustento y sentido cultural. Las boreales ocupan lugar central en la vida y cultura Iñupiat en términos del proceso de caza, la distribución comunal de la carne y otras partes del cuerpo, y los ceremoniales asociados con esto y otros eventos que sostienen el bienestar cultural, a todo lo cual denomino ciclo ballenero iñupiat. Para el presente estudio acuñé el término *cetaceousness* ["*cetaceonidad*"], como un híbrido entre *cetaceous* y *consciousness* [conciencia] que conecta el conocimiento humano con los cetáceos o ballenas. Utilizo este término para referirme a las interacciones hombre-ballena, en todos los niveles. Particularmente en Alaska, la *cetaceousness* es un proceso social y emocional de los Iñupiat para comunicarse con las ballenas. Con base en mi trabajo de campo etnográfico en Barrow y Point Hope, Alaska, entre 2004 y 2007, este estudio revela cómo se expresa y maneja en las prácticas iñupiat la incertidumbre colectiva acerca del entorno, y, por extensión, qué tan hondo penetra el calentamiento global en el núcleo cultural de su sociedad. Para hacerlo, ilustro diferentes aspectos de las relaciones iñupiat–ballenas boreales, o de la manera como la gente hace a las ballenas un rasgo central de sus vidas. Al influenciar la cosecha de las boreales y el terruño de los iñupiat, el cambio climático incrementa las incertidumbres ambientales que, a tiempo que amenazan las emociones humanas ligadas con la identidad, también las intensifican. Esta intensidad emocional se revela en la prevalencia de eventos y acciones tradicionales o de reciente invención relacionados con las ballenas, el número de personas involucradas, la frecuencia de su participación y el entusiasmo o sentimientos con los que ellos participan. Así pues, este estudio investiga cómo se expresaría y manejaría en las prácticas iñupiat la incertidumbre colectiva acerca del futuro del medio ambiente y, por extensión, qué tan hondo penetra el calentamiento global en el núcleo cultural de su sociedad. Mis descubrimientos demuestran las manera como los Iñupiat conservan y refuerzan su identidad cultural para sobrevivir a dificultades inesperadas en un entorno impredecible, reforzando su relación con las ballenas. *Palabras clave: adaptación, Alaska ártica, ballena boreal (Balaena mysticetus), cambio climático, identidad cultural, calentamiento global, geografía humanística, Iñupiat.*

The Iñupiat—the indigenous people of the North Slope Borough (NSB) of Alaska (Figure 1)—call themselves the "People of the Whales" (Figure 2). Bowhead whales (*Balaena mysticetus*) can weigh approximately one ton per 30 cm and live 130 to 150 years on average. Its flesh traditionally sustained Iñupiat health in an environment that is unsuitable for agriculture. Not surprisingly, Iñupiat depend on the bowhead for sustenance and cultural meaning. Through the hunting process, associated ceremonies, and the communal distribution of meat and other body parts, the bowhead remains central to Iñupiat cultural survival and worldview (Zumwalt 1988; Turner 1993; Brewster 2004; Sakakibara 2008, 2009). These ties between the Iñupiat and the bowhead have been threatened on many different levels. In this article, I explore specific ways in which climate change in the Arctic influences Iñupiat cultural institutions and practices that link people spiritually, materially, and politically with bowhead whales.

Several studies demonstrate the impact of climate change on land, sea, and human livelihoods; on the economics of human existence; and our human perceptions of environmental change (Riedlinger and Berkes 2001; Vincent, Gibson, and Jefferies 2001; Huntington 2002; Ford and Smit 2004; Fox 2004; George et al. 2004). Comparatively few studies, however, examine the impact of environmental change on the cultural activities associated with everyday life. The Iñupiat response to the impact of climate change has yet to be discussed extensively in the literature. To help fill this gap, I argue that by influencing the bowhead harvest and the Iñupiat homeland, climate change increases environmental uncertainties that both threaten and intensify human emotions tied to identity. I investigate how collective uncertainty about the future of the environment and their homeland is expressed and managed in Iñupiat cultural practices and, by extension, how deeply climate change penetrates the core of their society.

My research is not about the fragility of Iñupiat society or identity. Instead, it is a story about how the Iñupiat try to survive monumental challenges consciously and unconsciously by reinforcing their relationship with the whales. A whale-centric worldview based on the notion of collaborative reciprocity—what I call *cetaceousness*—helps people work toward retaining their identity throughout their associations with whales. Cetaceousness is a hybrid of *cetaceous* and *consciousness*, which links human awareness of and close bonding with cetaceans. The word reveals an attempt to capture the deep-rooted, spiritual, and material relationship the Iñupiat have with the whales. Among villagers of Point Hope, an origin story portrays the

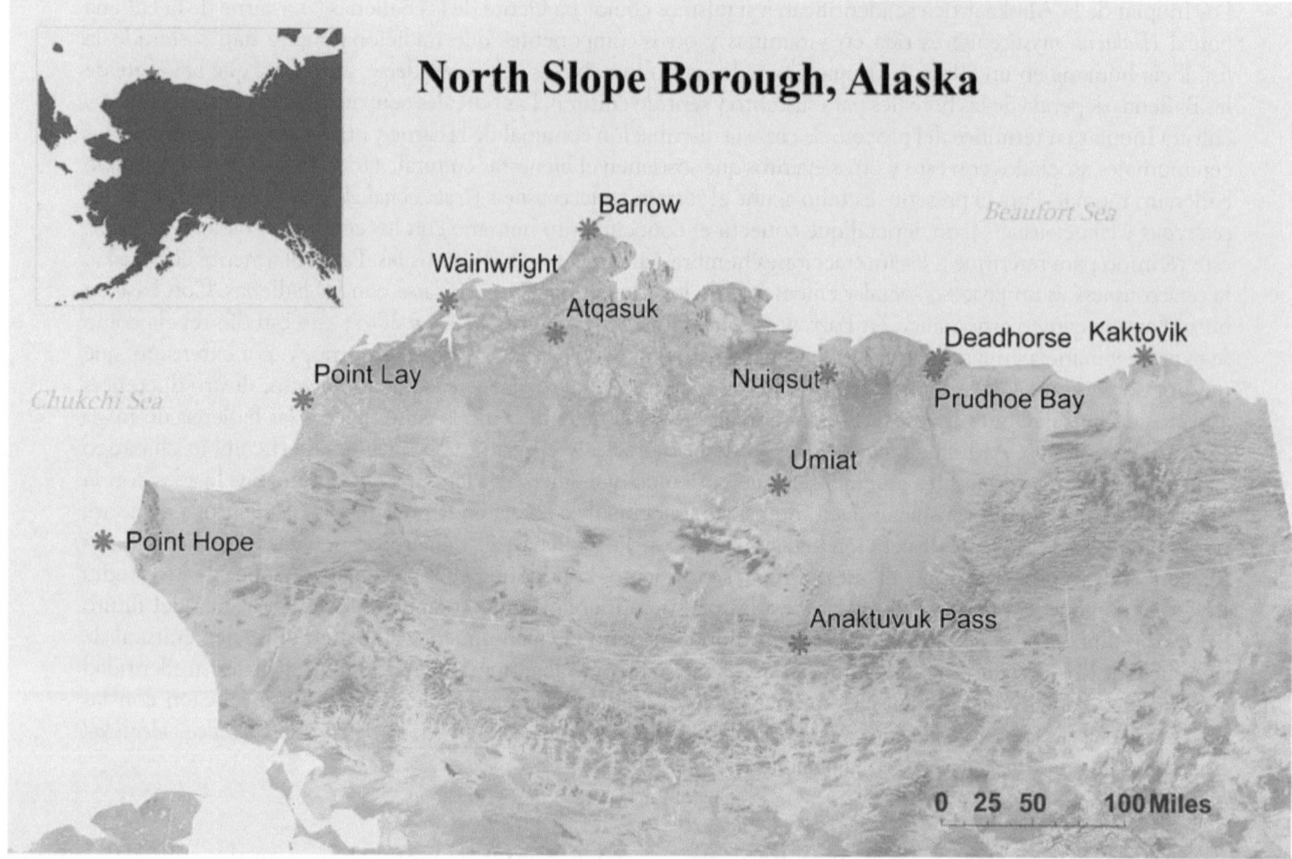

Figure 1. The North Slope Borough, Alaska. Courtesy of Jessica Jelacic, University of Cincinnati.

dramatic transformation of a bowhead into their ancestral homeland (Lowenstein 1992, 1993). To the villagers, then, their land was once a whale whose death coincided with the birth of a human habitat, which has led to an emotional kinship among the people, the land, and the whales (Reimer 1999). Significant changes to human–whale relations can cause a substantial shift in human well-being and cultural continuity.

Although my Iñupiat correspondents unanimously expressed the importance of human–whale integrity, I soon learned that there were no Iñupiat concepts or words directly equivalent to cetaceousness. During my conversations with villagers, I casually brought up the term and asked if they had other words or ideas associated with it in their cultural tradition or vocabulary. Elmer Frankson of Point Hope explained how they never had to have such a word because the whale had been historically integral to Iñupiat existence and "they always came back" (personal conversation, 29 August 2005). However, he continued: "We may need it [*cetaceousness*] now to have non-Eskimo people understand the value of the whales to our lives and urge them to think about our changing world." Ben Nageak of Barrow mentioned "[our] *whale*-ness," a concept I took to be similar to cetaceousness (personal conversation, 18 April 2005). The whale is deeply intertwined with their subsistence, as the following statements reveal: "Whales, that's who we are" (Caroline Cannon, Point Hope, personal conversation, 28 August 2005); "[W]e grow up with whales" (Irene Tooyak, Point Hope, personal conversation, 31 August 2005); and "[t]he whale-centeredness keeps us coming together in our land" (Jeslie Kaleak, Sr., Barrow, personal conversation, 2 July 2005). In this way, cetaceousness, as a unifying theme, becomes a foundation to assess current Iñupiat response to climate change through the ways people make whales a central feature of their lives.

Background: Climate Change and Cultural Survival in the Arctic Region

The Iñupiat world has been going through continuous change, particularly since indiscriminate American

Figure 2. The bowhead whale (*Balaena mysticetus*) and the People of the Whales. Photography by author. In Barrow, Alaska, Saavgak Crew captained by Clifford Okpeaha hunted and successfully landed a 15-m bowhead whale on 5 October 2005, which later enriched people's lives and facilitated celebrations in the community.

commercial whaling decimated the bowhead population by the 1920s. This affected the Iñupiat immediately but, then, even more significantly informed the International Whaling Commission's (IWC) attitude toward indigenous whaling in the 1970s. In effect, the IWC blamed and punished the victim. Following on the heels of the commercial whalers were the missionaries, and today most whaling communities identify as Christian, with Presbyterian denominations predominating in Barrow and Episcopalian in Point Hope. After the discovery of oil and the establishment of the NSB in 1972, further changes took place, such as environmental pollution and contamination (Kassam and The Wainwright Traditional Council 2000), nonrenewable resource extraction (Chance 1993), and nuclear threats and their aftermath (O'Neill 1995). All of these factors have affected Iñupiat livelihoods. In spite of all these recurrent impacts on traditional lifeways, a substantial percentage of Iñupiat maintain a balance between subsistence hunting and market interactions. However, the greatest challenge is perhaps yet to come. As the mass of the sea ice decreases, high waves erode their land and climate change as a whole increasingly begins to interfere with the Arctic ecosystem.

Environmental changes that might be regarded as minor elsewhere have significantly influenced the physical and cultural dimensions of life in the Arctic and are directly threatening indigenous populations. In late 2005, the Iñupiat representatives along with the Inuit Circumpolar Council (ICC) submitted a petition to the Inter-American Commission on Human Rights, an important agency of the Organization of American States. The petition described the violation of their human rights that had resulted from global warming related to greenhouse gas emissions from the United States. The ICC convention, held in Barrow in 2006, emphasized the immediate impact of global warming on native peoples and the anticipated human responses to environmental changes at both local and global scales. Following the petition and the convention, the Intergovernmental Panel on Climate Change (IPCC 2007) integrated indigenous perspectives of climate change into a more comprehensive global assessment. By considering indigenous experiences with climate change, international organizations are better able to incorporate how people are capable of adapting to the rapidly changing environment.

The Arctic region experienced a record sea ice minimum in June through September 2007, well below the previous sea ice minimum observed in September 2005. In February 2009, the amount of multiyear ice hit the lowest of the past 100 years (National Snow and Ice Data Center 2009). Sea ice provides an irreplaceable platform for Iñupiat subsistence activities, especially whaling, and buffers the coastline from winter storms. Prediction has it that all Arctic sea ice could disappear in the summer as early as 2030 if present trends continue (NASA Earth Observatory 2007). Increasingly unpredictable weather and climate patterns and environmental changes raise deep fears among Iñupiat. Such emotional reactions fuel cultural responses to global warming. Although the ancestors of today's Iñupiat have witnessed and withstood numerous stages of environmental change and cultural adaptation, many research participants that worked with me revealed that during no previous phase in their history were they forced to confront this "unbearable" type of emotional struggle and future uncertainty. As Elmer Frankson put it, "Our ground is melting and we are losing the land and the sea—the whales—altogether" (personal conversation, 29 August 2005).

Methods

This study is based on my fieldwork among NSB whaling communities between 2004 and 2007. My goal throughout my research was to investigate how Iñupiat maintain their physical and spiritual links with the bowhead whales in ways that sustain their cultural identity and help them cope with environmental change. Although based in Barrow (71°18′1″N 156°44′9″W;

population 4,500), I also spent a good deal of time in Point Hope (68°20′49″N 166°45′47″W; population 800), where I conducted a series of ethnographic interviews (individual, group, and questionnaires) with ninety-two participants from various generations, held formal and informal discussion forums with community members, and engaged in participant observation in seasonal and cultural events including subsistence activities (Sakakibara 2008, 2009). I was also able to participate in the ICC conference held in Barrow in July 2006 in which the northern indigenous communities played a significant role in representing whalers, the whaling tradition, and cultural survival. In the field, I learned about the people's conscious and unconscious efforts to adapt to the changing world (on both physical and political levels) by reinforcing their traditional relationships with the whales.

The notion of collaborative reciprocity is crucial for understanding Iñupiat cultural resilience. Central to this indigenous notion is the belief that humans and animals physically and spiritually constitute one other; that the soul, thoughts, and behaviors of animals and people interpenetrate in the collaboration of life (Fienup-Riordan 1990). This relationship with other living beings is fundamental to many indigenous moral philosophies (Momaday 1974; Nelson 1980; Basso 1996). For example, many northern indigenous groups believe that animals willingly give themselves to hunters in response to receiving respectful treatment as "nonhuman persons" in their own right (Nelson 1980; Fienup-Riordan 1990; Brewster 2004). Such relations are understandably influenced by animal behavior, distributions, and availability, all of which are influenced by climate and hence climate change. Just as humanistic geographers emphasize human relationships with animals and an emotional kinship with place and the environment in producing a cultural identity (Tuan 1974, 1976, 1984; Porteous 1990; Buttimer 1996; Bunkše 2004), collaborative reciprocity also should be understood as an adaptive strategy to engage Arctic life, and among Iñupiat hunters, cetaceousness serves as an example. Not only does it link human welfare with that of animals, but it affords people a means of expressing their concerns about environmental change and for coping with and adapting to such changes. This interrelationship is demonstrated in various public ways that are mutually reinforcing and include whaling organizations, food production, distribution, and consumption and in the styles of music, storytelling, and many other aspects of ceremonial and everyday life. Overall, throughout my fieldwork, ethnographic interviews, and participant observation, I explored how the people developed cetaceousness as a means to cope with unpredictable climate and environmental changes.

Climate Change and the Iñupiat Whaling Culture

The Iñupiat whaling cycle (Figure 3) is filled with meticulous preparations for whaling, whaling itself, and succeeding events and rituals to complete the full circle. Cetaceousness lies in the center as the circle of life. The Iñupiat whaling cycle also involves hunting other animals, year-round ceremonies, feasts, and many other social components that support intense human–whale relations.

"In our society, it is always the whale that brings us together," said Mae Ahgeak, whaling captain's wife (personal conversation, Barrow, 18 June 2005). The cultural survival and social ethics are all based on intimate relationships with the whale, and the whale symbolically and physically lies in the heart of human subsistence. Therefore, in everyday life, Iñupiat literally walk into the whaling cycle to sustain their lives, cultural values, and identity. In this context, *kiavallakkikput agviq* (the literal translation of this phrase is "standing by the whale," but it can also be interpreted as "into the whaling cycle") might be the closest Iñupiat phrase that embodies cetaceousness. The whale hunt is the highlight of this cycle but it is not the sole manifestation of cetaceousness.

Central to the act of whaling are *umialiks*, the whaling captains and their wives who own and operate *umiaks* (wooden frame boats covered by bearded seal skins used for whaling). Captains are mostly men and are respected community leaders. A captain organizes a crew, usually consisting of his family and relatives. Both men and women participate in the whaling cycle, and almost everyone belongs to at least one crew. *Umialiks* are responsible for nourishing crew members and sharing a

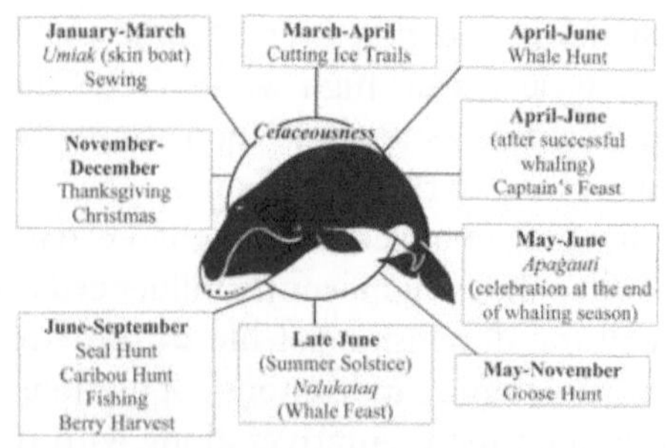

Figure 3. The Iñupiat whaling cycle.

whale with other villagers after a hunt. Every spring Iñupiat hunters camp out on the sea ice, which is the frozen sea that remains attached to land, patiently waiting for the arrival of whales. After a successful hunt, the hunters tie a heavy rope to the narrow point of the whale's immense tail and fasten all the *umiaks* and motorized boats to it. The cheers, cries, and calls of congratulations come out of radios, and people come from the village to help bring in the whale. When the whale finally reaches the camp, it is pulled up against solid ice. After establishing a block-and-tackle pulley system, the line is attached to the hands of approximately 100 men and women ready to pull the whale. The whale can be as long as twenty-three meters and weigh sixty tons. Butchering continues through the entire day and well into the night, until nothing remains except the voluminous whale's skull. Hunters slowly push the skull off the ice edge into the water to assure proper reincarnation of its soul so that it can return to them.

This whaling tradition, however, is being jeopardized by thinning sea ice, changing ocean currents that influence bowhead migration patterns, and new wind directions and climatic hazards. Some of these changes kill. Off the coast of Barrow, Iñupiat whalers of Kaleak Crew were celebrating their first successful spring hunt at their whaling camp on the sea ice shelf on 22 May 2005. The fifteen-meter bowhead whale was lifeless in the icy ocean, but the whalers were full of life in spite of their sleepless night spent paddling and towing. Their celebration was interrupted by the news that a missing youth had drowned in the Middle Lagoon as he drove through a heavy fog. He had been missing for a week, after suddenly disappearing on his way back from his uncle's whaling camp one night. Traditionally, mid-May is a time when hunters walk across or even ride snowmobiles on the fragile ice on the lagoons or the sea, but now the behaviors and conditions of the ice are changing. Despite the tragic report, the whaling crew worked quickly to butcher and remove the whale meat before the thinning sea ice would produce even greater tragedies.

In Iñupiat ways of living, traditions can often be modified and strengthened through adaptation. In recent years, as spring hunts experience difficulties, fall whaling has provided significant amounts of whale meat supplies for the entire North Slope region. The configuration of Point Barrow brings the bowhead migration near shore in both the northern and southern directions. In fall, when there is no shorefast ice attached to land, Barrow crews pursue whales with large outboard motor boats in the open water. Fall whaling does not happen when the quota is met during the spring hunt. For most other communities along the Arctic coast, fall whaling is not possible because the whales take a different migration route. Therefore, an unsuccessful spring hunt means few or no whales for an entire year for most of the whaling communities. Non-fall-whaling villages usually transfer their unfilled IWC-dictated annual quotas to Barrow so that Barrow whalers will send them harvested fall whales. In contrast to the spring whaling, fall whaling takes place on the open water as it precedes the late-fall freeze-up. As an example of Iñupiat adaptation, tradition and subsistence are often complemented with technology. Roy Nageak of Barrow, a whaling captain, told me about the use of a front-end loader that transports a whale from the beach to a butchering site after fall whaling: "The loader is newer than the origin of whaling, but we still treat the whale right this way" (personal conversation 1 October 2005). With the relocation of the butchering site further inland due to coastal erosion, the whalers would have had to drag the whale over dirt and gravel to the new site, in effect damaging the whale. "Technology helps us carry on our relationship with the whales. It doesn't hurt the whale or our tradition." Beverly Hugo of Barrow agrees: "This respect is what we have always treated the whales with" (personal conversation 2 October 2005). Tradition, technology, and adaptation are interrelated to embody the human appreciation of the whales. As Bodenhorn (2001) interpreted, it is neither the oldness nor newness that is appreciated in its own right among Iñupiat. Their traditions are never static, but it makes human subsistence possible and viable, especially when their environment and weather patterns are unpredictable. After each hunt, a cargo plane will carry whale meat and *muktuk* from Barrow to other communities. This form of sharing helps reinforce cetaceousness across the NSB, especially when any coastal villages suffer from harsh and unpredictable environmental conditions in the spring.

Coping with environmental changes has produced new catch patterns, particularly related to whale size. Point Hope whalers, who traditionally preferred to catch large whales, are now selecting smaller whales because they weigh less and are safer to butcher on the thinning ice. Barrow whaling captains also made decisions to go after smaller whales, even though the annual quota remains fixed, thus reducing the total amount harvested. The decisions people reach are based on a high awareness of environmental conditions but a belief that to not adapt, to not change, is to surely fail.

Recent shifts in climatic conditions influence both migration patterns of the bowhead whales and hunters' access to whales (Dixon 2003). The northward shifts in bowhead migration caused by the rise in sea water temperature requires whalers to use more fuel, gear, physical strength, and time (Eugene Brower, Barrow, personal conversation, 18 August 2005). Some whalers cannot afford to go whaling because of the sacrifices involved with this change. The migration of the ice-loving bowhead has also been influenced. With the warming sea water, the bowhead whales arrive earlier in northern Alaska in the spring and return from the Beaufort Sea later in the fall. Bowheads have evolved as ice whales, and it is unknown whether they could adjust to ice-free waters (Tynan and DeMaster 1997). Furthermore, the potentially ice-free Arctic Ocean will open the major routes for increased shipping, disturbing bowheads further still. On the other hand, gray whales are expanding their range as ice cover decreases (Reynolds, Moore, and Ragen 2005). "Gray whales?" Richard Glenn, whaling captain of Barrow in his forties responded: "No, we will only hunt the bowhead" (personal conversation, 21 April 2005). Iñupiat whalers are now obliged to make adjustments to their whaling cycle to accommodate these changes. Although changes in environment and climate have immense impacts on subsistence activities, hunting is just one manifestation of such. The following examples suggest other implications that have been revealed in the annual cultural cycle.

Discussion: Cetaceousness, the People of the Whales, and Climate Change

My humanistic studies on Iñupiat cultural resilience in the time of climate change reveal that Iñupiat make conscious and unconscious efforts to secure their cetaceousness through traditional expressive culture (Sakakibara 2008, 2009). Embracing their values of tradition and innovation, Iñupiat mobilize their cultural expressions as an elastic bond to preserve and develop their relationship with the whales. I have explored this bond by examining the spatial change in storytelling traditions in Point Hope before and after the 1977 village relocation due to coastal erosion, frequent storms, and flooding (Sakakibara 2008).[1] The other facet of my research examined climate change's impact on traditional Inupiat expressive culture, musical practices, performances, and emotional transformations among performers (Sakakibara 2009).[2] These case studies argue that Iñupiat efforts to maintain their relationship with the whale are agents of innovation and adaptation, that the Iñupiat are no miner's canary.

As Frankson's statement revealed earlier, the concept of cetaceousness has a global potential to appeal to nonindigenous populations who play a major role in the industrial activities that drive climate change. Currently, indigenous Arctic organizations, especially the ICC, the Alaska Federation of Natives (AFN), and the Alaska Eskimo Whaling Commission (AEWC), are building on their past strength to confront the environmental problems incurred by climate change. Cetaceousness can serve as a contemporary manifestation of northern indigenous efforts to better represent themselves in relation to colonial and nonindigenous influences from the south. The revival of Iñupiat identity in the 1970s has its origins in identity politics and ethnic sentiments associated with the struggle for whaling rights after the IWC tried to ban indigenous whaling. In response, Iñupiat whalers established the AEWC to cope with outsider idioms of whale species protection and environmental conservation. Currently, Iñupiat cultural identity has been represented by political organizations that successfully defend their right to whale. My interviewees often discussed the cultural importance of whale meat and blubber as symbols of their indigenous identity and as a metaphor for cultural survival. Cultural survival, in turn, provides the ideas and stability necessary to help mitigate the impacts of climate change. In this way, Iñupiat self-representation is both anchored in and reinforced by cetaceousness.

Iñupiat responses to climatic change link cetaceousness, whaling, and environmental conservation to human rights. Specifically, the ICC and the AEWC have begun working collaboratively to draw international attention to the human plight of the Arctic region and, as Sheila Watt-Cloutier, former Chair of the ICC, put it, to put "a human face on the global warming map" (ICC panel discussion on global warming, Barrow, 10 July 2006). Following the ICC's submission of a climate change petition to the Inter-American Commission on Human Rights, Watt-Cloutier used her time on the climate change panel at the ICC convention to show that human-induced global warming violated fundamental Iñupiat human rights, including health, physical integrity, security, subsistence access, and cultural well-being. Patricia Cochran, the succeeding ICC Chair, who also represents the Iñupiat, recently hosted the Indigenous Peoples' Global Summit on Climate Change, a five-day United Nations–affiliated conference with about 400 people from eighty nations attending in Anchorage, Alaska. The strategic turn from whaling rights to human rights seeks to force corrective measures to combat global warming in industrialized countries that have not signed the Kyoto Protocol, particularly the United States.

Despite some headway with the Inter-American court case, the Iñupiat continue to struggle to give global warming a human face. Collaborating with the NSB Department of Wildlife Management, the AEWC annually provides Arctic thoughts and voices on climate and environmental changes as observed by whalers. These messages are directed to non-Arctic regions in hopes of drawing attention to the Iñupiat homeland and cetaceousness. These reports typically reveal difficult hunting conditions and declining whale hunt efficiencies due to persistent ice-choked leads, dangerous ice conditions, strong ocean currents, and wind directions, all of which influence the success of the bowhead hunt (Suydam et al. 1993, 1996, 2003, 2006). This new form of self-representation seeks to bring the cultural plight of the Iñupiat to a wider audience but it relies on more recent political experiences. As Harry Brower, Jr. explains, "Our fight to keep the IWC from taking our bowhead whale away from us has taught us some very important lessons that we can use as we face global warming and [other] industrial threats" (ICC panel discussion on subsistence activities, Barrow, 12 July 2006). Using their own cetaceousness to raise awareness about themselves and bowhead whales, the Iñupiat—the People of the Whales—seek to extend their reciprocal relationship into the future to ensure the traditions of their past.

Conclusion

Through my fieldwork in two traditional whaling communities, I explored how climate change has influenced human–whale integrity in the Arctic. Recent and successful active mobilizations for whaling and now human rights are rooted in a deeply emotional relationship to the bowhead whale and the whaling cycle, what I have termed cetaceousness. Overall, my study exposed an Iñupiat cultural resilience that is founded on and simultaneously reinforces reciprocity and cetaceousness. To survive, Iñupiat have newly endowed their culture with the power to sustain their bond with the whales.

Reinforcement of cetaceousness also means a process of social change, as the late Pamiuq Elavgak of Barrow once remarked to me: "Whatever change happens, we shouldn't be disconnected [from our environment]. We should always be in the circle of the whale" (personal conversation, 12 September 2005). Iñupiat cultural resilience exemplifies an adaptive mechanism that is crucial for all human survival. The examples I present here show some of the complex pathways along which Iñupiat construct adaptive strategies to retain their bond with the whales and life more generally. Iñupiat adaptability is manifest through social action generated out of culturally informed innovation. Their subsistence is built on change; change and continuity are closely interrelated, and often coping with change facilitates human–environment integrity. I found that Iñupiat cultural practices and hope for survival converge in their tradition of being flexible and responsive to their surroundings, and this is where their cetaceousness begins to take a new turn. What characterizes Iñupiat cultural values is their strength in innovation and their faith in adaptability. As my Iñupiat collaborators continually emphasized, to keep whaling as the environment transforms around them is a way to strengthen their identity and nurture their survival. Inevitably, the future holds further changes and challenges, but Iñupiat relations with the whales continue to be an active agent in the process.

Acknowledgments

The author extends her gratitude for the financial assistance provided by the U.S. National Science Foundation Doctoral Dissertation Research Improvement Grant (No. 0526168. Geography and Regional Science Program and Arctic Social Science Program), logistical support by the Barrow Arctic Science Consortium and the North Slope Borough Department of Wildlife Management, and grants from the Center for Ethnomusicology and the Earth Institute, both at Columbia University, in addition to the Department of Geography and the Native American Studies Program at the University of Oklahoma. As a doctoral candidate in the Department of Geography, I was fortunate to be a student of Bob Rundstrom, my former advisor. I am grateful to Karl Offen for his personal and intellectual empowerment and fulfillment throughout my fieldwork and writing phases. I am also indebted to the inspiration and friendship provided by Wendy Eisner (University of Cincinnati) and Frederick E. Nelson (University of Delaware). Jessica Jelacic (University of Cincinnati) graciously shared her map of the North Slope Borough with me. Aaron A. Fox, Chair of the Department of Music at Columbia University, has also provided me with a strong rapport throughout our collaboration in Alaska. I would also like to extend my appreciation to the anonymous reviewers for their comments and suggestions. Last but not least, my deepest gratitude goes to the people of Point Hope and Barrow, Alaska, for their continuous encouragement and friendship throughout my fieldwork—*Quyanaqpak*. They are my teachers and are real people. Their valuable help and willingness to share the depth and breadth of their knowledge and experiences made my research possible, rewarding, and productive.

Notes

1. I described how environmental change is culturally manifest through tales of the supernatural, particularly spirit-beings or ghosts, by using Burch's (1971) prerelocation account and my postrelocation field notes as benchmarks to measure change. Point Hope villagers process environmental change through the storytelling tradition as a way to maintain their connection to a disappearing place. The changes are evident in the changing distribution of spiritual appearances, personal qualities, and encounter contexts. With their emotional and sensational appeal, supernatural stories give high visibility to places and a sense of place; as ghosts haunt places, the increasing number of supernatural tales, perceptions, and experiences set in the original settlement make the villagers' bond with the former village site, or Old Town, more tangible. I found how the settings of haunted landscapes have geographically reversed after the inhabitants moved from Old Town to New Town. In this context, storytelling is an expressive form of adaptation that reflects environmental uncertainties and the long-term well-being of the Iñupiat homeland. There is a particularly strong reciprocal link between a human sense of place and the presence of spirit-beings in their storytelling. For the Iñupiat, the content of stories and the very act of narrating them helps connect people to place, people to people, and ultimately people to whales. Stories serve to maintain a relationship and anchor the Iñupiat to their drowning home for old and young alike. The types of stories and modes of telling them reveal the villagers' uncertainty about the future based on their knowledge of the immediate past. As their environment becomes more unpredictable, the Iñupiat engage with new spirit-beings in their drowning home.
2. Among the people, "the whale is the drum and the drum is the whale" because the traditional drum membrane is often made of linings of whale livers, stomachs, or lungs (Sakakibara 2009). Influencing the volume and seasonality of successful whale hunting, environmental change heightens Iñupiat anxieties in three major ways: the availability of drum membranes made of the whale stomach skin and liver linings; disruption of the whaling cycle because an unsuccessful whaling season implies no celebrations and no musical occasions; and social disharmony and mutual distrust as a result. My work illustrated some of the ways singing, dancing, and drumming provide the means by which Iñupiat reconstruct the whaling cycle. In my interviews and participant observations, I found that Iñupiat music is both a means of adapting to environmental transformation and of indicating a transforming environment. In other words, music is not a miner's canary but rather an agent of innovation and adaptation. Drum performances and their recent resurgence in community events indicate increasing cetaceousness, not its demise. Specifically, the Iñupiat newly endowed their performance with an invitation for the whales to join their domain by reversing the human–whale relationship—traditionally, it was the whale that brought music and festivity to the people, but it is now the Iñupiat themselves who bring music to the whales to repair the broken whaling cycle as an act of collaborative reciprocity.

References

Basso, K. 1996. *Wisdom sits in places: Landscape and language among the Western Apache*. Albuquerque: University of New Mexico Press.

Bodenhorn, B. 2001. It's traditional to change: A case study of strategic decision-making. *Cambridge Anthropology* 22 (1): 24–51.

Brewster, K. 2004. *The whales, they give themselves: Conversations with Harry Brower, Sr.* Fairbanks: University of Alaska Press.

Bunkše, E. V. 2004. *Geography and the art of life*. Baltimore: Johns Hopkins University Press.

Burch, E. S., Jr. 1971. The nonempirical environment of the Arctic Alaskan Eskimos. *Southwestern Journal of Anthropology* 27 (2): 148–65.

Buttimer, A.1996. Geography and humanism in the late twentieth century. In *Comparison encyclopedia of geography*, ed. I. Douglas, R. Huggett, and M. Robinson, 837–59. London and New York: Routledge.

Chance, N. A. 1993. Sustainability, equity, and environmental protection. *Arctic Circle*. http://Arcticcircle.uconn.edu/NatResources/sustain.html (last accessed 10 January 2009).

Dixon, J. C. 2003. Environment and environmental change in the Western Arctic: Implications for whaling. In *Indigenous ways to the present: Native whaling in the western Arctic*, ed. A. P. McCartney, 1–24. Edmonton, Canada: Canadian Circumpolar Institute Press.

Fienup-Riordan, A. 1990. *Eskimo essays: Yup'ik lives and how we see them*. New Brunswick, NJ: Rutgers University Press.

Ford, J. D., and B. Smit. 2004. A framework for assessing the vulnerability of communities in the Canadian Arctic to risks associated with climate change. *Arctic* 57 (4): 389–400.

Fox, S. 2004. When the weather is *Uggianaqtuq*: Inuit observation of environmental change. Doctoral dissertation, University of Colorado, Boulder, Boulder, CO.

George, J. C. C., H. P. Huntington, K. Brewster, H. Eicken, D. W. Norton, and R. Glenn. 2004. Observation on shorefast ice dynamics in Arctic Alaska and the responses of the Iñupiat hunting community. *Arctic* 57 (4): 363–74.

Huntington, H. P., ed. 2002. *Impacts of changes in sea ice and other environmental parameters in the Arctic: Final report of the Marine Mammal Commission workshop, Girdwood, Alaska, 15–17 February 2000*. Bethesda, MD: Marine Mammal Commission.

Intergovernmental Panel on Climate Change (IPCC). 2007. Climate change 2007: Climate change impacts, adaptation, and vulnerability. In *Summary for policymakers: Working Group II Contribution of the Intergovernmental Panel on Climate Change Fourth Assessment Report*, ed. M. L. Parry, O. F. Canziani, J. P. van der Linden, and C. E. Hanson, 1–976. Cambridge, UK: Cambridge University Press.

Kassam, K.-A., and The Wainwright Traditional Council. 2000. *Passing on the knowledge: Mapping human ecology in Wainwright, Alaska*. Calgary, Canada: Arctic Institute of North America.

Lowenstein, T. 1992. *The things that were said of them: Shaman stories and oral histories of the Tikigaq people*. Berkeley: University of California Press.

———. 1993. *Ancient land: Sacred whale: The Inuit hunt and its rituals*. New York: Farrar, Straus and Giroux.

Momaday, N. S. 1974. Native American attitudes to the environment. In *Seeing with a native eye: Essays on Native American religion*, ed. W. Capp, 79–85. New York: Harper and Row.

NASA Earth Observatory. 2007. Record Arctic sea ice loss in 2007. http://earthobservatory.nasa.gov/Newsroom/NewImages/images.php3?img_id=17782 (last accessed 12 April 2009).

National Snow and Ice Data Center. 2009. Arctic sea ice news and analysis. http://nsidc.org/arcticseaicenews/ (last accessed 12 April 2009).

Nelson, R. K. 1980. *Shadow of the hunter: Stories of Eskimo life*. Chicago: University of Chicago Press.

O'Neill, D. 1995. *Firecracker boys*. New York: St. Martin's Griffin.

Porteous, J. D. 1990. *Landscapes of the mind*. Toronto: University of Toronto Press.

Reimer, C. S. 1999. *Counseling the Inupiat Eskimo*. Westport, CT: Greenwood Press.

Reynolds, J., S. Moore, and T. Ragen. 2005. Climate change and Arctic marine mammals—An uneasy glance into the future. Unpublished presentation. Barrow Arctic Science Consortium, Barrow, AK.

Riedlinger, D., and F. Berkes. 2001. Contributions of traditional knowledge to understanding climate change in the Canadian Arctic. *Polar Record* 37:315–28.

Sakakibara, C. 2008. "Our home is drowning": Climate change and Iñupiat storytelling in Point Hope, Alaska. *The Geographical Review* 98 (4): 456–75.

———. 2009. "No whale, no music": Iñupiaq drumming and global warming. *Polar Record* 45 (3): 1–15.

Suydam, R. S., J. C. George, C. Hanns, and G. Shefffield. 1993. Subsistence harvest of bowhead whales (*Balaena mysticetus*) by Alaskan Eskimos during 1992. International Whaling Commission Scientific Committee Report, Cambridge, UK.

———. 1996. Subsistence harvest of bowhead whales (*Balaena mysticetus*) by Alaskan Eskimos during 1995. International Whaling Commission Scientific Committee Report, Cambridge, UK.

———. 2003. Subsistence harvest of bowhead whales (*Balaena mysticetus*) by Alaskan Eskimos during 2002. International Whaling Commission Scientific Committee Report, Cambridge, UK.

———. 2006. Subsistence harvest of bowhead whales (*Balaena mysticetus*) by Alaskan Eskimos during 2005. International Whaling Commission Scientific Committee Report, Cambridge, UK.

Tuan, Y.-F. 1974. *Topophilia: A study of environmental perception, attitudes, and values*. New York: Columbia University Press.

———. 1976. Humanistic geography. *Annals of the Association of American Geographers* 66:266–76.

———. 1984. *Dominance and affection: The making of pets*. New Haven, CT: Yale University Press.

Turner, E. 1993. American Eskimo celebrate the whale: Structural dichotomies and spirit identities among the Inupiat of Alaska. *The Drama Review* 37 (1): 98–114.

Tynan, C. T., and D. P. DeMaster. 1997. Observations and predictions of Arctic climatic change: Potential effects on marine mammals. *Arctic* 50 (4): 308–22.

Vincent, W. F., J. A. E. Gibson, and M. O. Jefferies. 2001. Ice collapse, climate change, and habitat loss in the Canadian high Arctic. *Polar Record* 37 (201): 133–42.

Weller, G. 2004. Impacts of climate change. In *Encyclopedia of the Arctic*, ed. M. Nuttal, 945–46. London and New York: Routledge.

Zumwalt, R. L. 1988. The return of the whale: Nalukataq, the Point Hope whale festival. In *Time out of time: Essays on the festival*, ed. A. Falassi, 261–76. Albuquerque: University of New Mexico Press.

Benchmarking the War Against Global Warming

Douglas J. Sherman, Bailiang Li, Steven M. Quiring, and Eugene J. Farrell

We analyzed the HadCRUT3 reconstruction of the instrumental global temperature record for 1850 through 2008 to decompose thirty-year temperature trends into signal and noise components. The signal represents multidecadal trends and the noise represents annual variability about those trends. Historical estimates of temperature variability (e.g., noise) are used with seven temperature projections to simulate global warming time series. These trends include the 1979 through 2008 trend, four trends taken from the Intergovernmental Panel on Climate Change (IPCC) Fourth Assessment Report (AR4) simulations (A1B, A2, B2, and the constant composition commitment [CCC] scenarios) and trends representing approaches to 1.5°C and 2°C warming by 2100. Each series is simulated 1,000 times. The results are compared, statistically, to the current warming rate of 0.016°C year^{-1}. We calculate the time until those trends become statistically different from the trend observed over the most recent thirty-year period (1979–2008). The results indicate that it will probably be decades before distinct changes from the current warming rate become apparent. For the A1B scenario, only 25 percent of the simulations indicate difference by 2040. For the CCC, A2, 1.5°C, and 2°C scenarios, the 25 percent level is reached in about 2030, 2040, 2065, and 2075, respectively. Only about 10 percent of the B1 simulations indicate a difference before 2100. These results indicate that we should expect decades to pass before impacts of the war against global warming become apparent. *Key Words: Bayesian inference, Fourth Assessment Report, global warming, greenhouse gases, temperature change.*

我们通过分析 HadCRUT3 重建的 1850 年到 2008 年全球温测记录把 30 年的气温变化趋势分解成信号和噪声成分。此信号代表几十年的趋势，而此噪声成分代表那些趋势的年度变化。温度变化的历史估计（如噪音）和七次温度预测一起使用以模拟全球变暖的时间序列。这些趋势包括 1979 年到 2008 年的趋势，从代表政府间气候变化专门委员会（IPCC）第四次评估报告（AR4）模拟得来的四个趋势（A1B，A2，B2，和恒定成分承诺[CCC]场景），和到 2100 年 1.5°C 和 2°C 变暖的趋势。每个系列模拟了 1000 次。把结果与目前 0.016°C 一年的变暖速度进行了统计比较。我们计算了直到这些趋势与最近 30 年期间（1979 年至 2008 年）观察到的趋势具统计差异的时间点。结果表明，目前的气候变暖速率的显著变化趋势可能需几十年的时间才能表现出来。对于 A1B 情景，只有百分之 25 的模拟结果表明到 2040 年的区别。对于 CCC，A2，1.5°C 和 2°C 的情景，在约 2030 年，2040 年，2065 和 2075 年分别到达该百分之 25 的水平。只有约百分之 10 的 B1 模拟表明 2100 年之前的差异。这些结果表明，在对全球变暖的战争影响变得明显之前我们还将等待几十年。*关键词：贝叶斯推理，第四次评估报告，全球气候变暖，温室气体，温度变化。*

Analizamos la reconstrucción HadCRUT3 del instrumental del registro global de temperatura, de 1850 a 2008, para descomponer en componentes de señal y ruido las tendencias de la temperatura para lapsos de treinta años. La señal representa las tendencias multidecadales y el ruido representa la variabilidad anual en relación con esas tendencias. Los cálculos históricos de variabilidad de la temperatura (o sea, ruido) se utilizan con siete proyecciones de temperatura para simular series de tiempo de calentamiento global. En estas tendencias se incluye la registrada para 1979 a 2008, cuatro tendencias tomadas de las simulaciones (A1B, A2, B2 y la composición constante de los escenarios [CCC] de compromiso) del Informe de la Cuarta Evaluación (AR4) del Panel Intergubernamental sobre Cambio Climático (IPCC, sigla en inglés), más las tendencias que representan aproximaciones a un calentamiento entre 1.5°C y 2°C para el 2100. Cada serie es simulada 1.000 veces. Los resultados son comparados, estadísticamente, con la tasa de calentamiento actual de 0.016°C año-1. Calculamos el tiempo hasta cuando esas tendencias llegan a ser estadísticamente diferentes de las tendencias observadas en el período de treinta años más reciente (1979–2008). Los resultados indican que probablemente pasen décadas antes de que cambios conspicuos en la tasa de calentamiento actual lleguen a ser aparentes. Para el escenario A1B, solamente el 25 por ciento de las simulaciones muestran una diferencia para el 2040. Para los escenarios CCC, A2, 1.5°C y 2°C el nivel del 25 por ciento se alcanza alrededor del 2030, 2040, 2065 y 2075, respectivamente. Solamente cerca del 10 por ciento de las simulaciones B1 muestran una diferencia antes del año 2100. Estos resultados indican que debemos esperar que transcurran décadas antes de que los impactos de la lucha contra el calentamiento global se puedan notar. *Palabras clave: inferencia bayesiana, Cuarto Informe de Evaluación, calentamiento global, gases de invernadero, cambio de temperatura.*

During the twentieth century, global atmospheric temperatures rose at an average rate of 0.06°C per decade, although for the period from 1956 through 2005, the warming rate was nearly 0.13°C per decade (Intergovernmental Panel on Climate Change [IPCC] 2007). These changes have spurred intense, unprecedented environmental debates over the magnitude, causes, and future trends in atmospheric warming, leading to widespread declarations of "war against global warming" by the press, by politicians and nongovernmental organizations, and by bloggers and academics. This simple phrase provides the premise for this article. We accept that global atmospheric temperature has been rising and we entertain the analogy of a war being waged to counter human contributions to that increase. The purpose of this article is not to debate the degree to which anthropogenic activities contribute to global warming nor whether we are really in a war against global warming. Our objective is to analyze predicted temperature trends combined with simulations of observed annual temperature variability, to estimate how long it might be before we can recognize the outcome of this war. That distinction will occur at benchmarks when projected temperature trends become statistically different from the current trend. Our simulations are based on prior distributions established from historical records, without regard to the relative importance of natural and anthropogenic causes of variability.

War Against Global Warming

The phrase "war against global warming" has become commonplace (e.g., Oppenheimer 2003; Baron 2006; Pelosi 2007; Branson 2008; Nader 2008; Dimas 2009; Heritage Foundation 2009), frequently used to establish a prima facie basis for public policy arguments (although many think that phrase is inappropriate vis-à-vis human vs. human conflict). For example, the 28 April 2008 cover story in *Time* magazine (Walsh 2008; reproduced on the "Globalwarming.gov" Web site of the U.S. House of Representatives Select Committee on Energy Independence and Global Warming) offered a strategy for winning the war using the emissions control policy known as *cap-and-trade*.

Although the scientific debate over the degree of anthropogenic influences on global warming heated up in the 1970s (e.g., see historical reviews by Ramanathan et al. 1985; Kellogg 1987), much of the political controversy stems from the 1992 United Nations Framework Convention on Climate Change (UN 1992) wherein *climate change* was defined in Article 1 to refer only to "change in climate that is attributed directly or indirectly to human activity that alters the composition of the global atmosphere."

An objective of the UN Convention was to control the concentrations of greenhouse gases (GHG) as a means of minimizing human impacts on the climate system. The United State did not sign the Kyoto Protocol. Since that time, however, the national and international political and public will to impose restrictions on GHG emissions has increased, especially with regard to reducing CO_2 emissions, the focus of most declarations of war against global warming. The urgency for action is driven by predictions of catastrophic temperature change if GHG emissions are not reduced substantially and immediately (e.g., United Nations Development Program 2007). According to Stern (2008), current atmospheric concentrations of the six most prevalent GHGs are estimated to be the equivalent (CO_2e) of 430 ppm CO_2 and rising. If the CO_2e concentration reaches 450 ppm (predicted for 2011) there is a 78 percent probability that this would cause a 2°C warmer (than preindustrial) global equilibrium temperature, and at 550 ppm CO_2e there is a 24 percent chance of a 4°C increase (Stern 2008). More than 100 countries have committed to establishing policies to keep total warming, relative to preindustrial temperatures, to less than 2°C (e.g., Meinshausen et al. 2009; Rogelj et al. 2009). Meinshausen et al. (2009) argued that there is a 75 percent chance that this goal will be met if cumulative CO_2 emissions between the years 2000 and 2050 can be limited to 1,000 gigatons but noted that about one quarter of that total was emitted between 2000 and 2006. Rogelj et al. (2009) calculated that global GHG emissions in 2050 should be about 70 percent of 2000 levels to obtain a 25 percent chance of success with the 2°C target. Some nations, including many small island nations, are promoting a more aggressive target—a total of 1.5°C warming (Rogelj et al. 2009), with about half of that warming (0.76°C) already present (IPCC 2007). These emission and temperature targets are the basic objectives in the war against global warming and we presume that the attainment of a particular target, or benchmark, will represent victory for some constituency.

This research produces estimates of how long it might take to recognize whether a benchmark has been attained, relative to current rates of temperature change.

To accomplish this we first estimate the current rate of global atmospheric temperature change over the last thirty years to use as our baseline condition. It is against this observed trend that predicted rates of future temperature change are assessed. We then analyze a record of historical temperatures to derive empirical estimates of annual variability (noise) for every thirty-year trend (signal) in that record. These data are used to simulate seven temperature futures through the end of the twenty-first century and to estimate how long it might be before those trends become statistically distinct from the current trend. The point in time when that distinction occurs will not represent victory, but will signal that the tide has changed. All of our simulations begin by predicting the year 2009 temperature anomaly and using that value with the preceding twenty-nine years of historical data to establish a new trend.

Methods

The structure of our methods is conceptually straightforward: We (1) analyze historical temperature data in a series of thirty-year sets to quantify trends; (2) compute residual time series (annual variability) from those data sets; (3) assess the statistical characteristics of the ensemble residual data; (4) use the residuals and a Bayesian, predictive inference approach to simulate future residual time series; (5) use those residual series and smoothed temperature projections to produce simulated time series; and (6) calculate the time until the trends of those time series become statistically different from the current thirty-year trend. We consider the statistical differentiation to be a benchmark that establishes a magnitude and direction of climate change that is distinct from the current trend. The choice to establish the annual variability data to use with projected trends follows from Kim and North (1995, 2172), who stated that "one can add noise fluctuations to the greenhouse warming signal to mimic the true . . . warming signal."

Data

There are several data sets representing global mean atmospheric temperatures. These include the Goddard Institute for Space Studies Surface Temperature Analysis (e.g., Hansen et al. 2006), the HadCRUT3 series produced by the Hadley Centre's Climate Research Unit (e.g., Brohan et al. 2006; Rayner et al. 2006), and the National Climatic Data Center/National Oceanic and Atmospheric Administration's global surface temperature anomalies (e.g., Smith and Reynolds 2005). Although all three temperature data sets are based on overlapping sources, we chose the HadCRUT3 series for our analysis because it provides a record of surface temperature starting in 1850 and the data are accompanied by a comprehensive set of uncertainty estimates for data inhomogeneities associated with measurement and sampling error, temperature biases, biases due to changes in network density, and other sources of error as described by Peterson et al. (1998). They noted that even after careful quality control and application of bias adjustments, the net effect of these inhomogeneities is that the accuracy of all long-term records of global surface temperature (including HadCRUT3) is lower during the earliest and latest parts of the historical records. The HadCRUT3 record covers the period of 1850 through 2008, with averages computed over a $5^\circ \times 5^\circ$ grid using sea surface temperatures for marine regions. Temperatures are reported as anomalies relative to the thirty-year period from 1961 through 1990. Accuracy of the reconstructed temperatures is about $\pm 0.2^\circ$C for the 1850s, improving to about $\pm 0.05^\circ$C more recently (Folland et al. 2001).

Analysis

The World Meteorological Organization calculates climatological norms using decadally adjusted, thirty-year averages, such as the period from 1961 through 1990, which is the standard baseline for temperature anomalies. As such, there have been thirteen climate periods since 1850. However, from the perspective of a nonsentient atmosphere, the beginning of any thirty-year period is a chronological accident. Therefore, we analyzed every thirty-year period since 1850 by sliding a thirty-year window one year at a time until we reached 2008, yielding 130 climate periods for analysis. This approach is similar to the trend estimation procedure adopted by Barnett et al. (1998).

We used linear regression to fit a least squares line to each thirty-year time series, and derived their residuals. For each year in the record we obtain the set of all possible historical residuals, comprising up to thirty values, and there will be up to twenty-nine values for the following year's residual. This resulted in a total of 3,900 individual residual values whose population averages 3.4×10^{-17}°C with a standard deviation of 0.103°C. The maximum and minimum values are +0.343°C and −0.262°C, respectively. This data set represents every possible residual and sequence of residuals for any thirty-year period in the historical temperature record, representing the natural variability (or noise) of temperature about secular trends. This approach also ensures that many nonanthropogenic sources of short-term temperature changes, such as the Pacific Decadal Oscillation, El Niño-Southern Oscillation events, and volcanic eruptions, are represented in the data. To assess the stability of the residual series through time, we used linear regression to test for a trend in the magnitude of residual standard deviations. There was not a statistically significant relationship (slope $= -3.5 \times 10^{-6}$, $r^2 = 0.0001$, $p = 0.9$).

We simulate noise series using a Bayesian probability approach rather than using random samples drawn from a fitted probability (or spectral) density function (e.g., per Kim and North 1995). If we had used a frequentist approach and a large number of simulations, we would likely produce noise series unlike any represented in the HadCRUT3 data.

The Bayesian, predictive inference approach applies a mathematical framework on fixed, known data to quantify uncertainty about unknown data. This "data-centric" approach predicts values for future samples that will behave like observed samples, assuming that all forms of uncertainty are inherently quantifiable (Draper and Madigan 1997). This is an advantage over frequentist approaches that are based on fitting stringent distributions (such as the Gaussian) to data. Because the known (observed) and unknown (predicted) data sets have similar structures, any order of random variables is exchangeable and the statistical characteristics of the simulated data emulate those of the real data. We use the residual data described earlier to establish the Bayesian prior distribution needed for our simulations. Any autocorrelation in the temperature record is preserved by this approach—a necessity for developing the probability of a particular residual following another. This provides a mathematically sound procedure for simulation using Markov chains, for example.

We evaluated each thirty-year sequence of residuals in the following manner. First we paired the residual, R, for each year, t, and the residual for the following year, R_{t+1}, yielding twenty-nine pairs for each time series. We sorted the 3,900 residuals into class ranges (bins) of 0.02°C. For each R_t the related value of R_{t+1} is binned, resulting in the distribution depicted in Figure 1. Figure 1 also depicts the distribution of cases where the residual is increased, decreased, or stays the same. Every original R_t will be represented in a bin with a family of R_{t+1} residuals, each with an equal probability of being next. We use a random number generator to pick an

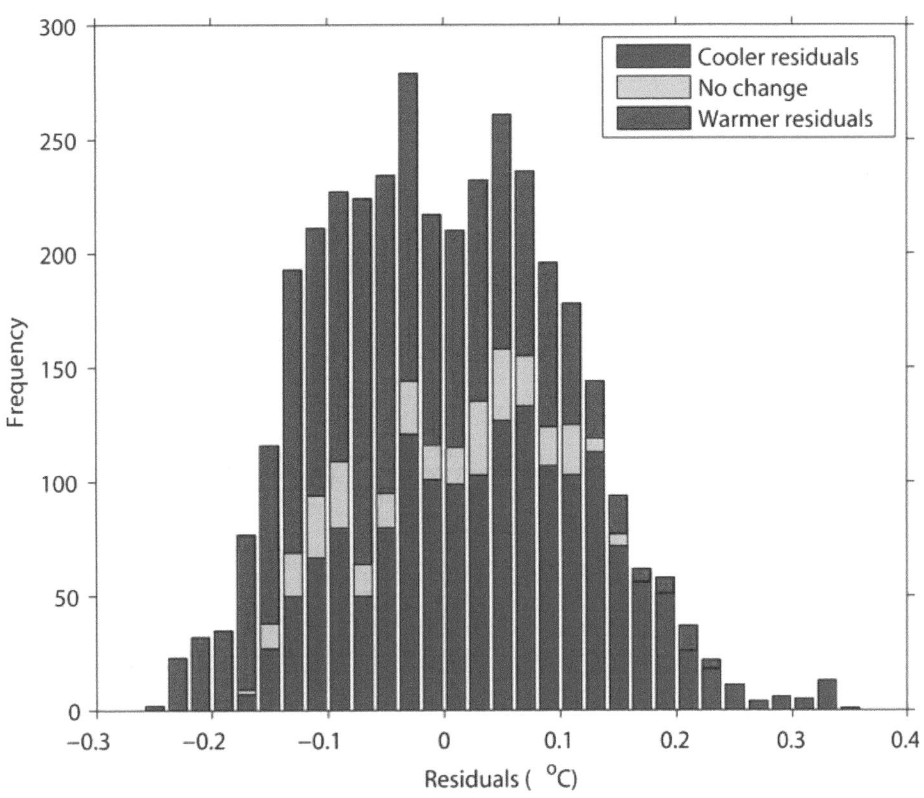

Figure 1. Distribution of the 3,900 residuals by increments of 0.02°C.

R_{t+1}, which becomes the next R_t in the series, and then the process is repeated. With this approach the magnitude and direction of change of temperature from year to year along the Markov chain are controlled completely by the distributions of historical fluctuations. All simulated residuals fall within the bounds of the actual data, and the probability of obtaining any particular value for R_{t+1} depends solely on its preceding R_t. For example, as a residual becomes more positive, it becomes increasingly likely that the subsequent residual will be cooler because the distribution of possible outcomes is adjusted according to historical precedence. If a residual is near 0°C, then there is an almost even likelihood of the next value being positive or negative.

We repeated the Markov chain simulation to get 1,000 residual time series of ninety-two years' length (through the year 2100) and compared the descriptive statistics with those of the original data. The mean, -7.6×10^{-6}, and the standard deviation, 0.103, of the simulated population were virtually identical to those of the historical data (with trivial differences due to rounding), thus meeting the exchangeability criterion for predictive inference. We consider these residual time series to represent the inherent variability in any thirty-year time series of mean global atmospheric temperature. This allows us to overlay the simulated series on any secular trend we choose.

Temperature Trends

We chose seven temperature projections (Figure 2) for evaluation. These trends include a linear continuation of the present trend, four trends taken from the IPCC Fourth Assessment Report (AR4) simulations, and trends representing approaches to 1.5°C and 2°C warming by 2100.

The "current" trend continues the 1979 through 2008 warming rate of 0.016°C per year^{-1}. The 95 percent confidence limits for that rate are 0.012 to 0.020°C year^{-1}. We picked four of the projected trends developed for the IPCC Third Assessment Report (TAR) and Fourth Assessment Report (AR4). The IPCC (2007) used several emission scenarios to drive General Circulation Model (GCM) simulations. The multiple emission scenarios provide a means to quantify the uncertainties associated with changes in GHG emissions based on estimates of economic growth, global relations, technological diffusion, environmental protection, and demographic change. We selected the IPCC emission scenarios A1B, B1, and A2. The temperature changes associated with these scenarios represent low (B1; 1.8°C warming by 2100), medium (A1B; 2.8°C warming by 2100), and high (A2; 3.4°C warming by 2100) examples of the ranges reported in AR4.

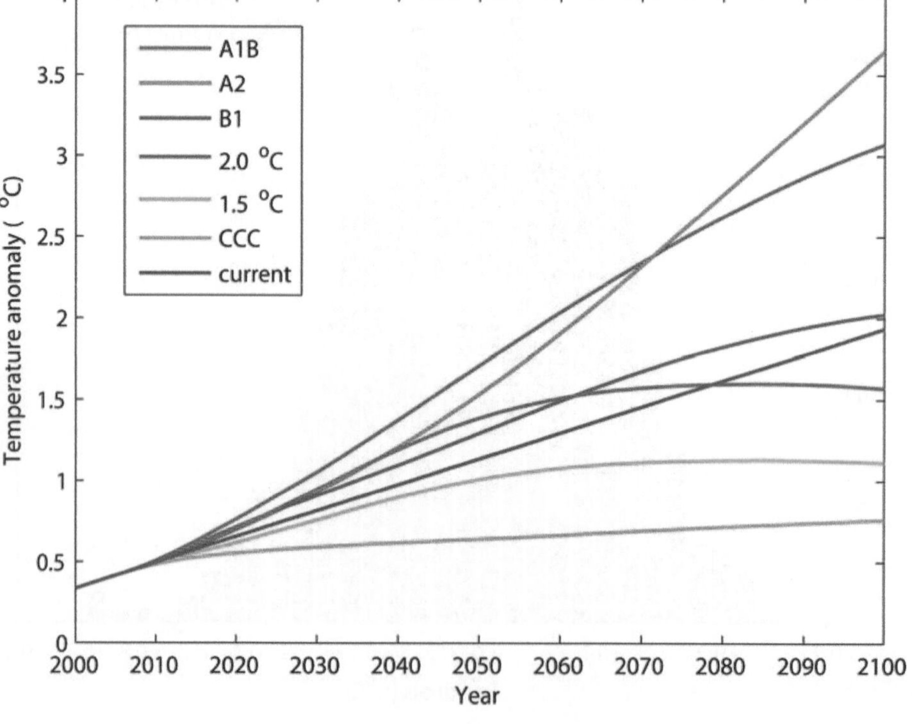

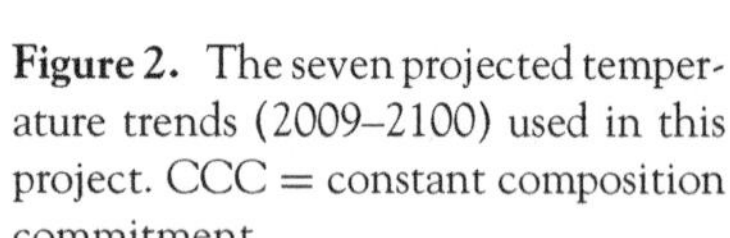

Figure 2. The seven projected temperature trends (2009–2100) used in this project. CCC = constant composition commitment.

The A1 scenarios are characterized by rapid economic growth, a population that peaks at 9 billion in 2050 and then declines, rapid introduction of new and more efficient technologies, and a world where disparities in income and lifestyle decline. The A1B scenario includes a balanced emphasis on all energy sources. The B1 scenario family is based on a convergent world with the same population growth as in A1 but with a rapid shift in the economy toward service and information-based sectors. The B1 scenario is ecologically friendly and features the introduction of clean and resource-efficient technologies. The emphasis is on economic, social, and environmental sustainability, including improved equity, but without additional climate initiatives. The A2 scenario family represents a more heterogeneous world with increasing population and regionally fragmented economic growth that is slower than in A1 or B2. The A2 scenarios also consider slower and more fragmented technological changes and improvements to per capita income. The fourth IPCC trend is the CCC curve based on predictions of temperatures if GHG concentrations were reduced to and held at their 2000 levels but with further warming of 0.1°C per decade for two decades because of the slow response of the oceans. Regardless of the scenario used, the best estimate GCM projections indicate that the magnitude of warming over the continents by 2030 is consistent (IPCC 2007, 12).

Several steps were necessary to adapt the four IPCC scenarios for this study. First, we digitized the curves in AR4 (IPCC 2007) Figure 10.4 over the period 2000 to 2100. Because one of the purposes of our study is to add short-term, Bayesian variability to temperature trends, it was necessary to remove that variability from the IPCC curves. We did this using a thirty-one-year moving average to obtain a smoothed data set for the years 2015 through 2085. Three steps were necessary to fill the time gaps. First, for 2000 through 2008 we used the current 0.016°C slope. Second, for 2009 to 2015 we interpolated using a uniform rate of change of slope to merge the 0.016°C trend with the respective AR4 trend to obtain a smooth transition of trends. Third, for the period 2086 to 2100, we used linear extrapolation.

The sixth and seventh trends are developed to approach total warming limits of 1.5°C and 2°C by 2100. Because there has already been 0.76°C of industrial-era warming, those limits become 0.74°C and 1.24°C from the present. We digitized the 2°C trend as presented in Meinshausen et al. (2009, Figure 2) and scaled the 1.5°C target against that.

A summary of the seven trends is presented in Table 1. Note that the year 2100 target temperatures do not match the temperatures indicated in Figure 2. This is because five of the seven target temperatures represent anomalies relative to the 1961 through 1990 mean temperature, whereas the 1.5°C and 2°C trends do not. The four AR4 curves are calibrated to merge with our temperature regression line in 2003, when the anomaly is already 0.384°C. The curves representing the 1.5°C (0.74°C remaining) and 2°C (1.24°C remaining) warmings are calibrated to merge with our data in 2008.

Benchmark Identification

Our assessment of trends is based on linear regression of thirty-year time series to find least squares trends and calculate the 95 percent confidence limits around those trends. We define a benchmark as the occurrence of the lower confidence limit of a particular trend exceeding the upper limit of the current trend (or any other trend of special interest) or when the upper confidence limit falls below the lower limit of the current trend. It is at

Table 1. Summary of temperature projection scenarios

Scenario	Temperature anomaly	Basis
A1B	+2.8°C by 2100 (AR4)	Rapid economic growth; A1B: balanced emphasis on all energy sources
A2	+3.4°C by 2100 (AR4)	A2: slower change; increasing global population
B1	+1.8°C by 2100 (AR4)	Ecologically friendly; introduction of clean, resource-efficient technologies
CCC	+0.76°C by 2100	Based on scenario that current greenhouse gas concentrations remain at 2000 levels
1.5°C Target	+1.5°C (0.74°C remaining) by 2100	Approach total warming limit of 1.5°C by 2100
2.0°C Target	+2.0°C (1.24°C remaining) by 2100	Approach total warming limit of 2.0°C by 2100
1979–2008	+1.5°C by 2100	Continuation of last thirty years trend from 2008 to 2100

Note: CCC = constant composition commitment.

those benchmarks (divergences) when the respective trends become statistically distinct.

There are two types of benchmark signals. There is a "false" signal that occurs when the noise in a time series causes a temporary divergence of confidence limit slopes that is reversed in a subsequent year. Second, there is the "true" signal that occurs when the divergence is not reversed before the year 2100. Note that the designation false is used only to designate the temporary nature of a threshold; as such a signal might indeed be a genuine precursor of a distinctive warming or cooling trend (relative to the 0.016°C year^{-1}). We are interested in both types of signals.

Projecting the Trends

We extrapolated future temperatures from 2008 using the simulations described earlier. The first extrapolation uses twenty-nine years of data from 1980 through 2008 with a predicted 2009 temperature (anomaly). The 2009 temperature is predicted by increasing the temperature from the 2008 value by an increment equal to the annual change rate of a particular warming scenario and then adjusting that value using a randomly generated R_{t+1}. A new thirty-year slope and its confidence limits are calculated. The confidence limits are compared to those from the 1979 through 2008 data. This procedure is performed 1,000 times for each scenario. The process is repeated for each succeeding year, using less and less historical data until, in the year 2038, all of the data are simulated. We continue these simulations until the year 2100.

It is important to note that all of the projected trends are nonlinear, other than that fit to the 1979 through 2008 data. For our analysis we approximate the curves in Figure 2 as a sequence of straight lines, each thirty years in length.

Results

An example of a set of simulated time series is presented in Figure 3. These simulations were randomly selected from the sets of 1,000 and are used to illustrate the magnitude of annual variability relative to actual warming trends when the residual-derived noise is added to the actual trends.

In this analysis we are interested in benchmarks of two types: when does a trend that warms faster than the current trend become statistically distinct and when does a trend that warms slower become distinct? In either case it is important to distinguish between false and true signals. We use the terms *warming* and *cooling* to represent temperature changes relative to the 1979 through 2008 rate. None of the projections that we evaluate represents actual cooling.

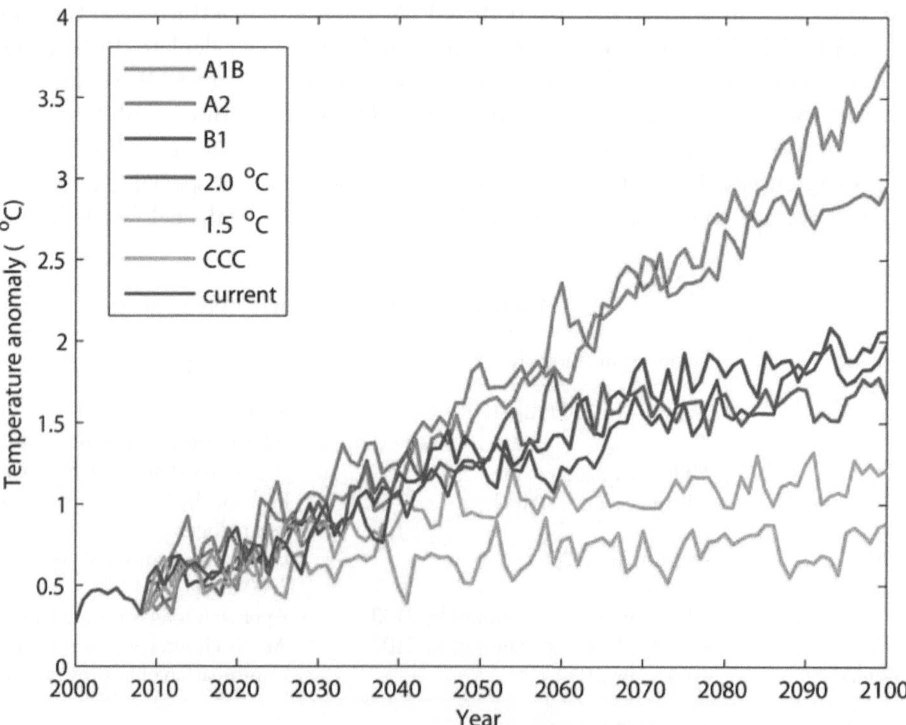

Figure 3. Seven random realizations of the "noise + signal" simulations. CCC = constant composition commitment.

Table 2. Distribution of false benchmarks from the 1,000 simulations of future temperatures

Scenario	False warming signals (year)			False cooling signals (year)		
	Min	Max	Median	Min	Max	Median
A1B	2020	2043	2032	—	—	—
A2	2021	2057	2039	—	—	—
B1	2021	—	2066	2057	—	—
2.0°C	2021	—	2036	2063	2091	2077
1.5°C	2040	—	—	2023	2084	2064
CCC	—	—	—	2022	2046	2026
Current	2035	—	—	2026	—	—

Note: Minimum (Min) is the first year when a false warming or cooling signal is observed in any of the 1,000 simulations and maximum (Max) is the last year when a false warming or cooling signal is observed in any of the 1,000 simulations. CCC = constant composition commitment.

A summary of the results for false signals is presented in Table 2. We expect only false signals, if any, from a continuation of the current trend. From the 1,000 simulations we found that the first false warming signal occurred in 2035 and a first false cooling signal in 2026. These cases are unusual, as the average simulation issues no such signal. The A1B trend issues a false warming signal as early as 2020 and as late as 2043, with the median in 2032. The A2 scenario leads to all cases becoming at least temporarily distinct before 2100, with maximum and median years 2057 and 2039, respectively. Neither A1B nor A2 produces a false cooling signal. The A2, B1, and 2°C trends all cause false warming signals in 2021. The 1.5°C simulations first indicate warming by 2040, but the CCC curve does not signal warming at all. False cooling signals are more common because three of the trends represent net cooling relative to an extrapolated 0.016°C year^{-1} warming rate. The typical duration of a "false" signal is one to five years, averaging about two years.

We report the occurrence of true benchmarks in a different fashion because in only the A2 scenario do all 1,000 simulations yield distinction from the current trend, where the first permanent crossing occurs in 2029 and the last in 2061, with a median date of 2044. Therefore, we report results according to the frequency with which simulations pass thresholds (diverge) by year (Figure 4). Net warming occurs under scenarios A2 and A1B. For the A2 curve, all cases are distinct by 2100 (indeed, by 2061), whereas only about 50 percent of the A1B cases are distinct from the current trend by 2100. For both scenarios, only about 25 percent of the time will a true change signal be generated by 2040.

The other four scenarios result in cooling thresholds, although none of them generate cooling signals for 100 percent of their cases. The first true signals for these trends occur in 2023 for CCC, 2051 for the 1.5°C target, 2067 for the 2.0°C, and 2092 for the B1 scenario.

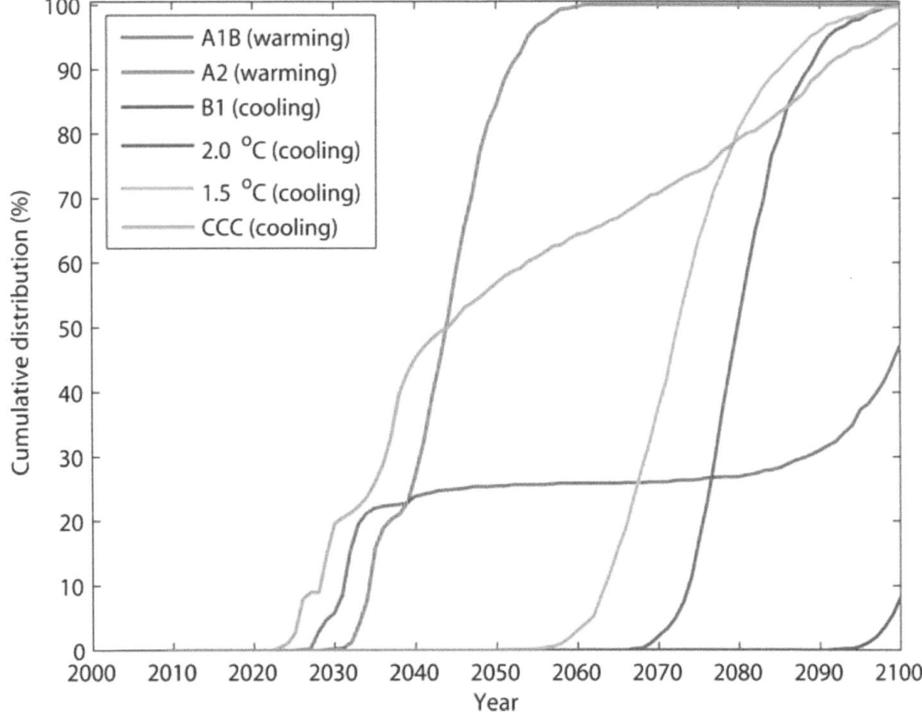

Figure 4. Cumulative distribution of true benchmark signals based on simulated temperature series and 95 percent confidence limits. CCC = constant composition commitment.

About 50 percent of the CCC simulations generate a true cooling signal by about 2050, but it takes until about 2070 and 2080 for the 1.5°C and 2°C scenarios, respectively. Less than 10 percent of the time will the B1 trend produce a statistically distinct change by 2100.

Discussion

The results indicate that it will be a long time, in a human context, before we can discern statistically significant differences between the current rate of atmospheric warming and the alternative warming scenarios. If we benchmark progress in 2030, we see that only two of the projections will signal a change in more than 5 percent of cases (Figure 4). By 2050, about 90 percent of the A2 cases, 30 percent of the A1B cases, and 55 percent of the CCC cases have signaled change. It will be difficult to recognize the occurrence of the B1 scenario as there will be minimal separation from the current trend by 2100.

Four factors could affect our results. First, we have made assumptions about the uniformity of the distribution of residuals through time. From the historical record we find that there is a mild dependence of the variability of residual sets on the slope of the least squares line. For the seventy-five cases where temperature trends were statistically significant ($p \leq 0.05$), the slope explained statistically about 37 percent of the changes in the standard deviation of residual sets ($r^2 = 0.37$, $p < 0.00001$). The range of standard deviations was only about 0.07°C to 0.12°C (and this range is the same for all 130 cases), however, so we ignored this difference in our analysis. Further, we assume that the distribution of noise (residuals) in the future will be similar to that found for the HadCRUT3 series.

Second, the results would be different if we chose 90 percent rather than 95 percent for the confidence limits around the regression lines (Figure 5). Using 90 percent confidence limits, attainment of 25 percent and 50 percent of the cumulative threshold cases occurs about five years earlier for the A2, 1.5°C, and 2°C scenarios, and about a decade earlier for the CCC projection. The changes are more extreme for the A1B scenario: The 25 percent attainment occurs in about 2025 instead of 2040, and the 50 percent level occurs by about 2035 rather than not at all.

Third, it is likely that some bias (although we believe it to be trivial) is introduced by our curve fittings and extrapolations to get the smoothed temperature projections. For the AR4 (IPCC 2007) curves in particular, the boundary conditions at 2100 made smoothing and extrapolating difficult. For example, the projections for the CCC and A2 trends are not modeled beyond 2100. The number of simulation models used to extend the A1B and B1 projections beyond 2100 changes from

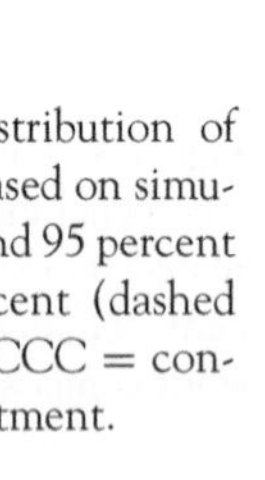

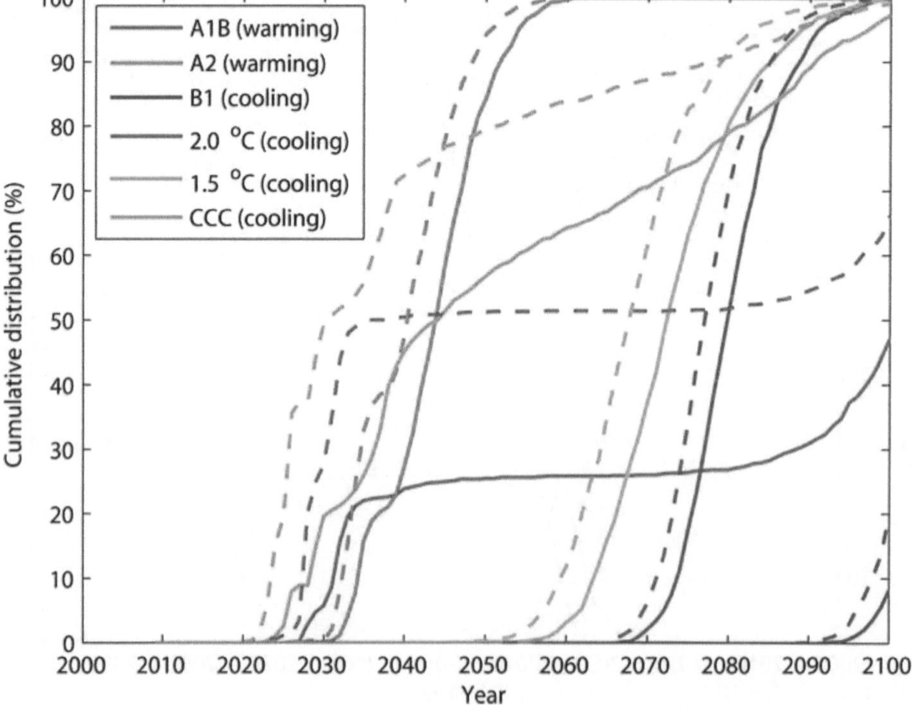

Figure 5. Cumulative distribution of true benchmark signals based on simulated temperature series and 95 percent (solid lines) and 90 percent (dashed lines) confidence limits. CCC = constant composition commitment.

twenty-one to seventeen and twenty-one to sixteen, respectively. This causes a temperature discontinuity at 2100, manifested as a negative offset in the trend—especially notable in the B1 projection.

Fourth, we have limited our regression analysis to linear methods, even in instances where trends appear to be clearly nonlinear and where other regression models might be more appropriate. In our evaluation of the 130 thirty-year periods that comprise the data set for trends, there was no single alternative method that reproduced trends better than the linear model. Further, we follow the examples of Kim and North (1995), the IPCC (2007), Easterling and Wehner (2009), and Knight et al. (2009), among others, who fit linear least squares trend lines to temperature series.

It is also important to note that the temperature projections that we consider, except the continuation of the current trend, are based on the assumption that warming will be governed solely by changes in GHG concentrations. This is consistent with the GCM models used in the A4R simulations. However, Feddema et al. (2005) and Pielke (2005) argued that land use and land cover change could also be significant drivers of climate change. Lean and Rind (2008) noted that there are other natural drivers of climate change, such as changes in solar activity, volcanic eruptions, and so on, although they are small relative to anthropogenic forcing.

Our results indicate that even if there is an aggressive international effort to reduce atmospheric concentrations of GHG immediately, it will be decades before the results become statistically discernable. If we require 90 percent certainty of change, there will only be a true signal of change 50 percent of the time by 2025 and only for the CCC projection. Although we have only presented formal comparisons of future trends with the current trend, there is similar ambiguity between other pairs of trends. For example, the A2 trend, predicting the fastest warming, is statistically indistinguishable from the 2.0°C trend until after 2050. This means that efforts to reduce GHG emissions and to sequester carbon must be sustainable for decades based on confidence in both the science that predicts that the efforts and in the policies that must be adopted and enforced. We can think of no such precedent in human history.

Conclusions

We have analyzed the surface temperature record for the period from 1850 through 2008 and extracted the natural temperature variability (noise) around linear trends. This variability was added, in sets of 1,000 simulations, to smoothed time series of seven temperature trends projected to the year 2100. Finally, we compared changes in trends relative to the 1979 through 2008 trend based on a consideration of the 95 percent confidence limits for the slopes of least squares lines. An appraisal of these results leads us to two main conclusions:

1. Decades might elapse before there are clear signals that projected warming trends can be distinguished from the present rate. For warming predicted under the AR4, A1B, B1, and the 1.5°C and 2°C warming scenarios, we note that less than half of the simulations produce a statistically significant signal before 2065.
2. The choice of the statistically based benchmark for distinguishing trends will impact substantially the length of time required to recognize whether climate control policies are succeeding. We have described a protocol for evaluating change using 95 percent and 90 percent confidence limits and cumulative distributions of simulations to demonstrate the characteristic lengths of time necessary for temperature projections to become distinct. However, from either a scientific or a policy perspective, there is no single, unambiguous benchmark criterion. Establishment of such criteria will likely be purpose driven, with some arguing for relatively lenient conditions. Others might argue for 90 percent or 95 percent confidence in trend discernment. The choice will be driven by the relative merits of making a Type I or Type II error. In either case, we are currently experiencing one of the fastest warming rates of the last century and a half, and there is every indication that warming rates will increase further unless remedial actions are taken. We anticipate that this will require a sustained commitment to stringent climate control policies for periods of decades or longer.

The phrase "war against global warming" is used by pundits, politicians, and advocates as a tool to focus attention and create a sense of urgency with the aim to influence the development of policies to mitigate the anthropogenic causes of global warming. Truly, however, it is not a war, and it will not become one. It will be a struggle. It will require unprecedented social, political, and economic commitments. Unlike a true war, however, we cannot anticipate victory, only, at best, a stalemate.

Revelle and Suess (1957, 19), in discussing the production of carbon dioxide from the burning of fossil fuels, wrote, "Thus human beings are now carrying out a large scale geophysical experiment of a kind that could not have happened in the past nor be reproduced in the future." That experiment was accidental. What we contemplate now is a purposeful, socioeconomically driven geophysical experiment of a magnitude unimaginable when Revelle and Suess made that prophetic statement a half-century ago—an experiment with outcomes unknowable for perhaps another half-century.

Acknowledgments

This work was partially supported by funding from the Stanford Development Group. We are grateful for the insightful comments offered by Seth Guikema, Gerald North, and two anonymous reviewers on an earlier version of the article. Their comments provided valuable guidance for our revisions.

References

Barnett, T. P., G. C. Hegerl, B. Santer, and K. Taylor. 1998. The potential effect of GCM uncertainties and internal atmospheric variability on anthropogenic signal detection. *American Meteorological Society* 11 (4): 659–75.

Baron, J. 2006. Thinking about global warming. *Climate Change* 77:137–50.

Branson, R. 2008. Addressing climate change: The United Nations and world at work. http://www.un.org/ga/president/62/ThematicDebates/statements/RichardBransonSpeech.shtml (last accessed 13 July 2009).

Brohan, P., J. J. Kennedy, I. Harris, S. F. B. Tett, and P. D. Jones. 2006. Uncertainty estimates in regional and global observed temperature changes: A new dataset from 1850. *Journal of Geophysical Research* 11: D12106. http://www.agu.org/pubs/crossref/2006/2005JD006548.shtml (last accessed 13 July 2010).

Dimas, S. 2009. World faces last chance to avoid fatal warming: EU. http://www.reuters.com/article/environmentNews/idUSTRE51Q22X20090227 (last accessed 13 July 2009).

Draper, D., and D. Madigan. 1997. The scientific value of Bayesian statistical methods. In *Trends and controversies*, ed. M. A. Hearst, 18–21. Berkeley: University of California.

Easterling, D. R., and M. F. Wehner. 2009. Is the climate warming or cooling? *Geophysical Research Letters* 36: L08706. http://www.agu.org/pubs/crossref/2009/2009GL037810.shtml (last accessed 13 July 2010).

Feddema, J. J., K. W. Oleson, G. B. Bonan, L. O. Mearns, L. E. Buja, G. A. Meehl, and W. M. Washington. 2005. The importance of land-cover change in simulating future climates. *Science* 310 (5754): 1674–78.

Folland, C. K., N. A. Rayner, S. J. Brown, T. M. Smith, S. S. P. Shen, D. E. Parker, I. Macadam, P. D. Jones, R. N. Jones, N. Nicholl, and D. M. H. Sexton. 2001. Global temperature change and its uncertainties since 1861. *Geophysical Research Letters* 28 (13): 2621–24.

Hansen, J., M. Sato, R. Ruedy, K. Lo, D. W. Lea, and M. Medina-Elizade. 2006. Global temperature change. *PNAS* 103 (39): 14288–293.

Heritage Foundation. 2009. The Heritage Foundation: Leadership for America. http://www.heritage.org/ (last accessed 13 July 2009).

Intergovernment Panel on Climate Change (IPCC). 2007. Summary for policymakers. In *Climate change 2007: The physical science basis. Contribution of Working Group I to the Fourth Assessment Report of the Intergovernmental Panel on Climate Change*, ed. S. Solomon, D. Qin, M. Manning, Z. Chen, M. Marquis, K. B. Averyt, M. Tignor, and H. L. Miller, 1–18. New York: Cambridge University Press.

Kellogg, W. W. 1987. Mankind's impact on climate: The evolution of an awareness. *Climate Change* 10:113–36.

Kim, K-Y., and G. R. North. 1995. Regional simulations of greenhouse warming including natural variability. *Bulletin of the American Meteorological Society* 76:2171–78.

Knight, J., J. J. Kennedy, C. Folland, G. Harris, G. S. Jones, M. Palmer, D. Parker, A. Scaife, and P. Stott. 2009. Do global temperature trends over the last decade falsify climate predictions? *Bulletin of the American Meteorological Society* 90:S56–S57.

Lean, J. L., and D. H. Rind. 2008. How natural and anthropogenic influences alter global and regional surface temperatures: 1889 to 2006. *Geophysical Research Letters* 35: L18701. http://www.agu.org/pubs/crossref/2008/2008GL034864.shtml (last accessed 13 July 2010).

Meinshausen, M., N. Meinhausen, W. Hare, S. C. B. Raper, K. Frieler, R. Knutti, D. J. Frame, and M. R. Allen. 2009. Greenhouse-gas emission targets for limiting global warming to 2°C. *Nature* 458 (909): 1158–62.

Nader, R. 2008. Ralph Nader challenges Al Gore to make his Obama endorsement count. http://www.votenader.org/media/2008/06/16/GoreObama/ (last accessed 13 July 2009).

Oppenheimer, M. 2003. After Iraq: Declare war on global warming. *New York Times* 8 April. http://www.nytimes.com/2003/04/08/opinion/08iht-edoppen_ed3.html (last accessed 13 July 2009).

Pelosi, N. 2007. House Speaker Nancy Pelosi calls the situation in Iraq "chaotic." *NPR News Morning Edition* http://www.npr.org/about/press/2007/013107.pelosi.html (last accessed 13 July 2009).

Peterson, T. C., D. Easterling, T. Karl, P. Y. Groisman, N. Nicholls, N. Plummer, S. Torok, et al. 1998. Homogeneity adjustments of in situ atmospheric climate data: A review. *International Journal of Climatology* 18:1493–517.

Pielke, R. A. 2005. Land use and climate change. *Science* 310 (5754): 1625–26.

Ramanathan, V., R. J. Cicerone, H. B. Singh, and J. T. Kiehl. 1985. Trace gas trends and their potential role in climate change. *Journal of Geophysical Research* 90 (D3): 5547–66.

Rayner, N. A., P. Brohan, D. E. Parker, C. K. Folland, J. J. Kennedy, M. Vanicek, T. Ansell, and S. F. B. Tett. 2006. Improved analyses of changes and uncertainties in marine temperature measured in situ since the mid-nineteenth century: The HadSST2 dataset. *Journal of Climate* 19:446–69.

Revelle, R., and H. E. Suess. 1957. Carbon dioxide exchange between atmosphere and ocean and the question of an increase of atmospheric CO_2 during the past decades. *Tellus* 9:18–27.

Rogelj, J., B. Hare, J. Nabel, K. Macey, M. Schaeffer, K. Markmann, and M. Meinhausen. 2009. Halfway to Copenhagen, no way to 2°C. *Nature Reports Climate Change* 3:81–83.

Smith, T. M., and R. W. Reynolds. 2005. A global merged land air and sea surface temperature reconstruction based on historical observations (1880–1997). *Journal of Climate* 18:2021–36.

Stern, N. 2008. The economics of climate change. *American Economic Review: Papers and Proceedings* 98:1–3.

United Nations (UN). 1992. *United Nations framework convention on climate change*. New York: United Nations.

United Nations Development Program. 2007. Making globalization work for us all: United Nations Development Program annual report 2007. New York: Office of Communications, United Nations.

Walsh, B. 2008. How to win the war on global warming. *Time* 28 April. http://www.time.com/time/covers/0,16641,20080428,00.html (last accessed 26 July 2010).

Regional Initiatives: Scaling the Climate Response and Responding to Conceptions of Scale

Melinda Harm Benson

In the absence of a coherent national policy to address global climate change, there was an emergence of new scales of environmental governance in the United States. They include regional initiatives that configure binding agreements between individual states within the United States that in some instances also involve cross-border collaborations with Canadian provinces. There are currently three such initiatives: the Northeastern Regional Greenhouse Gas Initiative, the Western Climate Initiative, and the Midwestern Regional Greenhouse Gas Reduction Accord. Combined, they include twenty-four states representing over half of the U.S. economy and four Canadian provinces representing almost three quarters of the Canadian economy. These initiatives are an interesting and unprecedented experiment in environmental governance. As such, they inform current conceptualizations of scale within human geography. They also evidence the need for geographers to enter into public debates on climate change governance and engage in a reconceptualization of the nature of sovereignty. *Key Words: climate change, governance, politics of scale, regional approaches, scalar fixity.*

基于缺乏一个连贯的国家政策以应对全球气候变化的情况，在美国出现了一种环境治理的新尺度。这些区域内的举措包括：配置美国各州之间具有约束力的协议，在某些情况下还涉及跨边界与加拿大各省的合作。目前有三个这样的举措：东北地区温室气体倡议，西部气候倡议，以及中西部地区温室气体减量协议。加在一起，它们包括占据美国经济半数以上的 24 州和几乎占据加拿大经济四分之三的 4 个加拿大省。这些措施是环境治理方面前所未有并引起广泛兴趣的实验。籍此，他们在人文地理学上引入了当前的概念化尺度。他们同时也提供了需求证据，以证明地理学者介入气候变化治理的公共辩论以及参与管理权性质的重新概念化的需要。*关键词：气候变化，管理，规模政治，区域办法，数量固定性。*

A falta de una política nacional coherente que enfrente el problema de cambio climático global, en Estados Unidos han emergido nuevas escalas de autoridad ambiental. Entre éstas se encuentran las iniciativas regionales que configuran acuerdos obligatorios entre estados individuales dentro de los Estados Unidos, las cuales a veces involucran colaboraciones transfronterizas con provincias canadienses. Actualmente existen tres de tales iniciativas: la Iniciativa Regional de Gases de Invernadero del Nordeste, la Iniciativa Occidental del Clima y el Acuerdo Regional de Reducción de Gases de Invernadero del Medio Oeste. Combinadas, estas iniciativas incluyen veinticuatro estados que representan más de la mitad de la economía americana, más cuatro provincias canadienses en donde se concentran casi tres cuartas partes de la economía canadiense. Estas iniciativas son un interesante experimento sin precedentes en términos de autoridad ambiental. En tal condición, las iniciativas reflejan las actuales conceptualizaciones de escala en geografía humana. También ellas evidencian la necesidad de que los geógrafos se comprometan en debates públicos sobre el gobierno de cambio climático y se comprometan en una reconceptualización de la naturaleza de soberanía. *Palabras clave: cambio climático, gobierno, políticas de escala, enfoques regionales, coherencia escalar.*

A scientific consensus that humans are causing climate change has existed for nearly a decade (Oreskes 2004). Although the environmental, economic, and social consequences of climate change are increasingly evident and alarming, the United States has been slow to take action and place a federal limit on greenhouse gas (GHG) emissions. In the absence of a coherent national policy in the United States, climate change efforts proliferated at other scales of governance. Cities, counties, states, universities, and even corporations have launched initiatives to reduce GHG emissions and employ mitigation and adaptation strategies (Betsill 2001; Hoffman 2006; Rappaport and Hammond 2007).

This article examines a new scale of governance produced in response to climate concerns. Described as "regional initiatives," they configure binding agreements between individual states within the United States

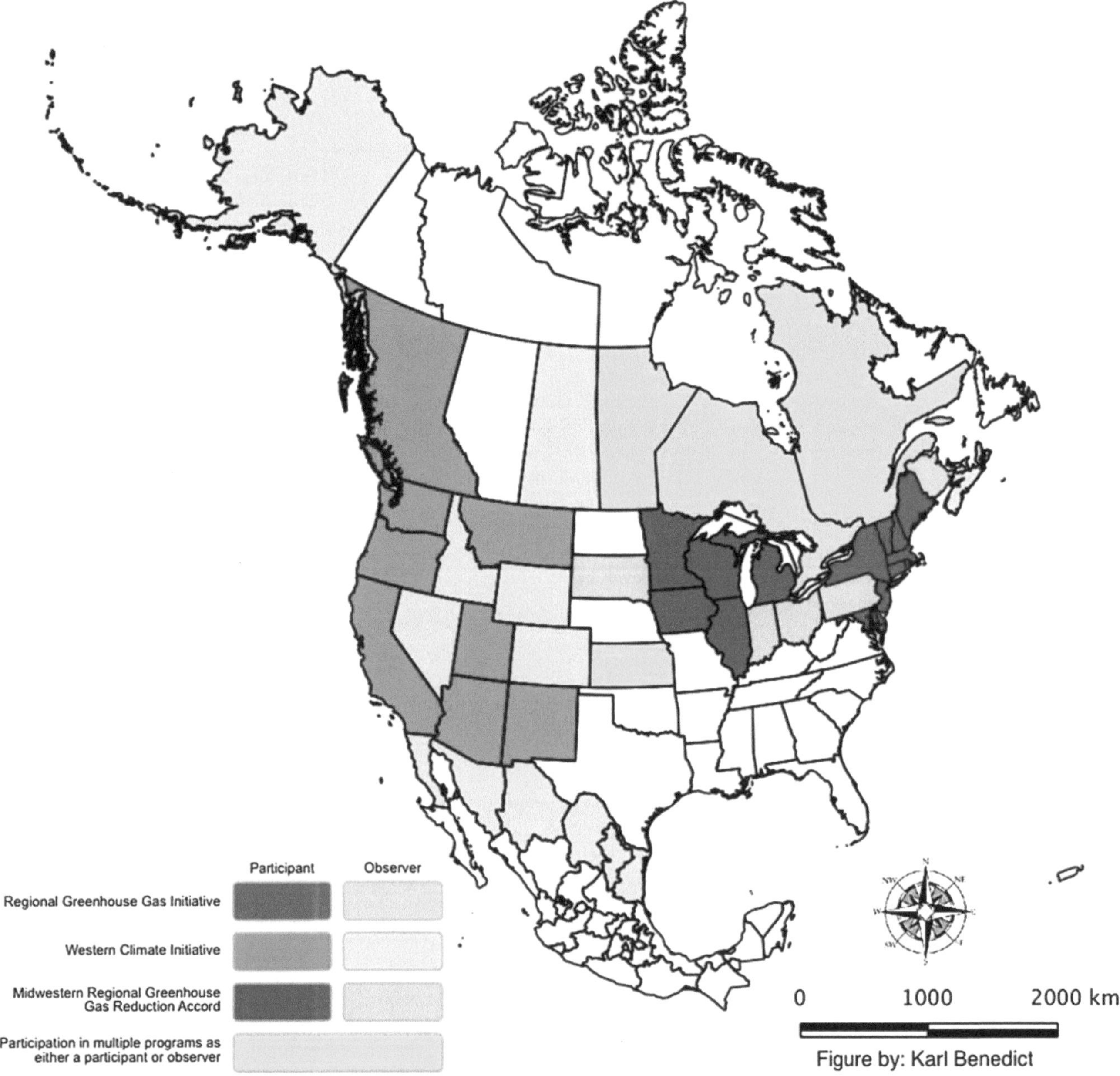

Figure 1. Regional climate initiatives.

that in some instances also involve cross-border collaborations with Canadian provinces. There are currently three such initiatives: the Regional Greenhouse Gas Initiative, the Western Climate Initiative, and the Midwestern Regional Greenhouse Gas Reduction Accord. Combined, these regional commitments include twenty-four states representing over half of the U.S. economy and four Canadian provinces representing almost three quarters of the Canadian economy (see Figure 1, Table 1).

Discussions regarding the relevance of regions and regionalism have a long and complicated history within geography (Hart 1982; Pudup 1988). In this instance, and as the map of these initiatives demonstrates, the term *region* is not descriptive and does not reflect any particular set of assumptions about place or space. Instead, the regional initiatives discussed here are a set of recently constructed, dynamic, and self-organizing relationships. They are an unprecedented experiment in environmental governance in the United States, which

Table 1. Comparison of regional initiatives

	Regional Initiative		
	Regional Greenhouse Gas Initiative	Western Climate Initiative	Midwestern Regional Greenhouse Gas Reduction Accord
Date commenced	December 2005	February 2007	November 2007
Participants	United States: Connecticut, Delaware, Maine, Maryland, Massachusetts, New Hampshire, New Jersey, New York, Rhode Island, Vermont	United States: Arizona, California, Montana, New Mexico, Oregon, Utah, Washington Canada: British Columbia, Manitoba, Ontario, Quebec	United States: Iowa, Illinois, Kansas, Michigan, Minnesota, Wisconsin Canada: Manitoba
Observers	United States: Pennsylvania Canada: Ontario, New Brunswick, Quebec	United States: Alaska, Colorado, Idaho, Kansas, Nevada, Wyoming Canada: Saskatchewan, Nova Scotia Mexico: Baja California, Chihuahua, Coahuila, Nuevo Leon, Sonora, Tamaulipas	United States: Indiana, Ohio, South Dakota Canada: Ontario
Goals	10 percent reduction of power-generated CO_2 emissions by 2018, cap reduces by 2.5 percent each year	15 percent below 2005 greenhouse gas levels by 2020	2020 target: 25 percent below 2005 greenhouse gas levels by 2020. 2050 target: 60–80 percent reduction below 2005 greenhouse gas levels by 2050[a]
Mechanisms/ regulatory tools	Mandatory cap and trade program put in place on electric power-based emission starting 1 January 2009	Mandatory cap and trade program scheduled to begin in 1 January 2012	1 January 2012[a]
Sectors regulated	Electric power generation	Electricity generation, including imported electricity; industrial and commercial fossil fuel combustion; industrial process emissions, gas and diesel consumption for transportation; residential fuel use	Electricity generation and imports; industrial combustion sources; industrial process sources, provided that credible measurement and monitoring protocols exist or can be developed; fuels serving residential, commercial, and industrial buildings not otherwise covered above; transportation fuels[a]
Percentage of total emissions represented by participants	10 percent of U.S. economy	73 percent of Canada's economy and 20 percent of U.S. economy.	25 percent of U.S. economy
Web site	http://www.rggi.org/home	www.westernclimateinitiative.org	http://www.midwesternaccord.org/

[a]Current Advisory Group recommendations under consideration.

has traditionally opted for either state or federal controls. They raise a critical issue of particular interest to geographers: one of scale (Liverman 2004). This article responds to the need identified by Sovacool and Brown (2009, 327) to "begin properly discussing scale as a vital aspect of policy intervention alongside the substance of climate policy" and also to the call made by past Association of American Geographers president Alexander Murphy (2006) for geographers to enter into debates of immediate relevance in the realm of public policy.

Geographers, perhaps because of their careful attendance to the importance of scale, are acutely aware of nature in which "geographical scale *is* political" (Taylor 1984, 7; Cox 1998; Marston 2000; Swyngedouw 2004). As the numerous accounts of the literature regarding the politics of scale and the rescaling of governance amply describe (Bulkeley 2005; Mansfield 2005; Norman and Bakker 2009), it is now widely accepted that scale is constructed, dynamic, and often contested (Leitner 2004). The consensus ends there. As both Moore (2008) and Neumann (2009) recently summarized, debates continue over the meaning of scale and its role in human geographic inquiry. Concerns that reflexive yet conflicting articulations of the "politics of scale" have led to "analytical blunting" of the concept (Brenner 2001) are met with calls to reinvent scale theory or even eliminate it as a concept in human geography (Marston and Smith 2001; Marston, Jones, and Wordword 2005).

Using scale as a way of understanding human relations can be tricky. Moore (2008) examined the ways in which engagement in scalar analysis often results in the reification of the very structures theorists seek to contest. Yet, despite this potential hazard, this article looks at these emerging regional initiatives and addresses both their theoretical and practical implications. It begins by examining the factors that gave rise to these regional efforts based on a review of the available materials on the initiatives, including their own publications, newspaper accounts, journal articles, gray literature, and unstructured interviews with participants. After outlining the basic composition and scope of each of the three regional initiatives, it assesses their relative efficacy and continued viability given the push for a single, national approach. It then addresses the implications of regional initiatives for current debates within human geography on the meaning of scale and how theoretical contributions from geographers might inform efforts to scale climate change responses.

Absence of Federal Leadership and the Emergence of Regional Initiatives

Many factors contributed to the vacuum that gave rise to these regional efforts. Three major influences are particularly worthy of discussion. First is the U.S. posture with regard to international agreements to address climate change. The United States initially signed the Kyoto Protocol, the agreement under the United Nations Framework Convention on Climate Change negotiated at the Earth Summit, held in Rio de Janeiro in 1992. However, neither the Clinton nor Bush administrations put it before the Senate for ratification. To date, the United States is not committed to any binding international agreements for GHG reductions.

The second major factor is President George W. Bush's position on climate change from 2000 to 2008. The Bush administration's National Energy Policy was released in May 2001, the same year the International Panel on Climate Change announced scientific consensus that humans were causing climate change (Oreskes 2004). The 170-page document dedicates a single page to climate change, noting the need for a better scientific understanding of its possible causes and potential solutions. Instead of leading a federal effort to address climate change, the Bush administration took the nation in the opposite direction: It accelerated fossil fuel extraction on federal lands and resisted efforts to regulate GHG emissions (Bryner 2002).

The third major influence was the failure of the existing regulatory system to address climate change at the federal level. In 1999, several states and environmental groups petitioned the Environmental Protection Agency (EPA) to regulate carbon dioxide emissions from motor vehicles as a "pollutant" under the Clean Air Act. The EPA rejected the petition, and that decision was subject to multiple court challenges. Ultimately, the Supreme Court granted certiorari and, in April 2007, the Court rejected the EPA's decision, ordering the EPA to begin a new analysis regarding whether it should regulate CO_2. Although the Obama administration is now moving forward, any actions taken are likely to be subject to further litigation (EPA 2009). Despite facing a pressing and enormous environmental challenge, the United States proved unable, or unwilling, to take meaningful action at the national level.[1]

When the federal system failed to respond to climate change, states began to act, both individually and as part of regional efforts. There are currently three regional climate agreements in the United States: the Northeastern Regional Greenhouse Gas Initiative (RGGI), the Western Climate Initiative (WCI), and the Midwestern Regional Greenhouse Gas Reduction Accord (MRGGRA; see Table 1). Participation in each of the regional initiatives is voluntary, and each initiative is open to new members. Once a state joins a regional initiative, it agrees to regulate activities within its borders to meet the commitments to reduce GHG emissions established by the regional initiative. States that do not want to be bound by the initiative but wish to participate in discussions while considering membership hold "observer" status. For example, several Mexican states currently hold observer status in the WCI.

All three regional initiatives are designed to set specific targets for reduction of GHG emissions, and all employ "cap and trade" approaches to meet those targets. Cap and trade programs are designed to regulate and decrease GHG emissions by first setting a "cap" on overall pollution based on reduction targets that are established (for example, the current proposed target for the WCI is 15 percent below 2005 GHG levels by 2020). These reductions are then achieved by regulating emissions through a "trade" system. Regulated emitters are either allocated or sold pollution "credits" that can be traded in a market-based program. Some companies might choose to reduce emissions substantially and then trade their remaining credits, and others might choose to buy additional credits rather than implement tougher pollution controls. Cap and trade programs are designed allow the market to decide how to most efficiently achieve pollution reduction goals. For example, under RGGI, there is a CO_2 emissions cap, and electric power generators in participating states are required to hold credits covering their emissions of CO_2. A market-based emissions auction was created by RGGI where electric power generators can buy, sell, and trade their CO_2 emissions credits.[2]

Regional Greenhouse Gas Initiative

RGGI was the first regional initiative and has made the most progress. With signatories from ten states, RGGI is the first mandatory cap and trade program in the United States for CO_2 and claims to be the largest carbon market in the world (RGGI 2009). RGGI began with a narrow focus. It addresses only power generation facilities and only CO_2, one of six GHGs under the Kyoto Protocol. This decision was strategic and helped the effort achieve initial success: Electric power generators are responsible for over a quarter of the overall CO_2 emissions from the region and is the most common GHG in the atmosphere. RGGI came into force in January 2009 and places an initial cap of 188 million tons, which represents a 10 percent reduction below 2009 levels during the first three-year compliance period from 2009 through 2011. This target represents a phased approach, starting with relatively modest reductions that will allow industry to begin to adjust to the new regulatory scheme and build confidence among participants. RGGI plans to stabilize power sector emissions over the first six years before initiating an emissions decline of 2.5 percent per year for the four years 2015 through 2018.

Western Climate Initiative

The WCI is younger, but it is more comprehensive in its approach. Committed members include eight states and four Canadian provinces. Together, they represent over 70 percent of the Canadian economy and 20 percent of the U.S. economy. WCI is currently in the process of developing a regional cap and trade program that will regulate not only carbon dioxide but also methane, nitrous oxide, hydrofluorocarbons, perfluorocarbons, and sulfur hexafluoride, with a goal of 15 percent below 2005 levels by 2020. WCI also covers more sectors, going beyond electricity generation to also include other industrial emissions, transportation, and residential fuel use. The cap and trade program is currently scheduled to launch in January 2012 when a three-year compliance period begins for facilities and other entities with emissions exceeding the threshold of 25,000 metric tons of CO_2 per year. Mandatory measurement and monitoring will commence January 1, 2010. WCI plans to begin regulating GHG emissions from transportation fuel in 2015. This more comprehensive strategy targets nearly 90 percent of the region's emissions.

Midwestern Regional Greenhouse Gas Reduction Accord

The MRGGRA is currently in the formative stages. Members currently include six U.S. states and one Canadian province. The original agreement signed in 2007 set an overall goal of establishing regional GHG reduction targets of 60 to 80 percent below current emissions levels through a multisector cap and trade system. Draft recommendations on the cap and trade program design were released in 2009, and the MRGGRA plans to begin implementation in 2010. The MRGGRA's advisory group recently made a series of recommendations for implementation that are currently under consideration by the member states, including a target of 25 percent below 2005 GHG levels by 2020 and a proposed set of measurement and monitoring protocols. It plans to begin enforcement in 2012.

Combined, these regional initiatives represent a tremendous amount of effort over the past several years to act in the absence of national leadership on climate change. They represent the first time states have joined forces to create mutually binding agreements to address a common environmental concern. Both the theoretical and practical implications of these regional approaches are now discussed.

Relative Efficacy and Continued Viability of Regional Approaches

Although the mere existence of these regional initiatives is interesting, their potential role in climate change governance, especially in comparison with a single, national approach, is worthy of examination. Sovacool and Brown (2009, 318) emphasized the "matching principle" as a central tenet for scaling policy initiatives to climate change. The matching principle dictates that the level of intervention should "match" the problem; that is, discrete issues should be addressed locally, international problems globally, and so forth. The problem comes when attempting to apply this concept. Climate change is a global phenomenon that manifests unique challenges on regional and local scales. From a contribution standpoint, climate change is clearly global in the sense that GHGs created anywhere in the world are felt everywhere. From an adaptation and mitigation perspective, more refined scales are needed to provide a meaningful assessment and response. These "scale dynamics" make the matching principle impossible to apply in any tidy way (Cash et al. 2006). Climate change has global, national, regional, and local implications. There is no "perfect match."

The regional approaches identified here have several of the strengths commonly identified with "local"-scale efforts (Sovacool and Brown 2009; Kates and Wilbanks 2003). First, because each initiative is unique, they promote a diversity of approaches, both in and among the regions. As such, they provide a basis for comparing the relative effectiveness of different approaches. For example, one of the main issues associated with a cap and trade system is how to distribute the allowances that will be traded under the cap: Should they be given away as "allowances" or auctioned off? Having more than one system in operation creates opportunities for experimentation.

Regional approaches are also relatively flexible. Because they fall outside the traditional frameworks of governance, they avoid many of the logistical and political hurdles that hamper national efforts. Rather than creating a situation in which states are competing for attention and resources, regional decision-making processes more closely resemble diplomatic negotiations. Perhaps the greatest testament to the relative flexibility of regional approaches is the pace at which the regional efforts have been able to move forward, especially when compared to the relative stagnation at the national level. Flexibility is always an asset when it comes to governance, but it is critical when it comes to addressing climate change. Our scientific understanding of climate change and its implications is constantly changing. This requires an ability to adjust policy approaches accordingly.

Regional approaches, to the extent that they loosely represent specific parts of North America, are also arguably in a better position to address many of the policy questions necessary to address adaptation and mitigation. Because states within regions tend to share similar concerns with regard to climate impacts, they are better able to focus attention on those issues. Electrical transmission lines provide one example. For the most part, transmission is already configured regionally. Regions are also arguably better suited to provide forums for approaching policy questions traditionally within the purview of state government, including state utility and land use regulation (Litz 2008).

There are also weaknesses associated with a regional approach, especially when compared to a national one. Although diversity has its strengths, it also results in a lack of uniformity. Lack of uniformity creates basic issues of fairness associated with some states bearing more of the climate burden than others, albeit voluntarily. Also relating to basic fairness is the problem of "leakage" or "spillover effects," situations in which pollution migrates to areas of the country with fewer regulatory controls. For example, in response to RGGI, coal-generated electricity production has escalated outside RGGI boundaries to meet demand for RGGI states (Sovacool and Brown 2009). As more regions begin enforcing their own cap and trade regimes, this will be less of a concern, but for the moment, there are inevitable inequities.

There are also issues associated with economies of scale. For cap and trade regimes, the general wisdom is that the bigger the market, the better in terms of the market's ability to generate competition and achieve the desired goal. Having several cap and trade systems operating at once allows for experimentation, but they arguably weaken overall effectiveness. Finally, there is the need for the United States to meaningfully participate in international agreements to address climate change. The current regional agreements in the United States include binding reductions by two national states: the United States and Canada. Although not insurmountable, negotiating with transboundary regional agreements would create challenges for international negotiations. There is no reason, however, why national commitments could not be met through initiatives representing more than one nation. The European Union provides one such example.

As of this writing, the future of regional initiatives is far from certain. There is currently a strong push at the federal level for a single, national climate change policy. Any national approach will likely derail the regional initiatives, either explicitly or implicitly due to the doctrine of preemption. Preemption is a constitutional mechanism that enforces the Supremacy Clause of the U.S. Constitution. It provides that, when there is a conflict between state and federal law, federal law governs (Klass 2008; Litz 2008). Current legislative proposals include a federal cap and trade program of GHGs that would likely "occupy the field" and render regional and state actions unenforceable. Current legislative proposals before Congress explicitly preempt state and regional efforts.[3] From a governance perspective, however, there is the question of whether regional efforts *should* be preempted. Each of these regional approaches is in a different stage of development, but together they have demonstrated one significant advantage over a federal approach: They are actually happening. RGGI is currently operational, and WCI and MRGGRA are both positioned to begin enforcement in the next few years. These advancements stand in sharp contrast with attempts to regulate GHGs on a national level. Although is it likely that there will be action on climate change at the federal level eventually, it is much less certain that the effort will be sufficiently responsive. Critics of current legislative initiatives note that, with their numerous loopholes, even some of the more aggressive proposals actually allow GHG emissions to continue to rise over the next two decades (Breakthrough Institute 2009).

Theoretical Implications: Beyond Nested Hierarchies and Capital Production

These regional initiatives have important implications for current discussions on the role of scale within human geography. The "scale debate" among theorists centers, at least in part, on conflicting views regarding how to properly conceptualize scale within contexts of governance. Whereas some conceive of scale as an essentially vertical hierarchical scaffolding of nested territorial units, others call for recognition of emerging, horizontal scales, often referred to as *networks*, which challenge traditional hierarchical accounts (Bulkeley 2005). Marston, Jones, and Wordword (2005) argued that dominant hierarchical conceptions of scale are so firmly embedded in the theory that attempts to incorporate horizontal and network theories are limited. Rather than continue to disagree over how scale should be operationalized, Marston and her colleagues encourage human geographers to jettison the concept and move toward "flat ontologies" that are reflective of complex, emergent spatial relations and avoid "vertical conceits" (422). This notion has been met with some alarm and also with numerous responses offering alternatives to the proposed exile. These include Kaiser and Nikiforova's (2008) call to engage in poststructural political genealogies of scalar knowledge as a way of intentionally assessing its utility and suggestions to draw experiences from other discourses, including identity politics (Moore 2008) and ecology (Sayre 2005).

The regional initiatives examined highlight several aspects of this debate. First, they confirm the need to go beyond embedded hierarchical conceptions of scale. Their emergence contests assumptions that scale can be seen as a series of "nested" relationships. As currently configured, they cut across traditional territorial boundaries and cannot accurately be termed simply subnational. Although they take on explicit and familiar roles of governance, they are not a set of enclosed, jurisdictional spaces (Cox 1998), and they do not reflect established notions of scalar relations. At the same time, regional initiatives cannot be viewed as networks. Networks, whether viewed more traditionally as conduits for the exchange of information and mechanisms for influencing the nation-state or viewed more radically as a move away from state-centered authority, are truly "horizontal" in the sense that they do not ground authority. By contrast, the regional initiatives discussed here create a new, self-organized sphere of authority reflecting interwoven and binding commitments. These regional initiatives provide further evidence of the nature in which both vertical and horizontal conceptualizations of scale are insufficient.

Next, these initiatives underscore another important concern regarding prevailing conceptualizations of scale. To date, much of the focus within the scalar literature is on the role of capital. "Preoccupied with questions of capitalist production, contemporary writing about scale in human geography has failed to comprehend the real complexity behind social construction of scale and therefore tells only part of a much more complex story" (Marston and Smith 2001, 233). This emphasis reflects political/economic geography's Marxist heritage and, although valuable, regional initiatives provide an opportunity to examine other ways in which new scales are currently being reconfigured and produced.

From this standpoint, the emergence of regional initiatives is particularly remarkable given that these new scalar configurations are subject to embodiment by the same political actors that occupy more traditional scalar configurations. These highly divergent geographical and political arrangements are employing many of the same mechanisms (i.e., cap and trade schemes) proposed by more traditional "fixed" scales of governance. In the end, this might prove problematic. Yet for the moment, these regional initiatives are, despite the familiar embodiments by political actors, moving forward with climate commitments where more fixed scales of governance have failed. These new approaches require an inquiry that is informed by, but not limited to, an examination of the role of capital.

As a related issue, the emergence of these regional initiatives reinforces McCarthy's (2005) call for more examinations of the production (as opposed to the politics) of scale as a way of exploring the meaning of scale. These new arrangements represent emerging spaces of engagement and exemplify the nature in which the scale of environmental governance is increasingly contextual (Cox 1998). In his discussion of what he describes as emerging, "post-sovereign" scales of environmental governance, Karkkainen (2004, 74) argued that "new governance arrangements represent a nascent, polycentric substitute for more familiar forms of sovereign authority." Using the Chesapeake Bay Program and the U.S.–Canadian Great Lakes Program as examples, Karkkainen identified three key characteristics of these emerging forms of environmental governance. First, they are nonexclusive in that they operate as multiparty collaborations and coimplementations in a manner that marks a departure from state-centric assumptions of authority. Second, they are nonhierarchical and rely on open-ended commitments by multiple parties that focus on establishing informal and formal modalities of cooperation rather than relying on a single authority to issue commands. Finally, they are post sovereign in the sense that they are not reliant on fixed jurisdictional lines but instead create a scale and scope needed to address the specific task at hand.

As with the initiatives studied by Karkkainen, these regional initiatives have many characteristics of an emerging environmental governance model. They are nonexclusive. Each initiative is open for membership across borders and jurisdictions. They are collaborative, with decisions based on consensus within each region. They are nonhierarchical, relying on open-ended commitments by multiple parties. They are post-sovereign in the sense that their boundaries are not defined by those of the nation-state but instead represent cross-border collaborations created by consensus. Finally, they are task-specific, formulated not along traditional, predetermined boundaries but organized instead by common interests to address a common problem.

Finally, these regional initiatives provide a cautionary example of the need to acknowledge both the emergence of new scales of governance and the ways in which they relate with more traditional scales of governance; that is, the nation-state. Although much of the scholarship examining environmental governance emphasizes the diminished role of the nation-state, others have cautioned against underestimating the continued viability and utility of the national scale as a dimension in scalar relations. Mansfield (2005, 461) examined the nature in which debates on globalization and rescaling "contribute to the idea that that national is less important today than in the past." She then argued that, although it is important not to privilege the national scale or provide it any particular ontological status, the national should be reintegrated as an important dimension in political and economic practices. Similarly, Norman and Bakker (2009) recently found that their examination of the rescaling of water governance across the Canada–U.S. borderland required them to reject dominant assumptions within the environmental governance literature that borders are more porous and less fixed.

The regional initiatives examined here are evidence of both the diminished capacity and relative fixity of the nation-state. They invite geographers to bring scale fixity more centrally into the debate over scale. McCarthy's (2005) conception of scale as a strategy in environmental governance is instructive in this respect. He examined the manner in which environmental nongovernmental organizations (NGOs) engage scalar configurations in ways that "often draw upon and reinforce the very scalar relationships they are attempting to contest or reconfigure" (749). These regional initiatives confirm McCarthy's observation that "scale jumping," although theoretically problematic, is common practice for NGOs and others seeking to influence practical outcomes in environmental policy.

Contributions from Theory: Escaping the "Territorial Trap"

This leads to the important issue of how current concepts of scale within human geography can inform pressing discussions regarding how to act in the face of global climate change. Perhaps the central contribution is something theorists generally seem to agree on: Scale is political. Although relatively fixed, the nation-state holds no particular ontological status. With regard to environmental governance, in particular, geographers agree on the nature in which "[i]n the globalizing 'post-national' era, new geographies of governance are emerging whereupon state capacities are being reorganized, territorially and functionally" (MacLeod and Goodwin 1999, 505). Examples are numerous, including both formal and informal production of subnational, regional governance in Europe, and transnational municipal networks (Bulkeley 2005; Ostergren 2005).

Although geographers might have successfully challenged the conceptual dominance of the nation-state in the literature, that conception is alive and well in the halls of Congress, within international climate negotiations, and even among proponents of regional initiatives, many of whom commonly characterize regional initiatives as stopgap measures to be taken before a coherent national approach is formulated. Bulkeley (2005, 878) noted that "within traditional accounts of global environmental politics, the notion of the state as the primary arena of political power remains unchallenged and there have been relatively few analyses of the changing nature of the state or of sovereignty." The result is what she referred to as the "territorial trap" in which "the naturalization of state space as the taken for granted demarcation of political power" (878).

Bringing scalar debates more centrally into the discussion on climate change has the potential for observations made by scalar theorists regarding the production and the politics of scale to facilitate a more critical view of the "territorial trap." Advancement of the theory further into the public debate has the potential to prompt a reconsideration of the nation-state as a privileged framework of governance. This would provide a significant contribution given that, although the nation-state does not hold any particular ontological status, on the material level its current relative fixity impedes the progress of emerging scales. Contributions from the discipline might move us collectively beyond traditional territorial assumptions of the nation-state as the primary actor in global environmental problem solving. More than a decade ago, Shafer and Murphy (1998) observed that pressing global environmental concerns highlight the limitations of the nation-state as an organizing principle and argued for a reconceptualization of sovereignty. If ever there was a time to engage in this reconceptualization, surely it is now.

Conclusion

Global climate change is the greatest environmental challenge ever faced. The failure of the United States to develop a coherent national policy to address climate change evidences the diminished capacity of the nation-state. Emerging from this void are regional initiatives to address climate change that configure voluntary, binding agreements to limit GHG emissions and promote adaptation strategies. They are an interesting and unprecedented experiment in environmental governance in the United States. Whether regional initiatives, if allowed to continue, would prove to be a more viable approach is a question that might never be answered. They will likely be preempted before being given a real opportunity to demonstrate their full capabilities. These initiatives inform current conceptualizations of scale within human geography and evidence the need to move beyond embedded hierarchical concepts of scale and toward a reconceptualization of sovereignty that is more reflective of new scales of governance and their relative capacities.

Acknowledgments

I would like to thank my colleagues Maria Lane, John Carr, and the members of the transdisciplinary research group at the University of New Mexico for their support of this project, as well as the anonymous reviewers who provided many constructive comments and suggestions.

Notes

1. An examination of the factors contributing to the nation's weakened position is beyond the scope of this article. For a summary of the literature, see Mansfield (2005).
2. A detailed explanation of cap and trade approaches is beyond the scope of this article. For more information, see Yamin (2005).
3. In addition to preemption, there is another potential constitutional constraint. The Compact Clause limits the ability of states to engage in an "agreement of compact" without the consent of Congress. Although historically the Compact Clause has been deemed violated only when such agreements "undermine the supremacy of the federal government" ("The Compact Clause and

the Regional Greenhouse Gas Initiative" 2007, 1958), some argue that regional initiatives to address climate change meet this threshold (Carothers 2006).

References

Betsill, M. M. 2001. Mitigating climate change in U.S. cities: Opportunities and obstacles. *Local Environment* 6 (4): 393–406.

Breakthrough Institute. 2009. Climate bill analysis, part 4: Emissions "cap" may let U.S. emissions continue to rise through 2030. http://thebreakthrough.org/blog/2009/05/climate_bills_offsets_provisio.shtml (last accessed 8 July 2009).

Brenner, N. 2001. The limits to scale? Methodological reflections on scalar construction. *Progress in Human Geography* 24 (4): 591–614.

Bryner, G. 2002. The national energy policy: Assessing energy policy choices. *University of Colorado Law Review* 73 (2): 341–412.

Bulkeley, H. 2005. Reconfiguring environmental governance: Towards a politics of scales and networks. *Political Geography* 24 (8): 875–902.

Carothers, C. 2006. United we stand: The interstate compact as a tool for effecting climate change. *Georgia Law Review* 41:249–54.

Cash, D. W., N. W. Adger, F. Berkes, P. Garden, L. Lebel, P. Olsson, L. Pritchard, and O. Young. 2006. Scale and cross-scale dynamics: Governance and information in a multilevel world. *Ecology and Society* 11 (2): 8.

The Compact Clause and the Regional Greenhouse Gas Initiative. 2007. *Harvard Law Review* 120 (7): 1958–78.

Cox, K. 1998. Spaces of dependence, spaces of engagement and the politics of scale, or: Looking for local politics. *Political Geography* 17 (1): 1–23.

Environmental Protection Agency (EPA). 2009. Proposed endangerment and cause or contribute findings for greenhouse gases under section 202(a) of the Clean Air Act. *Federal Register* 74:18886–910.

Hart, J. F. 1982. The highest form of the geographer's art. *Annals of the Association of American Geographers* 72 (1): 1–29.

Hoffman, A. J. 2006. Getting ahead of the curve: Corporate strategies that address climate change. Prepared for the Pew Center on Global Climate Change. http://www.pewclimate.org/global-warming-in-depth/all_reports/corporate_strategies/ (last accessed 8 July 2009).

Kaiser, R., and E. Nikiforova. 2008. The performity of scale: The social construction of scale effects in Narva, Estonia. *Environment and Planning D: Society and Space* 26:537–62.

Karkkainen, B. 2004. Post-sovereign environmental governance. *Global Environmental Politics* 4 (1): 72–96.

Kates, R. W., and T. J. Wilbanks. 2003. Making the global local: Responding to climate concerns from the ground up. *Environment* 45 (3): 12–23.

Klass, A. B. 2008. State innovation and preemption: Lessons from state climate change efforts. *Loyola of Los Angeles Law Review* 41 (4): 1653–719.

Leitner, H. 2004. The politics of scale and networks of spatial connectivity: Transnational urban networks and the rescaling of political governance in Europe. In *Scale and geographic inquiry: Nature, society and method*, ed. E. Sheppard and R. McMaster, 236–55. Oxford, UK: Blackwell.

Litz, F. T. 2008. Toward a constructive dialogue on federal and state Roles in climate change policy. World Resources Institute Solutions White Paper Series, World Resources Institute, Washington, DC.

Liverman, D. 2004. Who governs, at what scale and at what price? Geography, environmental governance and the commodification of nature. *Annals of the Association of American Geographers* 94 (4): 734–38.

MacLeod, G., and M. Goodwin. 1999. Space, scale and state strategy: Rethinking urban and regional governance. *Progress in Human Geography* 23 (4): 503–27.

Mansfield, B. 2005. Beyond rescaling: Reintegrating the "national" as a dimension in scalar relations. *Progress in Human Geography* 29 (4): 458–73.

Marston, S. A. 2000. The social construction of scale. *Progress in Human Geography* 24 (2): 219–42.

Marston, S. A., J. P. Jones, III, and K. Wordword. 2005. Human geography without scale. *Transactions of the Institute of British Geographers* 30:416–32.

Marston, S. A., and N. Smith. 2001. States, scale and households: Limits to scale thinking? A response to Brenner. *Progress in Human Geography* 25 (4): 615–19.

McCarthy, J. 2005. Scale, sovereignty and strategy in environmental governance. *Antipode* 37 (4): 731–53.

Moore, A. 2008. Rethinking scale as a geographical category: From analysis to practice. *Progress in Human Geography* 32 (2): 203–25.

Murphy, A. B. 2006. Enhancing geography's role in public debate. *Annals of the Association of American Geographers* 96 (1): 1–13.

Neumann, R. P. 2009. Political ecology: Theorizing scale. *Progress in Human Geography* 33 (3): 398–406.

Norman, E. S., and K. Bakker. 2009. Transgressing scales: Water governance across the Canada–U.S. borderland. *Annals of the Association of American Geographers* 99 (1): 99–117.

Oreskes, N. 2004. Beyond the ivory tower, scientific consensus on climate change. *Science* 306 (5702): 1686.

Ostergren, R. 2005. Concepts of region: A geographical perspective in regionalism. In *Regionalism in the age of globalism: Vol. 1*, ed. L. Hönnighausen, M. Frey, J. Peacock, and N. Steiner, 1–14. Madison: University of Wisconsin Press.

Pudup, M. B. 1988. Arguments in regional geography. *Progress in Human Geography* 12 (3): 369–90.

Rappaport, A., and S. Hammond. 2007. *Degrees that matter: Climate change and the university*. Cambridge, MA: MIT Press.

The Regional Greenhouse Gas Initiative (RGGI). 2009. Executive summary. http://www.rggi.org/docs/RGGI_Executive_Summary.pdf (last accessed 10 July 2009).

Sayre, N. F. 2005. Ecological and geographical scale: Parallels and potential for integration. *Progress in Human Geography* 29 (3): 276–90.

Shafer, S. L., and A. B. Murphy. 1998. The territorial strategies of IGOs: Implications for environment and development. *Global Governance* 4 (3): 257–74.

Sovacool, B. K., and M. A. Brown. 2009. Scaling the policy response to climate change. *Policy & Society* 27:317–28.

Swyngedouw, E. 2004. Scaled geographies: Nature, place and the politics of scale. In *Scale and geographic inquiry: Nature, society, and method*, ed. E. Sheppard and R. McMaster, 236–55. Oxford, UK: Blackwell.

Taylor, P. 1984. Introduction: Geographical scale and political geography. In *Political geography: Recent advances and future directions*, ed. P. Taylor and J. House, 1–7. London: Croom Helm.

Yamin, F. 2005. *Climate change and carbon markets: A handbook of emissions reduction mechanisms*. London: Earthscan.

Index

Page numbers in *Italics* represent tables.
Page numbers in **Bold** represent figures.

T - #0698 - 101024 - C0 - 279/216/18 - PB - 9781138852402 - Gloss Lamination